ISOSTASY AND FLEXURE OF THE LITHOSPHERE

Isostasy is a simple concept, yet it has long perplexed students of geology and geophysics. This fully updated edition provides the tools to better understand this concept using a simplified mathematical treatment, numerous geological examples and an extensive bibliography. It starts by tracing the ideas behind local and regional models of isostasy before describing the theoretical background, arguing that only flexure is in accord with geological observations. It then proceeds to describe the theoretical background, the observational evidence and the constraints that flexure has provided on the physical properties of the lithosphere. The book concludes with a discussion of flexure's role in understanding the evolution of the surface features of the Earth and its neighbouring planets. Intended for advanced undergraduate and graduate students of geology and geophysics, it will also be of interest to researchers in gravity, geodesy, sedimentary basin formation, mountain building and planetary geology.

A. B. WATTS is Professor of Marine Geology and Geophysics in the Department of Earth Sciences at the University of Oxford. He is a Fellow of the Royal Society; a recipient of the Murchison Medal of the Geological Society of London, the Arthur Holmes Medal of the European Geosciences Union and the Maurice Ewing Medal of the American Geophysical Union; and the author of more than 240 refereed publications.

ISOSTASY AND FLEXURE OF THE LITHOSPHERE

SECOND EDITION

A. B. WATTS
University of Oxford

Shaftesbury Road, Cambridge CB2 8EA, United Kingdom

One Liberty Plaza, 20th Floor, New York, NY 10006, USA

477 Williamstown Road, Port Melbourne, VIC 3207, Australia

314–321, 3rd Floor, Plot 3, Splendor Forum, Jasola District Centre,
New Delhi – 110025, India

103 Penang Road, #05–06/07, Visioncrest Commercial, Singapore 238467

Cambridge University Press is part of Cambridge University Press & Assessment,
a department of the University of Cambridge.

We share the University's mission to contribute to society through the pursuit of
education, learning and research at the highest international levels of excellence.

www.cambridge.org
Information on this title: www.cambridge.org/9781009278928

DOI: 10.1017/9781139027748

First published 2023

Printed in the United Kingdom by CPI Group Ltd, Croydon CR0 4YY

A catalogue record for this publication is available from the British Library.

ISBN 978-1-009-27892-8 Paperback

Additional resources for this publication at www.cambridge.org/isostasyandflexure2e.

The flexure of the lithosphere caused by the load of Miocene-Recent sediments in the Amazon deep-sea fan system, northeast Brazil. The sediment load has flexed the oceanic crust downwards by more than 2 km in offshore regions, and the continental crust upwards and downwards by up to 40 m in onshore regions. The flexures have had a significant impact on margin structure and stratigraphy offshore, and on landscape evolution and ecosystems onshore.

This book is dedicated to Mary and my family.

Contents

Preface to the Second Edition

It has been more than 20 years since the first edition of *Isostasy and Flexure of the Lithosphere* was published. Much has changed in the Earth and Space sciences during this time. Better monitoring of the land, oceans and atmospheres has fundamentally improved our knowledge of how Earth works while at the same time connecting us to some of the biggest challenges we face such as the human impact on climate, environment and sea level change. Nevertheless, phenomena such as isostasy remain important today, especially as they help us better understand what Earth has done in the past and what it might do again in the future.

This second edition of the book builds on the first edition, updating it to consider new data and ideas. The early development of isostasy has changed little, and so there have been few updates of Chapters 1 and 2 with the notable exception of the inclusion of Charles Darwin and his early work. When we read these two chapters through the eyes of today it is perhaps surprising how the early development of the field was dominated by such a small group of men working in just three countries, largely independently of each other. It is exciting, therefore, to see how interest in isostasy has changed during the past few decades as it has adapted and become more diverse, equal and international, and this bodes well, I think, for its future.

Other chapters have changed, some significantly. For example, Chapter 3 now includes a consideration of finite-difference methods and their applications to spatial variations in rigidity and flexure in three dimensions. Chapter 4 has been updated to show some recent results of glacial and inter-glacial, seamount and ocean island, river delta and trench-outer rise loading and unloading. Chapter 5 includes a discussion of the long-wavelength admittance, and Chapter 6 includes a discussion of models of rheology at lithospheric conditions. The most significant changes perhaps have been made to Chapters 7 and 8. In Chapter 7 the role of flexure in landscape evolution and dynamic topography has been explored, as has the idea of lithosphere memory and structural inheritance. In Chapter 8, new NASA mission topography and gravity data have been used to shed more light on planetary isostasy, and the postscript on Earth now includes an updated elastic thickness map and a discussion of gravity anomaly power spectra and their implications for the waveband of plate flexure.

Together, these changes have resulted in the addition of 66 new figures, the color enhancement of 221 figures, the update of 5 elastic thickness data tables, and the conversion of 17 scripts from Mathcad® to MATLAB®. Finally, 382 new sources that cover some of the many contributions to isostasy during the past two decades have been added to the references.

As was the case with the first edition, there are many people whom I would like to thank for their help. Foremost among them have been my close working colleagues: the late Evgenii Burov and Shijie Zhong, and especially Catherine Johnson, Simon Lamb, Rebecca Morgan, David Sandwell, Pål Wessel and Shijie Zhong who offered many helpful comments on earlier drafts of the chapters. I would also like to thank the new generation of students, especially Andrew Lin, Mohammed Ali, Cian Wilson, Rebecca Bell, David Close, John Hillier, Natalie Lane, Tom Jordan, Matt Rodger, Lara Kalnins, Guy Paxman, Dan Bassett, Johnny Hunter, Brook Tozer, Brook Keats and Rebecca Morgan, some of whom have gone on to positions in academia and made their own special contributions to isostasy. Finally, I thank Jue Wang at Washington University in St. Louis for her help with access to planetary topography and gravity data sets, David Sansom at Oxford for his help with re-drafting some of the figures and last, but not least, Matt Lloyd, Sarah Lambert and especially Susie Francis at Cambridge University Press for their patience, encouragement and help in seeing this project through.

A. B. Watts
Oxford,
September 2022

Preface to the First Edition

Isostasy is a concept that is fundamental to the Earth Sciences. It is based on the simple idea that the light crust is floating on the denser underlying mantle. Isostasy is a condition of rest that the crust and mantle would tend to in the absence of disturbing forces. The waxing and waning of ice sheets, erosion, sedimentation, and volcanism are some of the processes that disturb isostasy. The way that the Earth's crust and mantle respond to these disturbances helps us to constrain the long-term physical properties of the lithosphere and more complex geodynamical phenomena such as mountain building, sedimentary basin formation and the break-up of continents and the formation of new ocean basins.

The history of isostasy is a fascinating one. Ever since it was first defined in 1882, isostasy has been a subject of much debate among geologists, geodesists and geophysicists. Isostasy featured prominently, for example, in early major geological controversies such as the contraction theory and continental drift. By the early part of the last century, certain 'schools of thought' had emerged on the subject that pitched geologist against geophysicist and American against European. Even today, there is still much discussion about isostasy and the role that it may play in the context of a dynamic Earth.

The last book dedicated to isostasy was published in 1927 by the American geodesist, W. Bowie. Bowie's book appeared at a time when the geodesists were enjoying considerable success, having proved from gravity data the existence of isostasy. Yet, isostasy was a difficult concept for geologists to grasp. The book went a considerable way in explaining the relevance of local models of isostasy, such as Airy and Pratt, to geology. Unfortunately, it did so at the expense of the new ideas that had begun to emerge at the turn of the last century about the significance of regional rather than local schemes of isostasy. Although isostasy is discussed in subsequent books, notably Heiskanen and Vening Meinesz (1958) and Turcotte and Schubert (1990), it has not really been treated as a subject in its own right since Bowie's book.

The aim of this book is to put forward the modern ideas on isostasy that the outer layers of the Earth are strong and capable of supporting loads for long periods of geological time. The strength manifests itself in the flexing of the crust and mantle in the region of loads such as volcanoes and sediments. Some loads are large enough that they cause the crust and upper mantle to break, rather than flex. The book begins by tracing the history of isostasy from its roots at the turn of the last century to the present day. The observational evidence for flexure is then presented, using examples from glacial lakes, oceanic islands and seamounts, river deltas, and deep-sea trench outer rise systems, and its implications for the physical properties of the lithosphere are examined. Finally, the book considers the role that the phenomenon of flexure has played in our understanding the main surface features on the Earth and on other terrestrial planets.

The journey to the current book has been a long one and has involved many people. First, I would like to thank my 'mentors': Martin Bott, who first introduced me to the geological interpretation of gravity data, Manik Talwani, who taught me about marine gravity and geodesy, and Dick Walcott, whose work first stimulated my interest in the link between gravity, topography and flexure. The mid-1970s to mid-1980s was a busy time for flexure. Peter Buhl, Steve Daly, Bob Detrick, Neil Ribe, Jeff Weissel and especially Jim Cochran shared in a lot of this work. Finally, I would like to say a special thanks to my graduate students at Columbia and Oxford for their friendship, advice and never-ending enthusiasm. Those who have made their own special contributions to isostasy include John Bodine, Mike Steckler, Gary Karner, Julian Thorne, Uri ten Brink, Bernie Coakley, Pål Wessel, Walter Smith, Robin Bell, Andrew Goodwillie, Catherine Marr, Roxby Hartley, Rupert Dalwood and Jonathan Stewart.

This book was written during two sabbatical leaves, one from Columbia and the other from Oxford. I thank Vincent Courtillot at the Institut Physique du Globe de Paris and Wikki Royden at the Massachusetts Institute of Technology for their kind hospitality and for making available to me the facilities of these two fine institutions. Gloria Grace and the late Portia Takakjian at Columbia and Karen Krause and Claire Carlton at Oxford helped me to get the products of these sabbaticals into early drafts of some of the chapters.

The task of reading the chapters has fallen to current colleagues. I am very grateful to Eugvene Burov, Roger Scrutton, Jonathan Stewart, Shijie Zhong, and the 'marine group' at Oxford, Gareth Armstrong, Emily Black, Fiona Grant, David Haddad and Paul Wyer, for their critical comments and helpful advice. Finally, I thank Matt Lloyd at Cambridge University Press for helping see this project through.

A. B. Watts
Oxford,
April 2000

Acknowledgements

The publishers listed here are gratefully acknowledged for giving their permission to use quotations and original or redrawn figures in journals and books for which they hold copyright. The original authors are cited in the figure captions. Every effort has been made to secure permission to use copyrighted materials, and sincere apologies are rendered for any errors or omissions.

Publishers

Academic Press	
Quaternary Research	4.2, 4.3, 4.11
American Geophysical Union	
Geophysical Monograph Series	4.17
Journal of Geophysical Research	2.23, 2.29, 4.6, 4.20, 4.23, 4.37, 4.38, 4.39a, 5.31, Table 5.3, 7.8a, 7.14, 7.32, 7.34, 7.37, 7.52, 7.53, 7.54, 7.55, 7.57, 7.58, 7.59, 8.27, 8.28, 8.29, 8.30, 8.31, 8.32, 8.33, 8.44
Reviews of Geophysics and Space Physics	4.5
Tectonics	7.8b, 7.9b, 7.45, 7.47, 7.48, 8.4, 8.6, 8.9, 8.10, 8.11, 8.12, 8.17, 8.19, 8.20, 8.21, 8.22, 8.23
American Journal of Science	Quote – Chapters 2, 6, 2.3
Australian Journal of Earth Sciences	7.62
Birkhauser Verlag AG	
Pure and Applied Geophysics	6.22
Blackwell Science Ltd	
Geophysical Journal International	6.25
Center for Academic Publications, Japan	
Journal of Physics of the Earth	4.4
Dover Publications	Quote – Chapter 5
Elsevier Science	
Earth Planetary Science Letters	4.49

Institution	Reference
Tectonophysics	4.53
Geological Society of London	1.7, 2.28
John Wiley & Sons Limited	3.3, 4.10
Kluwer Academic Publishers	8.2, 8.3, 8.8, 8.13, 8.15
Macmillan Magazines Ltd	
Nature	7.61, 8.5
National Park Service of the United States	4.32
National Research Council Canada Press	
Canadian Journal of Earth Sciences	7.46
National Research Council of the United States	Quote – Chapters 4, 8
Springer-Verlag GmbH & Co. KG	7.55
The McGraw-Hill Companies	1.33, 2.18, Table 1.3, 1.4, 7.4
The Royal Society, London	
Catalogue of Portraits	1.5
Philosophical Transactions	1.6, 1.14, 4.13, 7.1
The University of Chicago Press	
Journal of Geology	2.4, 2.5, 2.16, Quote – Chapter 3, Table 2.3
Journal of the Franklin Institute	2.20
US Geological Survey Photographic Library	2.1
University of Michigan Press	3.7, 3.13, 3.16, 3.20, Table 3.2
Yale University Press	1.30

The authors listed here are gratefully acknowledged for giving permission to use original figures from journals or books or for providing photographic materials that have not been previously published.

Authors

Author	Reference
P. Cilli	4.27
N. Cordozo	MATLAB® script 3.24
R. W. Fairbridge	1.27, 1.28
D. Garcia-Castellanos	7.84
D. P. McKenzie	1.10, 1.13, 5.31, 8.24, 8.26
R. B. Owens	7.15
E. M. Parmentier	7.53
D. T. Sandwell	7.52, 8.22, 8.23
R. S. Scrutton	7.56
D. J. Shillington	4.63
D. E. Smith	8.14
S. C. Solomon	8.6, 8.8, 8.9, 8.10, 8.19, 8.20
U. S. ten Brink	7.57
R. I. Walcott	1.29, 1.31, 1.32, 4.9
P. Wyer	2.2

Data Sources

NOAA National Centers for Environmental Information **Marine Geology & Geophysics** **(Shipboard gravity, bathymetry, magnetics)** www.ngdc.noaa.gov/	**4.19, 4.29, 4.45, 4.57, 4.63, 7.60, 7.72, 7.74**
The Decade of North American Geology – Digital Data www.ngdc.noaa.gov/geomag/fliers/se-2004.shtml	4.7, 7.47, 7.71
CRUST 2.0, 1.0 **A new global crustal model at 2 × 2° and 1 × 1°** https://igppweb.ucsd.edu/~gabi/crust1.html	8.37, 8.39, 8.40
Digital isochrons of the world's ocean floor www.earthbyte.org/Resources/Agegrid/1997/digit_isochrons.html	5.20, 7.60 7.77
Marine gravity from satellite altimetry **Global topography** **(SRTM15, GEBCO 2020)** **Radar altimetry** **(Topography – V19.1, Gravity – V29.1, V31.1)** https://topex.ucsd.edu/	Book cover, 4.17, 4.54
National Geospatial-Intelligence Agency – Office of Geomatics **World Geodetic System 1984 (WGS 84)** **EGM2008** **Free-air gravity anomaly** https://earth-info.nga.mil/index.php?dir=wgs84&action=wgs84	4.54, 8.43, 8.44
Earth2014 Global relief model https://ddfe.curtin.edu.au/models/Earth2014/	8.43
Distribution of Pacific seamounts from Geosat/ERS-1 vertical gravity gradient data www.soest.hawaii.edu/pwessel/smts/	7.76
Planetary Data System **GRAIL (Moon), MESSENGER (Mercury), MOLA (Mars), Magellan (Venus) mission data** http://pds-geosciences.wustl.edu	8.11, 8.12, 8.17, 8.18, 8.21, 8.22, 8.23, 8.24, 8.26, 8.27, 8.34, 8.38
LALT Japanese Global DEM for the Moon www.science.org/doi/epdf/10.1126/science.1164146	8.10, 8.12
GeoMapApp **Swath bathymetry data** www.geomapapp.org	4.33
Global Central Moment Tensor (CMT) Project www.globalcmt.org/	4.32

Display Software

Generic Mapping Tools (GMT)
www.generic-mapping-tools.org/
Affinity Designer
https://affinity.serif.com/en-gb/

Tables and MATLAB® Files

CUP Server
www.cambridge.org/isostasyandflexure2e

Notation

D	Flexural rigidity
D_c	Depth of compensation (Pratt)
E	Young's modulus
f	Freeboard
f_e	Flattening of the Earth
f_l	Ratio of surface to sub-surface loading
g	Gravitational acceleration
G	Universal gravitational constant
k	Wavenumber
K	Curvature
ln	Natural logarithm (base e)
log_{10}	Common logarithm (base 10)
m_b	Earthquake body wave magnitude
M_e	Mass of the Earth
M_o	Bending moment
M_w	Moment magnitude of an earthquake
M_b	Body wave magnitude of an earthquake
P	Pressure
P_b	Line load
Q	Creep activation energy
r_e	Mean radius of Earth
R	Radius of regionality (Vening Meinesz)
R_g	Universal gas constant
T	Temperature
T_c	Thickness of zero elevation crust (Airy)
T_e	Elastic thickness
t_{sf}	Age of seafloor
V	Shear force

W_d	Water depth of deposition
y	Deflection (flexure)
Y	Backstrip
Z	Gravitational admittance
Z_{neck}	Depth of strength maxima
α	Flexural parameter (two-dimensional)
α_v	Coefficient of volume expansion
β	Flexural parameter (three-dimensional)
β_s	Stretching factor
γ^2	Coherence
δ	Deflection of the vertical
Δg	Gravity anomaly
Δ_{sl}	Sea level change
ε	Strain
η	Viscosity
θ_n	Fault hade
κ	Thermal diffusivity
κ_d	Denudational coefficient
κ_s	Subduing coefficient
λ_w	Wavelength
μ	Coefficient of friction
ν	Poisson's ratio
ρ	Density
σ	Stress
τ	Relaxation time
τ_s	Shear stress
ϕ	Geodetic latitude
ϕ_s	Porosity
ϕ_z	Phase of the admittance
φ	Isostatic response function
Φ	Astronomic latitude
φ_e	Flexural (elastic) response function
φ_v	Flexural (viscoelastic) response function

Units

Ma	Millions of years ago
Myr	Million years

1

The Development of the Concept of Isostasy

The keynote of isostasy is a working toward equilibrium. Isostasy is not a process which upsets equilibrium, but one which restores equilibrium.

(Chamberlin, 1931, p. 5)

1.1 Introduction

Isostasy is derived from the Greek words 'iso' and 'stasis' meaning 'equal standing'. The term is used to describe a condition Earth's crust and mantle tend to, in the absence of disturbing forces. In its simplest form, isostasy is the view that the lighter crust floats on the denser underlying mantle. It is an idealised state: a condition of rest and quiet. The transport of material over Earth's surface during, for example, the waxing and waning of ice sheets, the growth and decay of volcanoes and the deposition and erosion of sediments, disturbs isostasy and, in some cases, prevents equilibrium from being achieved. Seismic and gravity anomaly data suggest Earth's outermost layers generally adjust to these disturbances. One of the principal objectives of isostatic studies during the last two centuries has been to determine the temporal and spatial scales over which these adjustments occur. This information provides constraints on the physical nature of Earth's outermost layers, thereby improving our understanding of what drives more complex geodynamical processes such as mountain building, rifting and sedimentary basin formation.

The term isostasy was first coined in 1882, but there is evidence that questions concerning the equilibrium of the Earth's crust were being posed as far back as the Renaissance. Isostasy played a central role in the development of geological thought and featured prominently in some of the great controversies of the late nineteenth and early twentieth centuries such as the contraction theory, continental drift, and the permanence of the oceans and continents.

The discovery that the Earth's crust might tend to or be in a state of isostatic equilibrium is one of the most fascinating stories in the history of science. There were periods, for example, when it was accepted by one group of workers but rejected by another. There has also been considerable debate on which isostatic models best apply at a particular geological feature. These debates have led to some vigorous exchanges on isostasy in the literature and, on occasion, to the development of 'schools of thought', which divided geophysicists and geologists and North Americans and Europeans.

Today, isostasy still holds a central place in Earth Science. This is true despite a considerable body of work that shows Earth to be a dynamic planet that responds to loads over a wide range of spatial and temporal scales. Since isostasy is usually only concerned with how the crust and mantle adjust to shifting loads of limited spatial and temporal dimensions, it is only a 'snapshot' of these dynamical processes. Nevertheless, it is an important snapshot. By comparing the observed adjustments to models based on flotation, differential heating and cooling and bending of plates, we have learnt a considerable amount about Earth, its rheology, its composition and its structure.

In this introductory chapter, we will outline some of the key developments in the concept of isostasy. Special emphasis is given to the Airy and Pratt models of local isostasy. These models proved useful to the geodesists since they helped them in practical problems related to surveying. They were of less interest to geologists who struggled to incorporate the models into geological thought. The tussle between the geodesist and the geologist was an intriguing one that helps set the scene for later chapters.

1.2 First Isostatic Ideas

Some of the first ideas about the equilibrium of the Earth's outer layers originate with the engineer, artist and humanitarian Leonardo da Vinci (1452–1519). Translations of da Vinci's notebooks by Edward MacCurdy (MacCurdy, 1928; 1956) show that da Vinci gave considerable thought to how Earth might respond to shifts in loads over its surface. For example, the following quote (Delaney, 1940) illustrates how da Vinci thought the removal of sediment from a mountain might cause it to rise:

That part of the surface of any heavy body will become more distant from the centre of its gravity which becomes of greater lightness. The earth therefore, the element by which the rivers carry away the slopes of mountains and bear them to the sea, is the place from which such gravity is removed; it will make itself lighter . . . .. The summits of the mountains in course of time rise continually.

It was not, however, until 100 years later, when the first attempts were made to determine the Earth's shape, that it became possible to assess the equilibrium state of the continents and the oceans.

In the early eighteenth century, there were two main schools of thought concerning the shape of the Earth: an English and a French one. The English school, led by Isaac Newton (1642–1727), considered the Earth to be flattened at the poles, while the French school under Jacques Cassini (1677–1756) thought the Earth to be flattened at the equator. The Académie Royale des Sciences, under the direction of Louis XV, sponsored a team of scientists to go to different parts of the Earth to measure the length of a meridian degree to resolve the controversy. The first team, led by Charles Marie de La Condamine (1698–1758), made measurements in the region of the equator near Quito, Ecuador, while the second team, led by Pierre-Louis Moreau de Maupertuis (1698–1759), made measurements in the region of the Arctic Circle near Tornio, Finland.

The techniques used by Condamine and Maupertuis involved the measurement of the distance between two points of known position. The positions were determined astronomically by measuring the angle of elevation, Φ, between the pole star (Polaris) and the horizon, as indicated by level bubbles on an astrolabe. Level bubbles follow a surface, known as an 'equipotential surface', along which no component of gravity exists. The equipotential surface, which coincides with Earth's mean sea level, is known as the 'geoid', and so Φ is the angle between the pole star and the geoid. Because the direction of the pole star is normal to the equatorial plane, then it follows (Fig. 1.1) that Φ is also the angle between the normal to the geoid (i.e., the plumb-line direction) and the equatorial plane and, hence, the astronomical latitude at a point.

The distance between astronomical positions was determined by triangulation. In this technique, a network of triangles with vertices permanently marked on the Earth's surface are set up so that they connect the astronomical positions. One of the astronomical positions is then chosen as one vertex on the first triangle. If the length of one side of the triangle and its included angles are accurately measured, then it is possible to determine the distance between each vertex of the triangle. By extending the network of triangles to include the second astronomical position (Fig. 1.2), it is possible to estimate the total distance between astronomical positions from an accurate measurement of a length in the first triangle (which may be quite short) and the angles between vertices of all the other triangles.

The length of the meridian degree measured by Condamine was, as it turned out, much smaller than that measured by Maupertuis (Table 1.1). Furthermore, the length of the meridian degree on the Arctic Circle was greater, by about 900 m, than the length

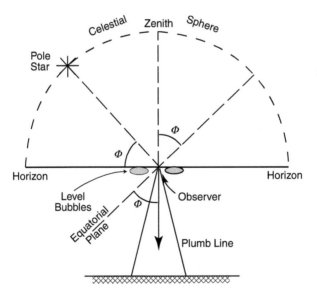

Figure 1.1 The determination of astronomical latitude, Φ, from observations of the pole star.

Table 1.1 *Summary of measurements of the length of a degree meridian according to the surveys led by Condamine and Maupertuis*

Nearest Town/City	Approximate Latitude	Length of a Meridian Degree (Toise*)	Difference from Paris (Toise*)
Tornio, Finland	66° (Arctic Circle)	57,525	+342
Paris (L'observatoire de Paris)	48° 50'	57,183	0
Quito, Ecuador	0° (Equator)	56,753	−430

* 1 Toise = 1.949 m.

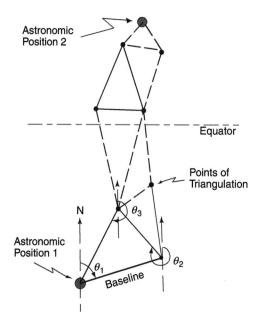

Figure 1.2 The measurement of the distance between two points by triangulation where astronomical latitude and longitude have been determined.

determined previously near Paris; and the length near the equator was smaller, by a slightly larger amount. These results convinced Condamine and Maupertuis that the Earth was indeed flattened at the poles, as suggested by Newton. The flattening, f_e, was estimated to be about 1/216.8 (Fig. 1.3). The measurements of Condamine and Maupertuis therefore solved the controversy of the overall shape of the Earth, although perhaps not quite in the way their sponsor, the Royal Court in Paris, had anticipated!

One member of Condamine's party though, Pierre Bouguer (1698–1758; Fig. 1.4), was not content, however, to let the matter rest there. Bouguer was puzzled by the consistency of the results because the measurements near the equator were obtained in the presence of much greater

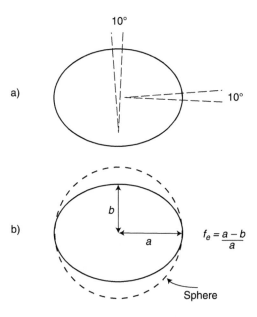

Figure 1.3 The flattening of the Earth, f_e, can be approximated by an ellipse with semi-major axis, a, and semi-minor axis, b.

Figure 1.4 Pierre Bouguer, the French mathematician and astronomer who accompanied Charles-Marie de La Condamine during the expedition of 1735 to Quito, Ecuador to measure the length of a degree meridian at the equator. (Pastel de Perroneau, eighteenth century; Musée du Louvre; Photo: Giraudon.)

topographic relief than those near the Arctic Circle. He surmised (Bouguer, 1749) that the mass of the mountains in the vicinity of Quito was sufficiently large that it should have caused the local plumb line to be deflected by as much as $1'\ 43''$ from the vertical.[1] Such a deflection would introduce errors into the astronomical positions because the elevation of a distant star is measured on a table with level bubbles so that the measurement is made with respect to the local plumb-line direction (Fig. 1.1). The astronomical positions were apparently not in error, however, leading Bouguer to conclude that the attraction of the mountains in the vicinity of Quito 'is much smaller than expected from the mass of matter represented in such mountains'.

A few years later, an Italian astronomer and mathematician with strong links to Croatia, Ruggero Giuseppe Boscovich (1711–87), provided an explanation of the problem that puzzled Bouguer. He said (Boscovich, 1755): 'The mountains, I think, are to be explained chiefly as due to the thermal expansion of material in depth, whereby the rock layers near the surface are lifted up. This uplifting does not mean the inflow or addition of material at depth, the void within the mountain compensates for the overlying mass.'

This passage is the first to use the term 'compensates'. Boscovich speculates that the mass excess of the mountain is compensated in some way by a mass deficiency at depth. Thus, the deflection of a plumb line near a mountain range may well be small, as Bouguer had suspected.

Boscovich, with broad interests beyond astronomy and mathematics, may, we speculate, have been influenced by a fellow Italian, the eminent scientist Luigi Ferdinando Marsigli (1658–1730). In 1728, 27 years prior to the appearance of Boscovich's paper, Marsigli illustrated in a beautiful set of drawings the difference in elevation and thickness of 'marine' (oceanic) and 'mountainous' (continental) crust. His beautiful, hemispherical, water-coloured and pen-drawn cross sections show a juxtaposed mountain-peak height and seafloor depth that appear to be in some sort of isostatic balance with a thicker crust beneath the mountains than beneath the oceans (Vai, 2006).

Little more appears to have been said on the matter for another 100 years. The statements made by Boscovich and the drawings of Marsigli on the compensation of mountains, as significant as they were, evidently had little impact on leading geologists of the time.

In the early 1800s geological thought in Europe was dominated by the contraction theory. According to this theory, the Earth's surface features were thought to have been the consequence of a gradual cooling of the Earth following its formation. Mountains were considered to be regions that had not cooled as much as ocean regions. The theory had its origins in the work of Gottfried Wilhelm Baron von Leibnitz (1646–1716) and René Descartes (1596–1650). Baron Jean Baptiste Joseph Fourier (1768–1830) subsequently measured the temperature gradient at shallow depths in the Earth, concluding that it was in accord with the predictions of the contraction theory. Therefore, Boscovich's statements on the thermal expansion of mountains may not have seemed all that inconsistent with the theory.

The eminent British geologist, Charles Lyell (1797–1875), was sceptical about the contraction theory. By 1833 he had completed his widely acclaimed book, *Principles of Geology* (Lyell, 1832–3), in which he proposed that the Earth's surface is continually

---

[1] Smallwood (2010) has pointed out that Bouguer grossly overestimated the predicted deflection of the vertical due to the mass of the mountains. Using a Digital Elevation Model (DEM) and local rather than whole Earth densities, he showed that the predicted deflection should be $\sim14\pm4''$, more than a factor of ~6 smaller than that estimated by Bouguer.

subject to periods of rest and change. He disagreed strongly with theories of catastrophes to explain geological events and with ideas forwarded by Leonce Elie de Beaumont (1798–1874) in France and Henry Thomas de la Beche (1796–1855) in England that geological processes, such as mountain building, were global events that occurred at similar times over widely separated regions. Lyell wrote: 'It is preposterous to imagine that just because they had similar trends the Allegheny and Pyrhenees mountain ranges could have been formed by the same catastrophic event.' Among Lyell's many influential friends was John Herschel (1792–1871, Fig. 1.5). In a letter addressed to Lyell, Herschel (1836) pointed out that he disagreed with the contraction theory. In his opinion, the outermost layer or 'crust' of the Earth was in some form of dynamic equilibrium with its underlying substratum or 'sea of lava'. He wrote 'the whole (crust) [floated] on a sea of lava'. According to Herschel, if the crust was loaded, say by sediments, it would sink, thereby causing the underlying lava to flow out from beneath the load and into flanking regions (Fig. 1.6).

Herschel's ideas on the equilibrium of the Earth's crust might never have been published had it not been for Charles Babbage (1790–1871), the mathematician and inventor of the computer, who had also befriended Lyell. Babbage decided to include the letter that Lyell had received from Herschel in a treatise he was writing. The treatise was prepared by Babbage, on his own initiative, as a sequel to the eight volumes of the 'Bridgewater Treatise', which had just been

Figure 1.5 John Frederick William Herschel, the son of the astronomer William Herschel who discovered the planet Uranus. Reproduced from the portrait on p. 161 of Robinson (1980) with permission of the Royal Society.

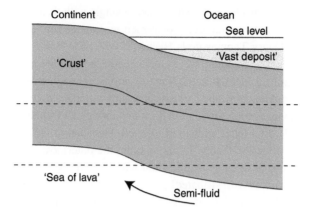

Figure 1.6 The adjustment of the crust to a 'vast deposit' by flow in the underlying 'sea of lava'. Reproduced from a figure in Herschel (1836) with permission of the Royal Society.

published using proceeds from Lord Bridgewater's estate. Babbage felt compelled to publish a ninth treatise because, in his opinion, a prejudice was emerging that the pursuit of science was unfavourable to religion. In a letter to Lyell, he pointed out that the ninth treatise would provide 'an opportunity to illustrate some of the magnificent examples of creation'.

In the ninth treatise Babbage included some observations (Babbage, 1847) he had made at the Temple of Serapis in Pozzuoli, Italy. Built towards the end of 200 AD, the temple had served as a spa (Fig. 1.7) for wealthy Romans. When Babbage visited the site in 1828, the temple's three remaining columns showed a dark encrustation about 4 m above their base. Above the dark encrustation, 2.5 m of the column had been perforated in all directions by a marine boring animal (*Modiola lithophaga*). This observation[2] suggested to Babbage that the temple had undergone a period of subsidence, followed by one of uplift. He attributed these movements of the crust to the action of heat, because of the location of the temple near the historically active volcano of Vesuvius. Babbage considered that the heating caused the crust to expand and contract locally and that these movements were in some way accommodated by movement in the underlying fluid lava. Thus, Herschel and Babbage agreed that the crust accommodated loads by lateral flow in a weak underlying substratum. While Lyell supported this view, it was strongly opposed by the supporters of the contraction theory, who believed that the subsidence and uplift was the consequence of thermal contraction and expansion on a global scale.

The ninth Bridgewater treatise was published in 1837, the same year that Charles Darwin (1809–1882) made a brief statement, to the Geological Society of London,[3] on some observations on the subsidence of ocean islands he had made during his circumnavigation of the world onboard HMS *Beagle*. Darwin, who was given a copy of Lyell's book by Captain FitzRoy at the start of the voyage, recognised three types of ocean islands: volcanic, coral and combinations of the two. It was his view (Darwin, 1842, p. 98, woodcut 4; p. 100, woodcut 5) that a submarine

[2] Sir Harold Jeffreys, a British geophysicist, was later so intrigued by this observation that he suggested (Crittenden, 1967, p. 279) a picture of the Temple of Serapis should hang in every Department of Geophysics as a reminder that movements of Earth's crust are not simple and may involve both subsidence and uplift at the same locality.

[3] Charles Darwin was elected to fellowship of the Geological Society in 1836 and early in 1838 became one of its 'secretaries'.

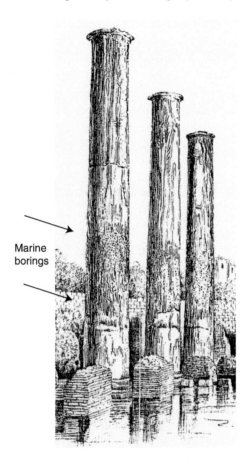

Marine
borings

Figure 1.7 The Temple of Serapis in Pozzuoli, Italy. The borings in the Roman-built columns were used by Babbage to infer that the crust had undergone subsidence followed by uplift. Reproduced from Babbage (1847) with permission of the Geological Society of London.

volcano would grow up on the seafloor to become a high island. Coral would grow in the shallow water fringing the volcano (e.g., as at Moorea in the Society Islands), and then as the volcano started to subside, the coral would attempt to keep pace and grow upwards, leaving a gap between a 'barrier reef' and the central volcanic island (e.g., as at Bora Bora in the Society Islands). Eventually, the island would sink below sea level and all that would remain is an atoll with a central lagoon (e.g., as at Aratika in the Tuamoto Islands). Darwin found direct evidence for subsidence at Santiago in the Cape Verde Islands where the volcano had 'disturbed' and 'bent' downwards a 'bright white' layer of limestone,[4] which Darwin supposed had been deposited horizontally and now underlay the volcano edifice (Darwin, 1844, p. 9, woodcut No. 2). He went on to propose that either the volcano had been uplifted and then subsided or that

[4] We now know that the volcanic evolution for Santiago ranges in age from 4.6 to 0.7 Ma and that the white limestone observed by Darwin is entirely Pleistocene in age (Johnson et al., 2012).

it had never been uplifted, which implied that the volcano somehow maintained its elevation despite the subsidence. Interestingly, from the viewpoint of isostasy, Darwin favoured the latter hypothesis.

It is tempting to speculate that the observations of Herschel and Babbage and, especially, those of Darwin relating to subsidence and uplift had set the scene for the next major development in our understanding of the science of the equilibrium of Earth's crust. However, it was not the geologists, but the geodesists who were to make the key observations that eventually led to the theory of isostasy.

1.3 The Deflection of the Vertical in India

The first measurements of the length of a meridian degree in the India sub-continent were carried out in 1840–59 by George Everest (1790–1866), who as Surveyor-General was charged with mapping the country. Everest's measurement techniques differed from those of the earlier surveys of Condamine and Maupertuis, as he considered astronomical as well as geodetic positions. Geodetic positions were computed at the vertices of each triangle by assuming the position at one vertex of the first triangle was known and then using the equation for the Earth's best-fitting reference ellipsoid to compute the other positions (Fig. 1.8). (Thus, instead of regarding the flattening of the Earth as an unknown, as the Condamine and Maupertuis parties had, Everest assumed it was by now well enough known to compute geodetic positions at points in between the astronomical positions.)

At points where both astronomical and geodetic positions are determined, it is possible to compare them. As Fig. 1.9 shows, astronomic and geodetic latitudes are referenced to a common surface: the equatorial plane. The astronomic position is defined as an angle between the equatorial plane and the local plumb-line direction, whereas the geodetic position is defined as an angle between the equatorial plane and the local normal to the Earth's best-fitting ellipsoid. The plumb-line direction does not necessarily follow the local normal to the ellipsoid because of disturbing masses in the Earth. The amount that the plumb line is deflected from the local normal is known as the 'deflection of the vertical'.

As part of his survey in India, Everest computed the geodetic position at a number of localities where he had already measured the astronomic position. He found that for two

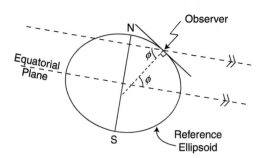

Figure 1.8 The determination of geodetic latitude, ϕ, from the theoretical formula for the Earth's best-fit reference ellipsoid.

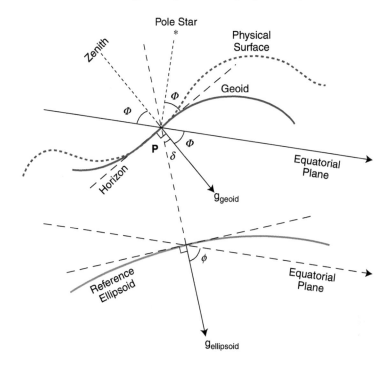

Figure 1.9 The determination of the deflection of the vertical, δ, from the difference between the astronomical latitude, Φ, and the geodetic latitude, ϕ, measured at a point **P** on the Earth's surface.

stations, Kaliana and Kalianpur, on the Ganges Plain to the south of the Himalaya, the latitude difference between them computed geodetically was 5.24″ smaller than that determined astronomically. Everest (1847) thought the discrepancy was unlikely to be in the astronomical position, which he considered to have been determined accurately, but in the geodetic position. He discussed the possibility the discrepancy was caused by the lateral gravitational attraction of the topography of the Himalaya (Everest, p. 179, 1st Hypothesis); but his preferred explanation was closure errors and an incorrect reference ellipsoid, which he then proceeded to 'disperse' throughout the triangulation surveys (Everest, p. 179, 2nd Hypothesis).

But a Caius College, Cambridge–educated mathematician, John Henry Pratt (1809–71: Fig. 1.10), who was Archdeacon of Calcutta at the time, disagreed with the so-called 'process of dispersion'. He believed the discrepancy could be directly related to the disturbing effects on the plumb line of the nearby Himalaya. As Bouguer had previously pointed out, the gravitational attraction of nearby mountains could locally perturb the direction of the plumb line, thereby introducing an error into the astronomic positions, since these positions are used to calculate the deflection of the vertical. Indeed, the largest discrepancies measured by Everest and his assistants during the survey of India were between Kaliana and Kalianpur, and Kalianpur was within 200 km of one of the highest peaks in the Himalaya. It was therefore somewhat surprising Everest did not favour an

Figure 1.10 John Henry Pratt one year after his appointment as Domestic Chaplain to Bishop Wilson of Calcutta. Portrait courtesy of D. P. McKenzie of Cambridge University.

explanation based on gravitational attraction of nearby topography, especially since he was well aware of its influence on the direction of the plumb line and astronomical observations from his earlier surveying experiences in the Cape Province, South Africa.[5]

In a paper read before The Royal Society of London on December 7, 1854, Pratt presented the results of his calculations of the gravitational attraction due to the Himalaya and its hinterland at Kaliana and Kalianpur. He estimated the attraction by dividing the mountains up into 'compartments', computing the gravitational attraction of each compartment, and then summing the results. The problem was determining the topography of the Himalaya and its hinterland. Since the region was largely unexplored, Pratt had to rely on interviews with travellers who had returned from the area!

Consider an elementary mass dm at M and a unit mass at P. The gravitational force of attraction, dF, between the masses is given by Newton's inverse square law:

$$dF = G\frac{dm}{d^2},\qquad(1.1)$$

where G is the universal gravitational constant and d is the distance between the two masses. In the case that M and P are points on the surface of a spherical Earth, then it follows (Fig. 1.11) that the component of the attraction at a station P due to the elementary mass at M in the direction of gravity is given by

---

[5] The author is grateful to Roger Bilham for pointing this out.

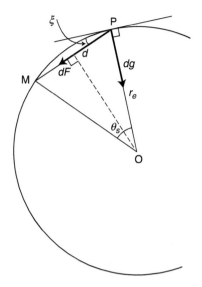

Figure 1.11 Pratt's determination of the gravitational attraction, dg, at a station P due to a mass dm at M.

$$dg = \frac{Gdm\sin(\xi)}{d^2},$$

where

$$\xi = \frac{\theta_s}{2}, \qquad d = 2r_e \sin\left(\frac{\theta_s}{2}\right)$$

$$dg = Gdm\left(\frac{\sin\left(\frac{\theta_s}{2}\right)}{4r_e^2 \sin^2\left(\frac{\theta_s}{2}\right)}\right). \tag{1.2}$$

Pratt used Eq. (1.2) to calculate the gravitational effect of the Himalaya at Kaliana and Kalianpur, and published the results in a 75-page paper (Pratt, 1855). By subtracting the gravity anomaly on the ellipsoid from the gravity anomaly due to the mass excess of the mountains, he calculated a plumb line deflection of 15.885″, more than three times the observed value (Everest, 1847, Fig. 1.12). Pratt was satisfied, despite problems with not knowing the detailed topography of the Himalaya, that he had correctly computed the effect of the mountains at Kaliana and Kalianpur. He concluded his paper by stating that he did not understand the cause of the discrepancy and that the problem should be investigated further.

1.4 Isostasy According to Airy

Shortly after Pratt's paper, George Biddell Airy (1801–92; Fig. 1.13), the Astronomer Royal, presented a paper to the Royal Society in which he offered an explanation for the discrepancy.

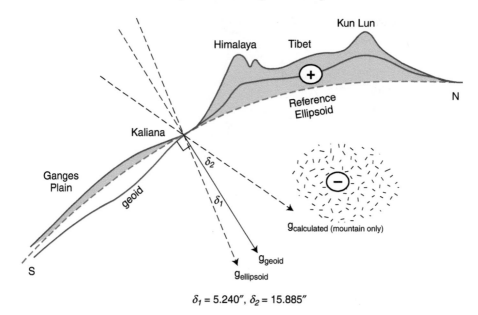

$$\delta_1 = 5.240'', \ \delta_2 = 15.885''$$

Figure 1.12 The deflection of the vertical at Kaliana, northern India. Deflection δ_1 is the deflection determined by Everest from astronomic and geodetic observations. Deflection δ_2 is the deflection calculated by Pratt due to the mass excess of the Himalaya, Tibet and Kun Lun above sea level. The shading schematically shows the mass deficiency that was inferred by Airy and, later, Pratt to underlie the mass excess.

Figure 1.13 George Biddell Airy, Professor of Astronomy at Cambridge University and Astronomer Royal. Portrait courtesy of D. P. McKenzie of Cambridge University.

Unlike Pratt, Airy was not surprised by the discrepancy. Indeed, he thought it should have been anticipated.

Airy's argument was based on his belief that the outer layers of the Earth consisted of a thin crust that overlay a fluid layer of greater density than the crust. He referred to the fluid layer as 'lava'. Airy compared the state of the crust lying on the lava to timber blocks floating on water. He wrote (Airy, 1855, p. 103):

the state of the Earth's crust lying upon the lava may be compared with perfect correctness to the state of a raft of timber floating upon water; in which, if we remark one log whose upper surface floats much higher than the upper surfaces of the others, we are certain that its lower surface lies deeper in the water than the lower surfaces of the others.

By using the analogy of icebergs, Airy suggested that a wide, flat-topped mountain or 'tableland' would be underlain by a less dense region, such that there would be a substitution of 'light crust' for 'heavy lava'. The effect on the local direction of gravity, he viewed, would depend on two actions: the positive attraction of the elevated 'tableland' and the negative attraction of the light crust. He argued that the reduction in attraction of the light crust would be equal to the increase in attraction of the heavy mass above, so that the total effect on the local direction of gravity would be small. This argument, which he was able to make in a three-page-long paper, provided a simple explanation for the observations of Everest.

Airy was quite cautious in applying his model, pointing out that it may not be entirely appropriate to consider all features on the Earth's surface as in a state of flotation. In the case of a tableland (e.g., Tibet), for example, Airy argued that it was unlikely to be supported in any way, other than by a lower-density crust that protruded into the underlying 'lava'. He argued that 'fissures' or 'breakages' would form at the edges of the tableland (Fig. 1.14). In the case of narrow features, such as Mt. Schiehallion in Perthshire, Scotland- where the Rev. Maskelyne (1732–1811) had previously used the deviation of the plumb line to estimate the mean density of the Earth, Airy suggested that his model may not apply, implying that the crust might be sufficiently strong to support this feature without fissure or breakage.

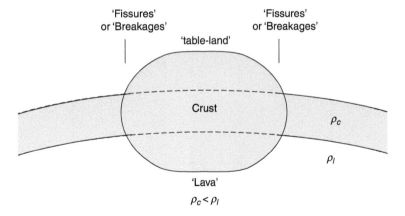

Figure 1.14 Airy's hypothesis of a crust that 'floats' upon 'lava'. Reproduced from fig. 2 of Airy (1855) with permission of the Royal Society.

1.5 Isostasy According to Pratt

There is little doubt that Airy's paper must have come as a surprise to Pratt. Pratt was a mathematician, and his approach to the problem was to carry out detailed, somewhat tedious calculations. Airy, on the other hand, approached the problem without mathematical reasoning, using simple physical concepts. Furthermore, there was a difference in the amount of work carried out; Pratt's paper was 75 pages in length, while Airy's contribution was only three pages!

In 1858, Pratt followed up on Airy's suggestions on mass excess and deficit, proposing his own model for the equilibrium of the Earth's outer layers. He criticised Airy's model on three main grounds (Pratt, 1859). First, the model was based on the assumption of a thin crust. Second, the model assumed that the crust was less dense than the underlying lava. Third, the model was not in accord with the prevailing contraction theory of the Earth. The first objection was based on a suggestion by a colleague of Pratt's at Cambridge, Mr. Hopkins, that the Earth's crust was at least 150 km thick. The second and third objections, however, were clearly a result of Pratt's adherence to the contraction theory.

In 1864 and 1870 Pratt presented two further papers on the subject to the Royal Society. In particular, he expanded on his view, based on the contraction theory, that the depressions and elevations of the Earth's surface were the product of thermal contraction and expansion (Pratt, 1864; 1871). He stated that the 'amount of matter in any vertical column drawn from the surface to a level surface below the crust is now and ever has been, approximately the same in every part of the Earth'. This statement implies (Fig. 1.15) that elevated regions were underlain by low-density rocks, whereas depressed regions were underlain by high-density rocks. Unlike Airy, however, Pratt did not speculate on what might cause different

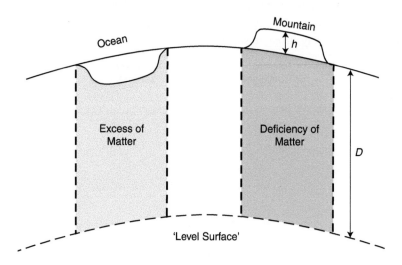

Figure 1.15 Pratt's hypothesis that mountains are underlain by low-density regions while oceans are underlain by high-density regions. At some depth, referred to as the 'level surface', masses are equal.

portions of the Earth's outer layers to be colder or hotter than others. But he recognised these regions as providing the compensation for surface depressions and elevations, a term that had not been used since Boscovich's study.

Despite Pratt's criticisms, Airy did not return to the discussion. The only occasion it seems was when Airy presented a lecture to the Cumberland Association for the Advancement of Literature and Science. A brief edited abstract of his lecture was published in *Nature* in 1878 (Anonymous, 1878). Apparently, Airy did not even refer to Pratt's model in his lecture: instead, he simply repeated the model that he had outlined to The Royal Society in 1855.

Pratt's contributions to isostasy were even more remarkable given that his colleague in Calcutta, Bishop Wilson, died in 1858 and so he had to take up an increasingly active role in the church. He was, by all accounts, a much admired archdeacon, and one of his greatest achievements was the establishment of funds to support the 'Hill schools' in the Himalaya and a missionary girl's school in Calcutta (Brown, 1872). The latter survives, and it is certainly fitting that the Memorial School, Jura Girja, one of the oldest and best known of the missionary schools, continues to recognise Archdeacon Pratt as its founder.

1.6 Fisher and Dutton on Isostasy

It is curious that the scientific discussions between Airy and Pratt during the 1850s were not really taken up by other workers at the time. The next development was not until 1881, when the Rev. Osmond Fisher published his book entitled *Physics of the Earth's Crust* (Fisher, 1881). Described by some as the first textbook in geophysics, the book pays little attention to the controversy between Airy and Pratt. Instead, its main purpose seems to have been to develop the arguments against the contraction theory, which was continuing to dominate geological thought at the time.

Fisher, like Herschel and Airy and Pratt before him, had his own views. He argued the crust and underlying substratum as being in some form of equilibrium: with the lighter crust floating on the denser substratum. He went on, however, to make an important new statement about how this equilibrium might be achieved. In his words (Fisher, 1881, pp. 275, 278): 'the crust must be in a condition of approximate hydrostatical equilibrium, such that any considerable addition of load will cause any region to sink, or any considerable amount denuded off an area will cause it to rise. ... the crust analogous to the case of a broken-up area of ice, refrozen and floating upon water'.

With these statements Fisher had captured the essence of what would later be called isostasy. According to Fisher, the crust obeyed Archimedes' principle. The weight of a floating block of crust of thickness B and density ρ_{block} that is floating in a fluid substratum of density ρ_{fluid} is then equal to the weight of the fluid displaced (Fig. 1.16):

$$B\rho_{block} = b\rho_{fluid}$$

$$\therefore \frac{b}{B} = \frac{\rho_{block}}{\rho_{fluid}}. \tag{1.3}$$

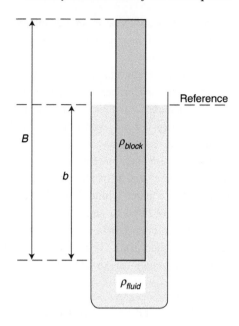

Reference

Figure 1.16 Archimedes' principle, which states that the weight of a block that floats in a fluid is equal to the weight of fluid that is displaced.

where b is the depth that the block is immersed in the fluid. Fisher used the analogy of a relatively light iceberg ($\rho_{ice} = 917$ kg m^{-3}) that floats in denser seawater ($\rho_{seawater} = 1{,}030$ kg m^{-3}) to estimate that the part of a crustal block that would be immersed in the fluid substratum would be about 9/10 of its total thickness. Using Maskelyne's preferred value of $\rho_{crust} = 2{,}750$ kg m^{-3}, this implies $\rho_{substratum} = 3{,}100$ kg m^{-3}, a plausible result if, as was thought in Fisher's time, the material beneath the crust consisted of lava.

The American army officer and geologist Clarence Edward Dutton (1841–1912, Fig. 1.17) was captivated by Fisher's book, writing a complimentary review of it in 1882 (Dutton, 1882) when Dutton was at the United States Geological Survey (Orme, 2007). Like Fisher, Dutton opposed the contraction theory, especially since it was based on vertical rather than horizontal movements and therefore offered no explanation for folding in mountain belts. In Dutton's words (p. 127): 'this hypothesis (interior contraction by secular cooling) is nothing but a delusion and a snare, and the quicker it is thrown aside and abandoned, the better it will be for geological science.'

Dutton's review of Fisher's book was noteworthy in another respect. The review contained the first reference to the term 'isostacy'. In Dutton's view, one of the 'fundamental doctrines' of Fisher's book was the notion that the broader features of the Earth's surface were simply those that were due to its flotation. This idea, he thought, should form an important part of any true theory of the Earth's evolution. He wrote in a footnote (Dutton, 1882, p. 289): 'In an unpublished paper I have used the terms isostatic and isostacy to express that condition of the terrestrial surface which follow from the flotation of the crust upon a liquid or highly plastic substratum ... isobaric would have been a preferable term, but it is preoccupied in hypsometry'.

Figure 1.17 Captain Clarence Edward Dutton, Corps of Ordnance, US Army, who first developed what Bowie (1927) later called the theory of isostasy. Reproduced from fig. 1 in Orme (2007).

It took Dutton a further seven years, however, to publish the paper that he referred to in the footnote. In this subsequent paper, Dutton (1889) pointed out that if the Earth was composed of homogeneous matter, its equilibrium figure would be a true spheroid of revolution; but if some parts were denser or lighter than others, its normal figure would no longer be spheroidal. Where the lighter material accumulated, he argued, there would be a tendency to bulge, and where the denser matter accumulated there would be a tendency to depress the surface. He wrote (Dutton, 1889, p. 53): 'For this condition of equilibrium of figure, to which gravitation tends to reduce a planetary body, irrespective of whether it is homogeneous or not, I propose the name isostasy.' Thus, the term 'isostasy' (note that Dutton now spelled the word with an 's' instead of a 'c') was born.

A major contribution of Dutton's work was to point out the relevance of isostasy to geology. Isostasy, he argued, explained the subsidence of a large thickness of shallow-water sediments and the progressive uplifts of mountain belts. Subsidence and uplift were, in his view, a result of gravitation restoring isostasy to regions that were disturbed on the one hand by sedimentation and on the other by erosion. Dutton, however, also recognised limitations with isostasy, particularly in its inability to explain regional subsidence and uplift such as observed in submerged oceanic plateaus and uplifted mountain platforms that were once at sea level. Isostasy, as defined by Dutton, gives no explanation for such permanent changes of the level surface. On the contrary, he realised that isostasy is a working towards equilibrium that implies a certain degree of 'protection' of the crust against 'massive lowering and raising'.

1.7 The Figure of the Earth and Isostasy

By the end of the nineteenth century, isostasy was still just an idea. It could not be proved by geological observations although, as Dutton pointed out, certain geological facts were in support of isostasy. The proof eventually came not from geology but from geodesy.

The science of geodesy may be considered as comprising two parts: geometrical geodesy and physical geodesy. In geometrical geodesy, triangulation techniques are used together with astronomical observations to determine information on the Earth's shape. Physical geodesy, on the other hand, uses the Earth's gravity field to determine its shape. The results of geometrical geodesy have been of immense practical importance in surveying on land.

The task of surveying the United States during the late 1890s was in the hands of the Coast and Geodetic Survey.[6] As a 'frontier' at this time, the western United States was experiencing booming land sales. It was important, therefore, that a network of monuments be established whose positions were accurately known. By the end of the nineteenth century, the Survey had carried out triangulation surveys along the Atlantic, Gulf and Pacific coasts. In addition, it had measured the astronomic positions at a large number of the triangulation stations. The triangulation of the country as a whole, however, was still not complete. When the triangulation surveys along the coasts were extended into the interior, it was found that significant gaps, overlaps and offsets occurred. These differences had to be eliminated if a national network of stations was to be established.

In 1899, John Fillmore Hayford (1868–1925) took charge of the Computing Division of the Survey and immediately set about the task of adjusting the different triangulation surveys. Early on in the work, Hayford realised that the source of error between the various surveys lay in the astronomical positions that were used to 'fix' the position of the monuments. The reason, he argued, was that these positions were referenced to the geoid, which, in turn, was influenced by the gravitational effect of the local terrain. He therefore corrected the astronomical positions for the gravitational effect of the local topography. The amount of work involved was enormous. This was because it was necessary to compute the effect of the topography above and below sea level, not only at the station, but also over an area within a radius of up to few hundred kilometres from the station point. In order to save time, Hayford constructed specially scaled templates for quickly reading topographic maps and employed a large number of workers (who he referred to in one publication as 'computers'!) to carry out the tedious calculations.

Hayford found that by including the topographic correction he substantially improved the misfit between the different triangulation surveys. However, he found an even better fit if it was assumed that the topography was in some form of isostatic equilibrium. The question was which of the models of compensation that were available at the time should be utilised, Airy's or Pratt's?

---

[6] This began as the Survey of the Coast. In 1836 it was renamed the Coast Survey in 1836 and the Coast and Geodetic Survey in 1878. Its present-day name is the National Geodetic Survey.

For reasons that are not entirely clear, Hayford chose Pratt's model rather than Airy's. Perhaps Hayford, like Pratt, supported the contraction theory. Alternatively, it may have been the convenience of computation (Jeffreys, 1926). In any event, Hayford's choice was a popular one with his colleagues and it set a trend that was to last for more than three decades in which the 'American school' of geodesy preferred the Pratt model for their isostatic studies.

Unfortunately, the Pratt model was not in any form that could be used in geodetic computations. Hayford therefore parameterised the Pratt model. In setting up the model (Fig. 1.18), he made the following assumptions:

- Isostatic compensation is uniform.
- The compensating layer is located directly beneath a topographic feature and reaches to the depth of compensation, D_c, where equilibrium prevails.
- The density of the crust above sea level is the same as the density of the crust at the coast.
- The density of the crust below sea level varies laterally, being less under mountains than under oceans.
- The depth of compensation is everywhere equal.

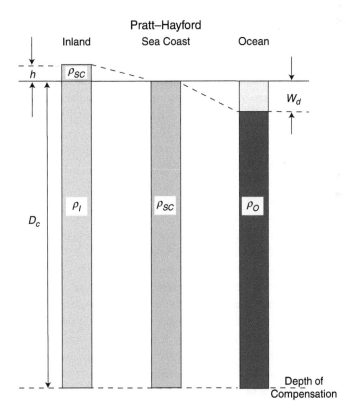

Figure 1.18 The Pratt–Hayford model of isostatic compensation. Reproduced from illustration 10 in Hayford (1909).

We can then write for the pressure P_I at the base of a column of rock beneath a mountain,[7]

$$P_I = \frac{force}{unit\ area} = \frac{mass}{unit\ area}g = \frac{density \cdot volume}{unit\ area}g,$$

$$P_I = D_c\rho_I g + h\rho_{sc}g,$$

and for the pressure P_{sc} beneath a seacoast column,

$$P_{sc} = D_c\rho_{sc}g.$$

Hence, for isostatic equilibrium we have

$$D_c\rho_I g + h\rho_{sc}g = D_c\rho_{sc}g,$$

$$\rho_I = \rho_{sc}\left(\frac{D_c - h}{D_c}\right). \tag{1.4}$$

The density of a column of rock beneath the ocean is then given by

$$\rho_o = \frac{\rho_{sc}(D_c - \rho_w)W_d}{(D_c - W_d)}. \tag{1.5}$$

If we assume $\rho_{sc} = 2{,}670$ kg m^{-3} and $D_c = 113.7$ km (the value that Pratt later argued best minimised isostatic gravity anomalies), then Eq. (1.4) shows that the mean density underlying a 3-km-high elevated region is about 3 per cent less than under the seacoast. Similarly, with $\rho_w = 1{,}030$ kg m^{-3} we find from Eq. (1.5) that the density beneath a 5-km-deep ocean is about 3 per cent more.

The Pratt model was used by Hayford to correct the triangulation surveys of the entire United States. He found that the model successfully reduced the discrepancies between overlapping surveys (Hayford, 1909). Moreover, the model helped reduce the discrepancy between geodetic and astronomic positions. Hayford found, for example, that the discrepancy in the deflection of the vertical was on average reduced to about 10 per cent of what it would have been if no isostatic model had been used. With this success, it is little surprise that the Pratt model continued to be widely used not only in the United States, but also elsewhere, such as India (Burrard, 1920) and Asia (Bowie, 1926). Curiously, there were few attempts to use the model in Europe. As Sidney Gerald Burrard (1860–1943), the Surveyor-General of India, wrote (1920, p. 57): 'The only countries so far that have calculated isostasy for their stations systematically are the United States and India. Europe has not touched the subject.' When in 1924 the Europeans did address the issue, however, it was not Pratt's model they turned to, but Airy's!

During the 1920s most geodetic work in Europe was being carried out in Finland. In 1924, while at the Superior Technical School in Helsinki, Weikko A. Heiskanen (1895–1971) considered the different types of isostatic models available at that time, choosing the Airy

[7] The columns in Figs. 1.18 and 1.19 assume rectangular prisms and a constant depth of compensation. This is the planar approximation. Kuhn (2003) expresses both the Pratt and Airy models in terms of a spherical approximation.

model (Heiskanen, 1931) ahead of Pratt's for his isostatic calculations. Like Hayford before him, he had first to put the model in a more precise form. In doing this, he assumed the following (Fig. 1.19):

- Isostatic compensation is uniform.
- The crust as a whole is floating in a 'sima-layer' according to Archimedes' principle.
- The compensating masses lie directly beneath mountains and oceans.
- The density of the crust is ρ_c everywhere and at every depth, and the density of the underlying 'sima-layer' is ρ_m everywhere and at every depth.
- The thickness of the crust T_c at zero elevation is constant everywhere.

As in Hayford's modification of the Pratt model, let us assume that the pressure at the base of a mountain column is equal to the pressure under a seacoast column. We can write for the pressure at the base of a column beneath the mountain, P_I, that

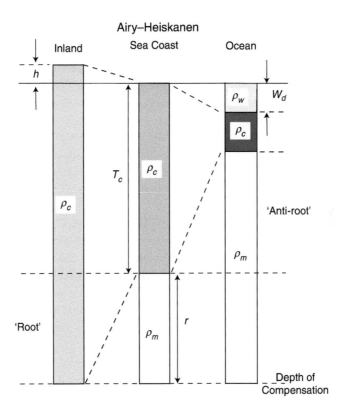

Figure 1.19 The Airy–Heiskanen model of isostatic compensation. Note the depth of compensation has been placed at the base of the deepest root. Unlike the Pratt model where the depth of compensation is at a fixed depth, the depth of compensation in the Airy model will vary spatially.

$$P_I = \rho_c g (T_c + h + r)$$

and for a seacoast column,

$$P_{sc} = T_c \rho_c g + r \rho_m g.$$

Assuming isostatic equilibrium, we can then write that

$$P_I = P_{sc},$$

$$r = \frac{h \rho_c}{(\rho_m - \rho_c)}.$$

(1.6)

Equation (1.6) shows that beneath a 3-km-high mountain the crust would 'project' into the mantle by 12.7 km, assuming values of ρ_c and ρ_m of 2,670 and 3,300 kg m^{-3}, respectively. This projection is called a root. The corresponding thinning of the crust beneath the oceans is called an anti-root.

By 1930, Finland had become one of the world's leading countries for work in physical geodesy. A major focus was to use gravity anomalies to compute the shape of the Earth. This proved particularly useful in poorly surveyed land regions, and at sea where it was difficult to use triangulation techniques. The only problem was that in order to compute the Earth's shape accurately it was necessary to know the gravity field everywhere. It was also important to use gravity anomalies that had been corrected for topography and its isostatic compensation (Lambert, 1930).

Exactly why Heiskanen chose the Airy model for his geodetic work in preference to Pratt's model is again not clear. In 1931, Heiskanen wrote (1931, pp. 113–14): 'the desire of helping isostatic research led W. Bowie during the Stockholm conference of the International Geodetic and Geophysical Union to suggest to me that I calculate complete isostatic tables based on Airy's hypothesis.'

W. Bowie (1872–1940), who had joined the US Coast and Geodetic Service as an assistant to Hayford in 1909, was a firm supporter of the Pratt model. Only three years prior to the Stockholm meeting he had written (Bowie, 1927, pp. 236–7): 'the Airy theory is untenable . . ... The Pratt theory is the only one so far that is sound.' So, was Bowie being provocative in his suggestion to Heiskanen that he use the Airy model? Heiskanen later clarified the situation. In 1966, Heiskanen wrote (1966, p. 777) that his reason for adopting the Airy model was because 'from the geophysical point of view, the Pratt–Hayford assumption did not seem suitable'.

Because of Heiskanen's work, the Airy model became well accepted throughout Europe. As a result, European and American geodesists were brought into a conflict with each other. The two schools of thought, however, were generally quite satisfied with the validity of isostasy. From their viewpoint, both the Pratt and the Airy models gave a satisfactory description of the density distribution of the outer layers of the Earth. A major question that remained was what was the geological significance, if any, of the different isostatic models?

1.8 Bowie's Illustration of Isostasy

Although a geodesist, Bowie was concerned with how geologists viewed isostasy (Bowie, 1927). He thought that geologists had difficulty in visualising isostasy, and so during his career he took several steps to try to explain its meaning. One of these, described in his book, involved setting up an experiment to demonstrate the simple physics that underlies the Airy and Pratt models of isostasy. The experiment was based on metal blocks that floated in an underlying fluid – materials that could be found in any physics laboratory.

Pratt isostasy is illustrated in Fig. 1.20a. Here, the blocks of different metals have been cast into prisms with the same cross-sectional area. Each prism has the same mass, but differs in length inversely with its density. If the prisms are placed in a vessel that is partly filled with mercury, each prism, because it has the same mass and cross-sectional area, displaces the same amount of mercury. The prisms exert the same pressure on the mercury, on which they rest, and so the lower surface of the prisms will be at the same level. We refer to the surface on which pressures are equal as the depth of compensation.

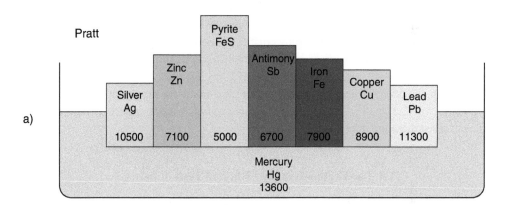

a)

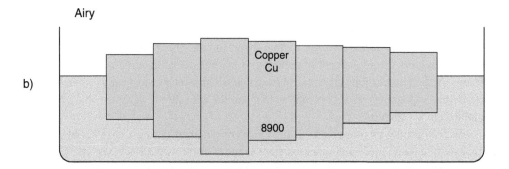

b)

Figure 1.20 Bowie's illustration of the (a) Pratt and (b) Airy models of isostasy using metal blocks floating in mercury. Reproduced from figs. 4 and 5 of Bowie (1927) with permission of E. P. Dutton, a division of Penguin Putnam Inc.

Airy isostasy is illustrated in Fig. 1.20b. Here, the prisms are made up of just one metal and, hence, are of equal density. In this case, not only do the longer prisms stand higher than the shorter ones but also, they project downwards further into the underlying mercury.

The depth of compensation is more difficult to define for the Airy model than the Pratt model. One way is to consider it as the base of the prism that projects the greatest depth into the underlying fluid. This prism stands the highest in the fluid. All other prisms will stand shorter by an amount that is in proportion to the height of their bases above the base of the highest prism. The depth of the lower surface of the highest prism will therefore be a surface of constant pressure, like the depth of compensation in the Pratt model. The main difference with the Pratt model is that for all prisms, except the prism that projects the greatest depth, the pressure is made up of the weight of both the prism and the mercury.

A common feature of both the Pratt and Airy models is that individual prisms are separated from each other such that they are able to move up or down without constraint from their neighbours. Under these conditions, a prism that has mass added to it would move downward, while one that had mass removed from it would move upward. The material beneath the prisms would accommodate these movements by flowing, like a fluid. Therefore, if mass was removed from one prism and added to another, the fluid would move out from under the loaded prism to beneath the prism that had been unloaded.

Bowie pointed out that in the Earth's crust there would, in fact, be great frictional and shearing resistance to any vertical movements arising from shifting of material from one region to another. However, he wrote (Bowie, 1927, p. 25): 'Since isostasy exists the crust must go up and down.' The main questions that remained were the extent to which it went 'up and down' and whether or not there were instances where the strength of the crust might, in fact, resist these movements.

1.9 The Earth's Gravity Field and Tests of Isostasy

The quantification of the Pratt model by Hayford and the Airy model by Heiskanen was of immense importance, not only for studies of the shape of the Earth, but also because they provided the means to test the extent of isostasy on a planetary scale. One test, for example, would be to calculate the gravitational effect of the models and compare it with measurements of the Earth's gravity field.

The gravity field depends on the Earth's shape and the distribution of masses within it. Isostasy involves the adjustment of the crust to the transfer of material across its surface. Therefore, departures of the gravity field from that which would be expected from models of the Earth's topography and its compensation should indicate the scale to which isostatic equilibrium is achieved.

The first gravity measurements in the US were obtained during the 1890s by George Rockwell Putnam (1865–1953), an assistant to the director of the US Coast and Geodetic Survey. The measurements were based on observations of the period of a swinging pendulum, T_p, which depends inversely on the local value of the acceleration due to gravity, g. The relationship between g and T_p can be derived by consideration of the concept of a fictitious mathematical pendulum, which is a dimensionless mass hanging on a weightless thread. If the pendulum length is l, we can write

$$g = \frac{4\pi^2 l}{T_P^2}. \tag{1.7}$$

In practice, several corrections have to be made to pendulum observations before they can be used to determine g. One of the most important is to correct for the movement of the stand that supports the pendulum. Other corrections that need to be applied are for temperature and pressure.

A practical difficulty was the long time that was required to make a gravity measurement. Putnam addressed this problem by using the new generation of light, portable pendulums, such as the quarter-second pendulum (Fig. 1.21). Unlike the early reversible pendulums which were of relatively long length, these relatively short-length pendulums were swung in a case in which the pressure had been reduced to one tenth of atmospheric using a portable air pump. By this means, the period of the more portable quarter-second pendulum's swing could be increased and the time therefore that it took to make a gravity measurement reduced. By 1894, Putnam had measured gravity along a transect of the United States from the east to the west coast (Putnam, 1895). In all, he acquired 26 measurements in 150 days. The highest measurement was obtained at Pike's Peak, Colorado, at 4,312 m above sea level (the United States Geological Survey reported the height as 4,302 m in 2002), where it was found that the mean value of gravity after swinging three quarter-second pendulums twice was 9.78094 m s^{-2}.

Putnam was not just concerned with the measurement of gravity. He was also interested in using gravity data to determine information about isostasy. But, first, it would be necessary to correct the observations for the shape and rotation of the Earth, the height of the station, and the mass that existed between a station and mean sea level.

To correct for the flattening of the Earth and its rotation, Putnam used a formula derived by the German geodesist Friedrich Robert Helmert that expressed the variation in gravity on an ellipsoid of revolution with a flattening, f_e, of 1/293.5. The derivation of this formula is based on Clairaut's first-order theorem and is obtained by solving Laplace's equation to a spheroidal boundary. Clairaut's theorem gives

$$g_{ellipsoid} = g_e(1 + \beta_e \sin^2\phi....), \tag{1.8}$$

where

$$\beta_e = \frac{5m}{2} - f_e,$$

$$m = \frac{\omega^2 a}{g_e}, \qquad f_e = \frac{a-b}{a}$$

and $g_{ellipsoid}$ is the value of gravity at some point on the ellipsoid, g_e is the value of gravity at the equator, ϕ is the latitude, ω is the angular velocity of the Earth's rotation, and a and b are

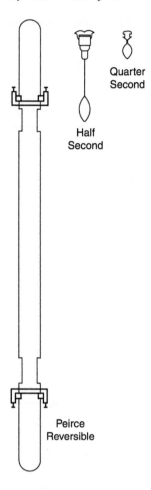

Figure 1.21 Comparison of the Peirce reversible pendulum that was swung in air with the half-second and quarter-second pendulums. The Peirce pendulum was used to make absolute gravity measurements at the Washington, DC gravity base station. The quarter-second pendulums, which were swung in a pressure-reduced case, were part of a portable measuring system used by George Rockwell Putnam to measure gravity in the United States during 1894. The complete apparatus consisted of three quarter-second pendulums that, together with their air pump, chronometer, and packing case, weighed 318 kg. Reproduced from fig. 3 in Putnam (1895).

the semi-major and semi-minor axes, respectively, that describe the ellipsoid that best fits the Earth's flattening. Helmert obtained the following:

$$g_{ellipsoid} = 9.78066(1 + 0.005243 \sin^2\phi\ldots), \qquad (1.9)$$

where $g_{ellipsoid}$ is in Gals (1 Gal = 1 m s^{-2} = 1,000 mGal). More widely used during the last century, however, have been the 1930 International Gravity Formula and the 1980 Geodetic

Reference System. The 1930 formula was based on $f_e = 1/297.0$ and a value of g_e derived empirically from terrestrial gravity measurements based on the Potsdam worldwide gravity network and is given by Heiskanen and Vening Meinesz (1958) as

$$g_{ellipsoid} = 9.780490(1 + 0.0052884 \sin^2\phi - 0.0000059 \sin^2 2\phi...).$$

The 1980 formula, on the other hand, was based on a satellite-derived $f_e = 1/298.257$ and a value of g_e corrected for the ~14 mGal difference between the older Potsdam and newer International Gravity Standardisation Net and is given, for example, by Li and Götze (2001) as

$$g_{ellipsoid} = 9.780327(1 + 0.0053024 \sin^2\phi - 0.0000058 \sin^2 2\phi...).$$

The variation of $g_{ellipsoid}$ with ϕ is illustrated in Fig. 1.22 where it is compared to the maximum range of gravity values measured on Earth's surface.

The correction for height of an observation station above mean sea level was computed using the 'Faye correction', later referred to as the free-air correction or *FAC*. This correction assumes no mass between the station and sea level. Then we can write

$$g_{geoid} = g_{observed} \pm FAC.$$

The plus sign represents an observation point above the geoid, and the minus sign a station below it. Now, we can also write

$$g_{observed} = \frac{GM_e}{(r_e + h)^2}, \qquad g_{geoid} = \frac{GM_e}{r_e^2},$$

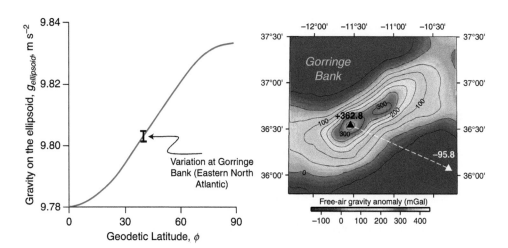

Figure 1.22 The variation in gravity over the surface of the Earth due to its shape and rotation. The bar shows the range of free-air gravity anomalies at the Gorringe Bank, which is associated with one of the largest variations in gravity (~459 mGal) on Earth.

where r_e is the mean radius of the Earth, M_e is the mass of the Earth and h is the height of the station above sea level. If we neglect higher-order terms in the expansion (i.e., h^2/r^2_e, h^3/r^3_e ...), then we can write

$$g_{observed} = \frac{GM_e}{r_e^2}\left(1 - \frac{2h}{r_e} + ...\right),$$

$$g_{geoid} = \frac{GM_e}{r_e^2},$$

$$FAC = g_{observed} - g_{geoid},$$

$$\therefore\ FAC = \frac{2h}{r_e}g_{geoid}.$$

(1.10)

With $g_{geoid} = 9.81$ m s^{-2} and $r_e = 6{,}370$ km, this gives, neglecting high-order terms, $FAC = 0.3080$ mGal m^{-1}. The difference in gravity on the geoid (which corresponds to mean sea level) and the reference ellipsoid is defined as the free-air anomaly, FAA.[8] We then have

$$FAA = g_{geoid} - g_{ellipsoid},$$

$$FAA = (g_{observed} \pm FAC) - g_{ellipsoid}.$$

(1.11)

Finally, in order to correct for the effect of mass distributions between the station and the geoid, Putnam used an ingenious method of constructing rings around a station point and computing their effects using expressions derived from the gravity effect of a cylinder. We can write the following formula for the gravitational attraction of elementary mass dm acting on a unit load at a station S above the axis of a vertical cylinder (Fig. 1.23):

$$df = \frac{Gdm}{d^2},$$

where d is the distance between S and dm. For the Cartesian co-ordinate system shown in Fig. 1.23,

$$d = \left(x^2 + y^2 + z^2\right)^{0.5}.$$

The component of df in the *direction* of gravity, dg, is given by

$$dg = df\frac{z}{d}$$

$$dg = \frac{Gdmz}{\left(x^2 + y^2 + z^2\right)^{1.5}}.$$

It is more convenient to replace x, y by polar co-ordinates r, θ. We can then write

---

[8] This definition does not include a correction (known as the 'indirect effect') for the gravity effect of the height difference between the geoid and the ellipsoid. Geoid anomaly maps show that these height differences reach ± 100 m and so the correction could be as high as ± 31 mGal. However, as Chapman and Bodine (1984) pointed out, the correction will only be significant at wavelengths > ~500 km. In local studies the correction appears as a constant offset and so is usually ignored in interpretation.

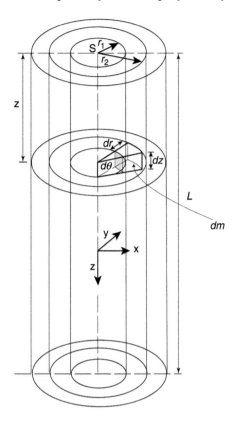

Figure 1.23 The calculation of the gravity effect of a vertical cylinder by consideration of the effect of a small, wedge-shaped element of mass.

$$r^2 = x^2 + y^2, \qquad dm = \rho r d\theta dz dr,$$

where ρ is the average density of the elemental mass. The gravity effect of a ring of the cylinder can be obtained by integrating the effect of an elemental mass over the radius and height of the ring. We then get

$$\Delta g_{ring} = \int_0^L \int_0^{2\pi} \int_{r_1}^{r_2} \frac{G\rho z}{(r^2 + z^2)^{1.5}} r d\theta dz dr,$$

$$\int_0^{2\pi} d\theta = 2\pi,$$

$$\int_{r_1}^{r_2} \frac{z}{(r^2 + z^2)^{1.5}} r dr = \left(\frac{z}{(r_1^2 + z^2)^{0.5}} - \frac{z}{(r_2^2 + z^2)^{0.5}} \right),$$

$$\Delta g_{ring} = 2\pi G\rho \int_0^L \left(\frac{z}{(r_1^2 + z^2)^{0.5}} - \frac{z}{(r_2^2 + z^2)^{0.5}} \right) dz,$$

$$\int_0^L \frac{z}{\left(r_1^2 + z^2\right)^{0.5}} \, dz = \left| \left(r_1^2 + z^2\right)^{0.5} \right|_0^L,$$

$$\int_0^L \frac{z}{\left(r_2^2 + z^2\right)^{0.5}} \, dz = \left| \left(r_2^2 + z^2\right)^{0.5} \right|_0^L, \qquad (1.12)$$

$$\Delta g_{ring} = 2\pi G\rho \left[\left(r_1^2 + L^2\right)^{0.5} - \left(r_2^2 + L^2\right)^{0.5} - r_1 + r_2 \right].$$

This is the form of the equation used by Putnam (1895, p. 44). So, given the height of a ring (e.g., above sea level), L, its inner radius, r_1, and outer radius, r_2, and density, ρ, it is possible to compute its gravity effect. The total correction was obtained by summing the contributions of individual 'rings' around a station. This correction is known as the Bouguer correction (*BC*).

For the special case that the topography extends at similar heights in all directions at a station, then

$$r_1 \to 0 \text{ and } r_2 \to \infty.$$

The Bouguer correction, *BC*, then becomes

$$BC = 2\pi G\rho L. \qquad (1.13)$$

Equation (1.13) is known as the 'Bouguer slab formula' and is widely used in gravity anomaly reduction and interpretation. The formula assumes a planar surface, which is valid for the Bouguer reduction in small survey areas. In larger areas, account should be taken of Earth's curvature. Qureshi (1976), for example, gives the formula for the gravitational attraction of a spherical cap of mass.

The Bouguer anomaly, BA, is the free-air anomaly corrected for the gravity effect of the masses above or below sea level. Hence,

$$BA = FAA - BC. \qquad (1.14)$$

Putnam calculated the free-air and Bouguer anomaly at the 26 stations in the western United States where he had obtained gravity measurements (Table 1.2). He found that the Bouguer anomaly values over the mountainous regions of Colorado were strongly negative, which indicated to him that there was a significant deficit of mass beneath these regions. He went on to suggest that this was the deficiency of mass that compensates for the excess of mass of the elevated regions. What intrigued Putnam was that the deficit was largest at Gunnison (2,340 m) in the central part of the Rockies, and not at Pikes Peak, the most elevated station. He attributed this to the existence of an excess of gravity in the vicinity of Pikes Peak, pointing out that this was also visible in the free-air anomaly over the region.

Even though Putnam's work was based on only 26 stations, it was an important landmark in the development of the concept of isostasy. For example, he argued (Putnam, 1895, p. 51) that 'the results of this series would therefore seem to lead to the conclusion that general continental elevations are compensated by a mass deficiency of density in the matter below

Table 1.2 *Summary of some of the measurements acquired by G. R. Putnam during the 1894–1895 trans-continental gravity survey of the western United States*

Station	Elevation (m)	Free-Air Anomaly (mGal)	Bouguer Anomaly (mGal)
Gunnison, CO	2,340	−7	−263
Pikes Peak, CO	4,293	226	−239
Salt Lake City, Utah	1,322	−53	−179

sea level, but that local topographical irregularities, whether elevations or depressions, are not completely compensated for, but are maintained by the partial rigidity of the earth's crust.' This is one of the earliest statements made concerning the spatial extent of isostasy in continental regions, and the first to hint isostatic equilibrium may not be complete everywhere.

By 1909, many gravity measurements had been obtained in other regions in the United States. To test the extent of isostasy more routinely at each station, Hayford devised a procedure similar to the one used earlier by Putnam. The main difference was that Hayford computed the gravity effect of isostatic compensation at each station using a model of isostasy, in his case Pratt. Another difference was that he divided the rings of a cylinder centred on each station into compartments (e.g., Fig. 1.24). It follows from the work of Putnam that the gravity effect of a compartment is given by

$$BC_{compartment} = G\rho(\theta'_1 - \theta'_2)\left[\left(h^2 + r_1^2\right)^{0.5} - \left(h^2 + r_2^2\right)^{0.5} - r_1 + r_2\right], \qquad (1.15)$$

where ρ is the density of the rock mass above sea level, h is the height of the compartment above sea level, r_1 is its inner radius, r_2 its outer radius, θ'_1 the azimuth of one side of the compartment and θ'_2 the azimuth of the other side. The gravity effect of each ring (or 'zone' as Hayford defined them) can then be obtained by summing the effects of individual compartments which could have different heights.

The gravity effect of the compensation, or isostatic correction (IC), can also be computed from these parameters. Using the Pratt–Hayford scheme for land areas gives

$$IC_{compartment} = G\rho\frac{h}{D_c}(\theta'_1 - \theta'_2)\left[\left(D_c^2 + r_1^2\right)^{0.5} - \left(D_c^2 + r_2^2\right)^{0.5} - r_1 + r_2\right]. \qquad (1.16)$$

The anomaly that was obtained by subtracting the BC and the IC from the free-air gravity anomaly was defined by Hayford as 'the new method anomaly'. We now know it as the 'isostatic anomaly'. The isostatic anomaly is given by

$$IA = FAA - BC - IC. \qquad (1.17)$$

In 1909, Hayford presented the results of his isostatic anomaly calculations for 56 stations in the United States. He found that the isostatic anomaly was smaller than either the free-air or the Bouguer gravity anomaly. Since the isostatic anomaly incorporates corrections not only

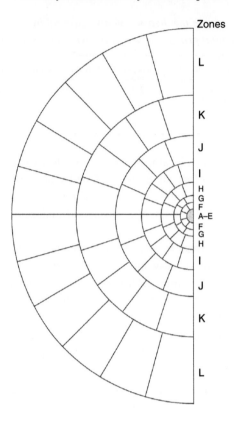

Figure 1.24 An example of one-half of a template used by John Fillmore Hayford to compute the gravitational effect of topography and its compensation at a gravity station. Each template would be drawn to a particular topographic map scale. The 'zones' are circles around a station, each of which is divided into compartments. The topography was averaged in each compartment and the gravity effect looked up in tables. Even with the tables, Hayford's assistants took an average of 17 hours to calculate the topographic and isostatic corrections at an individual station.

for the topography but also for the compensation, the magnitude of the anomaly provides some measure of the degree of compensation of a region: small-amplitude isostatic anomalies indicate that a geological feature is in some form of isostatic equilibrium, while large-amplitude anomalies indicate that a feature is either partially compensated or is uncompensated. Hayford argued that, since isostatic anomalies were generally small at all the stations, isostasy prevailed.

Hayford's work in the United States triggered an investigation into the state of isostasy in other regions. Using similar gravity reduction procedures, Bowie and others demonstrated that isostatic anomalies were also small at stations in southern Canada, India, Europe, western Siberia and islands in the Pacific Ocean. Bowie (1922, p. 49) wrote that 'it seems safe to assert that the theory of isostasy has been proven and that it may now be spoken of as the isostatic principle'. And by 1924 Heiskanen had completed a preliminary set of tables that could be used to compute the isostatic correction based on the Airy model. It therefore became possible in the United States and elsewhere to compare the isostatic anomalies based on the Pratt and Airy models. As Table 1.3 shows, the Bouguer gravity anomalies for stations in the mountainous regions are significantly reduced using both models of isostasy. The Pratt–Hayford model is associated with the smallest anomalies at two of the stations, while the Airy–Heiskanen model is associated with the smallest anomalies at the other two stations. Thus, gravity provides a test of isostasy, although it did not permit the early isostasists to distinguish between the different models.

Isostatic anomalies have proved of much value in determining the extent of isostatic equilibrium in a region. They also provide information on the depth of compensation in the Pratt model and the thickness of the crust at zero elevation in the Airy model. Hayford and Bowie (1912), for example, concluded that the best overall fit to gravity data from the United States was for a depth of compensation of 113.7 km. However, Heiskanen (1931) preferred a depth in the range of 20–40 km. The reason for the difference in depths is that these investigators used different models of isostasy. In the Airy model, the compensation takes the form of a relatively large density contrast at shallow depth, while in the Pratt model it is in the form of a small contrast extending to greater depths. If individual stations rather than the average are considered, however, large variations in estimates of the depth of compensation occur (e.g., Table 1.4).

The pioneering studies of Hayford, Bowie and Heiskanen were followed by several studies which used the isostatic gravity anomaly to assess the state of equilibrium of the Earth's crust and mantle. Early studies using either profiles (e.g., Fig. 1.25) or maps (e.g., Fig. 1.26) were based on the Pratt–Hayford scheme. Later, the Airy–Heiskanen scheme was used.

But, as pointed out earlier, gravity data were not able to distinguish between the two models of isostasy. The Pratt–Hayford model had certainly been used more widely than the Airy–Heiskanen model. However, as Jeffreys wrote in 1926 (p. 169): 'Convenience of computation, and perhaps tradition, rather than any physical probability, have been the chief reasons for the attention given to Pratt's hypothesis, and the lack of it to Airy's.' While there was therefore disagreement among the geodesists about the type of model to use in isostatic

Table 1.3 *Comparison of the isostatic anomaly computed using the Pratt–Hayford and Airy–Heiskanen models for representative stations in North America*

Station	Elevation (m)	Free-Air Anomaly (mGal)	Bouguer Anomaly (mGal)	Pratt–Hayford Isostatic Anomaly ($D_c = 113.7$ km) (mGal)	Airy–Heiskanen Isostatic Anomaly ($T_c = 40$ km) (mGal)
Gunnison, CO	2,340	35	−221	28	34
Pikes Peak, CO	4,293	224	−196	29	19
Salt Lake City, Utah	1,322	−15	−138	18	26
Truckee, CA	1,805	45	−154	−20	−4

Source: Based on data in table 7–1 of Heiskanen and Vening-Meinesz (1958).

Table 1.4 *Sensitivity in the Pratt–Hayford model to variations in the depth of compensation and in the Airy–Heiskanen model to variations in the thickness of the crust at zero elevation for a representative station in North America and Europe, respectively*

Pratt–Hayford: Pikes Peak, Colorado, USA	
Depth of Compensation (km)	Pratt–Hayford Isostatic Anomaly (mGal)
42.6	−204
85.3	+57
127.9	+32
Airy-Heiskanen: Sonnblick, Alps, Austria	
Thickness of crust at zero elevation (km)	Airy–Heiskanen isostatic anomaly (mGal)
20	−10
30	−24
40	−35

Source: Based on data from table 7–5 of Heiskanen and Vening-Meinesz (1958).

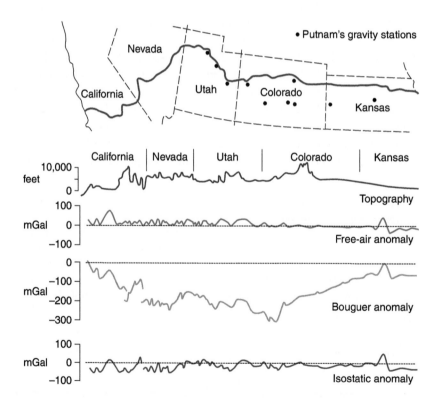

Figure 1.25 The first continuous trans-continental gravity anomaly and topography profile of the western United States. The purple solid line shows the profile, and the filled circles show the stations previously occupied by Putnam. Profile reproduced from plate 3 of Woollard (1943).

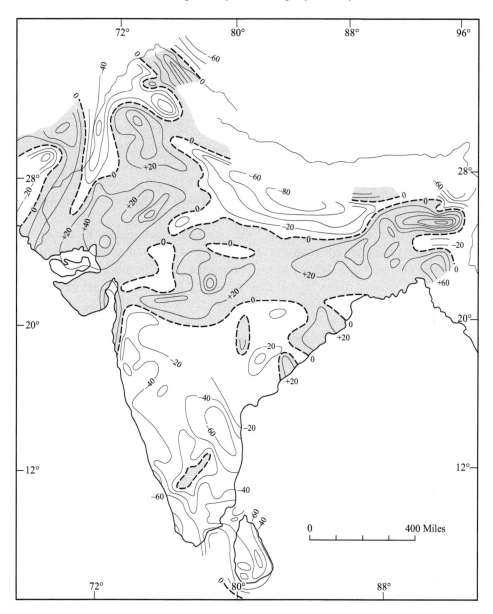

Figure 1.26 Isostatic gravity anomaly map of India based on the Pratt–Hayford hypothesis with $D_c = 113.7$ km. Green-shaded regions highlight positive isostatic anomalies. India has been considered the birthplace of isostasy because many of the early geodetic surveys were carried out there. Northern India, however, is associated with large-amplitude isostatic anomalies indicating a region out of isostatic equilibrium at least with regard to the Pratt model. Figure reproduced from fig. 9 of Daly (1939).

anomaly reductions, it was clear they thought that isostasy was an important doctrine that needed to be considered in geology. The problem was convincing the geologists. During the 1930s, for example, they appear to have been quite confused about the different models of isostasy and what information they implied, if any, about geological processes.

1.10 Isostasy and Geological Thought

The tests of isostasy that were carried out using gravity data convinced the geodesists that on the scales of continental masses, if not on the scales of mountain peaks and valleys, isostasy was operative. For example, Heiskanen wrote in 1931 (p. 111): 'isostasy is no longer a hypothesis but a principle positively verified, which must be taken into consideration in geological investigations'.

But geological thought during the early 1900s continued to be perplexed by isostasy. As Greene (1982) points out in *Geology in the Nineteenth Century*, the renowned Austrian geologist, Eduard Suess (1831–1914), declared himself a 'heretic in all regarding isostasy'. The American geologist Rollin Thomas Chamberlin (1931, p. 1) summarised the situation as follows: 'various geological observations and deductions, which the geologist regards as established facts, seem inconsistent with the isostatic theory as now applied by its leading advocates'. Chamberlin had little doubt about who to blame for the confusion (pp. 1–2): 'To the geologist the recognised geologic facts, which he can understand and appreciate, are vastly more convincing than mathematical interpretations based upon assumptions, some of which he does not understand, and others of which seem to him clearly at variance with actual earth conditions.' The principal problem was that the models used by the advocates of isostasy were static. Although the models enabled geodesists to calculate the disturbing effects on the gravity field of topography and its compensation, they provided little information on the origin of individual geological features. The geologists, on the other hand, based their work on careful field observations. By and large they were satisfied that they understood the processes that were responsible for the origin of the Earth's surface features and they did not find it necessary to take isostasy into account.

The following facts, for example, seemed to Chamberlin (1931) to contradict the theory of isostasy:

- The substantial amounts of shortening observed in mountain ranges could not be explained by simple vertical isostatic movements.
- Isostatic models could not explain the existence of peneplane surfaces without requiring large amounts of erosion.
- The substantial amounts of sediments in geosynclines could not be explained by isostatic processes alone, and so other factors must be operative.
- The pattern of rebound of the continents following melting of the Pleistocene ice sheets did not follow the pattern of the ice load very closely.

Also puzzling to Chamberlin (1931) were the large variations in the depth of compensation that were required by Hayford and Bowie to minimise isostatic anomalies. Chamberlin questioned whether the depth of compensation was a physical entity. The query was understandable. All isostatic models tend to an equilibrium state. In an Airy model, the depth of compensation varies such that if crust is removed or added to, say by erosion or sedimentation, then isostatic equilibrium will be restored as the uniform density crust responds by thinning or thickening. But in a Pratt model, the depth of compensation is fixed and it is not immediately obvious exactly how equilibrium would be restored following a disturbance. In Chamberlin's view, geological processes were complex, and he saw no reason why the process responsible for compensation of the Earth's surface features should not be likewise.

Chamberlin's views were shared by a number of other geologists at the time, notably Shepard (1923), King-Hubbert and Melton (1930) and de Lyndon (1932). Shepard (1923, p. 208), for example, wrote: 'In spite of the apparently indisputable evidence that the earth's crust is in almost perfect isostatic adjustment, geologists have been slow to accept the theory of isostasy.' He pointed out the difficulties of explaining the peneplanation of mountain ranges and their subsequent uplift in terms of isostatic models. Shepard attempted to explain these difficulties, however, by introducing other forces, such as lateral compression and uplift associated with volcanism. King-Hubbert and Melton (1930, p. 695), on the other hand, were not as condescending as Shepard. They wrote that 'the theory of isostasy must, for the present, be regarded as resting upon a not too secure foundation and is hardly trustworthy for use as a major premise in present discussions of Earth problems'.

The other difficulty, raised by de Lyndon (1932), concerned structural styles of mountain belts. He argued that the most striking features of the Uinta mountains (Colorado Rockies) were the over-turned anticlines and low-angle thrust faults which together proved that compression, not isostasy, was responsible for orogeny. The principle of isostasy should be considered in his view, but it was a minor, not a major contributor to Earth's topography.

We learn from these discussions that despite the usefulness of isostasy to the geodesists, the concept remained a difficult one for the geologists to accept. Geologists such as Chamberlin recognised that isostasy was rooted in careful physical measurements and that it was supported by simple physical reasoning. But they were disappointed that despite all the hard work that Hayford, Bowie, Heiskanen and other geodesists had put in to quantify the Airy and Pratt models, the concept of isostasy could not be of more use in the solution of geological problems.

Not all geologists shared this disappointment. As Harry Fielding Reid, a contemporary of Bowie and Chamberlin, cautioned (1922, p. 317): 'Geologists often ask too much of the principle of isostasy. When they find that it will not explain all earth movements, they think it is not a true principle.'

1.11 Nature's Great Isostatic Experiment

A few years after Chamberlin's paper was published, Reginald Aldworth Daly (1871–1957), a Canadian geologist, published *The Changing World of the Ice Age*. In the book Daly (1934) describes the pattern of uplift and subsidence that followed loading of the North

America and Fennoscandia crust by the great Pleistocene ice sheets. These patterns provide important information on the physical nature of the crust and its underlying mantle.

About 18 ka, a large portion of the Northern Hemisphere was covered by thick ice sheets, the largest of which were centred on Hudson Bay in Canada and on Sweden. There is evidence that the ice sheets were on the crust for a sufficiently long time that a state of isostatic equilibrium prevailed. Then, about 12.5 ka, the ice sheets began to melt. The melting upset the equilibrium, releasing large amounts of water into the surrounding oceans. The adjustment of the crust and upper mantle to shifts in ice and water loads is the subject of glacial isostasy.

The history of glacial-isostatic movements is recorded in raised (Fig. 1.27) and sub-merged (Fig. 1.28) beaches. By dating shells in the beach deposits using radio-carbon techniques it is possible to reconstruct the uplift and subsidence history of a region that has undergone a glaciation or de-glaciation event. Figure 1.29 shows, for example, the movements of a 6-ka surface that, prior to the melting event, was a horizontal surface at or near sea level. The figure shows that those regions that were loaded by the ice sheet are rising, while peripheral areas are sinking.

Daly, who had built on the earlier contributions of Thomas Jamieson and Nathaniel Shaler (Wolf, 1993), examined the pattern of subsidence and uplift around the large Pleistocene ice sheets and concluded that they sank for two main reasons: elasticity and plasticity. He argued that the crust and upper mantle would respond to the weight of an ice

Figure 1.27 A view to the south of Richmond Gulf (Hudson Bay, Canada) showing the lichen-covered raised beaches. Photo courtesy of the late R. W. Fairbridge of Columbia University.

Figure 1.28 A drowned river valley or 'ria' along the south-west coast of England. Photo courtesy of the late R. W. Fairbridge of Columbia University.

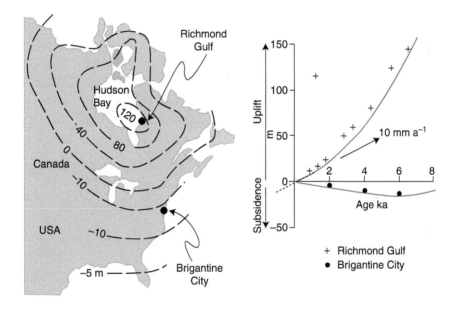

Figure 1.29 The pattern of glacio-isostatic rebound that followed the retreat of the Laurentide ice sheet. The graph shows the uplift measured at Richmond Gulf and the subsidence measured at Brigantine City (New Jersey, USA). Data reconstructed with permission from fig. 1 of Walcott (1973) and figs. 5 and 10 of Walcott (1972b).

sheet as an elastic material, such that if the ice load were removed through melting, then the crust and upper mantle would rebound elastically. If, however, the ice was on the crust long enough, then there would be flow such that the stronger crust would be 'basined' by the weaker underlying mantle. On melting, the crust would initially respond elastically, followed by a slow response as material in the mantle flowed back from flanking regions towards the loaded region.

In Daly's view, the areas peripheral to an ice sheet were critical to understanding the mechanisms of adjustment. He believed, for example, that material in the mantle would flow out from beneath the load and into flanking regions. This pattern of flow would up-arch the strong crust in flanking regions, producing a peripheral bulge. On melting, the bulges would collapse, forming troughs adjacent to the uplift area (Fig. 1.30).

Unfortunately, Daly was unable to corroborate these views in observed subsidence and uplift data. He therefore proposed a hypothesis, known as the 'punching hypothesis', which he thought better explains the observations. According to this hypothesis, ice sheets are large enough loads that they 'indent' the strong crust, setting up vertical shear stresses in the mantle that underlies the ice sheet. Eventually, mantle materials respond to these stresses by flowing. Because the whole mantle supports the ice sheet, at least momentarily, then on melting the whole mantle flows. A consequence of the punching hypothesis is that upon loading the flanking areas are dragged down but are eventually uplifted, whereas upon unloading the flanking areas are uplifted, then they subside.

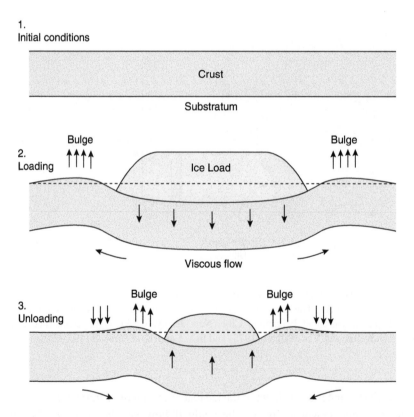

Figure 1.30 The 'viscous bulge' hypothesis to explain the pattern of uplift and subsidence of a deglaciated region. Reproduced from fig. 69 of Daly (1934) with permission of Yale University Press.

A strong crust, as Haskell had earlier pointed out, is not actually required to explain the observed uplift and subsidence patterns. He developed a formula, based on the motion of a viscous fluid under a surface load (Haskell, 1935), in which the time, t, to reach full (i.e., asymptotic) isostatic compensation is given by (Haskell, 1937, eq. 5)

$$t \sim \frac{20\eta}{\rho g w},$$ (1.18)

where g is gravity, w is one-half width of a surface load, and η and ρ are the viscosity and density of the fluid respectively. Substituting Haskell's preferred values of $w = 1{,}000\,\mathrm{km}$, $g = 9.81\,\mathrm{m/s^2}$, $\rho = 3{,}000\,\mathrm{kg\ m^{-3}}$ and $t = 18\,\mathrm{ka}$ into Eq. (1.18) yields a viscosity, required to explain the rise of the Fennoscandia area after melting of the ice, of $\sim 8.35 \times 10^{20}$ Pa s. Therefore, flow in the mantle occurs naturally, and a strong crust is not required to explain observations. Haskell favoured Daly's punching hypothesis, however, pointing out that the only way to produce the bulges would be to restrict the flow to a narrow, viscous channel beneath the region of the load.

An important feature of Daly's hypothesis is that it involves some form of flow in the mantle that underlies the crust. As such, the hypothesis is in accord with isostasy. But isostasy as envisaged by Pratt and Airy is based on two fundamental assumptions:

- that features on the Earth's surface – no matter how small – are compensated for either by changes in crustal thickness or by lateral changes in density of the crust and mantle.
- that the crust is 'floating' in a state of hydrostatic equilibrium with its surroundings, such that the mantle that underlies the crust has little or no strength.

The first assumption cannot be easily tested using the pattern of isostatic adjustments in North America and Fennoscandia. The reason for this is that the ice sheets that loaded these regions were so large that they caused significant yielding of the crust, largely uninhibited by its physical properties. Only perhaps in localised regions was the mass of the ice sheet sufficiently small not to cause yielding. Since these loads, both large and small, have largely disappeared, it is not possible to examine what relationship existed, if any, between the size of the ice sheet and the degree of yielding of the underlying crust.

Daly was, however, able to test the second assumption. His approach (Daly, 1939) was to use the information that was present in the gravity anomaly concerning the stresses that existed in the Earth's outermost layers. He found, for example, that the former Fennoscandia ice sheet is associated with a free-air gravity anomaly low of about −20 mGal. Daly used the amplitude of the gravity anomaly as a measure of the compensation and hence the amount of ice that had originally been added to the crust. He found, using the Bouguer slab formula, that the equivalent height of an ice load of density $917\,\mathrm{kg\ m^{-3}}$ is 520 m.

Once the load had been estimated, Daly was able to apply the theory of George Howard Darwin, the son of Charles Darwin, for the deformation of an elastic half-space to calculate the maximum stress difference required to support that load (Darwin, 1908). Darwin showed that for a periodic load of the form

$$\rho g h \cos\left(\frac{2\pi z}{\lambda_w}\right), \qquad\qquad \rho g h \sin\left(\frac{2\pi z}{\lambda_w}\right),$$

the stress-difference at a depth z is given by

$$\sigma = 2\rho g h \left(\frac{2\pi z}{\lambda_w}\right) e^{\frac{-2\pi z}{\lambda_w}}, \qquad\qquad (1.19)$$

where λ_w is the wavelength of the load. The maximum stress difference, σ_{max}, is given by

$$\sigma_{max} = \frac{2\rho g h}{e},$$

which occurs at

$$z = \frac{\lambda_w}{2\pi}.$$

The diameter of the negative free-air gravity anomaly over Fennoscandia is approximately 1,500 km, which corresponds to a wavelength of ~2 × 1,500 = 3,000 km. With values of $h = 520$ m, $\lambda_w = 3,000$ km and $\rho = 917$ kgm^{-3} in Eq. (1.19), we get for the maximum stress difference, $\sigma_{max} = 3.4$ MPa at a depth $z = 480$ km. The stress difference at 113.7 km, the depth of compensation according to the proponents of the Pratt model does not exceed 0.7 MPa. Thus, in the view of Daly, the assumption of a mantle with little or no strength, as suggested by the isostatists, was fully justified.

The gravity anomaly over a glaciated region is useful in one other regard, that of estimating the time that it would take for the deep flow to come to an end. Over the Laurentide ice sheet, for example, there is a close correlation (Walcott, 1970c) between the free-air gravity anomaly and topography on the scale of a 1° × 1° 'square' (Fig. 1.31). Assuming this anomaly is due to warping of the crust by the ice sheet, then it is possible to calculate how far the depressed region must rise for the free-air anomaly to vanish. The equation we use is (e.g., Case 3, Fig. 1.32)

$$\Delta g = 2\pi G y (\rho_{crust} - \rho_{air}) + 2\pi G y (\rho_{mantle} - \rho_{crust}),$$

$$\Delta g = 2\pi G y \rho_{mantle},$$

$$y = \frac{\Delta g}{2\pi G \rho_{mantle}}. \qquad\qquad (1.20)$$

With $\rho_{mantle} = 3,300$ kg m^{-3} and $\Delta g = -50$ mGal, Eq. (1.20) yields $y = 360$ m. Therefore, the Hudson Bay region of Canada has a further 360 m to rise before equilibrium is re-established. At the 10 mm a^{-1} uplift rate that is observed today (Fig. 1.29) this will take some 30–40 ka, which is about a factor of two longer than that calculated earlier by Haskell (1937).

Thus, the study of glacial isostatic adjustment has provided two important pieces of information on isostasy: one concerns the magnitude of the stresses that are supported by the mantle, and the other the timescales over which the phenomenon of isostatic adjustment occurs.

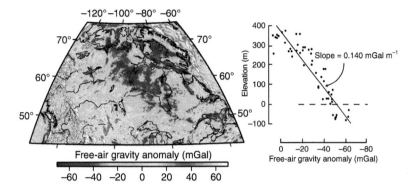

Figure 1.31 The free-air gravity anomaly over Canada and the northern United States and its relationship to the topography. The map is a downsampled 10 × 10 minute grid derived from the combined satellite and terrestrial gravity field grid (V31.1) of Sandwell et al. (2019). The gravity-topography slope plot is reproduced with permission from fig. 4 of Walcott (1970c).

1.12 Success of the Airy and Pratt Models of Isostasy in Explaining Crustal Structure

Despite the studies of Daly and Haskell, most geologists in the late 1930s still paid little attention to isostasy. Perhaps they thought the spatial scales of glacial-isostatic adjustment were too large to be of interest to geological processes such as erosion and sedimentation which generally occur over much narrower tracts. Alternately, they thought that the timescales of glacial-isostatic processes were too short to be of interest to the long timescales of geological processes.

One exception was the Scottish-born American geologist Andrew Cowper Lawson (1861–1952), who had long championed the concept of isostasy. He published (Lawson, 1938, 1942) the first of several articles on the response of the crust and upper mantle to sediment loads in large river deltas. He showed, contrary to the opinions of Chamberlin, that substantial thickness of sediments could accumulate at a river delta simply by the isostatic adjustment to sedimentary loads.

The main 'users' of isostasy during the 1930s, however, continued to be geodesists. In Europe, efforts were being focused on determining the patterns of uplift in glaciated regions more precisely using geodetic techniques. In the United States, on the other hand, a new initiative was begun to use seismic techniques to determine the structure of the crust and upper mantle – one which was to have an immediate impact on isostasy.

Until the early 1950s, the gravity field had provided the principal means with which to test isostasy. But it soon became clear that gravity, due to its fundamental non-uniqueness, was limited in its ability to distinguish between the different isostatic models. Furthermore, the 'best-fit' depths of compensation determined by isostasists were strongly model dependent. Even though isostatic anomalies could be significantly reduced by changing the depth of compensation, the depths varied substantially according to whether the Airy or Pratt models were used.

Isostasy and Flexure of the Lithosphere

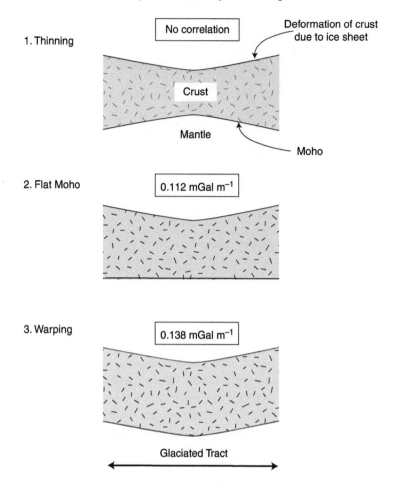

Figure 1.32 Expected correlation between free-air gravity anomaly and topography for different models of crustal structure. Note how close the calculated correlation between gravity and topography for the warping model is to the observed correlation in Fig. 1.31. Reproduced with permission from fig. 3 of Walcott (1970c).

A more satisfactory test of isostasy is seismic data on the structure of the crust and upper mantle. The Pratt and Airy models, for example, imply quite different crustal and mantle structure beneath an elevated mountain and a deep ocean basin. The Airy model predicts that these features would be compensated by changes in the thickness of a uniform-density crust, whereas the Pratt model predicts that they will be compensated by lateral changes in the density of the crust and mantle. The result is that the two models imply quite different depths of compensation. Isostatic gravity anomalies in the United States suggest, for example, a depth of compensation of about 40 km if an Airy model is used or a depth of 113.7 km for a Pratt model.

Following World War II, George P. Woollard (1908–79) and Maurice Ewing (1906–74) began an extensive program of seismic refraction measurements using explosive sources and geophones or hydrophones onshore and offshore the United States. The main result of these studies was to show that the crust is generally thickest beneath the continents and thinnest beneath the ocean basins. Figure 1.33 shows, for example, a compilation of the crustal thickness along a profile extending from Hawaii in the west, across the continental United States, to offshore France and Portugal in the east. The profile provides striking evidence that changes in topography are accompanied by changes in crustal thickness and, hence, that the Airy model provides a better description in general of the crustal structure than the Pratt model does.

It was clear from the earliest seismic measurements, however, that the relationship between topography and crustal thickness was not a simple one. The continental crust, for example, is actually thinnest in the western United States where the elevation is greatest, and thickest in the eastern United States where the elevation is smallest. Furthermore, in some oceanic regions, there are large lateral variations in crustal velocities that would not be predicted by the Airy model.

To better understand the departures from Airy isostasy, Worzel and Shurbet (1965) used gravity anomaly and seismic refraction data to construct a so-called 'standard' oceanic and continental column. The continental column includes adjustments for the elevation above sea level and variations in the depth of the interface between crust and mantle. The ocean column was constructed in a similar way except there was no elevation correction applied.

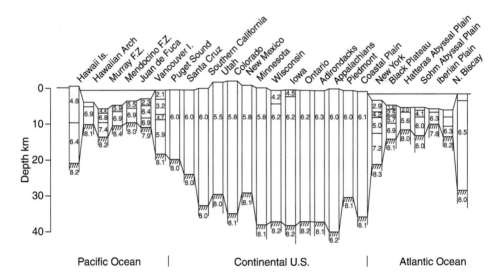

Figure 1.33 Seismic refraction profile of the crust and upper mantle structure from Hawaii Island in the west, across the mainland United States, to offshore France and Portugal. The profile shows striking support for the Airy–Heiskanen model of isostasy with mountains underlain by thick crust and oceans underlain by thin crust. The numbers in the columns are P wave velocities in km s^{-1}. Reproduced from fig. 7–5 of Heskanen and Vening-Meinesz (1958).

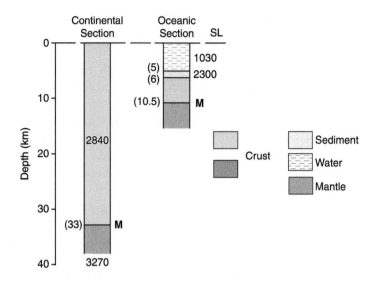

Figure 1.34 'Standard' columns of the continental and oceanic crust. Unbracketed numbers are densities in kg m^{-3} and bracketed numbers are depths below sea level (SL) in kilometres. Reproduced from fig. 2 of Worzel and Shurbet (1965).

Figure 1.34 shows the standard columns that they adopted. Both columns are in balance with each other such that the free-air gravity anomaly over them would average to zero.

The standard columns (Worzel and Shurbet, 1965) have been widely used to test the state of isostasy at particular geological features. One example is shown in Fig. 1.35. The figure shows a compilation of the seismically constrained crustal structure across the margin offshore eastern Canada (Keen et al., 1975). The dashed line in the figures is the depth to the M-discontinuity in the transition zone as predicted assuming the Worzel and Shurbet standard column and the Airy model of isostasy. The figure shows a reasonably close fit between the seismic structure and the predictions of the Airy model, although there are departures in detail.

It would therefore appear from these discussions that seismology provides overwhelming support for isostasy. The crust is indeed lighter than the mantle that it overlies and, in this sense, is in a state of flotation. The question that remains is the 'architecture' of the crust. Is it characterised by lateral density changes in a crust of uniform thickness as predicted by a Pratt model or by changes in the thickness of a uniform-density crust as suggested by Airy? The evidence from seismic refraction data seems to favour the Airy model. However, as we will see in later chapters, the Pratt model, as well as other models, still have considerable merit.

1.13 Bloom's Test of Isostasy

In 1967, A. Bloom (1928–) extended the earlier work of Daly and others on glacial isostasy by considering the effects that the transfer of water from the ocean to the ice sheets and the

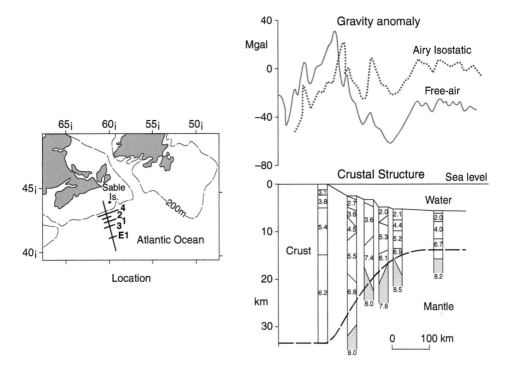

Figure 1.35 Gravity anomalies and crustal structure at the transition between oceanic and continental crust at the continental margin offshore Nova Scotia, Canada. The predicted Moho based on an Airy–Heiskanen model of isostasy is shown as a dashed line. Data reproduced from figs. 1 and 5 of Keen et al. (1975).

release of water from the ice sheet to the oceans would have on the isostatic balance of the stable part of the Earth's crust (Bloom, 1967). By examining the heights of former shorelines, he was able to test the concept of isostasy; in particular, to test the timescales at which isostatic adjustments might take place.

Bloom considered a simple model (Fig. 1.36) of a stable continent with a gently sloping shelf juxtaposed to an ocean basin. Initially, the shoreline is in equilibrium such that a notch will be cut at some position, say S_1. If sea level then falls by h, a second notch will be cut at a point lower down on the shelf slope. The second notch will not necessarily be h below S_1, however, because the ocean basin adjusts to the removal of water by uplift. The uplift causes water to spill over from the ocean basin and re-flood the shelf. Where the second notch is cut depends on how fast the crust adjusts to the removal of the water load. For example, if the adjustment is slow, then a notch will be cut first at S_2, which is a height h below S_1, and then at S_3, which is a height y above S_2. In the case of a fast adjustment, the sea level surface will proceed directly to S_3 and there will be no intermediate notch cut at S_2.

The heights between the old and new shorelines can be calculated by assuming that a column in the ocean is in local isostatic balance before and after unloading. Following Fig. 1.37, we then have

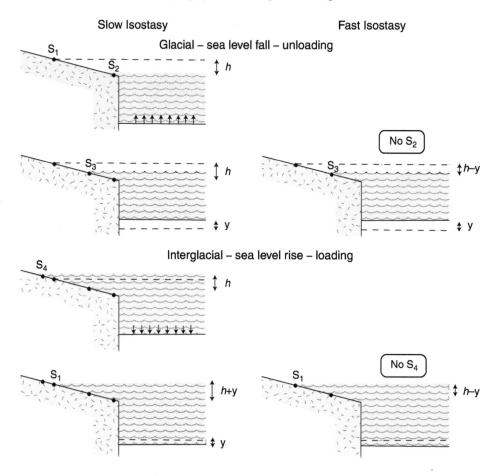

Figure 1.36 Schematic diagram showing the relationship between the position of a 'notch' and a sea level rise and fall. The diagram shows two cases that vary according to whether isostatic adjustment is slow or fast.

$$h\rho_{water}g + w_d\rho_{water}g + T_c\rho_{crust}g = w_d\rho_{water}g + y\rho_{mantle}g$$

$$\therefore \; y = h\left(\frac{\rho_{water}}{\rho_{mantle}}\right). \qquad (1.21)$$

Equation (1.21) shows that if sea level fell by an amount $h = 108$ m, the ocean basin will rise due to unloading by an amount $y = 33$ m, assuming $\rho_{water} = 1{,}030$ kg m^{-3} and $\rho_{mantle} = 3{,}330$ kg m^{-3}. The height between the shorelines and where a notch is cut will depend on how fast isostasy proceeds. If the uplift proceeds more slowly than the removal of water, then the distance between the upper and lower notches will be 108 m. If, however, the uplift proceeds rapidly, then the height between the shorelines is reduced to $h - y = 75$ m.

The height that sea level appears to an observer on the continent is called the freeboard. We see from Fig. 1.37 that the freeboard, f, associated with a sea level fall is given by

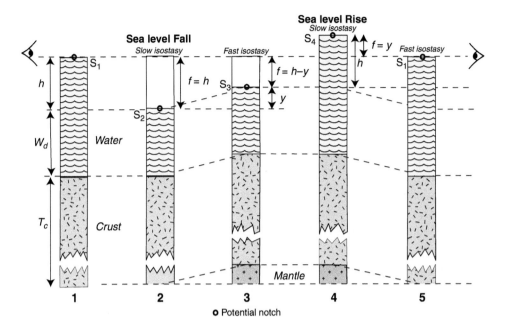

Figure 1.37 The freeboard, f (i.e., the sea level as it would appear to an observer on the continent), during a sea level fall and rise and its relationship to the rate of isostatic adjustment to changing water loads.

$$f_{slow\ isostasy} = h, \qquad f_{fast\ isostasy} = h - y \qquad (1.22)$$

Usually, during glacial and inter-glacials a sea level fall is followed by a rise. A rise following the fall in the simple model in Fig. 1.37, for example, will cause the shoreline to move to a point above S_3. If the rise is similar in magnitude to the fall and isostasy proceeds slowly enough, then the shoreline will move first to S_4 where it might cut a notch. If, however, isostasy is fast, then the shoreline will move directly to S_1 and there will be no intermediate notch at S_4.

These considerations suggest that the height between the upper and lower shorelines following a sea level fall and rise will depend on how rapidly isostasy proceeds in relation to the time that it takes for the water to withdraw from and flood onto the shelf. In the case of a 108 m rise in sea level, the height between the shorelines will be 108 m for slow isostasy and $108 - 33 = 75$ m for fast isostasy.

Sea level changes due to changes in ice volume will be recorded not only on the gently sloping shelves of the continents, but also on the steep slopes of oceanic islands. Oceanic islands are attached, however, to the oceanic crust, and so they respond directly. During a sea level fall and rise, for example, ocean islands rise and fall with the underlying ocean basin so that there will be no intermediate shorelines S_3 and S_4. Thus, ocean islands provide only limited information on how fast or slow isostasy proceeds. However, the height

Table 1.5 *Comparison of sea level on the continents and oceanic islands*

Glacial Maximum	Observed Sea Level Rise on Ocean Islands (Cathles, 1975) h (m)	Calculated* Sea Level Rise on Continents Based on Fast Isostasy $h - y$ (m)	Observed Sea Level Rise on Continents (Morner, 1969) h (m)
Late Wisconsin (~50 ka)**	109	77	70–90
Early Wisconsin (~1,000 ka)**	130	91	85–90

* Assumes $\rho_w = 1,030 \, \text{kg m}^{-3}$ and $\rho_m = 3,330 \, \text{kg m}^{-3}$.
** Ages refer to the approximate culmination of glaciation.

between the preserved shorelines on oceanic islands corresponds to S_1 and S_2 and therefore they have the greatest potential to determine the actual sea level change.

These considerations suggest that, if the isostatic model assumed by Bloom is correct, the height of continental shorelines should provide information on the rate of isostatic adjustment.

We compare in Table 1.5 the sea level rise that followed the last major glaciations as they are observed on oceanic islands and continents. The table shows that the sea level rise as observed on oceanic islands is significantly higher than that observed on the continents. This suggests that isostatic adjustment has been relatively fast, since otherwise the heights between shorelines on the continents would have been much higher. This is confirmed in the table, where good agreement is shown between the observed sea level rise and the predicted rise based on Bloom's model with fast isostasy.

We know from the last glacial and inter-glacial that seas can rise and fall in relatively short periods of time. Therefore, we may conclude that the data from shorelines suggest that isostatic adjustments are rapid and are essentially complete within a few tens of thousands of years.

1.14 Summary

The history of the development of ideas about isostasy is a fascinating one. The concept was discovered during the late 1880s following the acquisition of new data as part of a geodetic survey of northern India. Initially, there was much controversy about the form of isostatic compensation at different geological features, especially among geologists. The geodesists, however, demonstrated with mathematical rigour that isostasy prevailed to a notable degree. But opinions concerning the types of isostatic model that should be used were divided, even amongst the geodesists. By the 1930s, two schools of thought had emerged: an American school, which favoured the Pratt model, and a European one that preferred the Airy model. During the 1940s and 1950s the predictions of the two isostatic models had begun to be compared with observations of seismic structure of the crust and mantle in continental and oceanic regions. These data showed that the crust was thicker beneath the continents than beneath the oceans, a result that is in accord with the predictions of the Airy model.

Despite the geodetic and seismic evidence, however, the concept of isostasy continued to perplex the geologists. The main problem centred on the Airy and Pratt models and what these models implied about the physical nature of the crust and upper mantle. To many geologists these models were inconsistent with the known geological facts, ignoring as they did the strength of the crust and upper mantle. In the next chapter we explore the development of the ideas that eventually led to the notion that the outer layers of the Earth had a finite strength.

2

Isostasy and Flexure of the Lithosphere

The gravity anomalies prove with great conclusiveness the existence of regional isostasy to a notable degree.

(Barrell, 1919b, p. 304)

2.1 Introduction

In Chapter 1, the concept of isostasy was traced from its earliest beginnings. It was shown that the Airy and Pratt models satisfied the needs of the geodesists who were concerned with minimising closure errors in triangulation surveys. One of the models, Airy, was also successful in explaining the seismic structure of the oceanic and continental crust. Despite these successes, however, there was still a widespread dissatisfaction with the Airy and Pratt models among the leading geologists of the time.

The Airy and Pratt models imply that geological features – no matter what their size – are compensated locally. Compensation is achieved by a thickening of a crust of uniform density, by lateral changes in density of a uniform-thickness crust, or by some combination of these factors. While some geologists had few difficulties with this, others argued that the strength of the crust and upper mantle was sufficient that it might prevent certain surface features from being completely compensated.

At the turn of the twentieth century, a new way of thinking about isostasy emerged. Rather than considering isostatic compensation as just the consequence of flotation or differential cooling, the view began to take hold that it was the topography-forming process that determined the compensation. The essence of this idea was that geological features such as ice sheets, volcanoes and river deltas were loads imposed on the surface of the crust, which then responded in some way. In the view of Gilbert, Putnam and Barrell, for example, the response involved a strong crust rather than a weak one, as implied by the local models of isostasy.

The work of these early geologists was significant because they attempted to give a physical meaning to compensation – at a time when the geodesists were quite content to use the Airy and Pratt models without even questioning what these models implied about the physical nature of the crust. It was these geologists who laid the foundations of what we now consider to be 'modern isostasy'.

2.2 Gilbert and the Strength of Earth's Crust

One of the earliest discussions of the importance of considering the strength of Earth's crust was by the American geologist G. K. Gilbert (1843–1919). In 1889, Gilbert (Fig. 2.1) published a short article entitled 'The strength of the Earth's crust'. As he saw it, there were two ways by which the Earth's surface features could be characterised: rigidity and isostasy. He wrote (Gilbert, 1889, p. 25): 'Mountains, mountain ranges, and valleys of magnitude equivalent to mountains, exist generally in view of the rigidity of the Earth's crust; continents, continental plateaus, and oceanic basins exist in virtue of isostatic equilibrium in a crust heterogeneous as to density.'

In 1890, Gilbert published what is now widely regarded as a classic study of Lake Bonneville, a former Late Pleistocene lake that occupied the site of the present Great Salt Lake in the Basin and Range Province of the western United States. He deduced that, at its maximum, Lake Bonneville was up to 300 m deep and had occupied an area of some 10^5 km^2. Remarkably, the shorelines of the lake were 'notched' both in the surrounding mountainous terrains as well as in some volcanic 'islands' within the lake (Fig. 2.2). By careful observation, Gilbert was able to map the ancient shorelines for distances of up to a few hundreds of kilometres. He found that the shorelines were not flat but were warped upwards in the central part of the former lake by as much as 39 m above the lake's edge.

Figure 2.1 Grove Karl Gilbert, former Chief Geologist of the US Geological Survey, aged 55 years. (Photo: Courtesy of the US Geological Survey Photographic Library. Portrait 129)

a)   b)

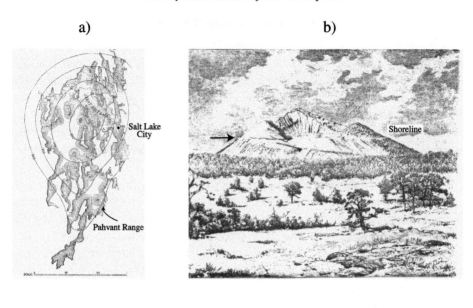

c)

Figure 2.2 Pahvant Butte, a volcanic cone in the Sevier Basin in Utah. A shoreline of Lake Bonneville has been cut into the old part of the cone. (a) Map showing location of Pahvant Butte. (b) Sketch of Pahvant Butte from the north. (c) Photograph courtesy of P. Wyer. (a) and (b) are reproduced from plate XL of Gilbert (1890).

Gilbert attributed the warping of the Lake Bonneville shorelines to isostatic adjustment of the crust following the removal of the water load as the former lake drained through openings to the north and south. To determine how the crust responded to unloading, he used an Airy model of isostasy to calculate the uplift of the crust. Assuming a state of isostatic balance existed before and after water loading, Gilbert calculated that the removal of a 300-m-deep lake would result in an uplift of 90 m assuming $\rho_{water} = 1{,}030$ kg m^{-3} and $\rho_{mantle} = 3{,}330$ kg m^{-3}. This amount of uplift, he found, was approximately 2.5 times larger than the observed value.

As Gilbert was well aware, local models of isostasy ignore the strength of the crust. He therefore suggested that the observed uplift was less than expected because the crust was able to resist the uplift due to its intrinsic strength.

In a footnote to his 1890 paper, Gilbert used the observed uplift to estimate the thickness of the mechanically strong part of the crust. Unfortunately, he did not detail his calculations. He did mention, however, an 'engineers' formula', which he used to argue that the observed uplift of the former Lake Bonneville was compatible with a model in which the crust behaved as a 50-km-thick elastic beam.

The calculation by Gilbert assumes the crust in the vicinity of Lake Bonneville was in isostatic equilibrium prior to and following removal of the water load. When the lake formed, isostasy was disturbed, and material flowed out from beneath the water load to the surrounding regions. On unloading, however, the crust returned to its equilibrium, pre-deformation configuration. Gilbert therefore ignored any effects that the viscosity of the substratum may have had on the pattern of uplift.

In 1894, Gilbert was asked to accompany G. R. Putnam on a trans-continental gravity survey of the United States. His charge was to describe the geology where Putnam had made his gravity measurements so that a better account could be given of the factors, such as rock density, that were needed to reduce and interpret the data. In the spirit of scientific barter, Gilbert provided his geology field notes to Putnam and in return Putnam provided Gilbert with the basic facts of the gravity measurements. Both scientists may have been a little surprised, however, to find that by 1895 each had written a separate interpretation of the same data set!

As it turned out, Gilbert and Putnam had reached rather different conclusions. Like Putnam, Gilbert believed that the gravity measurements between Ithaca, New York and Denver, Colorado indicated that the interior plains of the United States were approximately in isostatic equilibrium. But he differed with Putnam concerning the interpretation of gravity data at stations that were occupied west and east of the interior plains. Whereas Putnam had concluded that the major departures from isostasy were limited to a few mountain stations, Gilbert believed that the entire Rocky and Appalachian Mountains and Wasatch plateau system departed significantly from isostasy. He wrote (Gilbert, 1895a, p. 74): 'The Appalachian and Rocky mountains and the Wasatch plateau all appear to be of a nature of added loads, the whole mass above the neighbouring plains being rigidly upheld.' In a second paper concerned with gravity anomalies, Gilbert modified this state-ment to reflect the fact that features wider than mountains may be supported locally. He wrote (Gilbert, 1895b, p. 333): 'While the new data thus indicate that the law of isostasy does not obtain in the case of single ridges of the size of a large mountain range, they agree with all other systems of gravity measurements in declaring the isostasy of the greater features of relief.'

Although Gilbert was active in research for a further 18 years, these remarks were to be his last on the strength of the Earth's crust. In 1913, he published a paper that, although concerned with isostasy, makes no reference to his previous work concerning the importance of crustal rigidity. Instead, the paper includes numerous references to the work of the geodesists, Hayford and Bowie. He wrote (Gilbert, 1913): 'The success of the (isostatic) hypothesis in reducing anomalies of the vertical and anomalies is held – properly, as I think – to show that isostatic adjustment in the Earth's crust is nearly perfect.' Why Gilbert changed his mind about the strength of the crust is not clear. By 1913, after a long spell as an administrator in Washington, he may have wearied somewhat of the problem. Alternatively, he may have been persuaded – like many of his colleagues – by the views of the geodesists, who had shown so convincingly that local models of isostasy could account for the closure errors in triangulation surveys and that it was therefore not necessary to take into account the strength of the crust.

2.3 Isostasy According to Barrell

If Gilbert had been persuaded by the geodesists to give up his views on the importance of the strength of the crust, they had little effect on one of his younger contemporaries, J. Barrell (1869–1919).

During 1914 and 1915, while an associate professor at Yale University, Joseph Barrell (Fig. 2.3) published a remarkable series of eight papers (1914a through 1914g) under the general title of *The strength of the Earth's crust*. In the papers, Barrell advances the view that the Earth's crust was a good deal stronger than the early advocates of isostasy, such as Hayford and Bowie, would care to admit. While he supported the idea that the larger areas of the Earth's surface, those of 'continental and oceanic proportions', were in isostatic equilibrium, he believed that smaller features, such as those that enter into the 'composition of the landscape', were supported by the rigidity of the crust. Between these two end-members were features such as mountain ranges, plateaus and basins. Barrell viewed these features as the result of horizontal forces, which had later been modified by erosion and sedimentation.

There is little doubt that Gilbert's work was a strong influence on Barrell. The latter adopted (Barrell, 1914d, p. 304) *The strength of the Earth's crust* for the collective title of his papers in recognition of the work of the same title published by Gilbert in 1889. But Barrell extended the work much further and did not seem as much in awe as Gilbert was of the geodesists. He also sought an answer to the question raised by Gilbert (1913) about the spatial extent of the isostatic anomalies mapped by Hayford and Bowie, and whether they resulted from vertical differences in the scheme of compensation or were the consequence of real departures from isostasy.

In Paper 1, Barrell (1914a) reviews the lines of geologic evidence that can be used to test the degree to which the crust resists vertical stresses. He includes as evidence mountain ranges and volcanic cones, the shifting of loads due to climatic change (e.g., Lake Bonneville), erosion cycles and the sediment loads that make up large river deltas. Because river deltas seemed to him best adapted to quantitative studies, Barrell chose them for a more detailed study.

It was Barrell's view that the Earth's crust was strong enough to support river deltas, such as the Niger and Nile, without being deformed. His argument was based on the fact that

Figure 2.3 Joseph Barrell, Professor of Geology at Yale University, aged 49 years. Portrait from Anonymous (1919). Reproduced with permission of the *American Journal of Science*.

there was no evidence in the available seafloor maps, at the time,[1] for a trench-like depression around the flanks of the deltas that might reflect deformation. Using the maps, he was able to estimate that the volumes of sediment trapped in the Nile and Niger deltas were 8.9×10^4 km^3 and 1.2×10^5 km^3, respectively. These volumes implied that the masses of the two deltas were 2.2×10^{17} kg and 3.0×10^{17} kg, respectively, assuming $\rho_{sediment} = 2{,}500$ kg m^{-3} (the value used by Barrell). It was Barrell's view that the crust was strong enough to support loads of this size. He (Barrell 1914a, p. 48) concluded as follows: 'These deltas point to a measure of crustal rigidity capable of sustaining to a large degree the downward strains due to the piling-up and overthrusting of mountains built by tangential forces, or those resulting from the load of sediments in areas of deposition, or those upward strains produced by the erosion of plateaus previously uplifted toward isostatic equilibrium.' Barrell was aware, however, that while the load of river deltas could be supported, there were areas of the Earth's crust that might yield under load. For example, he believed that island arc-trench systems were localities where loads might be large enough that there would be significant deformation of the crust. He wrote (p. 45): 'It appears as though the mountain ranges had been piled too high by tangential forces, and, by virtue of the partial rigidity of the crust, had depressed the neighbouring ocean bottoms.' In his next four papers (Papers 2–5), Barrell concerns himself with gravity anomaly

---

[1] The first edition of the GEneral Bathymetric Chart of the Oceans (GEBCO) was published in 1903 based on >12,000 lead-line soundings and funding from Prince Albert of Monaco, so it is quite likely that Barrell had access to such maps.

interpretation. He was particularly interested in Hayford's so-called 'new method' anomalies, which were the differences between observed and calculated gravity values at a gravity station, corrected for Pratt isostasy. While Hayford found the isostatic reduction useful because it helped minimise the closure errors in geodetic surveys, Barrell was interested in the information that the anomalies might provide on the physical nature of the crust and mantle. He regarded Hayford's new method anomalies, for example, as a measure of the difference at a station between the actual physical condition of the Earth's crust and a highly idealised one.

In Paper 2, Barrell (1914b) carefully re-examines the geodetic evidence that supports the principle of isostasy. He agreed with Hayford and Bowie (1912) that the Pratt model of isostasy reduced the discrepancy between observed and calculated anomalies at nearly all the US gravity stations. He disagreed, however, that the magnitude of the discrepancies was significant enough that it favoured the Pratt model over any other type of compensation model. Barrell advanced two arguments. Firstly, he pointed out that Hayford's new method anomalies formed well-defined positive and negative anomaly 'belts' that extended over large regions of the United States. He argued that the existence of such anomalies, which reached magnitudes that ranged from –93 mGal near Seattle to +33 mGal near Olympia in Washington State, indicated significant departures from Pratt isostasy. Secondly, he pointed out that when the isostatic corrections are calculated using a model in which the compensating mass about a station is distributed regionally rather than locally, there is no significant reduction in the differences between observed and calculated anomalies. In 18 stations in mountainous regions, for example, the mean difference based on local compensation was 18 mGal, compared with 18, 17 and 20 mGal when the compensation was extended to the outer limit of zone K (18.8 km), zone M (58.8 km) and zone O (166.7 km), respectively. In Barrell's opinion, the difference between a local and a regional model of compensation was insignificant, especially when the uncertainties in the measurements and their reductions were considered.

Barrell concluded that Pratt's was not the only model that could explain the discrepancies between observed and calculated gravity anomalies over the United States. He argued that a model in which the compensation was extended out to a distance of at least 100 km could explain the discrepancies equally well. In fact, he made this point with the Pikes Peak, Colorado station of Hayford and Bowie (1912), where he showed that a model of regional compensation improved the discrepancies, reducing them from 19 to 2 mGal (Table 2.1). Barrell interpreted this result as indicating that the crust was sufficiently rigid to support mountains without significant yielding. A similar conclusion, based on different reasoning, had been reached 18 years earlier by Putnam and Gilbert.

Table 2.1 *Comparison of the isostatic anomalies based on a local compensation model to a model with regional compensation out to Hayford zones K, M and O*

Station	Elevation (m)	Local Compensation (0 km)	Zone K (18.8 km)	Zone M (58.8 km)	Zone O (166.7 km)
Pikes Peak, CO	4,293	+19	+11	+6	+2

Papers 3–5 (Barrell, 1914c, 1914d, 1914e) are concerned with the interpretation of the gravity anomalies determined by Hayford and Bowie (1912) using the Pratt model. In his view these anomalies are (p. 307) 'mostly the measure of real loads, due to irregularities in density in the crust, with little or no relation to the topography formed by erosive processes.' Barrell describes at great length methods to interpret these anomalies in terms of density distributions in the crust. He uses simple gravity models to estimate the likely depths of the anomalies, concluding that a significant portion of the density differences must occur at relatively shallow depths in the crust. Barrell concludes Paper 5 with the following statement: 'The excesses and deficits of mass . . . will be a more accurate measure of the capacity of the rigid crust to carry without viscous yielding loads which have borne through geological time, hidden loads whose magnitudes in many regions appear to mask by contrast the present relief between mountains and valleys.' With these statements Barrell introduces the concept of buried loads. Thus, not only had he clearly recognised the occurrence of surface loads on the crust, such as river deltas, but also he used the departures from a local isostatic compensation model to infer other, hidden, loads in the crust that had little or no surface expression.

In Paper 6, Barrell (1914f) returns to the question of the information gravity anomalies provide about isostasy. He points out that the null hypothesis of 'no isostasy' would introduce large systematic 'errors' in the computations of the deflection of the vertical and the gravity anomaly. He agreed with Hayford and Bowie that on the scale of the oceans and continents, isostatic equilibrium exists to a notable degree. Barrell interpreted this as indicating that there must be a mechanism involved, such that when regions are disturbed by erosion and sedimentation there is a transfer of material that restores equilibrium. Erosion involves a rising region, whereas sedimentation implies sinking. When the stresses involved in erosion and sedimentation exceed the strength of the crust, the excess or deficiency of pressure from the heavy or light area is transmitted to an underlying zone of weakness. In Barrell's words (Barrell 1914f, p. 658): 'This deep zone is in turn the hydraulic agent which converts the gravity of the excess matter in the heavy column into a force acting upwards against the lighter column . . . . By this means even the continental interiors are kept in isostatic equilibrium with the distant ocean basins.' Barrell referred to the strong outermost layer of the Earth as the 'lithosphere', and the weak underlying layer as the 'asthenosphere' (Fig. 2.4). In doing so he was the first to use the term asthenosphere. Lithosphere, however, appears to have been used previously by Dana (1896) to describe the solid earth, in contrast to the hydrosphere and atmosphere, and, by I. D. Lukashevich,[2] to describe the rigid outermost part of the Earth.

Barrell had specific views on how the combined lithosphere–asthenosphere system could explain the erosion–sedimentation 'cycle'. Suppose a plateau 1,000 km wide in a continental interior is separated from a region of sedimentation by an intermediate area of similar horizontal dimensions. Let an erosion cycle cause the removal of 0.5 km of material from the continental interior, which is then deposited over an equal area of the seafloor. Assuming the width of the section is 1 km, then a total of 500 km^3 has been removed by erosion. To restore the mass below the continental interior, a similar volume of material must flow in past

---

[2] M. G. Kogan (personal communication). 'Lithosphere' was used in a paper entitled 'Mechanics of the Earth's crust', which was presented by the scientist and revolutionary, I. D. Lukashevich, to the Imperial Mineralogical Society on November 13, 1907. I am grateful to G. D. Karner for a translation of the paper published by Lukashevich the following year in St. Petersburg.

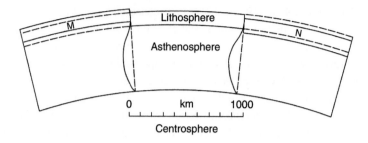

Figure 2.4 Schematic vertical section of the outer layers of the Earth, showing the nature of the 'undertow' in the asthenosphere that is necessary to restore isostatic equilibrium after modification of the lithosphere by uplift and erosion. Reproduced from fig. 14 of Barrell (1914f) with permission of the University of Chicago Press.

the dashed line from the seaward direction. Barrell estimated that the amount of lateral movement would be about 0.83 km, if the depth of the weak zone were 600 km. Thus, the amount of flow is small, even for a large continental area that has been significantly reduced by erosion.

In Paper 7, Barrell (1914g) returns to the main theme of Papers 2–5 that the 'new method' gravity anomalies represent hidden loads that are maintained by the rigidity of the crust. He outlined a method whereby these anomalies could be used to determine the stress-differences that are required to support them. Barrell proposed a two-part approach. First, calculate the equivalent loads on the surface that are implied by the gravity anomalies. Second, use the equations of G. H. Darwin to estimate the stress differences associated with these loads, assuming the Earth responds to these loads in a similar way as would an elastic half-space.

Consider, for example, Barrell's interpretation of the gravity anomalies that were mapped by Hayford and Bowie (1912) at Seattle and Olympia in the north-western United States. The 'new method' anomaly at Seattle and Olympia is +33 and –93 mGal, respectively, so the difference between the stations is 126 mGal. Using Eq. (1.13) with $\Delta g = 126$ mGal and $\rho = 2,670$ kg m^{-3} gives an equivalent load height of 1,123 m. Since Seattle and Olympia are about 80 km apart, the load has a full wavelength of $2 \times 80 = 160$ km.

Barrell used Darwin's equations (Darwin, 1908) to calculate the stress differences associated with mountains and valleys. But, in the process, he had assumed no isostatic compensation and that all topographic relief on the Earth's surface was supported by its internal rigidity.

This fact had limited the applicability of Darwin's equations by previous workers such as Gilbert. But Barrell believed Darwin's equations could be applied, provided that they were based on Hayford's 'new method' anomalies. The reason for this was that he viewed these anomalies as the consequence of the rigidity of the Earth's outermost layers and, hence, a direct measure of it.

In Paper 6, Barrell wrote (1914f, pp. 656–7) 'that although the relations of continents and ocean basins show with respect to each other a high degree of isostasy, there is but little such adjustment within areas 200 to 300 km in diameter . . .. Individual mountains and mountain ranges may stand by virtue of the rigidity of the crust.'

It is clear, with this statement, why Barrell used Hayford's 'new method' anomalies. In the reduction of these anomalies, the so-called 'high degree of isostasy' shown by the continents and oceans had already been accounted for by Hayford's use of the Pratt model. The anomalies themselves, however, provided a measure of the departures from what Barrell regarded as an idealised state of local isostasy. In Barrell's view, the departures did not represent uncertainties in the Pratt model's assumptions, due, for example, to variations in the depth of compensation. Rather, they represented buried loads in the crust that were supported by its intrinsic strength. Thus, in the way he saw the problem, Barrell was quite justified in applying Darwin's elastic half-space model to Hayford's new method anomalies.

We presented Darwin's equation in Section 1.11 when discussing the work of Daly. Substituting $h = 1,123$ m, the equivalent load height for Seattle and Olympia, $\lambda_w = 160$ km and $\rho = 2,670$ kg m^{-3} in Eq. (1.19) gives a maximum stress difference of 22 MPa at a depth of $z = 25$ km. At $z = 120$ km, Barrell's preferred thickness of the lithosphere, the value of the stress difference is only 2.5 MPa, which is an order of magnitude smaller than the stress difference at $z = 25$ km.

Figure 2.5 summarises the stress profiles calculated by Barrell for Seattle and Olympia in addition to the results from isostatic gravity pairs at the Niger Delta, Hawaii and the regions covered by icesheets. We see from these profiles that it is the relatively short-wavelength loads that are supported at shallow depths whereas long-wavelength loads are supported at greater depths. The largest stresses are all in the lithosphere, with the asthenosphere being comparatively free of stress. Both the lithosphere and asthenosphere, however, are involved in the support of surface loads.

In Paper 8, his final contribution, Barrell (1914h) sought to explain isostasy in terms of the mechanical behaviour of rocks. The evidence from isostasy suggests the existence of a zone of little strength at depth in the Earth. Barrell pointed out that rigidity is strictly a measure of the stiffness of the crust, whereas a very different quantity, the limit of elastic yielding, or the beginning of flow, is a measure of its strength. He saw the weak zone as (p. 441) 'solid rock, sodden with occluded gases, giving mobility to the growing fluid and ready to play their part in assisting recrystallization'. He also discussed the results of isostasy in terms of the evidence from tides and earthquakes, concluding that they do not show evidence for an asthenosphere. They only show that it is not fluid and that it is not unlike other parts of the Earth in its elastic properties.

There is little doubt that with these eight papers, Barrell had laid the foundations of what we might now call 'modern isostasy'. The papers are notable for their elegant words, thoroughness, and breadth. Barrell had considerable foresight. He seems, for example, to have even anticipated plate tectonics! In Paper 7 he wrote, when describing the development of a load (p. 42), 'the crust above would come to act to some extent as a bending plate, the stresses within it would increase, chiefly within the upper and lower portions'. Barrell died in 1919, at the relatively young age of 49 years. At the time of his death, the editors of the *American Journal of Science* were in possession of two further papers on isostasy (Barrell, 1919a, 1919b). His colleagues were generous in their praise of his work (Anonymous, 1919; Willis, 1919), describing him as one of the leading young geologists of his generation and

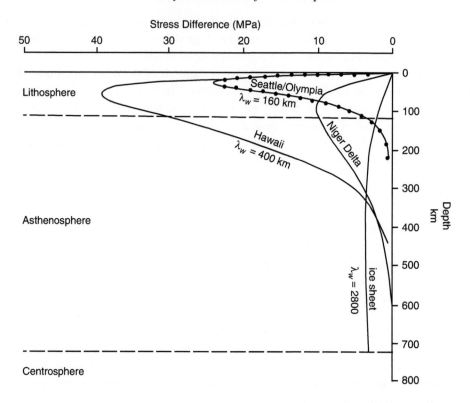

Figure 2.5 Stress differences for harmonic loads on an elastic Earth, the loads representing departures from local isostatic models. Reproduced from fig. 18 of Barrell (1914g) with permission of the University of Chicago Press.

one of its brightest hopes for the future. Yet, it is curious that Barrell's work has not been well cited. This is true not only in Europe, but also in America. For example, Bowie's influential review article on isostasy (Bowie, 1922) makes no mention of Barrell's work.

2.4 Bowie's Criticism of Barrell

Bowie did eventually refer to Barrell in his book (Bowie, 1927), published a few years later. Curiously, he did not comment on his interpretation of the 'new method' gravity anomalies, a topic to which Barrell had devoted so much of his time. His focus instead was on the origin of river deltas.

Barrell had made the statement that large river deltas such as the Nile and Niger were essentially supported by the rigidity of the crust. Bowie disagreed, pointing out (Bowie 1927, p. 86) that the crust should yield in response to such large loads that were emplaced on the Earth's surface.

To follow the reasoning behind Bowie's argument, consider a vertical column of the crust and mantle, which is initially unloaded and then is re-loaded by sediment (e.g., Fig. 2.6). In

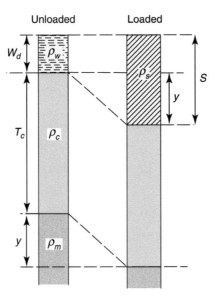

Figure 2.6 Schematic diagram showing a hypothetical column of the crust and mantle before and after loading by sediments.

Barrell's view, the sediments would not cause the underlying crustal rocks to yield by any notable degree. Therefore, if a depth of water, W_d, were available for sedimentation, then the final thickness of sediments would also be W_d, the increased load being supported by the strength of the crust. But, in Bowie's view, the crust would yield to the extra load according to the principle of isostasy. Therefore, a greater thickness of sediment could accumulate than the available water depth.

According to Barrell, the maximum depth of the sediments in the Niger delta is about 3 km. Bowie considered that the crust was not sufficiently strong to support this amount of sediment and calculated, using isostasy, the expected amount of yielding. Assuming isostatic equilibrium prior to and following loading, we can write:

$$\rho_w W_d g + \rho_c T_c g + \rho_m yg = \rho_s W_d g + \rho_s yg + \rho_c T_c g$$

$$\therefore \; y = \frac{W_d(\rho_s - \rho_w)}{(\rho_m - \rho_s)}. \tag{2.1}$$

Using $W_d = 3$ km, $\rho_m = 3{,}150$ kg m^{-3}, $\rho_s = 2{,}500$ kg m^{-3} and $\rho_w = 1{,}030$ kg m^{-3}, Bowie obtained $y = 6.9$ km.

Thus, in Bowie's view the crust would yield to a 3-km sediment load by about 6.9 km. The total thickness of sediments that would accumulate would then be $3 + 6.9 = 9.9$ km. Bowie did not believe this to be excessive at a river delta, especially when he knew that thicknesses of more than 10 km of clastic sediments existed in the Appalachian, Himalaya and other mountain belts.

Irrespective of the actual thickness of sediment beneath the Niger, Bowie had made an important point. If the thickness was 3 km, for example, it could still be explained along the lines of Bowie's argument. The water depth W_d that would be required to explain this thickness of sediments can be obtained from the relation

$$S = y + W_d, \tag{2.2}$$

where S is the total thickness of sediments. Substituting Eq. (2.2) in Eq. (2.1) we get for the water depth, W_d,

$$W_d = \frac{S(\rho_m - \rho_s)}{(\rho_m - \rho_w)} \tag{2.3}$$

which with $S = 3$ km and the densities as defined previously gives $W_d = 0.92$ km. Thus, the crust would yield to a load by 2.08 km. The total thickness of sediment would be 0.92 + 2.08 = 3 km.

Bowie's calculations were carried out around the concept of a 'base' (Fig. 2.7). The base divided the sediment wedge into two parts: a region above which is the load that has been added, and a region below which is the response of the mantle to that load. Bowie was

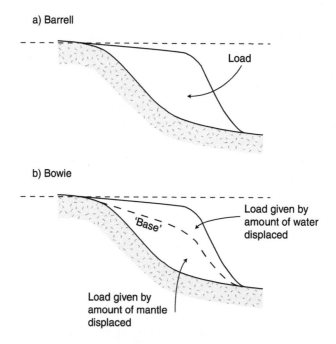

Figure 2.7 Schematic diagram illustrating the contrasting views of Barrell and Bowie on how a sedimentary wedge at the edge of a water-filled basin is supported by the crust and mantle. (a) Barrell's view that the entire wedge represents the 'driving' load, which is supported by the crust without yielding. (b) Bowie's view that only a portion of the wedge above 'base' acts as the 'driving' load, the remainder being accommodated by yielding.

intrigued by the possibility that the base may vary during the evolution of a sedimentary basin. The concept enabled Bowie to deduce that there were two main factors that controlled the thickness of sediment in a sedimentary basin: one that defined the position of the base and, hence, the magnitude of the loads acting on the crust, and the other that defined the response of the crust to these loads. The critical factor according to Bowie was the position of the base, since large thicknesses of shallow-water sediments would not be possible if the only force that was acting on the crust to cause subsidence was the load of the sediment. He believed that the movement of the base was due to a cause other than the load of the sediments, namely some form of 'contraction' of the underlying crust and mantle material.

But the essence of Bowie's criticism was not the thickness of sediments below the Nile delta. Rather, it concerned the results of recent gravity measurements that had been obtained by the Coast and Geodetic Survey in the Mississippi river delta region. Bowie pointed out that if the entire delta was an additional load on the crust that was supported by its rigidity, as implied by Barrell, then the delta should be associated with a large-amplitude positive gravity anomaly. If, on the other hand, the delta was isostatically compensated in the manner suggested by Bowie, then there would be a competition between the gravity effect of the excess mass of the load and the mass deficiency of the compensation. The net result would be small-amplitude gravity anomalies.

In the case of the Niger, we can estimate the approximate magnitude of the gravity anomaly according to each hypothesis using Eq. (2.1) and the thickness previously estimated. On Barrell's hypothesis, the gravity anomaly is given by

$$\Delta g = 2\pi G(2,500 - 1,000)3,000 = +189 \text{ mGal}.$$

But on Bowie's hypothesis, the gravity anomaly is a consequence of two loads: one given by the weight of water displaced, and the other by the weight of the mantle displaced. Using Eqs. (2.2) and (2.3), we have estimated the values of W_d and y corresponding to the total sediment thickness of 3,000 as 920 m and 2,080 m, respectively. The gravity anomaly due to the displaced water is given by

$$\Delta g_{water} = 2\pi G(2,500 - 1,000)920 = +56.8 \text{ mGal},$$

and the gravity anomaly due to the downward displaced crust and mantle is

$$\Delta g_{crust+mantle} = 2\pi G\left[(2,800 - 2,500)2,800 + (3,150 - 2,800)2,800\right] = +56.8 \text{ mGal}.$$

Since sediment is denser than the material it displaces (water), whereas the crust is less dense than the material it displaces (mantle), the total gravity anomaly based on this hypothesis is

$$\Delta g_{total} = \Delta g_{water} - \Delta g_{crust+mantle}.$$

Instead of finding large-amplitude positive gravity anomalies over the Mississippi delta, Bowie found anomalies that ranged in amplitude from −32 to +19 mGal with a mean of −7 mGal. This observation indicated that the delta was very nearly in complete isostatic equilibrium. To Bowie, as well as to many other geodesists at the time, this was compelling evidence that Barrell was wrong. The Earth's crust could not sustain large loads such as river deltas without significant yielding.

Bowie never wavered from the view that local compensation was much nearer the truth than regional compensation. In his book, Bowie (1927) repeats the analysis he had made in 1917 of the average gravity anomalies over mountain ranges. He found that regional compensation with a radius of 59 km around a station satisfies the data as well as local compensation, but a regional scheme of compensation out to a distance of 167 km did not fit quite so well.

But, had Bowie missed the point? The question posed by Barrell (1919b, p. 306) was not about which isostatic models, local or regional, would best satisfy the data over a particular geological feature. Barrell was satisfied that there were observations that could be used in support of both models. His question was more fundamental: how much nearer the truth was local than regional compensation?

2.5 Putnam and Local versus Regional Compensation

One researcher who was in a good position to address Barrell's question was one of Bowie's colleagues at the Coast and Geodetic Survey, George Rockwell Putnam (1865–1953). By 1895, Putnam, with the help of Gilbert, had obtained some of the first gravity and density measurements in the western United States. He also made some of the first statements on the extent of isostasy in the continents, hinting quite strongly that isostasy may not be complete everywhere (Putnam, 1895).

In 1912, Putnam described a method to investigate isostasy by considering the gravity anomalies at pairs of stations from the same general locality but differing considerably in elevation. He chose the method because he thought that the effects of incomplete compensation would be most apparent in the data from station pairs.

Putnam thought the compensation around each station would be more spread out than local models would predict, and he termed the phenomenon 'regional compensation'. To model the compensation, he averaged the surface elevation within a certain radial distance or 'zone' of the station and applied the appropriate compensation scheme to this average elevation (Fig. 2.8). He found that this method eliminates or greatly reduces the extreme isostatic anomalies in the elevated regions as found by Hayford.

The effect was particularly noticeable for the Mauna Kea, Hawaii and Honolulu, Oahu pair of stations (e.g., Table 2.2). As Fig. 2.9 shows, increasing the area of regional compensation reduces the isostatic anomaly differences. If the compensation is extended to 360 km, then the differences between the two stations approach zero.

Putnam's subsequent publications continued to emphasise the importance of regional compensation. He reminded his readers (Putnam, 1922, 1929) that Hayford had divided the area around each gravity station into very small compartments and had assumed a local scheme of compensation for each compartment. The first zone is a cylindrical column 2 m in radius and extending downward 113.7 km, and this column is assumed to be in complete isostatic equilibrium. As Putnam pointed out, this could not possibly be the true situation.

Table 2.2 *Comparison of differences in isostatic anomalies for pairs of stations with local and regional compensation*

Station pairs	Elevation (m)	Difference Pratt Isostatic Anomaly (mGal)	Difference Regional Isostatic Anomaly (mGal)		
			Zone K	Zone M	Zone O
Mauna Kea, Hawaii	3,981	131	118	105	68
Honolulu, Oahu	6				
Pikes Peak, CO	4,293	28	20	16	12
Colorado Springs, CO	1,841				

Source: Reproduced from data in Putnam (1912).

Station

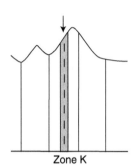
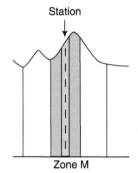
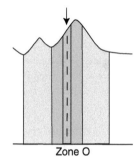

Figure 2.8 Putnam's scheme for the reduction of isostatic anomalies. The colour-shaded regions show the extent of the crust and mantle used to calculate the gravity effect of the compensation. Zone K is the narrowest and assumes that compensation is local. Zone O is the widest and assumes the compensation is regional. Putnam showed that the difference in the gravity anomaly between pairs of stations, say one on a high elevation and the other on a low elevation, was greatly reduced if the compensation at each station was assumed to extend out to Zone O.

By the late 1920s and early 1930s, Putnam had emerged as one of the strongest proponents of regional isostasy in the United States. He stated that there was 'no question that the load of a mountain, for example, is distributed beyond the area of its base'. He recognised, however, that it was difficult from gravity anomaly data alone to determine the lateral extent of the compensation (Putnam, 1930). Nevertheless, he was convinced that regional isostasy existed to a notable degree. Moreover, he recognised the link between regional compensation and the strength of the crust. He wrote: 'Regional compensation corresponds to a crustal strength capable of bearing considerable loads, and in this respect is more consistent than is local isostasy, with visible evidence on the Earth's surface.'

Putnam's advocacy of regional isostasy was even more surprising given that he worked under Hayford while at the Coast and Geodetic Survey. Since assuming the directorship in

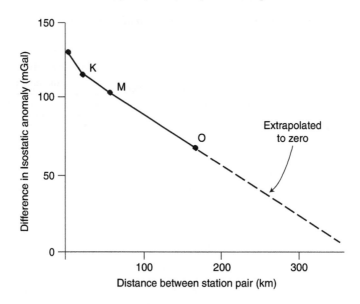

Figure 2.9 Diagram illustrating how the difference between pairs of isostatic anomalies at Moana Kea, Hawaii and Honolulu, Oahu, are reduced by increasing the area over which regional isostasy is assumed to occur. Based on data published by Putnam.

1909, Hayford had channelled much of the survey's efforts and resources into the construction of tables, based on the Pratt local model of isostasy, that could be used to compute the gravity effect of topography and its compensation. Putnam's results on regional isostasy therefore raised the question of the usefulness of Hayford's tables. We do not know what the relations were like between the two men but, in 1910, only one year after Hayford became director, Putnam left the survey to take up a position as the first Commissioner of Lighthouses (Kearney, 2011). Despite his new activities, which included the writing of books on lightships and lighthouses, nautical charts and navigation at sea, Putnam continued to work on isostasy, publishing his last three papers on the subject (Putnam, 1922, 1929 and 1930) as an affiliate of the Department of Commerce. At the end of his career, he became involved in the League of Nations, an international organisation created after the end of World War I as part of the Treaty of Versailles.

2.6 Vening Meinesz and the Radius of Regionality

The next significant development came with the work of the Dutch geophysicist, Felix Andries Vening Meinesz (1887–1966). Vening Meinesz (Fig. 2.10) was a graduate of Delft Technical University in civil engineering who had begun his research career by making pendulum gravity measurements on the unstable soils of the Zuyder Zee. During the 1920s, he persuaded the Royal Netherlands Navy to install his pendulum apparatus on a submarine so that gravity measurements could be obtained at sea (Vening Meinesz, 1929, 1941a,

Figure 2.10 Felix Andries Vening Meinesz, Hoofd-Curator van het Koninklijk Nederlands Meteorologisch Instituut, with a pendulum apparatus. Photo courtesy of the late B. J. Collette of the Vening Meinesz laboratorium, Rijksuniversiteit, Utrecht.

1941b, 1948). The first measurements were made in the Dutch East Indies region around Java and Sumatra. The remainder were acquired during transits between Holland and the Dutch East Indies in the Indian and Pacific Oceans.

Following his early work on instrumentation, Vening Meinesz turned his attention to the reduction of gravity data. In discussing the procedures that had been previously used by Hayford and Bowie to correct for isostatic compensation, he pointed out (Vening Meinesz, 1931) two main problems. First, the procedures were extremely tedious. Second, the models used made no statement on exactly how compensation was achieved at an individual topographic feature.

In a paper entitled 'Une nouvelle methode pour la réduction régionale de l'intensité de la pesanteur', Vening Meinesz introduced a new method for the reduction of gravity data. In discussing the development of the method, Vening Meinesz appears to have been influenced by Putnam. Referring to Putnam's insistence on using a regional rather than a local compensation model, Vening Meinesz wrote (1931, p. 34): 'M. Putnam n'a cessé d'attirer l'attention sur cette question.' Like Putnam, Vening Meinesz believed that the compensation of the Earth's surface features was not necessarily local, as Hayford and Bowie had assumed, but instead might be regional. Unlike Putnam, however, Vening Meinesz regarded topography as a load on the surface of the crust (Fig. 2.11). In this view, the load would cause the crust to bend, in much the same way as would an elastic plate that overlies a weak inviscid substratum. If the load was large enough, it would force the rigid crust downwards into the substratum, causing a low-density root similar to that predicted by the Airy model. The main difference with the Airy model, however, was that the root would be shallower and broader because the load was now supported, at least in part, by the rigidity of the crust.

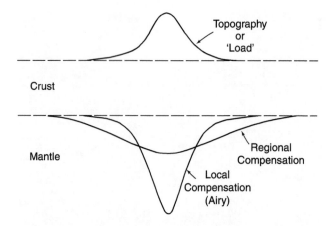

Figure 2.11 Schematic diagram illustrating the local and regional models for compensation of a topographic load. In contrast to Airy-type local compensation, there is less of a depression beneath a topographic feature or load than in regional compensation and the compensation is more spread out.

An elastic plate model had, in fact, already been invoked a few years earlier by Jeffreys (1926) to account for the response of the crust to folding on the one hand, and denudation and re-deposition on the other. Jeffreys, who was critical of Bowie's book and his adherence to local models of isostasy (Jeffreys, 1928), argued the crust would bend in response to these loads, modelling it as a two-dimensional beam of infinite length. He showed that the departures from local compensation occurred within a distance, x, from the edges of the load given by

$$x < \frac{\pi}{2a},\tag{2.4}$$

where

$$a = \sqrt[4]{\frac{3g\rho}{Ed^3}}$$

and g is average gravity, ρ is the density of the load, E is the Young's modulus of the beam and d is the beam thickness. Jeffreys, in the absence of any other information, assumed $d = 70$ km. He also assumed that $E/\rho = 2 \times 10^{11} \mathrm{m^2 s^{-2}}$ and $g = 9.8 \mathrm{~m~s^{-2}}$, from which it follows that $x < 180$ km.

Vening Meinesz required a bending model that could be applied to gravity data around an individual station. He therefore needed a solution of the elastic plate equation in three rather than two dimensions as had been used by Jeffreys. The solution he opted for was the one derived by Hertz (1884), who had applied it to the problem of the deformation of ice caused by the weight of a skater. Hertz gives the solution for the deflection of an elastic plate, y, by a concentrated load, P, as

$$y = \frac{P}{2\pi\beta^2(\rho_{mantle} - \rho_{infill})g} \left\{ \begin{array}{l} \left(\dfrac{r}{\beta}\right)^2 \dfrac{1}{2^2} \ln\left(\dfrac{r}{\beta}\right) - \left(\dfrac{r}{\beta}\right)^6 \dfrac{1}{2^2.4^2.6^2} \left(\ln\left(\dfrac{r}{\beta}\right) - \dfrac{5}{6}\right) \\[3mm] -1.1159\left(\left(\dfrac{r}{\beta}\right)^2 \dfrac{1}{2^2} - \left(\dfrac{r}{\beta}\right)^6 \dfrac{1}{2^2.4^2.6^2} + \right) \\[3mm] +\dfrac{\pi}{4}\left(1 - \left(\dfrac{r}{\beta}\right)^4 \dfrac{1}{2^2.4^2} + \left(\dfrac{r}{\beta}\right)^8 \dfrac{1}{2^2.4^2.6^2.8^2} - \cdots\right) \end{array} \right\}, \quad (2.5)$$

where ρ_m is the density of the material under the plate, ρ_{infill} is the density of the material infilling the deflection, r is the radial distance from the load P and β is the three-dimensional flexural parameter, which is given by

$$\beta = \left[\frac{D}{(\rho_m - \rho_{infill})g}\right]^{0.25}, \quad (2.6)$$

where D is the 'flexural rigidity'. The maximum deflection, y_{max}, occurs beneath the load (i.e. at $r/\beta = 0$) and is given, according to Eq. (2.5), by

$$y_{max} = \frac{P}{8\beta^2(\rho_m - \rho_{infill})g}. \quad (2.7)$$

The bending profile or flexure curve[3] that corresponds to Eq. (2.5) is shown in Fig. 2.12. The curve shows a 'flexural depression' immediately beneath the load. The depression extends into the region beyond the load because of the rigidity of the plate. At radial distances greater than 3.887β, however, the curve shows a 'flexural bulge' that flanks the flexural depression. A bulge is a characteristic feature of models in which an elastic plate overlies an inviscid substratum.

Vening Meinesz used a modified form of Hertz's flexure curve to compute the form of the compensation. Basically, he computed a set of tables that summarised the total gravity effect of the compensation at a station for different assumed values of the parameter β. The parameter β is a measure of the flexural rigidity of the plate and determines both the amplitude and wavelength of the compensation. The distance at which the flexure projects to zero was referred to by Vening Meinesz as the 'radius of regionality', R, which he defined as

$$R = 2.905\beta \quad (2.8)$$

Vening Meinesz corrected each of his submarine pendulum measurements for the effects of regional compensation. Figure 2.13 shows, for example, the geometry of the crust that he would have assumed to calculate the isostatic anomaly at a station that was centred on the axis of a submarine volcano. The calculations were laborious, and a complete set of his tables was not published until 1941.

---

[3] A MATLAB® file, chap2_hertz.m, to calculate the flexure of a circular plate by a concentrated load using the equations of Hertz (1884) is available from the Cambridge University Press server.

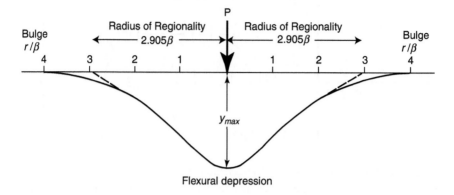

Figure 2.12 The flexure of an elastic plate by a concentrated load P based on the formula of Hertz; r is the radial distance from the load and β is a measure of the flexural rigidity of the plate. Vening Meinesz defined the value of $r = 2.905\beta$ as the radius of regionality, R.

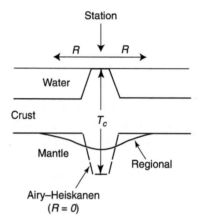

Figure 2.13 Schematic diagram illustrating the radius of regionality, R, at an oceanic island load.

Figure 2.14 illustrates the gravity data obtained by Vening Meinesz aboard the submarine 'K13' in the vicinity of Oahu in the Hawaiian Islands; the first such measurements to be obtained over the sea areas flanking a large submarine volcano. The red solid line shows the observed free-air gravity anomaly. The anomaly reaches a maximum value of about +210 mGal over the flanks of the volcano and minimum values of about −100 mGal in adjacent regions. The other lines show two isostatic anomaly profiles: one based on the Airy model and the other on Vening Meinesz's regional model. Both models significantly reduce the magnitude of the free-air anomaly. However, the Vening Meinesz model appears to reduce the free-air anomaly by a much greater amount than the Airy model.

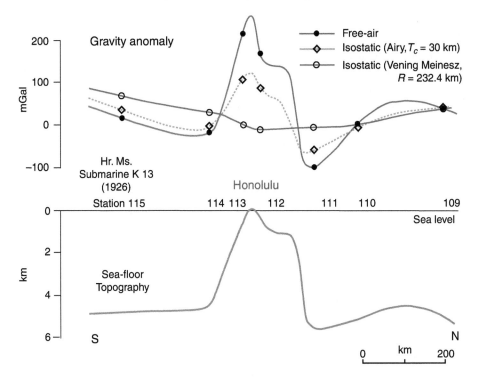

Figure 2.14 Gravity anomaly and topography profiles of the Hawaiian Ridge in the region of Oahu. Based on measurements acquired by Vening Meinesz on board Submarine K13 of the Royal Netherlands Navy. The Airy and Vening Meinesz isostatic anomaly were obtained by subtracting the combined gravity effect of the topography and the compensation from the observed free-air gravity anomaly. Only the Vening Meinesz model accounts for the free-air anomaly, suggesting that the ridge is supported by a regional rather than a local scheme of compensation.

Despite this most convincing demonstration of the importance of regional isostasy, Vening Meinesz appears to have been quite cautious, treating his model as just another reduction procedure that 'geodesists may wish to consider'. In his last major work on the subject, Vening Meinesz wrote (1948; p. 13): 'The (bending) curve . . . seems acceptable, even if we should not feel too sure that the crust really behaves as a purely elastic plate.' Vening Meinesz did not return to the question of regional isostasy for another 10 years. The next time he raised it again was in the book he co-authored with Heiskanen (Heiskanen and Vening Meinesz, 1958). Again, he seems quite cautious about the applicability of his model, mentioning, somewhat tersely, that (p. 141) '[t]he regional isostatic system is a modification of the Airy–Heiskanen floating theory'. The reason for his reluctance to emphasise one isostatic model over another is found a little later in the same book. He wrote (p. 142) that 'isostatic compensation occurs in various parts of the earth in different ways, so it is impossible to find a system which would correspond everywhere to the actual conditions'. Although Vening Meinesz had provided the first proof of regional isostasy, he never claimed the concept (the late B. J. Collette, personal communication). He considered that both local and regional models of isostasy had a role. The Pratt–Hayford model was in accord with the development of mountain ranges by vertical expansion of the Earth's crust, while the Airy–Heiskanen model was acceptable at the transition between continents and oceans. Finally, regional isostatic models were in accord with topography brought about by erosion, sedimentation, volcanism or folding of Earth's upper crustal layers.

It was certainly diplomatic of Vening Meinesz to give each model of isostasy a contributing role. But was his basic premise true: that no one model stood out any more than another?

2.7 Gunn and the Principle of Isobary

At about the same time that Vening Meinesz was working on his model for regional isostasy, the American physicist and inventor Ross Gunn (1897–1966) was working on his own scheme. There is little evidence in their publications that the two scientists, one working in Europe and the other in the United States, had any contact with each other. Vening Meinesz, for example, does not refer to Gunn in any of his publications. Similarly, Gunn does not reference any of Vening Meinesz's work, except as a source for his submarine gravity measurements south of Java.

During 1937 and 1949, Gunn (Fig. 2.15) published six sole-author papers in which he outlined his views on isostasy. Like Barrell and Putnam before him, Gunn

Figure 2.15 Ross Gunn. Graduate of Yale University in Physics, Research Physicist at the US Naval Laboratory in Washington, DC, and Research Professor of Physics at The American University. Reproduced from Abelson (1998).

questioned the results of the geodesists, who had persisted in emphasising local models of isostasy, presenting instead the argument that surface features were supported by regional isostasy. In Gunn's view (Gunn, 1943a, p. 65), 'the presumed approach of the Earth's crust to exact isostatic adjustment has been somewhat exaggerated in the past and ... some of the corrections applied to geodetic data have been carried out to an accuracy beyond those values actually justified. Large sections of the earth do tend toward a precise regional isostatic equilibrium.' The main focus of Gunn's research was to examine the role that the strength of the lithosphere might play in explaining the departures from local models of isostasy. He was the first to examine systematically the extent of regional isostasy at mountain ranges (Gunn, 1937), volcanic islands (Gunn, 1943b), continental margins (Gunn, 1944), and island arcs and trenches (Gunn, 1943a, 1947).

Gunn (1943a) stated that local isostasy could not exist everywhere – the strength of the crust would prevent it. He argued that there are two ways to consider isostasy: ideal and regional. Like Bowie before him, Gunn illustrated the two schemes with reference to prisms of crustal material floating in a weak underlying substratum. 'Ideal isostasy', in Gunn's view, implies that an individual vertical prism is self-supported and no friction whatever exists between it and neighbouring prisms. The mass/unit area of the floating prism is on average equal to the mass/unit area of the underlying substratum that it displaces, so that different thicknesses or densities of the floating prism do not produce significant differences in the gravity anomaly. 'Regional isostasy' (Fig. 2.16), on the other hand, assumes that the prisms are in contact with each other such that friction limits the adjustment of individual prisms. If a load is applied, then each prism that is loaded is forced down, but now, because each prism is mechanically attached to adjacent prisms, they are also forced down while the load is being supported by vertical shear stresses. The mass/unit area of each floating prism will not now be equal to the mass/unit area of the underlying substratum displaced, because material has been forced out from below adjacent prisms even though no load was applied to them. Isostasy is still achieved, however, when the system of prisms is considered, since just enough substratum material is displaced to carry the applied load.

Although regional isostasy involves the support of loads over large areas, Gunn pointed out that both ideal and regional isostasy could be explained in terms of the equilibrium of an individual prism – provided that the vertical shear stresses that act on the side of the prism are considered. In the case of ideal isostasy, for example, we can write

$$P = Mg = \text{constant}, \tag{2.9}$$

where P is the pressure at the depth of compensation and M is the average mass/unit area of the prism between the depth of compensation and the upper surface of the prism. In the case of regional isostasy, the pressure at the depth of compensation is made up of two components: namely, the average mass/unit area of the prism between the depth of compensation and the surface of the prism; and the vertical shear stress S_v that is imposed on the prism by the intrinsic strength of the lithosphere. Equation (2.9) then becomes

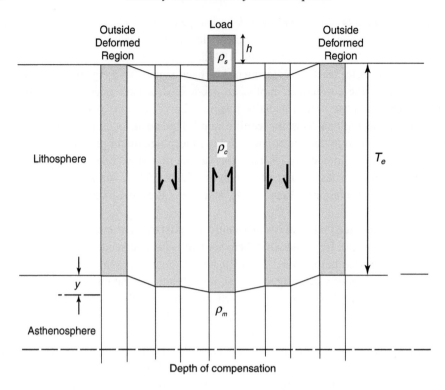

Figure 2.16 Regional isostasy according to Gunn. Surface loads are supported by vertical shear stresses over a broad region of a strong elastic plate.

$$P = Mg + S_v = \text{ constant.} \tag{2.10}$$

When the lithosphere adjusts to the condition represented by Eq. (2.10), Gunn referred to it as being in 'isobaric equilibrium', since 'isobary'[4] implies equal pressure. Isobary is therefore a state of mechanical and hydrostatic equilibrium of the outer layers of the Earth, in contrast to ideal isostasy, which only describes a state of hydrostatic equilibrium. Of course, Eq. (2.10) shows that isobary reduces to ideal isostasy in regions where the vertical stresses are negligible.

The vertical shear stress, S_v, plays an important role in regional isostasy. As Gunn pointed out, if S_v acts on a freely moving prism, it will displace substratum material. Equilibrium will be restored when

$$S_v = M'g, \tag{2.11}$$

where M' is the average mass/unit area of the displaced material. If we neglect high-order terms, then the first-order gravity anomaly due to this change in mass is

$$\Delta g = 2\pi G M'. \tag{2.12}$$

---

[4] The term 'isobary' is no longer in use, probably because of confusion with 'isobar', which is in wide use in meteorology.

Combining Eqs. (2.11) and (2.12) gives

$$\Delta g = 2\pi G \frac{S_v}{g} \tag{2.13}$$

so that if the vertical shear stress, S_v, can be calculated, then it should be possible to evaluate the gravity anomalies due to isobary.

In order to apply his ideas to geological problems, Gunn used the elastic plate model. We do not know, but it appears Gunn derived his own set of equations for the deflection of an elastic plate by distributed loads. The equations were of the form used earlier by Jeffreys, although he does not reference him. They differed from those used by Vening Meinesz because he thought the mathematical complications imposed by three dimensions would make his investigations too 'laborious and impractical'. Besides, he viewed many of the geological features he was likely to study as essentially two-dimensional in nature.

The basic equation for the deformation of an elastic beam, y, overlying an inviscid fluid is given (e.g., Gunn 1943a) by

$$D \frac{d^4 y}{dx^4} + (\rho_m - \rho)gy = 0, \tag{2.14}$$

where ρ_m is the average density of the substratum, ρ is the density of material infilling the deflection and D is the flexural rigidity of the plate given by

$$D = \frac{E T_e^3}{12}, \tag{2.15}$$

where T_e is the elastic thickness of the plate and E is Young's modulus. Gunn defined a parameter, b, given by

$$b = \sqrt[4]{\frac{(\rho_m - \rho)g}{4D}}. \tag{2.16}$$

Gunn (1943a) gives the solutions to Eq. (2.14) for a line load that is applied to a 'continuous' and a 'fractured' or 'broken' plate. He recognised that in both applications, b was an important parameter that controlled the shape of the flexure (Fig. 2.17). He therefore used geological observations to try to estimate b.

The first application by Gunn was to the belt of negative gravity anomalies mapped earlier by Vening Meinesz, south of Java. Gunn attributed these anomalies to the 'down-bending' of the crust due to 'tangential stresses acting in the lithosphere'. Using Vening Meinesz's data (Fig. 2.18), he estimated the average width of the negative anomaly belt to be about 170 km. He attributed the anomalies to a down-bending of a fractured plate by an end load (Fig. 2.19). The width of the flexure, x_b, due to a line load that acts on the edge of a fractured plate is given by

$$x_b = \frac{\pi}{2b}. $$

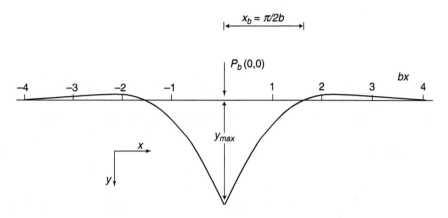

Figure 2.17 The 'fractured' or 'broken' plate model of Gunn.

If $x_b = 170$ km, then $b = 9.2 \times 10^{-8}$ cm^{-1} or $1/b = 108.7$ km. By matching the positive anomalies that flank the negative anomaly belt, Gunn obtained two other estimates of b south of Java of 9.0×10^{-8} cm^{-1} and 7.8×10^{-8} cm^{-1}.

In another application, Gunn considered the Nile and Niger deltas modelled earlier by Barrell. Using Barrell's data, Gunn estimated $b = 7.85 \times 10^{-8}$ cm^{-1} for these features.

Gunn recognised that b provided fundamental information about the physical properties of the lithosphere, terming it the 'lithospheric constant'. One of the goals of his work was to use b, together with the gravity anomaly, to try and determine the degree to which a particular load on the Earth's surface might be compensated at depth.

To follow Gunn's argument, let us first consider the gravity anomaly that would be expected for a block load of height h and density ρ_L that is emplaced on the surface of an infinitely rigid lithosphere. The gravity anomaly is given by

$$\Delta g_{rigid} = 2\pi G \rho_L h. \tag{2.17}$$

If, on the other hand, the plate was not rigid but deformed under the weight of the load, then the gravity anomaly is the difference between the positive gravity anomaly due to the load, and the negative anomaly due to the depressed lighter crustal material. In this case,

$$\Delta g_{total} = \Delta g_{rigid} - \Delta g_{compensation}$$

$$\Delta g_{total} = 2\pi G \rho_L h - 2\pi G (\rho_m - \rho_L) y, \tag{2.18}$$

where y is the amount of deformation, ρ_m is the average density of the substratum and the density of the crust is equal to the density of the load.

Next, we need to consider the response of this lithosphere to loads of simple shapes. Gunn showed, for example, that the maximum deflection of a continuous elastic plate due to a rectangular load that is emplaced on its surface is given by

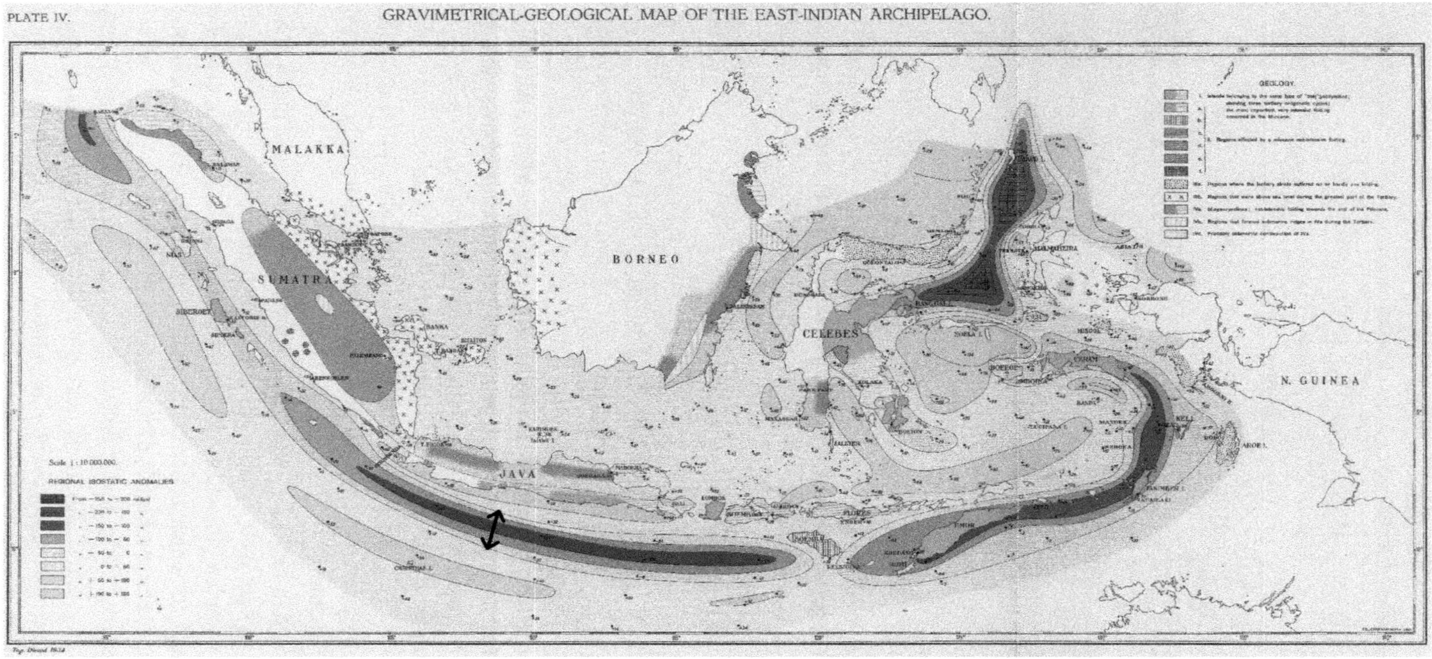

PLATE IV.   GRAVIMETRICAL-GEOLOGICAL MAP OF THE EAST-INDIAN ARCHIPELAGO.

Figure 2.18 Regional isostatic gravity anomalies south of Java that Gunn used to estimate the value of the lithospheric parameter, *b*. Positive anomalies are yellows. Negative anomalies are in reds. The double-headed arrows delineate the Java trench. Reproduced from fig. 10-C of Heiskanen and Vening Meinesz.

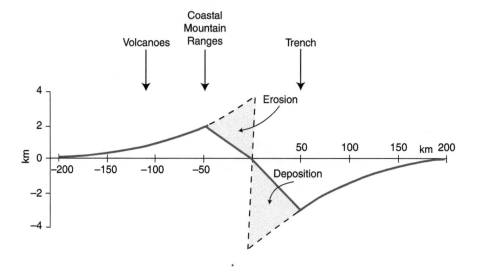

Figure 2.19 Gunn's model for the origin of 'ocean deeps' and 'mountains' by tangential (compressional) stresses which lead to overthrusting, failure along shear planes and flexing.

$$y = \frac{\rho_L h}{\rho_m - \rho_L}\left\{1 - e^{-bm}\cos(bm)\right\}, \tag{2.19}$$

where m is the half-width of the load. Substituting Eq. (2.19) in Eq. (2.18), we get

$$\Delta g_{total} = 2\pi G \rho_L h (e^{-bm}\cos(bm)). \tag{2.20}$$

The importance of Eq. (2.20) is that it provides a measure of the degree to which a particular geological feature on the Earth's surface is compensated. The equation shows that, as $bm \to 0$ (i.e., either the load is narrow or the plate is strong), then $\Delta g_{total} \to 2\pi G \rho_L h$, which is the case of uncompensation where the load is supported by an infinitely rigid lithosphere. Also, if $bm \to \infty$ (i.e., the load is either wide or the plate is weak), then $\Delta g_{total} \to 0$, which is the case of perfect compensation where the load is entirely supported by the weak underlying substratum.

Gunn's results are illustrated in Fig. 2.20, which shows the fraction of a load that will be supported by a strong lithosphere as a function of bm. The figure shows that isostatic compensation is essentially complete for values of $bm > 1.4$. Hence, he argued that the main factors controlling the extent to which a particular geological feature is compensated depends on its width and the elastic properties of the underlying plate.

Gunn was quick to realise, of course, that if b could be constrained from observations, then it would be possible to use Eq. (2.20) to estimate the width that a geological feature needs to be in order to be compensated. Moreover, if both the width and b were known, then it would be possible to estimate the degree of compensation.

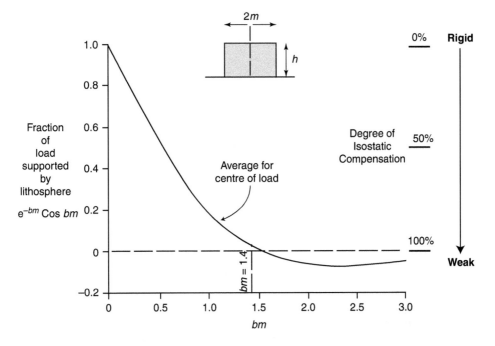

Figure 2.20 Gunn's illustration of how the lithosphere will support loads of different width. Reproduced from fig. 2 of Gunn (1943a) with permission of the University of Chicago Press.

Using Gunn's preferred value of b, based on the average value of the Java trench and the Niger and Nile delta results, for example, we get

$$m > \frac{1.4}{b} \qquad \therefore m > \frac{1.4 \times 10^8}{8.4}\,\text{cm} \qquad \therefore m > 160\ \text{km}.$$

Therefore, the Java and Niger/Nile value of b implies that a geological feature would have to be at least $2 \times 160 = 320$ km wide before it would be completely compensated. Narrower features would be only partially compensated.

This estimate for the size that a load would need to be in order to be fully compensated differed markedly from that of previous workers. It was Bowie's view, for example, that geological features as narrow as 110 km were fully compensated. As Fig. 2.20 shows, in order for this to be the case, then $1/b = 39.3$ km, which is a factor of three smaller than the value estimated by Gunn at the Java Trench and Nile and Niger deltas.

Gunn's final paper on regional isostasy was published in 1949. Although mainly a review, the paper included a table that summarised in a succinct way the various types of equilibrium that may be achieved by the lithosphere (Table 2.3).

In Gunn's view, the principle of isobary accounts for the way the crust responds to loads, irrespective of the size of the load. Bouguer and isostatic anomalies were of limited use, in

Table 2.3 *Comparison of the various types of isostatic equilibrium*

Assumed Character of Lithosphere	Sufficient Approximation for Actual Earth When Load Is:	Resulting Character of Equilibrium	Lithospheric Constant, b	Type of Gravity Anomaly	Principal Term in the Gravity Anomaly
Rigid	Small and concentrated	Mechanical	Zero	Bouguer	$2\pi G\rho h$
Plastic	Uniform and larger than 330 km	Hydrostatic	Infinite	Isostatic	Zero
Elastic	Unrestricted	Mechanical and hydrostatic	$8.4 \times 10^{-8}\,\mathrm{cm}^{-1}$	Isobaric	$2\pi G(\rho h - \rho_0 y)$

Source: Reproduced from table I in Gunn (1949) with permission of the University of Chicago Press.

his view, because they misrepresented the response of the lithosphere as either too rigid (Bouguer) or too weak (isostatic). They were therefore 'end-members' to his more general scheme of isobary.

With his 1949 paper, Gunn essentially concludes the work begun by Gilbert, Barrell, Putnam and Vening Meinesz on the importance of considering the strength of Earth's outermost layers. It is interesting to speculate on what it was that stimulated Gunn's interest in isostasy. He published his work while he was a technical advisor at the US Naval Research Laboratory and Director of the US Air Force Weather Bureau Cloud Physics Project in Washington, DC (Gunn, 1962; Abelson, 1998). In 1945, two years after his paper on gravity anomalies at continental margins was published, Gunn was cited by the Secretary of the US Navy for his work on the development of the first atomic submarine! Isostasy must therefore have been somewhat of a sideline to his day-to-day work as a senior government scientist. However, isostasy, we can perhaps speculate, allowed him to pursue his interest in the application of physics to problems of the Earth and, maybe, helped to provide him with a much needed respite from the pressures of wartime America.

2.8 Isostasy and Plate Tectonics

Vening Meinesz and Gunn died within two months of each other in 1966. The following spring, a series of papers were presented at the Annual Meeting of the American Geophysical Union (AGU) in Washington, DC, that were to have a profound impact on the way that geologists viewed Earth's evolution. The papers were not based on gravity or geodesy data or even on considerations of isostasy, but instead on marine magnetic anomaly and earthquake data. These data, which were presented in >70 abstracts at the meeting, confirmed the earlier ideas of continental drift and seafloor spreading and suggested that the mid-oceanic ridges were the sites of the creation of new oceanic crust. At the end of 1967, McKenzie and Parker (1967) showed that earthquake slip vector data in the north-west

Pacific Ocean could be explained if the Pacific had rotated as a rigid plate about an Euler pole centred at latitude 50° N, longitude 85° W and was presently being consumed at the Kuril and Japan trenches beneath Eurasia. And in 1968, the results from seafloor bathymetry, marine magnetic anomalies and earthquake seismology were combined by Morgan (1968) into the general theory of 'plate tectonics'.

Plate tectonics considers the outer layers of the Earth to comprise six large, rigid plates (Fig. 2.21) and several smaller plates that are in motion with respect to each other (Fig. 2.22). Three main types of plate boundaries are recognised: those where plates diverge, converge or slide past each other. The driving mechanism of plate tectonics is not known, but it is generally agreed that the plates are driven by some form of thermal convection in the underlying mantle.

One of the fundamental tenets of plate tectonics is that deformation is concentrated at the plate boundaries and that the interiors of the plates behave rigidly on long geological timescales. These assumptions are the basis for kinematic studies, which attempt to reconstruct the relative positions of the continents and the distribution of the intervening oceans through geological time.

The proponents of regional isostasy had, of course, already addressed the question of the rigidity of the Earth's outer layers long before the development of plate tectonics. Yet, it is curious that the work of Gilbert, Barrell, Putnam, Vening Meinesz and Gunn were so seldom cited in the geophysical literature of the time. It is true that some of the terms adopted in plate tectonics, such as 'lithosphere' and 'asthenosphere', were borrowed from the early isostatists. And Le Pichon et al. (1973) make several references to Barrell, Gunn and Vening Meinesz in *Plate Tectonics* (the book was dedicated to Gunn). Nevertheless, regional isostasy appears to have played no more than a passing role in the development of plate tectonics.

But plate tectonics did not suddenly develop from the new ideas that emerged from a single meeting. It was built on the much earlier concepts of continental drift and seafloor spreading. As we will now see, isostasy played a significant role in the development of these concepts.

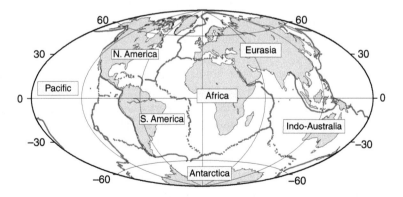

Figure 2.21 The distribution of Earth's seven major plates. Plate boundaries are shown in red and are defined by a mid-ocean ridge, a deep-sea trench or a transform fault.

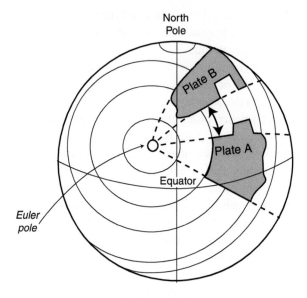

Figure 2.22 Plate tectonics assumes that the plates behave rigidly on long geological timescales such that their relative motions can be described by a series of motions about one or more poles of rotation.

The theory of continental drift is generally attributed to the German meteorologist, A. Wegener (1880–1930). In 1915, he published the first edition of the now classic book *Die Enstehung der Kontinente and Ozeane*. In this book (an English translation of the 1962 edition was published in 1966), Wegener develops the idea that the continents on either side of the Atlantic Ocean were once joined, and that they drifted apart into their present positions. Wegener carefully considered isostasy, concluding that the concept was in accord with continental drift. In particular, he appealed to the fluidity of the weak layer required in isostatic models to help explain how the 'keels' of continents might be able to move through the mantle. He did not express any preference between the different types of isostatic models because they all implied some form of flow in the weak layer that provided the mechanism for the restoration of isostatic movements that followed loading and unloading of the crust and mantle.

The early isostatists, however, were not as accepting as Wegener and were initially critical of continental drift theory. Bowie, for example, found Wegener's hypothesis puzzling. In his 1927 book, he suggests that the primary cause of the distribution of oceans and continents was vertical, not horizontal, movement. The most important processes involved in shaping the Earth, he argued, were surface processes such as evaporation, precipitation, erosion and sedimentation. The 'driving forces' to these phenomena are the maintenance of isostatic equilibrium by sub-crustal flow, and the expansion and contraction of the crust due to 'changes of temperature of the crustal matter'.

Bowie therefore appears to have been more in favour of the 'permanence' of the continents and oceans, and so it seems he belonged to the school of the early 'fixists' along with, for example, B. Willis, R. T. Chamberlin and W. Bucher. Certainly, he did not accept the notion advanced by Wegener and other 'mobilists' such as E. Argand, A. Holmes and A. L. du Toit that large continents could fragment, disperse and leave behind large 'floating' islands in their wake.

Yet, a few years after his book was published, Bowie (1929) had modified his views. While reviewing ideas on Earth's origins, he concluded that a planetesimal theory is one that is in accord with the known isostatic facts. In particular, the detachment of the Moon from the Earth during its early history disrupted the outer granitic layers of the Earth, causing it to 'fragment' into several blocks. Bowie saw no difficulty with the idea that these blocks could drift around freely in the fluid layer that he and other isostasists believed to underlie the crust.

The supporting evidence for continental drift came from the similarity of the coastlines on the opposing sides of the ocean basins, as well as petrologic, stratigraphic, structural and palaeontologic similarities between continents that are now widely separated. Wegener's was certainly not a view of permanence, and he disagreed vigorously with workers such as Dana, Heim and Suess who, like Bowie, still found favour with the contraction theory.

One worker who shared Wegener's views was the British geologist, Arthur Holmes (1890–1965). He considered continental drift plausible and sought the mechanisms to explain it. Holmes (1931) proposed that upwelling convection currents deep in the mantle would eventually break up a floating continental 'block' and form a new ocean at an 'island', which would then push the separate blocks apart and onto a downwelling convection current at a 'geosyncline'. The 'geosyncline' was a popular concept in the 1940s through the 1960s (e.g., Kay, 1947), when it was used to describe the facies, lithology and thickness of sediments that accumulated at continental margins.

In the US, the Princeton academic and wartime Navy officer, Harry Hammond Hess (1906–69) appears to have agreed with Holmes, although there is no evidence of any contact between them. Hess, however, had access to new echo sounder technology that had been installed on the troop-carrying ships under his command in the Pacific Ocean, and he would have undoubtably been more familiar with what was on the seafloor and what lay beneath it than Holmes.

In Hess's view, the Earth's crust was in horizontal as well as vertical motion. The mid-oceanic ridges were the outlets for upwelling material in the mantle, and the deep-sea trenches the sites where newly created oceanic crust descended back into the mantle (Hess, 1962). Support for his idea came from consideration of the composition of the crust and mantle. Hess believed the mantle to be made up of peridotite, which is serpentinised on its ascent to the surface at ridges where it forms the oceanic crust. The serpentinised crust then moves horizontally towards the trenches, where it is reheated as it descends into the mantle. By this means, the water that produced the serpentinisation at mid-oceanic ridges would be released and recycled back into the mantle.

A short time after Hess's paper, another American geologist, Robert Sinclair Dietz, published a similar hypothesis, dubbing it seafloor spreading (Dietz, 1961). Unlike Hess, however, Dietz attached less significance to the crust. He believed that it was the outer, rigid layer of the Earth, or lithosphere, that moved as a single entity, not just the crust. He also maintained, as Barrell had before him, that the lithosphere was a strong layer that was underlain by a weaker layer, the asthenosphere, which allowed the lithosphere to be in horizontal motion above it.

The idea of rigid 'blocks' in horizontal motion provided the Canadian geologist John Tuzo Wilson (1908–93) with an explanation for the sharp offsets that had been discovered, for example, by Bruce Heezen, Marie Tharp and colleagues in the Equatorial Atlantic Ocean, on the crests of the mid-ocean ridge. Wilson argued (1965) the offsets were faults that traced the relative motion between adjacent oceanic crustal blocks. The offsets were not 'strike-slip' faults such as are seen on land, but were a new class of fault, which he termed 'transform faults'. In Wilson's view, transform faults represented slip between two blocks of oceanic crust that were created as a result of seafloor spreading; the direction of slip being opposite to that which would be expected from the offsets observed in the seafloor of the mid-ocean ridge crest.

The ideas of Holmes, Hess, Dietz and Wilson on the tectonic significance of sea-floor features such as mid-oceanic ridges, deep-sea trenches and transform faults, together with the publication by Walter Clarkson Pitman III and James Ransom Heirtzler of the magnetic anomaly 'magic profile' acquired on board R/V *Eltanin* cruise 19 to the Pacific-Antarctic Ridge (Pitman and Heirtzler, 1966), were the catalysts for the development of plate tectonics. The main architects of the new theory were arguably McKenzie and Parker (1967), Morgan (1968) and Le Pichon (1968) who, among many other contributions, made the first attempts to quantify the motions of the major plates on a sphere.

The development of plate tectonics during the late 1960s had a profound impact on the Earth Sciences. As with most scientific paradigms, it was followed by a period of testing. Initially, effort was focused on plate kinematics since knowledge of the relative motions of all the plates was needed to construct a self-consistent model for global plate motions. This was followed during the mid-1970s by a focus on the physical and chemical properties of the plates. Of particular importance to understanding the driving mechanism of plate tectonics was the answer to the question: how thick is the lithosphere?

The principal method to determine the thickness of the lithosphere is seismology. The seismic P and S waves that are generated by controlled sources and earthquakes, for example, can be used to constrain the velocity structure of the Earth's outermost layers.

Seismic studies (e.g., Grand and Helmberger, 1984) show that the P- and S-wave velocities increase with depth. At a certain depth, however, the S-wave velocity decreases. P waves pass through all types of media, but S waves only traverse materials that can support shear stresses. In other words, P and S waves are passed by solids that are strong and deform by bending, but only S waves are unable to pass through liquids. The decrease in S-wave velocity has therefore been interpreted as defining the boundary between the strong lithosphere above and the weaker asthenosphere below.

As Fig. 2.23 shows, the decrease in S-wave velocity is most obvious in oceanic regions.[5] The most extensive data set that exists on the thickness of oceanic lithosphere has come from surface-wave phase velocities that have been 'regionalised' into zones of different age

---

[5] A MATLAB® file, chap2_shear_wave_profile.m, which plots profiles of S-wave velocity versus depth in oceanic (and continental) regions using the S-wave gridded data set of Shapiro and Ritzwoller (2002) is available from the Cambridge University Press server.

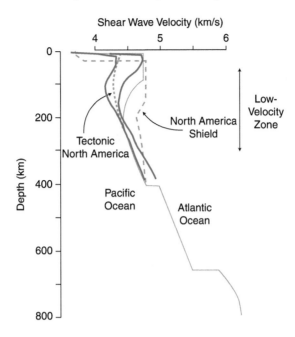

Figure 2.23 Plot of shear-wave velocity against depth based on surface-wave studies of tectonic North America, shield North America, the north-west Atlantic Ocean and the Pacific Ocean. Reproduced from data in Grand and Helmberger (1984) (green lines) and Ritzwoller et al. (2004) (red lines). Copyright by the American Geophysical Union.

(e.g., Leeds et al., 1974; Forsyth, 1975). These data show that the seismic thickness of oceanic lithosphere increases with age, from about 10 km for 20-Ma-old oceanic crust to about 80 km for 70-Ma-old crust (Fig. 2.24). Results are in accord with long-range seismic refraction data (e.g., Asada and Shimamura, 1975), which suggest that old oceanic lithosphere is about 90 km thick.

One problem with using seismic velocities to estimate the thickness of the lithosphere is anisotropy. This causes *P*- and *S*-wave velocities to vary with azimuth. Under the oceans, the fast direction is usually in the direction of seafloor spreading, which suggests the orientation of minerals such as olivine is controlling the elastic properties and, hence, the velocity structure of the lithosphere. Regan and Anderson (1985), for example, estimated that the seismic thickness of old oceanic lithosphere when corrected for the effects of anisotropy may only be 50 km, half the value estimated in previous studies. More recent studies (e.g., Nishimura and Forsyth, 1989; Shapiro and Ritzwoller, 2002), however, are in better agreement with the early studies.

A decrease in *S*-wave velocity is more difficult to identify in the continents than in the oceans (Fig. 2.23). Hence, there is still much uncertainty about the seismic thickness of continental lithosphere. Lerner-Lam and Jordan (1987), using fundamental mode surface-wave data, suggest that, on average, the thickness of continental lithosphere exceeds 220 km beneath the Archean and Proterozoic cratons. This has been confirmed in more recent studies based on higher overtones (e.g., Priestley and Debayle, 2003). The thickness is now

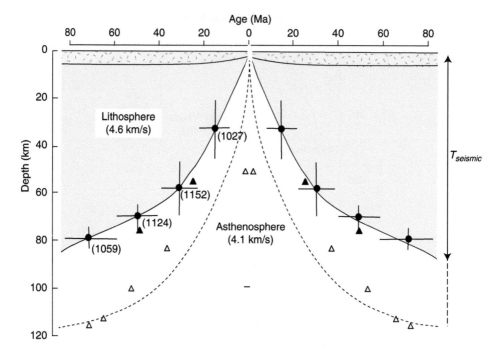

Figure 2.24 Seismic thickness of the oceanic lithosphere based on regionalised surface-wave studies of Leeds et al. (1974), Forsyth (1975) and Nishimura and Forsyth (1989) with the addition of the depth to the top of the low-velocity zone of Ritzwoller et al. (2004) (open triangles) and the depth to the lithosphere/asthenosphere boundary (filled triangles) of Kawakatsu et al. (2009).

considered to be less in Phanerozoic orogenic regions, where it approaches that of old oceanic lithosphere, and in active extensional regions such as the Basin and Range, where it is most similar to that of young oceanic lithosphere.

The thickness of the lithosphere can also be estimated from observations of sea-floor depth and heat flow. Oceanic lithosphere, for example, is created at a mid-oceanic ridge by the intrusion of hot molten material that cools, solidifies and subsides with age. As the lithosphere moves away from the ridge it loses heat (principally to seawater), becomes denser and subsides. By comparing the predictions of thermal models to observations of the depth of the seafloor and heat flow measurements (e.g., McKenzie, 1967; Sclater and Francheteau, 1970; Parsons and Sclater, 1977), it has been possible to estimate the parameters that define thermal cooling models.

Figure 2.25 illustrates the temperature structure of oceanic lithosphere based on the cooling-half space model of McKenzie (1967) and the cooling plate model of Parsons and Sclater (1977). The main difference between the models[6] is in the boundary conditions assumed at the base of the cooling plate. Both models predict that the seafloor depth and depth to individual

---

[6] A MATLAB® file, chap2_thermal_models.m, to calculate the temperature structure of the cooling half-space and plate cooling models of McKenzie (1967) and Parsons and Sclater (1977) and compare it to bathymetry in the North Pacific Ocean is available from the Cambridge University Press server.

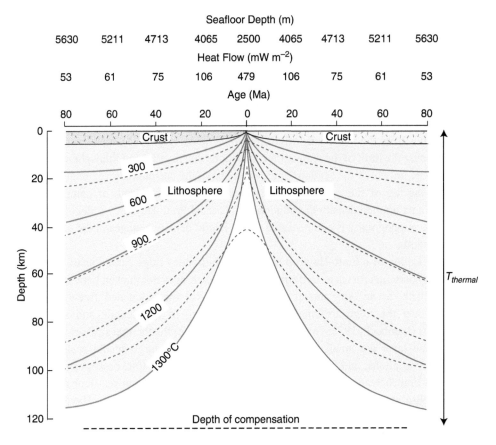

Figure 2.25 Thermal structure of the oceanic lithosphere based on the cooling plate model of Parsons and Sclater (1977). Numbers above the *x*-axis show the predicted seafloor depth and heat flow based on the model. The red solid lines show the model depth to the 300, 600, 900, 1,200 and 1,300°C isotherms, and the dashed red lines the same isotherms based on McKenzie et al. (2005) which consider the temperature dependence of the thermal conductivity.

isotherms increase with age as the oceanic lithosphere cools. After about 80 Ma, the two models start to depart. The cooling half-space model continues to subside with age, but subsidence in the cooling plate model is arrested as the lithosphere begins to 'feel' its lower thermal boundary layer. Oceanic lithosphere then approaches its equilibrium thermal thickness, which Parsons and Sclater (1977) estimate is about 125 km. McKenzie et al. (2005) recently modified the cooling plate model to include a temperature-dependent thermal conductivity, and the resulting isotherms are compared to those of Parsons and Sclater (1977) in Fig. 2.25.

Irrespective of the thermal model used, a comparison of Figs. 2.24 and 2.25 shows the thermal thickness of oceanic lithosphere may well exceed its so-called seismic 'lid' thickness. The ocean bottom seismometer data of Kawakatsu et al. (2009), for example, shows that the base of the seismic lithosphere is given approximately by the depth of the 1,100°C oceanic isotherm. This temperature is of the order of values measured in erupting lava lakes in Hawaii and to temperatures obtained by pyroxene geothermometry on mantle-derived ultra-mafic xenoliths. Solomon (1976) suggested therefore that the base of the seismic lithosphere may correspond to the mantle solidus. If this is the case, then the decrease in S-wave velocity may reflect a zone of partial melting in the mantle.

A basic assumption of the thermal cooling models is that an isostatic balance is maintained between the relatively cold, dense, excess mass of a mid-ocean ridge with the relatively hot, light, mass deficit of the region that underlies it. The depth of compensation, which extends from the mid-ocean ridge crest to its flanks (Fig. 2.25), then has a physical meaning, being the limit of the temperature variations due to crust and lithosphere cooling. Such a model, in which the seafloor is elevated above the surrounding regional sea-floor depth and the top of the asthenosphere is elevated above the depth of compensation, resembles in its implied density distribution the classical Pratt–Hayford model of isostasy, where surface topography is compensated for by lateral density changes in the underlying crust and mantle.

The application of a Pratt–Hayford model to a mid-ocean ridge can be justified by the facts that (a) oceanic crust is of uniform thickness (i.e., it does not thicken between a ridge crest and flank, as an Airy model would predict); (b) the calculated free-air gravity anomalies over the ridge are small in amplitude, long in wavelength, and similar to observations (Lambeck, 1972; Pearson and Lister, 1979); (c) the thermal stresses generated in the elastic layer are small and can be ignored over the large spatial scales of the temperature variations; and finally (d) the range of thermal plate thicknesses estimated by Sclater and Francheteau (1970) and Parsons and Sclater (1977) from sea-floor depth and heat flow observations (70,125 km) are of the order of those derived by Hayford (e.g., 113.7 km) from his global analysis of the compensation depths required to minimise Pratt–Hayford isostatic gravity anomalies.

The success of thermal cooling models at accounting for seafloor depth away from a mid-ocean ridge led to their application in other geological settings, most notably at fracture zones, which are fossil transform faults. Fracture zones are characterised by abrupt changes in seafloor depth and thermal structure where old, cold, deep regions may be juxtaposed to young, hot, shallow regions. Louden and Forsyth (1976) and Sibuet and Veyrat-Peinet (1980) showed, for example, that the seafloor depth and thermal structure of a fracture zone yields a characteristic free-air gravity anomaly positive–negative 'couple' with the positive anomaly over the young, hot, shallow region, and the negative anomaly over the old, cold, deep region, which averages to zero over the length of a profile, as would be expected for what is an isostatic model.

When considering the thermal structure of the lithosphere and its more general application, it is useful, as Cochran (1981) has shown, to isostatically balance a column of the crust and lithosphere at a mid-ocean ridge with that of zero-elevation continental crust. He found, assuming thermal parameters similar to those derived by Parsons and Sclater, a linear temperature gradient in the continental crust and sub-crustal mantle, and a linear gradient in the oceanic crust and an isothermal (1,330°C) sub-crustal mantle, the isostatically balanced columns illustrated in Fig. 2.26. The columns are equivalent to the crust and mantle 'standard columns' discussed in Chapter 1 but differ through their consideration of the thermal structure. The isostatically balanced lithospheric columns have been used in

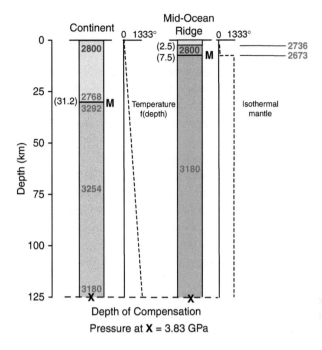

Figure 2.26 Cochran's 'standard' crust and lithosphere columns for a zero-elevation continent and a mid-ocean ridge. The red bold numbers show the calculated density structure based on the temperature structure shown, a zero-temperature crust and mantle density of 2,800 and 3,330 kg m^{-3} respectively, and a coefficient of volume expansion of 3.4×10^{-5}°C^{-1}. The assumed depth to Moho, top of the oceanic crust and base of the thermal lithosphere are shown in brackets in kilometres.

modelling the gravity anomaly, especially in those tectonic settings where a thermal origin has been suggested, such as in sedimentary basins in rifted continental margin and interior plate settings.

In 1978, William Fulton Haxby (1950–2006) and Donald Lawson Turcotte (1932–) introduced the term 'thermal isostasy' to describe those models that integrate the thermal structure of a plate with an isostatic condition. Haxby and Turcotte (1978) used the fact that an isostatically balanced thermal model has a dipole density structure and derived expressions for the geoid anomaly associated with the cooling and subsidence of a newly formed oceanic plate at a mid-ocean ridge. They showed the geoid anomaly, ΔN, to be a linear function of seafloor age, t, at least out to ~80 Ma as given by

$$\Delta N = \frac{2\pi G \rho_{lo} \alpha_v T_m k}{g} \left[1 + \frac{2\rho_{lo} \alpha_v T_m}{(\rho_{lo} - \rho_w)\pi} \right] t,$$

where G = gravitational constant, g = gravitational acceleration, α_v is the coefficient of volume expansion, T_m is temperature of the isothermal mantle and ρ_{lo} and ρ_w are the densities of zero-degree mantle and water respectively. By calculating the geoid anomaly associated with isostatically compensated seafloor topography and comparing it to the observed geoid anomaly derived from satellite altimetry, Haxby and Turcotte (1978) were able to estimate the depth of compensation associated with heating of plates by mantle plumes at mid-plate swells, constraining it to depths of ~100 km. Sandwell (1982) and Nakiboglu and Lambeck (1985) extended the work of Haxby and Turcotte by considering moving plates, vertically stratified rheologies and temperature perturbations in the lithospheric crust and mantle.

Several workers have attempted to reconcile the seismic and thermal estimates into a single model for the evolution of oceanic lithosphere. The fact that observations of heat flow and seafloor depth away from a mid-ocean ridge are better explained by a cooling plate than a cooling half-space model suggests that some form of heating is provided to the base of the plate to arrest its cooling and maintain it at a constant temperature. One possible source of the additional heat is some form of small-scale convection beneath the plate, as has been proposed by Richter and Parsons (1975), Haxby and Weissel (1986) and Huang and Zhong (2005) among others.

Parsons and McKenzie (1978) proposed that the thermal thickness of the lithosphere comprises two layers: an upper layer where heat is lost conductively, and a lower layer where heat is lost convectively. They place the boundary between the two thermally defined layers at the depth of the 1,100°C oceanic isotherm (Fig. 2.27) and speculated that it may also correspond to a mechanical boundary that separates the rigid part of the lithosphere, which is moving with the plates, from the weak (viscous?) underlying asthenosphere which accommodates the movement.

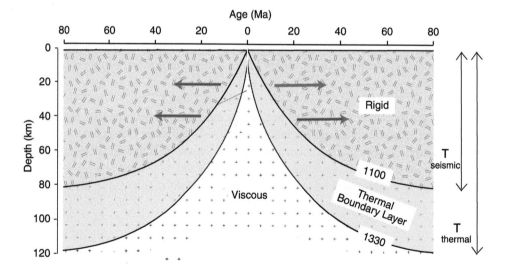

Figure 2.27 Diagram illustrating the concept of the oceanic lithosphere as a thermal and mechanical boundary layer. The arrows delineate that part of the lithosphere that moves with the plates. Reproduced with modification from Parsons and McKenzie (1978).

Seismic and thermal estimates of the thickness of the lithosphere are more difficult to reconcile in the continents. The lack of a systematic relationship between thermal age and topography and heat flow and the uncertainties in the amount of crustal heat production and thermal conductivity complicate the determination of a suitable thermal model for continental lithosphere. Sclater et al. (1980) suggest that the oceans and continents may have a similar thermal thickness of 125 km, the continents being the equilibrium portion of the lithosphere that has not been destroyed by subduction. However, as Lerner-Lam and Jordan (1987) point out, such a thickness is inconsistent with seismic studies which suggest that the thickness of the continental lithosphere is, on average, about twice that of oceanic lithosphere.

Recent estimates of seismic thickness have been based on surface wave tomography in combination with temperature estimates from ocean plate cooling models and garnet peridotite nodules (Priestley and McKenzie, 2013). The thickest lithosphere (up to 250 km) is associated with the cratonic cores of continents, which at the time of the last supercontinent assembly (e.g., Fig. 2.28) formed an arcuate region of thick lithosphere that extended from the southern into the northern hemisphere (McKenzie et al., 2015). Thicknesses decrease over adjacent regions, notably the mobile belts that flank cratons.

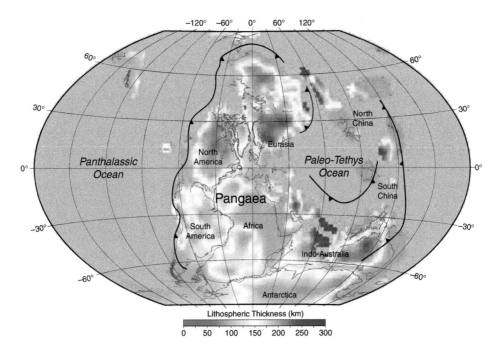

Figure 2.28 Seismic thickness of the lithosphere superimposed on a re-assembly of Pangaea at ~290 Ma. The colour scale reflects the thickness of the lithosphere derived from surface wave tomography in combination with temperature estimates from plate cooling models (Priestley and Mckenzie, 2013). Pangaea was reconstructed using the Euler rotations of Mathews et al. (2016). Reproduced from fig. 1 in Watts et al. (2018).

Nevertheless, thicknesses are still significant, reaching 100 km and higher over Europe and large tracts of the African and Arabian plates.

Seismic data therefore generally support the idea, advanced by the early isostasists, of a thick, strong, rigid lithosphere that overlies a weak asthenosphere. The limitation here is that these data only define the response of Earth's outermost layers to loads that are of short duration (i.e., a few seconds to a few hundred seconds). What is of interest to plate tectonics, however, is the behavior of the lithosphere on the much longer timescales of geological processes, such as those associated with continental assembly and dispersal and the opening and closing of ocean basins.

2.9 Walcott and Flexure of the Lithosphere

Three years after the AGU 1967 Spring meeting that played such an important role in the development of plate tectonics, Richard Irving Walcott (1934–) published a series of papers on the flexure of the lithosphere. In these papers, he examined the evidence for flexure at a number of different loads on Earth's surface, including ice sheets and glacial lakes (Walcott, 1970c), sediments (Walcott, 1972a) and oceanic islands and seamounts (Walcott, 1970b). By comparing predictions based on simple elastic plate models with the observed topography and gravity anomalies in the region of these loads, Walcott was able to estimate the flexural rigidity of the lithosphere in a wider range of tectonic settings than had previously been possible.

Walcott's work was an important milestone in the development of regional isostasy. While it extended the earlier work of Vening Meinesz and Gunn, it was, nonetheless, timely. It directly addressed one of the fundamental assumptions of plate tectonics: namely the idea that the plates behave rigidly on long geological timescales and that they could use in kinematic reconstructions. Although both Vening Meinesz and Gunn had estimated the flexural rigidity, it was Walcott who first determined precisely how rigid the plates were and the extent to which their long-term mechanical properties varied spatially and temporally.

Walcott (1970a) showed that the elastic thickness (which is determined by the flexural rigidity) of the lithosphere in the region of large loads in the plate interiors was in the range of 5–114 km. More importantly, he argued that a correlation existed between the elastic thickness and the load age: the largest values of elastic thickness correlated with the shortest duration loads, such as the glacial lakes, while the smallest values were associated with the longest duration loads, such as the oceanic islands. Walcott therefore surmised that the elastic thickness depends on the age of the load, and that the wide range of values reflected some form of load-induced stress relaxation in the lithosphere.

This result was interesting because the model that he, and Gunn and Vening Meinesz before him, had used was an elastic plate model. Such a model implies that loads are supported by the strength of the plate indefinitely. Walcott argued, however, that a better model might be a viscoelastic rather than an elastic plate. A viscoelastic plate is initially elastic and then becomes more viscous as the age of the load increases. Walcott found (Fig. 2.29) that the observed estimates of the elastic thickness, which were from continents

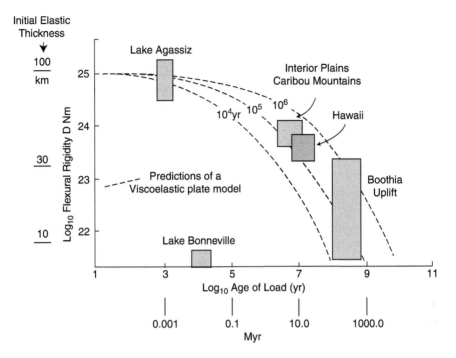

Figure 2.29 The relationship between flexural rigidity and age of a load. Reproduced from fig. 5 of Walcott (1970a), copyright by the American Geophysical Union.

as well as oceans, could be explained by a Maxwell viscoelastic plate model with an initial elastic thickness of about 100 km and a relaxation time of 10^5 a.

At the time of Walcott's study there were few other estimates of the thickness of the lithosphere, but the evidence based on seismic and thermal considerations was certainly consistent with estimates of up to about 100 km for the elastic thickness of the lithosphere.

Walcott's study, however, raised several questions about the long-term mechanical behaviour of the lithosphere. In particular, what is the physical meaning of the elastic thickness and to what extent does it depend on load age as opposed to other factors such as the geological setting and thermal age? The answers to these questions are of fundamental importance to plate tectonics, and it is no surprise, perhaps, that Walcott's work triggered a new wave of interest in flexure studies that was to last throughout the 1970s and into much of the following decade.

2.10 Summary

The idea that loads on the Earth's surface may be regionally rather than locally supported can be traced back to the work of Gilbert and Barrell at the turn of the twentieth century. Their work was carried out at a time when the US Coast and Geodetic Survey was promoting the use of local, rather than regional, models of isostasy. The geodesists, with the notable exception of Putnam, showed that regional models of isostasy were not required to explain geodetic data.

While Gilbert appears to have been influenced by the geodesists, his younger contemporary, Barrell, was not. Barrell challenged the conclusions of the geodesists concerning local models of isostasy, invoking instead the idea that topography was supported regionally by the lateral strength of the lithosphere. Following Barrell's premature death, there was a vacuum, and the new ideas of regional isostasy gradually gave way again to local models of isostasy, which continued to be championed by the geodesists, notably Hayford, Bowie and Heiskanen.

The history thereafter is a punctuated one with 'bursts' of activity followed by long periods of quiet. The work centred on the activities of a few individuals: Vening Meinesz, Gunn and Walcott. Vening Meinesz and Gunn were contemporaries, but there is little evidence of any contact between them.

It is perhaps surprising that regional isostasy (Fig. 2.30) played such a small role in the development of plate tectonics. Although terms such as 'lithosphere' and 'asthenosphere' can

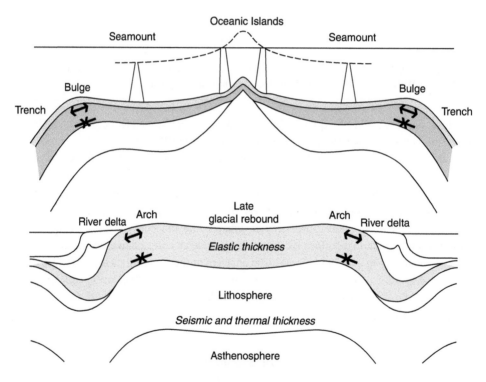

Figure 2.30 Schematic diagram illustrating the structure of the Earth's outer layers, from the viewpoint of regional or flexural isostasy. Upper diagram – Oceans. Lower diagram – Continents. In later chapters we will show that the elastic thickness is approximately one third of the seismic thickness. The arrows schematically illustrate the locations of the maximum extensional and compressional bending stress in the flexed plate.

be traced back to Barrell, the foundations of plate tectonics were built instead on continental drift and seafloor spreading. The main disciplines involved were marine magnetics and earthquake seismology. Gravity, which had played such a key role in proving isostasy, had little or no role. It was only after plate tectonics that gravity was to really come into its own as a means to constrain the crustal and mantle structure of the plate boundaries initially and then the long-term thermal and mechanical properties of the plates themselves.

3

Theory of Elastic Plates

It is sufficiently accurate, therefore, to assume that the lithosphere is a strong, elastic sheet of sufficient uniformity that it obeys well-established laws of elasticity and that it is everywhere supported on a weak underlying magma that, given adequate time, will adjust itself to hydrostatic equilibrium.

(Gunn, 1949, pp. 267–8)

3.1 Introduction

In the previous two chapters, the historical development of the concept of isostasy was presented. The Airy, Pratt and flexure models of isostasy provide a simple means with which to examine the extent of isostatic equilibrium at individual geological features. The Airy and Pratt models, however, assume that topography – no matter what its size – is locally compensated.

The flexure or elastic plate model, as developed by Jeffreys, Vening Meinesz, Gunn and Walcott, is also based on assumptions. The most important is that during loading and unloading, the crust and upper mantle behave as a perfectly elastic material. It is true that at relatively low temperatures and pressures all rocks behave elastically. However, rocks are also brittle at these temperatures and pressures, so they may fracture if the applied loads are sufficiently large. Another problem is that at high temperatures and pressures many rocks deform plastically.

The use of an elastic plate model may seem, therefore, a gross over-simplification of the way that the crust and mantle behave on the long timescales of geological processes. Yet, perhaps because of its simplicity, the model has proved a popular one with geologists. It provides a reference model against which a wide range of geological and geophysical observations can be compared and allows the derivation of a consistent set of parameters that describe the thermal and mechanical properties of the tectonic plates. In this chapter, we develop a simple theory that serves to illustrate the physics that underlies the elastic plate model.

100

3.2 Linear Elasticity

We begin with the assumption that the stresses that develop in an elastic plate during bending are linearly proportional to the strain. This assumption is important in two regards. First, it greatly simplifies the mathematics. Second, it allows use of the so-called principle of superposition. This principle allows the flexure due to a load to be computed from the sum of its component loads. Furthermore, it is known from experimental studies of a wide range of materials that the assumption of linear elasticity gives good results, at least for small strains.

Consider first a stressed material, and the right-handed, orthogonal, Cartesian co-ordinate system x, y, z and the alternative x_1, x_2, x_3 system as illustrated in Fig. 3.1. For any stress vector there will be three scalar components, σ_1, σ_2, σ_3, of the stress that refer respectively, to the x_1, x_2, x_3 axes. If σ_1, σ_2, σ_3 are the principal stresses and ε_1, ε_2, ε_3, the principal strains, then stress and strain are related (e.g., Jaeger, 1969, p. 55) by

$$\left\{ \begin{array}{l} \sigma_1 = (\lambda + 2G) \cdot \varepsilon_1 + \lambda\varepsilon_2 + \lambda\varepsilon_3 \\ \sigma_2 = \lambda\varepsilon_1 + (\lambda + 2G)\varepsilon_2 + \lambda\varepsilon_3 \\ \sigma_3 = \lambda\varepsilon_1 + \lambda\varepsilon_2 + (\lambda + 2G)\varepsilon_3 \end{array} \right\}, \tag{3.1}$$

where G and λ are certain properties of the material known as Lamé's parameters:

G is the modulus of rigidity, which is defined as the ratio of the shear stress to the shear strain in simple shear.

λ is the bulk modulus, which is defined as the ratio of the hydrostatic pressure to the dilatation it produces.

We also define the following:
E is Young's modulus and is defined as the ratio of tension or compression to extension in a solid that is under axial compression or tension and that is unrestricted laterally:

$$E = \frac{G(3\lambda + 2G)}{(\lambda + G)}.$$

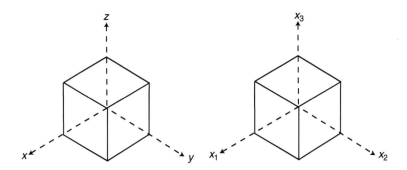

Figure 3.1 The right-hand, orthogonal, Cartesian co-ordinate system showing the x, y and z axes, and the alternative x_1, x_2 and x_3 axes.

v is Poisson's ratio and is defined as the ratio of lateral extension to longitudinal extension in a solid that is under axial compression or tension and that is unrestricted laterally:

$$v = \frac{\lambda}{2(\lambda + G)}.$$

Equation (3.1) gives the stress in terms of the strain. Solving this equation and using the preceding relations for E and v gives

$$\begin{cases} \varepsilon_1 = \dfrac{1}{E}\sigma_1 - \dfrac{v}{E}\sigma_2 - \dfrac{v}{E}\sigma_3 \\[2mm] \varepsilon_2 = \dfrac{-v}{E}\sigma_1 + \dfrac{1}{E}\sigma_2 - \dfrac{v}{E}\sigma_3 \\[2mm] \varepsilon_3 = \dfrac{-v}{E}\sigma_1 + \dfrac{v}{E}\sigma_2 + \dfrac{1}{E}\sigma_3 \end{cases}. \tag{3.2}$$

A widely used boundary condition for the materials of the crust and mantle is that of uniaxial strain. According to this condition, there is strain in the direction of one axis, but not in the other two axes; that is, $\varepsilon_2 = \varepsilon_3 = 0$, $\varepsilon_1 \neq 0$. Substituting these values in Eq. (3.1) gives

$$\sigma_1 = (\lambda + 2G)\varepsilon_1, \qquad \sigma_2 = \sigma_3 = \lambda\varepsilon_1$$

for the principal stresses and

$$\sigma_2 = \sigma_3 = \frac{v}{(1-v)}\sigma_1 \tag{3.3}$$

for the relationship between them.

In geological applications, σ_1 is usually assumed to be the stress that acts in the vertical direction (i.e., in the direction of gravity), while σ_2 and σ_3 are orthogonal horizontal stresses (Fig. 3.2). If the vertical stress that acts on a cube of rock is due to the pressure of the overlying overburden, then

$$\sigma_1 = \rho g h,$$

where h is depth of burial of the cube and ρ is density of the rock.

If the stressed cube of rock can move in the vertical direction, but is unable to move in any other direction (i.e., the condition of uniaxial strain applies), then it follows from Eq. (3.3) that

$$\sigma_2 = \sigma_3 = \frac{v}{(1-v)}\rho g h.$$

Assuming values of $h = 5$ km, $\rho = 2{,}840$ kg m^{-3} and $v = 0.25$, then we get $\sigma_1 = 13.93$ MPa and $\sigma_2 = \sigma_3 = 4.64$ MPa. Therefore, the stress acting on the cube in the horizontal plane is approximately one-third of that which acts in the vertical direction.

3.3 Cylindrical Bending

To model flexure, we need a simple theory that is appropriate to the bending of plates that have large horizontal dimensions, but which are thin in a vertical direction. The flexure of

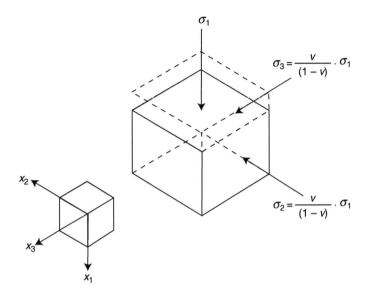

Figure 3.2 Stress vectors acting across each of the co-ordinate planes of a cube of rock resolved into their three components parallel to the x_1, x_2 and x_3 axes. The stresses have been computed assuming the condition of uniaxial strain. In this condition, there is a displacement in the x_1 direction, but no displacements in the x_2 and x_3 directions. Stress σ_1 causes the displacement in the x_1 direction. Stresses σ_2 and σ_3 are called into play to prevent displacements in the x_2 and x_3 directions.

thin plates can be calculated analytically by requiring them to be in equilibrium under all the forces that are acting on them. Unfortunately, in most geological situations it is not possible to specify all these forces.

Timoshenko (1958b, p. 78) has shown that if a plate that is loaded at its edges is three times longer than it is wide, then the resulting flexure is cylindrical in form (Fig. 3.3). Thus, any distortion of the plate during bending is limited to the extreme edges of the plate. This means that it may not be necessary to use a plate to model the deformation of the Earth's crust and mantle. A beam of unit width can be considered instead, if consideration is given to the stresses and strains that act on the beam as a consequence of its being part of a larger plate.

3.4 Flexure of Beams

The flexure of elastic beams has been a subject of much study, especially by mechanical engineers. Several good textbooks already exist on the subject, notably those by Timoshenko (1958a, 1958b), Nadai (1963) and Hetényi (1979). The books by Nadai and Hetényi are noteworthy because of the attention they give to the deformation of beams that overlie weak foundations. These cases have many practical applications, including the determination of the stresses that develop in dams, railway tracks and ship's hulls.

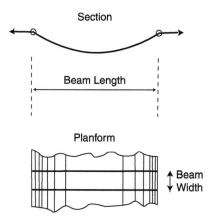

Figure 3.3 Cylindrical bending occurs if (a) the length of a plate is large in comparison with its width, (b) the acting forces do not vary along the length of the plate and (c) only the part of the plate at a sufficient distance from the ends is considered. Reproduced from fig. 52 of Timoshenko (1958b) with permission of John Wiley & Sons Limited.

3.4.1 Stresses

Let us first consider the stresses that develop in an elastic beam that is deformed by a turning moment, M_o, applied to opposite ends. Using the co-ordinate system illustrated in Fig. 3.4, the turning moments deform the beam into a convex downward shape.

We make the assumption that the deformation takes place in such a way that (a) the depth into the beam of any horizontal surface and (b) any vertical cross-section of the beam are the same before and after bending. It follows then from Fig. 3.4 that triangles OCD and DEF contain one common angle and three parallel sides, so they are similar. The strain of any fibre in the beam in the x direction, ε_x, is given by

$$\varepsilon_x = \frac{\text{change}}{\text{original length}} = \frac{EF}{CD} = \frac{DF}{OD} = \frac{y_f}{r}, \tag{3.4}$$

where r is the radius of curvature and y_f is the distance from the neutral surface (a surface on which there is no strain) to a fibre in the beam.

A beam may be treated as a 'cut-out' of a larger plate by applying a plane stress boundary condition. Such a condition implies that the beam is stressed in the x and z directions but is only free to move in the y direction. The stress in the x direction arises from the bending while the stress in the z direction acts to prevent strain in this direction, thereby simulating the effect of the surrounding, much larger, plate.

The conditions of plane stress can be written as

$$\sigma_2 \neq 0, \quad \sigma_1 \neq 0, \quad \sigma_3 = 0$$

or

$$\sigma_x \neq 0, \quad \sigma_z \neq 0, \quad \sigma_y = 0. \tag{3.5}$$

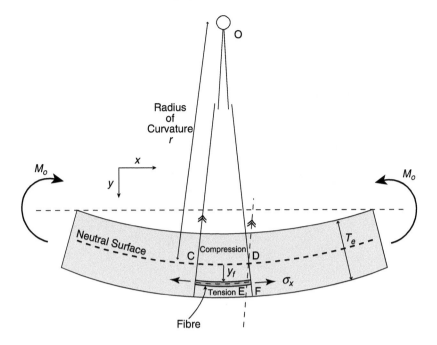

Figure 3.4 The bending of an elastic beam by an applied bending moment, M_0. The shaded region shows how bending causes a fibre lengthwise strip in the lower part of the beam to undergo a longitudinal tensile stress, σ_x. There is an equivalent longitudinal compressive stress induced in the upper part of the beam. The neutral surface is the surface within the beam that experiences neither tensile nor compressive stresses and, hence, undergoes no strain.

Substituting Eq. (3.5) in Eq. (3.2) gives

$$\left.\begin{array}{l} E\varepsilon_x = \sigma_x - v\sigma_z \\ E\varepsilon_z = \sigma_z - v\sigma_x \\ E\varepsilon_y = -v(\sigma_x + \sigma_z) \end{array}\right\}. \tag{3.6}$$

But $\varepsilon_z = 0$, so Eq. (3.6) reduces to

$$\left.\begin{array}{l} \sigma_x = \dfrac{E\varepsilon_x}{(1 - v^2)} \\ \sigma_z = \dfrac{E\varepsilon_x v}{(1 - v^2)} \end{array}\right\}. \tag{3.7}$$

Substituting Eq. (3.4) in Eq. (3.7) gives

$$\left.\begin{array}{l} \sigma_x = \dfrac{Ey_f}{r(1 - v^2)} \\ \sigma_z = \dfrac{Ey_f v}{r(1 - v^2)} \end{array}\right\}. \tag{3.8}$$

3.4.2 Shearing Force and Bending Moment

The stresses produced in the beam during bending must balance the moment, M_o, that is applied to the ends of the beam. In most geological applications, however, it is more convenient to replace M_o by a vertical shear force V and a bending moment M that acts in the plane of any cross section of the deformed beam.

The bending moment can be estimated by considering the stresses that act across any cross section of the beam. We can write (e.g., Fig. 3.5) as follows for the force dF that acts on an element of the cross section of the beam that has a height dy and unit width:

$$\text{force} = \text{stress} \times \text{area},$$

$$dF = \sigma_x dy 1.0,$$

$$dF = \left(\frac{E y_f}{r(1 - v^2)}\right) dy.$$

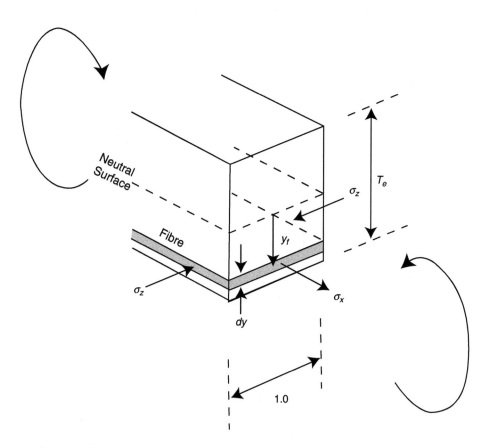

Figure 3.5 Cross section showing the stresses that act on an element of a beam of height dy, unit width and distance y_f from the neutral surface.

The moment, dM, that acts on the element about the neutral surface of the beam is given by

$$moment = force \times distance,$$

$$dM = dFy_f,$$

$$\therefore \ dM = \left(\frac{Ey_f^2}{r(1-v^2)}\right)dy.$$

The total bending moment is obtained by integration of all the elemental moments. Hence

$$M = \frac{E}{r(1-v^2)}\int_{T_e0.5}^{-T_e0.5} y_f^2 dy,$$

where T_e is the elastic thickness of the plate,

$$\therefore \ M = \frac{-ET_e^3}{12(1-v^2)}\frac{1}{r}$$

or

$$M = -\frac{D}{r}, \tag{3.9}$$

where D is the flexural rigidity of the plate and given by

$$D = \frac{ET_e^3}{12(1-v^2)}. \tag{3.10}$$

The radius of curvature in the x, y co-ordinate system illustrated in Fig. 3.4 is given by

$$\frac{1}{r} = \frac{\left[\frac{d^2y}{dx^2}\right]}{\left[1+\left(\frac{dy}{dx}\right)^2\right]^{1.5}},$$

which for small slopes approximates to

$$\frac{1}{r} \rightarrow \frac{d^2y}{dx^2}. \tag{3.11}$$

Substituting Eq. (3.11) in Eq. (3.9) gives

$$M = -D\left[\frac{d^2y}{dx^2}\right]. \tag{3.12}$$

In the co-ordinate system used here, the flexure is concave up and d^2y/dx^2 is negative. Therefore, M will be positive. This is consistent with engineering convention (e.g., Timoshenko, 1958a, p. 75) that M is positive if it acts in a clockwise direction on the left-hand side of a vertical cross section in the beam.

Once the bending moment has been defined, it is a relatively easy matter to compute the shear force, V, that acts on any vertical cross section of the deformed beam.

Let us consider an element of the beam of width dx that is bounded by two vertical cross sections, mn and m'n' (Fig. 3.6). If the element is of negligible mass, then the shear force that acts on opposing sides of the element will be equal. The bending moments, however, will not be equal. The increase in bending moment, dM, is given by the moments of the couple represented by the two (equal and opposite) vertical shearing forces; that is,

$$dM = Vdx$$

or

$$\frac{dM}{dx} = V. \tag{3.13}$$

If additional loads act on the surface and base of the beam element, then the shear forces on opposing sides of the element will change. Assume, for example, that a force q/unit area is applied to the surface of the element while a force p/unit area is applied to its base (Fig. 3.7). The force balance on the element is obtained by summing all the downward and upward acting forces and is given by

$$qdx1.0 + (V + dV) - pdx1.0 - V = 0,$$

$$\therefore \quad \frac{dV}{dx} = (p - q) = \frac{d^2M}{dx^2},$$

$$\frac{d^2M}{dx^2} - (p - q) = 0. \tag{3.14}$$

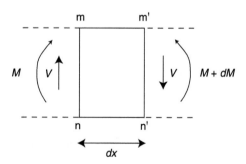

Figure 3.6 The bending moments and shearing forces that act on an element of a beam of width dx.

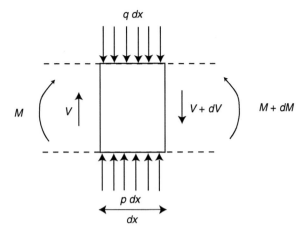

Figure 3.7 The bending moments and shearing forces that act on an element of a beam of width dx, which is loaded by a force q/unit area on the surface and by a force p/unit area from below. Reproduced from fig. 2 of Hetényi (1979) with permission of the University of Michigan Press.

Substituting Eq. (3.12) in Eq. (3.14) gives

$$D\frac{d^4y}{dx^4} + (p - q) = 0. \tag{3.15}$$

3.4.3 Winkler Foundation

In geological applications, we are usually concerned with loads of a certain shape that act on the surface and the base of the crust (Fig. 3.8). It is therefore more convenient to replace the forces q/unit area and p/unit area by a load of a specified height and density. Consider, for example, the material that infills the deflected part of the beam. If the infill material completely fills the deflection and its density, ρ_{infill}, is greater than that of the material it displaces, then the downward acting force is given by

$$\begin{aligned} q &= \text{force/unit area} \\ &= \text{mass} \times \text{acceleration/unit area} \\ &= \text{density} \times \text{volume} \times \text{acceleration/unit area} \\ \therefore q &= \text{density} \times \text{height} \times \text{acceleration} \\ &\text{and } q = \rho_{infill} y g. \end{aligned} \tag{3.16}$$

Finally, consider the foundation of the beam. If the foundation exerts an upward force that acts on the base of the bent part of the beam – due, say, to the buoyancy of the material displaced – then

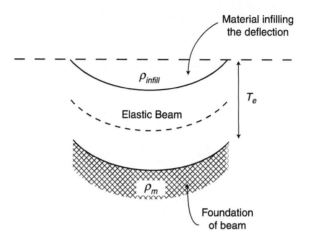

Figure 3.8 The deflection of an elastic beam that is loaded by a downward acting force due to sediments, for example. The beam is acted on its base by an upward load due to buoyancy of the foundation and on its surface by a downward load due to material that infills the deflection.

$$p = \rho_m yg. \tag{3.17}$$

A foundation that acts with a force that is proportional at every point to the deflection is known as a Winkler foundation[1] (Hetényi, 1979). Examples of such foundations include the ballast that supports railway lines and the water displaced by a ship's hull. In both cases, it is sufficient to assume that the foundation behaves as an inviscid fluid (i.e., a fluid with zero viscosity). Substitution of Eq. (3.16) and Eq. (3.17) in Eq. (3.15) gives

$$D\frac{d^4y}{dx^4} + (\rho_m - \rho_{infill})yg = 0, \tag{3.18}$$

which is the general equation for the deflection of an elastic beam overlying an inviscid substratum. Equation (3.18) is an example of a homogeneous fourth-order differential equation, the solution of which can be obtained using the method of quadratics. It can be verified, for example, by substitution, that the solution of Eq. (3.18) is of the following form:

$$y = e^{\lambda x}(A_c\cos \lambda x + B_c\sin \lambda x) + e^{-\lambda x}(C_c\cos \lambda x + D_c\sin \lambda x), \tag{3.19}$$

where A_c, B_c, C_c and D_c are the four arbitrary constants of integration; λ is a parameter that determines the amplitude and wavelength of the deformation and is given by

$$\lambda = \left[\frac{(\rho_m - \rho_{infill})g}{4D}\right]^{0.25}. \tag{3.20}$$

[1] After E. Winkler who first introduced the term in 1867.

3.5 Beams of Unlimited Length

So far, consideration has only been given to beams of finite length. While this is a frequently occurring situation in mechanical engineering, we are mostly concerned in flexure studies with beams that extend for very great distances on either side of where a particular load is emplaced. Equation (3.19) therefore needs to be solved for the special cases of loads that are emplaced on beams of infinite and semi-infinite length.

3.5.1 Infinite Beams

Consider first the case of a line load P_b/unit width that is applied at a point O (0,0) on a beam of infinite length (Fig. 3.9). The load, P_b, is referred to as a line load because it is concentrated in the x direction, but it extends along the unit width of the beam in the z direction. The following boundary conditions apply:

Condition 1: $x \rightarrow \pm \infty$, $y \rightarrow 0$. This condition ensures that there is no deflection at the far edges of the beam. The condition is met if terms in $e^{\lambda x}$ in Eq. (3.19) vanish. This occurs if $A_c = B_c = 0$. In this case, Eq. (3.19) reduces to

$$y = e^{-\lambda x}(C_c \cos \lambda x + D_c \sin \lambda x). \tag{3.21}$$

Condition 2: $x = 0$, $dy/dx = 0$. This condition ensures symmetry in the deflection on either side of where the load is applied. This condition is met if the first derivative of the deflection curve is zero at $x = 0$. It follows from Eq. (3.21) that

$$\frac{dy}{dx} = C_c\left(-\lambda e^{-\lambda x}\cos \lambda x - \lambda e^{-\lambda x}\sin \lambda x\right) + D_c\left(-\lambda e^{-\lambda x}\sin \lambda x + \lambda e^{-\lambda x}\cos \lambda x\right).$$

At $x = 0$ this reduces to

$$\frac{dy}{dx} = -C_c\lambda + D_c\lambda,$$

which is only zero if $C_c = D_c$. Equation (3.21) therefore becomes

$$y = e^{-\lambda x}D_c(\cos \lambda x + \sin \lambda x). \tag{3.22}$$

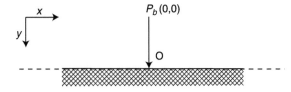

Figure 3.9 A line load, P_b/unit width, applied at a point O (0,0) on the surface of a beam of infinite length.

The remaining integration constant, D_c, can be determined by considering the force balance that acts on the surface and base of the deflected beam. Such a balance exists when the downward-acting loads are equal to the sum of all the upward-acting loads. We then have

$$P_b 1.0 + 2\rho_{infill}g \int_0^\infty y\,dx = 2\rho_m g \int_0^\infty y\,dx, \qquad (3.23)$$

where the loads on the left-hand side of the equation represent the downward-acting driving and infill loads on the surface of the beam, and the load on the right-hand side is the upward-acting load due to the buoyancy of the inviscid substratum. The factor of 2 in Eq. (3.23) ensures that all loads on either side of P_b are considered.

Re-arranging Eq. (3.23) yields

$$P_b = 2(\rho_m - \rho_{infill})g \int_0^\infty y\,dx.$$

Substituting for y from Eq. (3.22) and integrating gives

$$P_b = \frac{2(\rho_m - \rho_{infill})gD_c}{\lambda}, \qquad (3.24)$$

and substitution of Eq. (3.24) into Eq. (3.22) gives

$$y = \frac{P_b \lambda}{2(\rho_m - \rho_{infill})g} e^{-\lambda x}(\cos \lambda x + \sin \lambda x). \qquad (3.25)$$

The form of Eq. (3.25) can best be illustrated by considering some limiting values. For example,

(a) $x = 0$, $y = y_{max} = \frac{P_b \lambda}{2(\rho_m - \rho_{infill})g}$,

(b) $y = 0$ when $(\cos \lambda x + \sin \lambda x) = 0$. This occurs at $x = 3\pi, 7\lambda/4\pi \,...,$

(c) Positions of maxima and minima occur when $dy/dx = 0$. Differentiating Eq. (3.35) with respect to x gives

$$\frac{dy}{dx} = \frac{-P_b \lambda^2}{(\rho_m - \rho_{infill})g} e^{-\lambda x} \sin \lambda x \qquad (3.26)$$

which is zero when $x = 0, \pi/\lambda, 2\pi/\lambda \,...$. The value of y at $x = 0$ has already been evaluated. This is y_{max}. The value of y at $x = \pi/\lambda$ is therefore

$$y_{x=\pi/\lambda} = y_{max}e^{-\pi}(\cos \pi + \sin \pi)$$

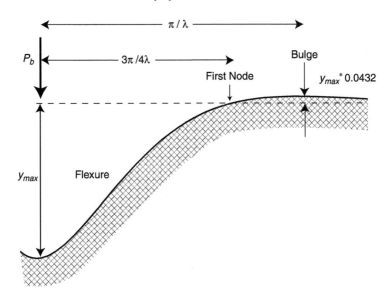

Figure 3.10 The flexure of a beam of infinite length by a line load P_b/unit width.

or

$$y_{x=\pi/\lambda} = -y_{max}0.0432.$$

The limiting values indicate that the deflection (Fig. 3.10) has the form of an exponentially damped cosine and sine function. There is a downward deflection, the 'flexural depression', for $0<x<3\pi/4\lambda$, and an upward deflection, the 'flexural bulge', for $3\pi/4\lambda<x<7\pi/4\lambda$. At $x=3\pi/4\lambda$ there is neither a depression nor a bulge and this point is referred to as the 'first node'. The parameter λ therefore controls both the amplitude and the wavelength of the flexure curve.

The parameter λ, for example, in Eq. (3.20) is determined by the elastic properties of the beam. In engineering applications, λ is usually a known quantity, while in geology it is an unknown that needs to be estimated from observations. Unfortunately, λ is not used consistently in the geological literature. Some workers, for example, prefer to express it in terms of the two-dimensional flexural parameter, α, which has units of distance and is given by

$$\alpha = \frac{1}{\lambda} = \left[\frac{4D}{(\rho_m - \rho_{infill})g} \right]^{0.25}.$$

Others prefer the flexural rigidity, D, or the elastic thickness, T_e. Table 3.1 summarises the different parameters that have been used, together with their equivalent values of λ.

The discussion thus far has been in terms of the flexure of the beam. It is also of interest to determine the bending stresses that develop in the beam because of bending (e.g., Fig. 3.11). We have from Eq. (3.8) and Eq. (3.11), for example, that

Table 3.1 *The main elastic parameters that have been used to describe the long-term mechanical properties of the lithosphere*

Description in Original Paper	Symbol in Original Paper	Equivalent λ	Reference
–	α	$\alpha/(1-v^2)^{1/4}$	Jeffreys (1926)
Lithospheric constant	b	$b/(1-v^2)^{1/4}$	Gunn (1943a)
Flexural parameter	α	$1/\alpha$	Walcott (1970a)
Flexural rigidity	D	$\left((\rho_m-\rho_{infill})g/4D\right)^{1/4}$	Walcott (1970a)
Elastic thickness	T_e	$3g(\rho_m-\rho_{infill})(1-v^2)g/T_e^3$	Walcott (1970a)

$$\sigma_x = \frac{E}{(1-v^2)}\frac{d^2y}{dx^2}y_f, \tag{3.27}$$

where y_f is the distance to a particular horizontal fibre in the beam. It follows from Eq. (3.25) that the second derivative of the deflection profile is

$$\frac{d^2y}{dx^2} = \frac{-P_b\lambda^3}{(\rho_m-\rho_{infill})g}e^{-\lambda x}(\cos(\lambda x)-\sin(\lambda x)). \tag{3.28}$$

Substituting Eq. (3.28) in Eq. (3.27) gives

$$\sigma_x = \frac{E}{(1-v^2)}\frac{-P_b\lambda^3}{(\rho_m-\rho_{infill})g}e^{-\lambda x}(\cos(\lambda x)-\sin(\lambda x)). \tag{3.29}$$

We can use Eq. (3.29) to evaluate the bending stresses that act within the beam. The maximum stress occurs in the uppermost and lowermost parts of the beam. For example, if we choose to evaluate the stress on the uppermost surface of the beam, then $y_f = -T_e/2$ and

$$\sigma_x = \frac{ET_eP_b\lambda^3}{2(1-v^2)(\rho_m-\rho_{infill})g}e^{-\lambda x}(\cos\lambda x-\sin\lambda x). \tag{3.30}$$

The form of the stress profile is best illustrated by consideration of the limiting values. For example,

(a) $x=0$, $\sigma_x = \sigma_{max} = \frac{ET_eP_b\lambda^3}{2(1-v^2)(\rho_m-\rho_{infill})g}$,

(b) $\sigma_x = 0$ when $(\cos\lambda x - \sin\lambda x) = 0$. This occurs at $x = \pi/4\lambda, 5\pi/4\lambda, 9\pi/4\lambda, \ldots$,

(c) Positions of maxima and minima occur when $d\sigma_x/dx = 0$ Differentiating Eq. (3.30) with respect to x gives

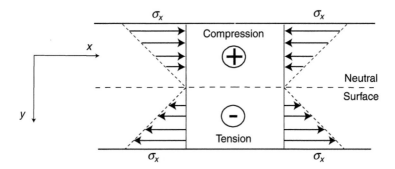

Figure 3.11 Fibre stresses in a bending beam. The upper half of the beam is compressed and the fibre stress is positive. The lower half of the beam is extended and the fibre stress is negative. The fibre stress is zero on the neutral surface.

$$\frac{d\sigma_x}{dx} = \frac{-ET_eP_b\lambda^4}{(1-v^2)(\rho_m-\rho_{infill})g}e^{-\lambda x}\cos\lambda x,$$

which is zero when $x = 0, \pi/2\lambda, 3\pi/2\lambda \dots$. We have already evaluated the value of σ_x at $x = 0$. This is σ_{max}. The value of σ_x at $x = \pi/2\lambda$ is therefore

$$\sigma_{x=\pi/2\lambda} = \sigma_{max}e^{-\pi/2}\left(\cos\left(\frac{\pi}{2}\right) - \sin\left(\frac{\pi}{2}\right)\right)$$

or

$$\therefore \ \sigma_{x=\pi/2\lambda} = -\sigma_{max}0.208.$$

Like the flexure, the stress profile is determined by λ (Fig. 3.12). Stresses in the upper part of the beam are positive for $0 < x < \pi/4\lambda$, forming a broad region of compression that flanks the load. The maximum compressive stress is at $x = 0$. In contrast, stresses are negative for $\pi/4\lambda < x < 5\pi/4\lambda$, forming a broad region of tension. The maximum tensile stress occurs at $x = \pi/2\lambda$, which is also the point of greatest curvature of the deflected beam.

The discussion so far has been concerned with line loads. In most geological applications, however, loads are distributed, sometimes over wide regions. The flexure due to a distributed load can be calculated by integrating a line load element over the actual width of the load.

Consider, for example, the case of a rectangular load that is placed on the surface of the beam. First, the line load P_b/unit width needs to be replaced by a distributed load q_x/unit area; that is,

$$P_b1.0 = q_xdx1.0.$$

The incremental deflection, dy, due to the incremental load q_xdx is therefore

$$dy = \frac{q_xdx\lambda}{2(\rho_m-\rho_{infill})g}e^{-\lambda x}(\cos\lambda x + \sin\lambda x).$$

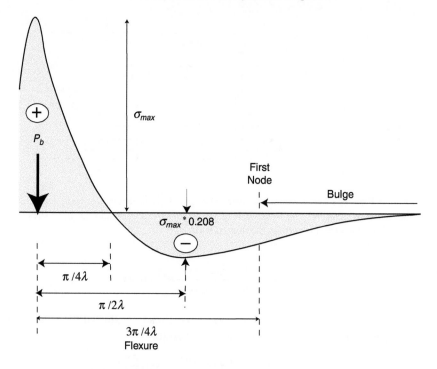

Figure 3.12 Maximum fibre stresses in the upper half of a bent beam of infinite length. Positive stresses are compressive. Negative stresses are tensile.

The total flexure is obtained by integration.

For example, consider a point C (0,0) that is directly beneath the load (Fig. 3.13a). The total flexure is

$$y = \frac{q_x\lambda}{2(\rho_m - \rho_{infill})g}\left[\int_0^b e^{-\lambda x}(\cos\lambda x + \sin\lambda x)dx + \int_0^a e^{-\lambda x}(\cos\lambda x + \sin\lambda x)dx\right], \quad (3.31)$$

where the limits of integration a and b represent the distance from C to the left and the right edge of the load respectively. Evaluating the integrals in Eq. (3.31) gives (e.g., Hetényi, 1979, p. 15)

$$y = \frac{q_x}{2(\rho_m - \rho_{infill})g}\left(2 - e^{-\lambda b}\cos\lambda b - e^{-\lambda a}\cos\lambda a\right). \quad (3.32)$$

For a point to the left of the distributed load (i.e., Fig. 3.13b) it follows that

$$y = \frac{q_x}{2(\rho_m - \rho_{infill})g}\left(e^{-\lambda a}\cos\lambda a - e^{-\lambda b}\cos\lambda b\right), \quad (3.33)$$

and for a point to the right (i.e., Fig. 3.13c),

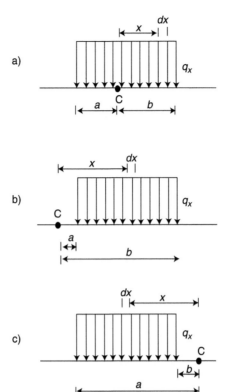

Figure 3.13 Diagram showing the deflection due to a distributed load on the surface of a beam can be calculated at (a) points beneath, (b) to the left and (c) to the right of the load. Reproduced from fig. 8 of Hetényi (1979) with permission of the University of Michigan Press.

$$y = \frac{-q_x}{2(\rho_m - \rho_{infill})g}\left(e^{-\lambda a}\cos(\lambda a) - e^{-\lambda b}\cos(\lambda b)\right).\qquad(3.34)$$

Figure 3.14 illustrates the flexure of a beam due to two distributed loads: one narrow (i.e., width = 25 km) and the other wide (i.e., width = 300 km), computed using Eqs. (3.32) to (3.34).

3.5.2 Semi-infinite Beams

So far, we have only considered the case of an infinite beam. Of particular interest in geological applications are cases where loads are applied on or near a break in a plate. Such a situation can arise at a fault scarp where one crustal block has been vertically offset from another. If the offset is two-dimensional (i.e., it extends for a great distance in the

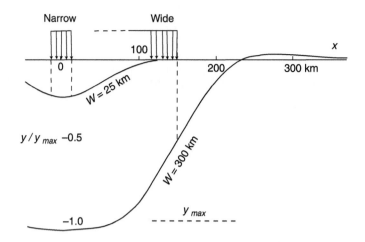

Figure 3.14 Deflection of a beam due to a narrow (i.e., 25 km) and wide (i.e., 300 km) distributed load. The deflections have been normalised to the value expected for Airy isostasy and assume $\lambda = 1/54\,\mathrm{km}^{-1}$ (which corresponds to $T_e = 20\,\mathrm{km}$), $E = 100$ GPa and $\sigma = 0.25$.

$\pm z$ direction), then we can approximate the behaviour of the plate by a semi-infinite beam that has a free-end and is infinite in one direction.

The deflection of semi-infinite beams is most easily obtained by first considering the moments and forces that are generated during bending of an infinite beam. By 'nulling' the bending moment and shear force at the free-end it is then possible to derive an expression for the deflection of a semi-infinite beam.

Consider, for example, the case of the loading of the infinite beam illustrated in Fig. 3.15a. The line load P_1/unit width causes a shear force and bending moment to be set up in the plane of any cross-section of the beam. The shear force, V_A, and the bending moment, M_A, help to maintain the mechanical continuity of the beam on either side of A. If, however, a shear force, P_o, and bending moment, M_o, are applied to the beam that nullify V_A and M_A (Fig. 3.15b), then the beam would become mechanically discontinuous, or break, at A. The values of P_o and M_o that are needed to produce such a condition are called (Hetényi, 1979) the end-conditioning forces.

A balance of bending moments and shearing forces at A then gives

V_A+*vertical shearing force at A due to* M_o+*vertical shearing force at A due to* $P_o=0$

$$V_A + V_{M_o} + V_{P_o} = 0$$

$M_A +$ *moment at A due to* $M_o +$ *moment at A due to* $P_o = 0$

$$M_A + M_{M_o} + M_{P_o} = 0. \tag{3.35}$$

We begin by evaluating the deflection, bending moments and shear forces that arise in an infinite beam due to P_o.

The flexure due to P_o is

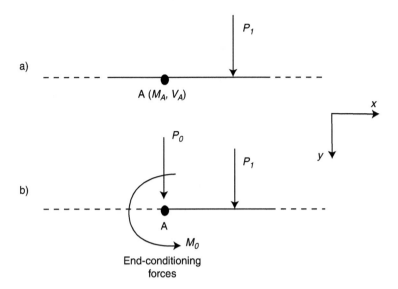

Figure 3.15 Diagram illustrating the moment, shearing force and end-conditioning forces, M_o and P_0, at A due to a line load P_1/unit width on an infinite beam.

$$y = \frac{P_o \lambda}{2(\rho_m - \rho_{infill})g} e^{-\lambda x}(\cos \lambda x + \sin \lambda x). \tag{3.36}$$

The bending moment due to P_o is (see Eq. 3.12)

$$M_{P_o} = -D\frac{d^2 y}{dx^2}. \tag{3.37}$$

Differentiating Eq. (3.36) twice with respect to x gives

$$\frac{d^2 y}{dx^2} = \frac{-P_o \lambda^3}{(\rho_m - \rho_{infill})g} e^{-\lambda x}(\cos \lambda x - \sin \lambda x). \tag{3.38}$$

Substituting Eq. (3.20) into Eq. (3.38), and Eq. (3.38) into Eq. (3.37) gives

$$M_{P_o} = \frac{P_o}{4\lambda} e^{-\lambda x}(\cos \lambda x - \sin \lambda x). \tag{3.39}$$

The shear force due to P_o is

$$V_{P_o} = \frac{dM_{P_o}}{dx},$$

$$V_{P_o} = -\frac{P_o}{2} e^{-\lambda x}\cos \lambda x. \tag{3.40}$$

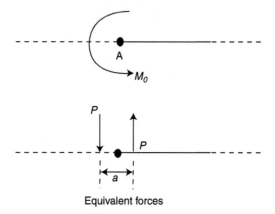

Equivalent forces

Figure 3.16 M_o can be considered as a limiting case of two opposing vertical forces P/unit width a distance a apart. Reproduced from fig. 6 of Hetényi (1979) with permission of the University of Michigan Press.

Next, we need to evaluate the deflection, bending moment and shear force due to M_o. The easiest way to do this is to consider M_o as the limiting case of two opposing vertical forces, P, that are applied to the beam a distance a apart (Fig. 3.16). The deflection due to the downward vertical force is

$$y_{down} = \frac{P\lambda}{2(\rho_m - \rho_{infill})g} e^{-\lambda x}(\cos \lambda x + \sin \lambda x). \tag{3.41}$$

The deflection due to the upward vertical force, which is offset a distance a from the downward force, is then

$$y_{up} = \frac{-P\lambda}{2(\rho_m - \rho_{infill})g} e^{-\lambda(x+a)}(\cos \lambda(x+a) + \sin \lambda(x+a)). \tag{3.42}$$

The total deflection is

$$y_{total} = y_{up} + y_{down}$$

or

$$y_{total} = \frac{-Pa\lambda}{2(\rho_m - \rho_{infill})g} \left[\frac{e^{-\lambda(x+a)}(\cos \lambda(x+a) + \sin \lambda(x+a))}{a} - \frac{e^{-\lambda x}(\cos \lambda x + \sin \lambda x)}{a} \right].$$

$$\tag{3.43}$$

We now follow Hetényi (1979, p. 14) and assume that as a approaches zero, P simultaneously approaches the value of M_o (i.e., $Pa \rightarrow M_o$). Equation (3.43) then simplifies to

$$y_{total} = \frac{-M_o\lambda^2}{(\rho_m - \rho_{inf\,ill})g}e^{-\lambda x}\sin\lambda x. \tag{3.44}$$

The bending moment and vertical shear are given by the successive derivatives of y_{total}. Hence,

$$M_{M_o} = \frac{M_o}{2}e^{-\lambda x}\cos\lambda x \tag{3.45}$$

and

$$V_{M_o} = \frac{-M_o\lambda}{2}e^{-\lambda x}(\cos\lambda x + \sin\lambda x). \tag{3.46}$$

Consider now that the point A is at $x = 0$. Equations (3.39), (3.42), (3.45) and (3.46) then simplify to

$$M_{P_o} = \frac{P_o}{4\lambda}, \quad V_{P_o} = \frac{-P_o}{2}, \quad M_{M_o} = \frac{M_o}{2}, \quad V_{M_o} = \frac{-M_o\lambda}{2}.$$

Substitution in Eq. (3.35) then gives

$$V_A = \frac{M_o\lambda}{2} + \frac{P_o}{2}, \quad M_A = -\left(\frac{M_o}{2} + \frac{P_o}{4\lambda}\right).$$

Consider now Fig. 3.17, which shows the case of a load P_b/unit width on the free-end of a semi-infinite beam. In this case,

$$V_A = P_b, \qquad M_A = 0.$$

Therefore,

$$M_o = \frac{-2P_b}{\lambda}, \qquad P_o = 4P_b \tag{3.47}$$

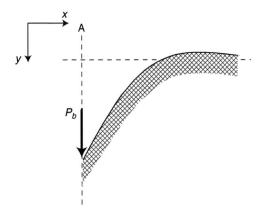

Figure 3.17 The deflection of a semi-infinite (broken) beam by a load P_b/unit width at A.

where M_o and P_o are the end-conditioning forces that are required to produce the condition of the free-end. The deflections for these forces are obtained from substitution of Eq. (3.47) into Eq. (3.44),

$$y_{M_o} = \frac{M_o \lambda^2}{(\rho_m - \rho_{infill})g} e^{-\lambda x} \sin \lambda x$$

$$\therefore \ y_{M_o} = \frac{-2P_b \lambda}{(\rho_m - \rho_{infill})g} e^{-\lambda x} \sin \lambda x,$$

$$(3.48)$$

and Eq. (3.47) into Eq. (3.36),

$$y_{P_o} = \frac{P_o \lambda}{2(\rho_m - \rho_{infill})g} e^{-\lambda x} (\cos \lambda x + \sin \lambda x)$$

$$\therefore \ y_{P_o} = \frac{2P_b \lambda}{(\rho_m - \rho_{infill})g} e^{-\lambda x} (\cos \lambda x + \sin \lambda x).$$

$$(3.49)$$

The total deflection, y, is obtained by summing Eq. (3.48) and Eq. (3.49):

$$y = \frac{2P_b \lambda}{(\rho_m - \rho_{infill})g} e^{-\lambda x} \cos \lambda x.$$

$$(3.50)$$

As before, the form of the deflection can be determined by considering limiting conditions. For example,

(a) $x = 0$, $y_{max} = \frac{2P_b \lambda}{(\rho_m - \rho_{infill})g}$,

(b) $y = 0$ when $\cos \lambda x = 0$. This occurs at $x = \pi/2\lambda$, $3\pi/2\lambda$, ...,

(c) Positions of maxima occur when $dy/dx = 0$. Differentiating Eq. (3.50) with respect to x gives

$$\frac{dy}{dx} = \frac{-2P_b \lambda^2}{(\rho_m - \rho_{infill})g} e^{-\lambda x} (\cos \lambda x + \sin \lambda x)$$

which is zero when $\cos \lambda x \rightarrow -\sin \lambda x$; that is, when $x = 3\pi/4\lambda$, $7\pi/4\lambda$, ... The value of y at $x = 3\pi/4\lambda$ is given by

$$y_{x=3\pi/4\lambda} = -y_{max}0.0671.$$

Figure 3.18 illustrates the form of the flexure. As in the case of the infinite beam, the semi-infinite beam is depressed immediately below the load and is uplifted in flanking regions. The main differences between the two cases are that the width of the depressed region is smaller (i.e. $0 < x < \pi/2\lambda$ compared with $0 < x < 3\pi/4\lambda$), the distance

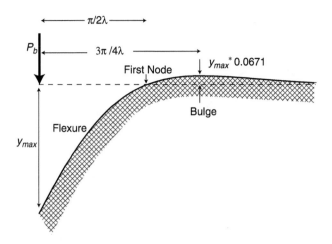

Figure 3.18 The flexure of a semi-infinite beam by a line load P_b/unit width.

to the crest of the bulge is shorter (i.e. $3\pi/4\lambda$ compared with π/λ) and the amplitude of the bulge is larger (~6.7 per cent of the maximum flexure compared with ~4.2 per cent).

The stresses in the beam can be determined from

$$\sigma_x = \frac{E}{(1-v^2)} \frac{d^2y}{dx^2} y_f.$$

Differentiating Eq. (3.50) with respect to x gives

$$\frac{d^2y}{dx^2} = \frac{4P_b\lambda^3}{(\rho_m - \rho_{infill})g} e^{-\lambda x} \sin \lambda x.$$

Substituting and evaluating the stress at $y_f = -T_e/2$ then gives

$$\sigma_x = \frac{E}{(1-v^2)} \frac{2P_b\lambda^3 T_e}{(\rho_m - \rho_{infill})g} e^{-\lambda x} \sin \lambda x. \qquad (3.51)$$

The form of Eq. (3.51) can be illustrated by considering some limiting values. For example,

(a) $\sigma_x = 0$ when $\sin \lambda x = 0$. This occurs at $x = 0, \pi/\lambda, 2\pi/\lambda, \dots$.

(b) Positions of maxima occur when $d\sigma_x/dx = 0$. Differentiating Eq. (3.51) with respect to x gives

$$\frac{d\sigma_x}{dx} = \frac{E}{(1-v^2)} \frac{-2P_b\lambda^4 T_e}{(\rho_m - \rho_{infill})g} e^{-\lambda x} (\sin \lambda x - \cos \lambda x)$$

which is zero when $\sin \lambda x = \cos \lambda x$. This occurs when $x = \pi/4\lambda, 5\pi/4\lambda, \dots$. The maximum value of σ_x occurs at $x = \pi/4\lambda$ and is given by

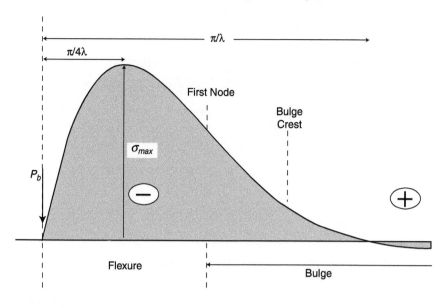

Figure 3.19 Maximum fibre stresses in the upper half of a flexed beam of semi-infinite length that has been deformed by a downward acting line load. Negative stresses are tensile. Positive stresses are compressive.

$$\sigma_{max} = \frac{E}{(1 - v^2)} \frac{P_b \lambda^3 T_e}{(\rho_m - \rho_{infill})g} 0.644.$$

The stresses in the uppermost part of a flexed semi-infinite beam are illustrated in Fig. 3.19.

The final case is the flexure caused by distributed loads on semi-infinite beams. This can be evaluated by integration. Hetényi (1979), for example, gives the solutions for the case of a rectangular load on the end of a semi-infinite beam.[2]

3.6 Hetényi's Functions

It can be seen from Eqs. (3.25), (3.30), (3.50) and (3.51) that the deflections and stresses associated with both infinite and semi-infinite beams can be expressed in terms of four main functions. These are as follows:

$$A_{\lambda x} = e^{-\lambda x}(\cos \lambda x + \sin \lambda x)$$
$$B_{\lambda x} = e^{-\lambda x} \sin \lambda x$$
$$C_{\lambda x} = e^{-\lambda x}(\cos \lambda x + \sin \lambda x)$$
$$D_{\lambda x} = e^{-\lambda x}\cos \lambda x.$$

[2] A MATLAB® file, chap3_hetenyi_plate_break.m, which uses the Hetényi functions to calculate the flexure due to a rectangular load whose left-hand edge is placed at the break of a semi-infinite beam, is available from the Cambridge University Press server.

These functions, which we will refer to here as the 'Hetényi functions', are easily evaluated, and so it is a simple matter to compute the flexure of infinite and semi-infinite beams by line loads.

3.7 Beams of Variable Rigidity and Restoring Force

The solutions thus far have been for beams of uniform elastic properties and two-dimensional loading. While there are several geological situations where such models are appropriate, cases exist where the elastic properties vary along the length of the beam and loads are three-dimensional rather than two-dimensional.

3.7.1 Continuous Variation

It has so far been assumed (e.g., Eq. 3.9) that the flexural rigidity, D, is constant along the length of a deflected beam. However, the rigidity along the length of a beam may vary because of changes in the cross-sectional area due to width and thickness variations. An example of a varying flexural rigidity is a wedge-shaped beam, the width of which varies linearly with x (Fig. 3.20). Because of the variable width, the beam will be subject to a corresponding linearly varying restoring force.

While it is possible to derive analytical expressions for the deflection of wedge-shaped beams (Hetényi, 1979), the solutions are complex and not easy to apply to geological loads and unloads. The main problem is the requirement of polar rather than Cartesian co-ordinates. A consequence of this, which has geological applications, is that the problem can be reduced to consideration of the deformation of a circular plate supported on an elastic foundation and subjected to a concentrated load.

3.7.2 Circular Plates

Analytical solutions for the flexure of a circular plate that overlies an inviscid fluid substrate by a concentrated load have been given by Hertz (1884), Brotchie and Silvester (1969) and

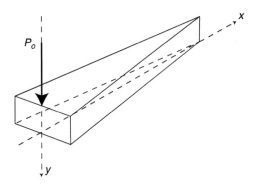

Figure 3.20 A wedge-shaped beam subject to a concentrated load P_0. Reproduced from fig. 88 of Hetényi (1979) with permission of the University of Michigan Press.

Hetényi (1979). Hertz's solutions were applied by Vening Meinesz to the calculation of regional isostatic anomalies (see Section 2.6), while Brotchie and Silvester's solutions have been widely applied, especially to concentrated loads such as volcanoes, seamounts and small ice sheets.

Brotchie and Silvester (1969) showed that the general equation for the deflection of an elastic, thin, spherical shell can be written as

$$\nabla^4 w + \frac{1}{\beta^4} w = \frac{q}{D}, \tag{3.52}$$

where w is the radial displacement, q is the applied load, β is the three-dimensional flexural parameter defined in Eq. (2.6) and ∇^4 is the biharmonic operator in the surface coordinates of the shell.

The solution of Eq. (3.52) for a concentrated load P is given by

$$w = \left(\frac{P\beta^2}{2\pi D}\right) \text{kei}\left(\frac{r}{\beta}\right), \tag{3.53}$$

where r is radial distance from the load and kei (along with ker, bei and ber) is a Bessel–Kelvin function of zero order.[3]

The solution of Eq. (3.52) for a disc-shaped load of height h, radius R_d and density ρ_{load} is given, for example, by

$$w = \frac{h\rho_{load}}{(\rho_m - \rho_{infill})}\left[\left(\frac{R_d}{\beta}\right)\text{ker}'\left(\frac{R_d}{\beta}\right)\text{ber}\left(\frac{R_d}{\beta}\right) - \left(\frac{R_d}{\beta}\right)\text{kei}'\left(\frac{R_d}{\beta}\right)\text{bei}\left(\frac{r}{\beta}\right) + 1\right] \tag{3.54}$$

beneath the disc load (i.e., $r < R_d$) and

$$w = \frac{h\rho_{load}}{(\rho_m - \rho_{infill})}\left[\left(\frac{R_d}{\beta}\right)\text{ber}'\left(\frac{R_d}{\beta}\right)\text{ker}\left(\frac{R_d}{\beta}\right) - \left(\frac{R_d}{\beta}\right)\text{bei}'\left(\frac{R_d}{\beta}\right)\text{kei}\left(\frac{r}{\beta}\right)\right] \tag{3.55}$$

outside the disc load (i.e., $r > R_d$). kei$'$, ker$'$, bei$'$ and ber$'$ are the first derivatives of the Bessel–Kelvin functions.

The flexures of a circular plate by a narrow disc load ($R_d = 12.5$ km) and a wide disc load ($R_d = 150$ km) are illustrated in Fig. 3.21.

We now return to the question of the flexure of a plate with variable elastic properties. Hetényi (1979) gives an analytical solution for the flexure along the x axis of a wedge-shaped beam by a concentrated load. He also extends the solution of a wedge-shaped beam to the more general case of the deflection of a circular plate by uniformly distributed loads, giving analytical solutions for the deflection of beams of variable-restoring force (Hetényi, 1979, p. 108) and flexural rigidity (Hetényi, 1979, p. 112).

---

[3] These functions can be obtained from the NAG Library. The MATLAB® file, chap3_hertz_kei.m, which compares the Hertz (Eq. 2.5) and Brotchie and Sylvester solutions for the flexure due to a concentrated load, is available from the Cambridge University Press server.

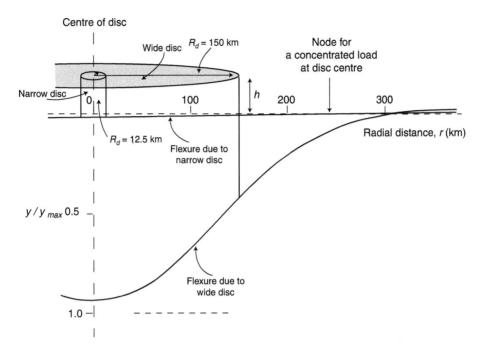

Figure 3.21 The flexure of a circular plate by 'narrow' and 'wide' disc loads. The calculations are based on the Brotchie and Sylvester (1969) analytical solution with $T_e = 20$ km,
$\rho_{load} = \rho_{infill} = 2,800$ kg m^{-3} and $\rho_m = 3,270$ kg m^{-3}.

Unfortunately, Hetényi (1979) only gives solutions for the flexure in the immediate vicinity
of a distributed load. His solutions are therefore difficult to apply to the lithosphere, where we
are concerned with the deflection beneath both the load and the regions that flank the load.

To overcome these difficulties, most workers now use numerical rather than analytical
methods to calculate the flexure of variable-rigidity beams and plates. Foremost among
these methods is the finite-difference method. In principle, this method (e.g., Conte and De
Boor, 1965) involves the simultaneous solution of a number of linear equations, each
equation being a finite-difference approximation of a general equation. The equations for
the flexure of elastic plates, together with their finite-difference approximations, are given
in Bodine (1980), Stewart (1998), Contreres-Reyes and Osses (2010b) and Hunter and
Watts (2016)[4] for two-dimensional loads and in Van Wees and Cloetingh (1994), Stewart
(1998), Ventsel and Krauthammer (2001), Wyer (2003) and Li et al. (2004) for three-
dimensional loads.

We compare the two-dimensional Bodine, and Contreres-Reyes and Osses finite-
difference methods with the Hetényi solutions in Fig. 3.22. The agreement is close, as
seen by small Root Mean Square (RMS) difference between the two finite-difference and
the Hetényi solutions which are < 0.5 per cent.

---

[4] The MATLAB® file, chap3_flex2d_fd.m, which computes the flexure of an arbitrary-shaped two-dimensional load on an elastic
plate with spatially varying rigidity is available from the Cambridge University Press server.

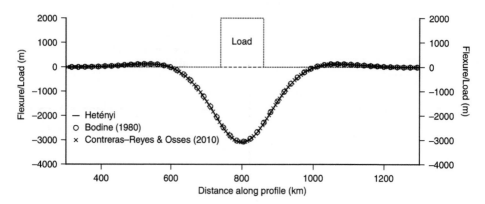

Figure 3.22 Comparison of the 2D Hetényi analytical solution to the finite-difference method of Bodine (1980) and Contreras-Reyes and Osses (2010b). The calculations assume a 120-km-wide, 2-km-high load that displaces water, load, displaced material, infill and mantle densities of 2,650, 1,030, 2,650 and 3,330 kg m^{-3} respectively, and a uniform T_e of 20 km.

The advantage of finite-difference methods is that they are well suited to cases where the flexural rigidity and equivalent elastic thickness and the restoring force vary spatially. They have been used, for example, to simulate a plate break by setting $T_e = 0$ at the plate break and a finite value of T_e elsewhere. Figure 3.23 shows a close agreement between the semi-infinite (broken) plate analytical solutions of Hetényi (1979) and the finite different solutions of Bodine (1980).

In the past few years, several three-dimensional finite-difference codes have been made available to the Earth Science community. Figure 3.24 shows an example of the use of the MATLAB® code of N. Cardozo, which in turn is based on the equations in Ventsel and Krauthammer (2001). Other finite-difference codes that combine the thermal and mechanical properties of the plate have been developed by Garcia-Castellanos et al. (1997) in two dimensions and Hinojosa and Mickus (2002) in three dimensions and applied by them to geological problems.

3.8 Thin- versus Thick-Plate Flexure

The discussion thus far has been based on a simple approximate theory for the bending of elastic plates, known as 'thin-plate' theory. Such a theory assumes that the beam (or plate) thickness is small compared to the radius of curvature, perfectly elastic, and the component of stress, σ_y, in the vertical direction is small compared to the other stress components and may be set to zero.

Comer (1983) derived analytical expressions for the deflections and stresses due to loading of an elastic plate of arbitrary thickness and compared them with those for thin-plate solutions. Comer's so-called thick-plate solution is exact for small strains and linear boundary conditions, in the absence of gravitational forces. He showed that for narrow loads

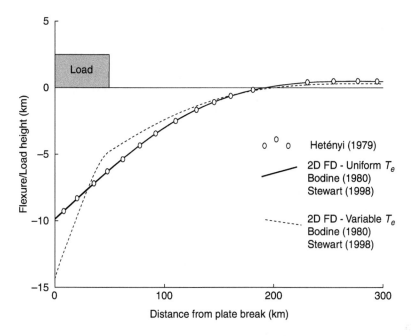

Figure 3.23 The deflection of a semi-infinite beam by a rectangular end load. The calculations are based on the analytical solution of Hetényi (1979) and the numerical (finite-difference) solutions of Bodine (1980) and Stewart (1998). $T_e = 25$ km, $\rho_{load} = \rho_{infill} = 2,800$ kg m^{-3} and $\rho_m = 3,330$ kg m^{-3}. There is an excellent agreement between the two solutions. The dashed lines show an example of a variable T_e case, for comparison, where $T_e = 5$ km beneath the load and 25 km elsewhere.

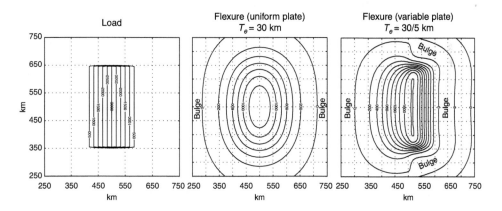

Figure 3.24 The flexure of a circular plate by an elongate triangular load of maximum height 2,500 m, width 150 km and length 300 km. The calculations are based on the MATLAB® code 'Flex3D' copyright of N. Cordozo which, in turn, is based on eq. 3.83 of Ventsel and Krautmann (2001). The left-hand panel shows the load; the middle panel, the flexure for a uniform $T_e = 30$ km; and the right-hand panel, the flexure for a variable case in which $T_e = 30$ km for horizontal distance, $x < 500$ km and $T_e = 5$ km for $x > 500$ km. Other parameters are densities of the load, infill and mantle, of 2,700 and 3,330 kg m^{-3} respectively. Note the decrease in wavelength of the flexure and the migration of the bulges towards the load in the low T_e region of the variable case.

the width of oceanic seamounts and wide loads the width of planetary volcanoes there is a close agreement between the predictions of thin- and thick-plate theory. The main departures were confined to the region immediately beneath the loads, where thin-plate theory may underestimate the flexure by as much as 5 to 10 per cent.

Wolf (1985) agreed with the conclusions of Comer (1983) that, for most geological applications, the deflections and bending stresses based on thin-plate theory were usually good approximations. However, he pointed out some inconsistencies in Comer's thick-plate analysis. He developed an 'improved' thick-plate model (Wolf, 1985), which included the effect of the gravitational body forces. This model showed that the differences between thin- and thick-plate solutions are typically even smaller than deduced by Comer. The closest correspondence was for long- rather than short-wavelength loads.

While the debate between thin and thick plate continues (e.g., Zhou, 1991), and although flexures at the surface and base of a loaded plate may well differ for geological reasons, it appears that a thin plate is a satisfactory approximation to model the flexure of most geological loads. Hence, there is good reason to believe that estimates of the elastic parameters (e.g., flexural rigidity, elastic thickness) of the lithosphere that have been published to date are not in need of significant revision and can be used in rheological and other studies with some confidence.

3.9 Spherical versus Flat Earth

We have discussed thus far the deformation of elastic plates on a flat earth and so have ignored any effects of sphericity. A number of workers (e.g., Franke, 1968) have pointed out the importance of considering sphericity, for example, in controlling the shape of Earth's surface features such as island arcs and their associated deep-sea trenches. Timoshenko and Woinowsky-Krieger (1959), among others, have presented the basic equations for the deformation of spherical shells, but as Tanimoto (1997) point out, these equations do not include the effects of buoyancy forces and so have limited applicability to Earth. Tanimoto (1998) showed that when buoyancy forces are included, the deformation at island arcs and trenches is quite well approximated by a flat Earth assumption. However, stresses may not be well predicted along-strike, and Tanimoto (1998) speculates that spherical effects may be an important source of seismicity at plate boundaries. Despite this, it seems that spherical shell theory has limited application on Earth, the surface of which comprises many small plates, but may be important in the interiors of large plates (e.g., Pacific) and on one-plate planets, as has been pointed out by Beuthe (2008).

3.10 Summary

Simple analytical solutions exist for the deflection by line loads of thin elastic beams that overlie an inviscid fluid substrate. Deflections for more-complex rectangular- and triangular-shaped loads can be evaluated by integration. The deflection of a beam that varies in thickness along its length is best solved, however, by numerical rather than analytical methods. Foremost among them are finite-difference techniques and both two- and three-dimensional codes are now widely available. Thin-plate theory and a flat Earth are good approximations to make in most geological modelling applications.

4

Geological Examples of the Flexure Model of Isostasy

Isostasy could not be proved by geologic observations, although they might suggest it.

(Reid, 1932, p. 39)

4.1 Introduction

In Chapter 2, it was argued that of the various types of isostatic models only one, the flexure model, is in general accord with geological observations. The flexure model has several advantages over other isostatic models. It permits the ideas of the response of the crust and mantle to loading and unloading to be discussed and allows considerations to be made about how much the crust and mantle might bend and what conditions might eventually lead it to break. These ideas are difficult to conceive of when using, for example, the Airy and Pratt models.

The proof of isostasy has traditionally been based on geodetic rather than geologic observations. The geodesists Hayford and Heiskanen, for example, argued that because free-air gravity anomalies were generally small over continental and oceanic regions, the Earth's crust and mantle must be in some form of isostatic equilibrium. It was clear, however, that because of ambiguities in gravity anomaly interpretation, it was not always possible to use gravity to determine unequivocally which model of isostasy might apply. The geodesists therefore sought to adopt their own models, with the Airy and Pratt models finding the most favour.

The flexure model has not been widely used by geodesists. Early proponents such as Barrell, Vening Meinesz and Gunn argued that the flexure model was in accord with geological observations. However, with the notable exception of Vening Meinesz, they failed to go the next step and set up quantitative procedures to correct the free-air gravity anomalies for the gravitational effects of a flexural-type compensation. This limited the application of the model by the geodesists, who earlier had so enthusiastically set up such procedures for the Airy and Pratt models.

Although the flexure model failed to find favour with the geodesists, it was the only model of isostasy that from its inception held the attention of geologists. One of its appeals was that it could explain a wide range of seemingly unrelated geological phenomena. For example, Barrell, Vening Meinesz, Gunn and Walcott showed that lithospheric flexure was manifest in features as diverse as former late-glacial lakes, river deltas, seamounts and oceanic islands, and deep-sea trench systems. These workers considered not only the traditional indicators of isostasy, such as gravity and topography, but also other observations, such as seismic reflection and refraction, and uplift and subsidence data.

In this chapter, we discuss those geological observations that have played a key role in establishing the phenomenon of flexure. Like the early advocates, we will restrict our discussion to a few simple unloading and loading examples. By limiting the discussion in this way, it is hoped that we will be able to prepare the ground better for Chapter 7, which discusses at greater length the role of flexure in more complex loading situations such as at sedimentary basins, rift valleys and orogenic belts.

It is helpful, when discussing flexure at different geological features, to use a 'reference' model against which observations can be compared. In this chapter, we will use the model discussed in Chapter 3 of an elastic plate that overlies a weak inviscid substratum. This is not to imply that such a model best describes the long-term mechanical properties of the lithosphere. To the contrary, we will argue in later chapters that the strength of the crust and mantle is limited and that only a portion of the lithosphere may behave elastically. Nevertheless, the elastic plate model has proved remarkably robust. It is mathematically simple, has been widely used in the Earth Sciences and has, as we will see in later chapters, allowed a first-order assessment to be made about parameters such as the flexural rigidity and the elastic thickness of the lithosphere and how these parameters might vary spatially and temporally.

4.2 Glacio-Isostatic Rebound

During the past 2–3 Ma, much of the Northern hemisphere was covered by ice. The last major ice sheet advance across North America and Fennoscandia, the Wisconsin event, occurred about 20–25 ka. The load of the ice had a profound effect on the underlying crust and mantle. Regions that were loaded were subject to subsidence, while flanking regions experienced uplift.

By about 16 ka, global temperatures had risen sufficiently for the ice sheets to start to melt and begin their retreat towards the poles. The retreat resulted in a major shift in the centre of mass of the ice load, which, in turn, influenced the underlying crust and mantle. Broadly speaking, regions that were once depressed by the ice load started to rise while regions that were once uplifted began to subside. The precise nature of this response provides information on the physical properties of both the lithosphere and the underlying asthenosphere.

The time that it takes for the lithosphere to respond to the load shifts depends, for example, on how quickly the materials of the asthenosphere can flow back from beneath the unloaded region into the loaded one. By comparing the decay time of the topography of North America and Fennoscandia with predictions of simple models, it has been possible to estimate the viscosity structure of the mantle.

The relaxation time, τ, associated with the deformation of a viscous sphere of viscosity η and density ρ is given (e.g., Peltier, 1974; Cathles, 1975) by

$$\tau = \frac{\eta}{\rho g n r_e} \left(2n^2 + 4n + 3\right), \tag{4.1}$$

where g is average gravity, r_e is the mean radius of the Earth and n is the spherical harmonic degree of the dominant term in the deformation.

The deformation of North America and Fennoscandia extended over a roughly circular region of radii 1,500 and 550 km which correspond to wavelengths of ~6,000 km and ~2,200 km and values of n of 6 and 16, respectively. According to the uplift data of Sauramo (1955) and Hillaire-Marcel and Fairbridge (1978), the relaxation time, τ, for Fennoscandia and North America is about 8.0 ka and 1.7 ka respectively. Substituting these values into Eq. (4.1) with $\rho = 3{,}000$ kg m^{-3}, $g = 9.81$ m s^{-2} and $r_e = 6{,}371$ km gives

$$\eta_{fennoscandia} = 2 \times 10^{22} \text{ Pa s and } \eta_{north\ america} = 6 \times 10^{22} \text{ Pa s.}$$

An important question is the role the lithosphere plays in modifying the patterns of post-glacial rebound. Intuitively, the lithosphere would be expected to modify the pattern: a thick, rigid lithosphere would respond differently, for example, from a thin, weak one. In practice, the large size of the ice loads, together with the relatively short times that these loads are present on the crust, means that the pattern of rebound is determined more by the asthenosphere rather than the lithosphere.

One of the first to argue for a lithosphere in models of glacial-isostatic rebound was R. K. McConnell. In 1965, he extended the calculation of the relaxation time to models of an elastic layer that overlies one or more viscous layers and calculated the spectra of the uplift data from Fennoscandia and compared it to observations. In 1968, he showed that the Fennoscandia spectra was characterised by two prominent peaks at wavelengths of 475 and 1,800 km. By fitting exponentially decaying functions to data in the vicinity of these peaks, he argued that the short-wavelength components of the deformation responded more quickly (relaxation time ~10^3 yr) than did the long-wavelength ones (~10^4 yr). This result is contrary to what would be expected for a model in which the Earth is considered as a viscous sphere of constant viscosity and density which, as Eq. (4.1) implies, predicts that as wavelength increases, then the relaxation time should decrease, not increase. McConnell (1968) found that a 120-km-thick elastic plate, with a rigidity modulus (see Section 3.2) of 6.5×10^{10} N m^{-2} that overlies a viscous mantle in which the viscosity decreases from 4.1×10^{20} Pa s near the base of the elastic layer to 2.7×10^{20} Pa s between 220 and 400 km depth and then increases to $> 6.8 \times 10^{21}$ Pa s below 1,200 km, yields a decrease in relaxation time at short wavelengths and a nearly constant increase in relaxation time at long wavelengths, and this explains the twin-peaked Fennoscandia spectral data well.

Walcott (1973), however, pointed out some problems with McConnell's analysis. One was that he neglected to account for the ability of the viscous layers to deform elastically. For example, we know that on the short timescales of P and S waves, the Earth responds elastically to very great depths in the mantle. According to Walcott (1973), the elastic effects are large enough that McConnell's estimates of mantle viscosity may be as much as a factor of two too high. A more serious concern, however, was that his model could not explain the present-day rates of uplift of Fennoscandia. These data required much greater relaxation times. Since long relaxation times are a feature of the long-wavelength deformation, Walcott (1973) attributed the discrepancies to uncertainties in McConnell's analysis at these wavelengths. By only considering 800-km-long profiles, he had, in effect, limited the study to relaxation spectra for wavelengths < 1,600 km.

The uncertainties with McConnell's analysis mainly concerned his estimate of the mantle viscosity structure. Walcott (1973) agreed with McConnell that an elastic layer was needed to explain the relaxation data, pointing out that such a layer was an efficient means of decreasing relaxation times due to its ability to store stresses for long periods of time. Walcott's preferred model is distinguished from McConnell's mainly by its thicker low-viscosity zone (e.g., Fig. 4.1).

Since McConnell's and Walcott's work, there has been a succession of improved models. Of particular note was the work of Peltier (1974) and Cathles (1975) who based their studies on spherical viscoelastic layered models. Such models combine the properties of elastic and viscous spheres: they respond initially as an elastic material, but if the load is applied for

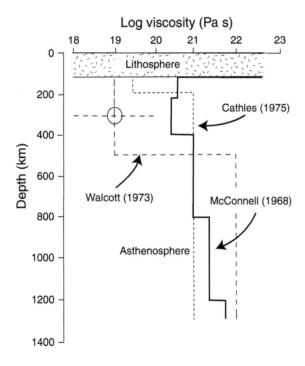

Figure 4.1 Models with an elastic layer overlying viscous layers used to fit glacial uplift phenomena. Based on fig. 12 in McConnell (1968), fig. IV-69 in Cathles (1975) and fig. 3 in Walcott (1973).

long enough, they behave as a Newtonian viscous fluid. The special case of a Maxwell viscoelastic material is characterised by four parameters: density, elastic compressibility, initial elastic rigidity and viscosity. Usually, the first three of these parameters can be determined from seismology. One objective of glacio-isostatic rebound studies is to determine the fourth of these parameters, the viscosity.

Both Cathles and Peltier used an impulse response function approach (see e.g., Section 5.2) to compute the deformation of a viscoelastic spherical Earth for various load shapes. In each case, the loads consist of two parts: the ice sheets themselves and the meltwater that they release. One of the main difficulties of post-glacial rebound studies is that the geometry of these loads is not known precisely. It is possible, however, to provide some constraints. For example, sea level curves deduced from dating of shallow-marine and near-shore deposits in continental shelves (e.g., Emery and Garrison, 1967) can be used to estimate the total mass of the ice sheets, while geomorphological observations (e.g., scour marks) can be used to constrain the area over which they extended. By assuming the load geometry, it is possible to compute the deformation of the Earth associated with different viscosity profiles in the mantle. The deformation manifests itself in several ways. The most obvious is through changes in apparent sea level at coastlines. One of the most important results of the work of Peltier, Cathles and colleagues was to show that following the last major deglaciation, the entire Earth's surface deformed in a series of upwarps and downwarps away from where the main ice loads were applied. This 'ripple' effect produced apparent sea level curves that varied from place to place over the globe.

Figure 4.2 shows a map of the five principal zones of deformation predicted to have followed instantaneous melting of the ice at 16 ka (Clark et al., 1978). In Zone 1, the sea level curve is monotonic and shows a continuous fall in regions that were once ice loaded. Equivalently, the land has been rising in this region. In Zone 2, the sea level change is again monotonic, but the sense is reversed and there has been a continuous rise in sea level. Equivalently, the land has been sinking. The deformation is not limited to Zones 1 and 2 but extends for great distances away from the main loads. When we consider the predicted deformation in its entirety, Zones 1, 3 and 5 are regions of sea level fall while Zones 2 and 4 are regions of sea level rise.

Broadly speaking, the observed pattern of sea level changes for the past 16 ka is consistent (e.g., Fig. 4.3) with the predictions of the spherical viscoelastic layered models. The fit to the far-field data is particularly good at oceanic islands. For example, most oceanic islands in the central and west Pacific show emergence of a few meters since 4–6 ka, which is in close agreement with the model predictions.

The fit of the near field is not, however, as good. In particular, the predicted emergence in Zone 1 is much greater than the observed. Peltier (1980) attributed the discrepancy to a 'trade-off' in errors in ice load history and the viscosity profile in the mantle. He showed that he could reconcile the discrepancy with a lower mantle viscosity and thinner ice sheets that melt earlier.

Iwasaki and Matsu'ura (1982) showed that some of the discrepancy could be explained by the introduction of an elastic layer. Using a similar loading history as used by Peltier and Andrews (1976), these workers showed that the near-field data could be explained if a 100-km-thick

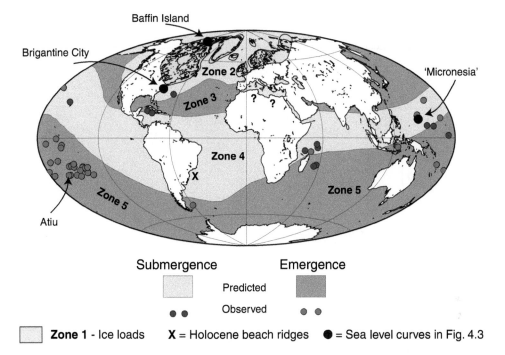

Figure 4.2 Illustration of the five sea level zones assuming instantaneous uniform melting of Northern Hemisphere ice at 16 ka. Sea level curves are predicted to be similar within each zone. Blue fill shows Zone 1. Red and blue filled circles show islands that according to Nunn (1994; tables 8.5 and 8.6) experienced emergence and submergence, respectively during the middle (to late) Holocene. Black filled circles show location of the sea level curves in Fig. 4.3. Reproduced from fig. 5 of Clark et al. (1978) with permission of Academic Press.

elastic layer was introduced into the model. A thinner layer would predict too large an uplift, while a thicker layer would produce too small an uplift.

The importance of including an elastic layer was demonstrated by Iwasaki and Matsu'ura in their study of Lake Bonneville. This approximately 350-m-deep Late Pleistocene lake once occupied much of the topographic depression that characterises the present-day Great Salt Lake in Utah. The water load was narrower and more concentrated than either of the Laurentide or Fennoscandia ice sheet loads, and therefore the effects of the lithosphere, if any, are more likely to be apparent here. Data compiled by Crittenden (1963) and Iwasaki and Matsu'ura (1982) showed that a plate with T_e of 30–40 km could explain well the pattern of uplift that followed the draining of the lake (Fig. 4.4). This range of elastic thickness estimate was two to three times less than the value preferred by McConnell and Walcott for North America and Fennoscandia.

Recent studies of glacial isostatic rebound have included both an elastic layer and spatial variations in its thickness and, hence, flexural rigidity. Latychev et al. (2005) and Whitehouse et al. (2006), for example, have shown that a discontinuity in T_e across an

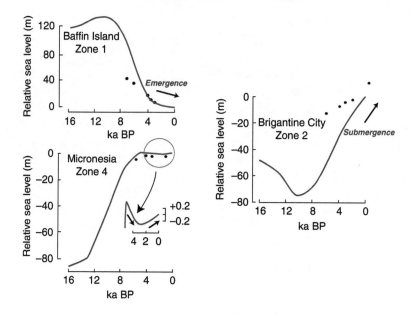

Figure 4.3 Representative sea level curves from sites within Zones 1, 2 and 4 of Fig. 4.2. Sites within Zones 1 and 2 show emergence and submergence respectively. Sites within Zone 4 show emergence and then submergence in the past 4 kyr. Reproduced from figs. 16, 18 and 20 of Clark et al. (1978) with permission of Academic Press.

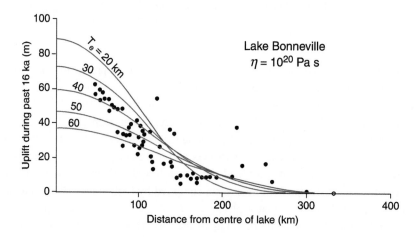

Figure 4.4 Comparison of observed and calculated uplift associated with former Lake Bonneville. Observed data are based on the height of a 16-ka shoreline. Calculated curves are based on a model with a viscous mantle ($\eta = 10^{20}$ Pa s) and an elastic lithosphere with T_e in the range of 20–60 km. Reproduced from fig. 24 of Iwasaki and Matsu'ura (1982) with permission of the Center for Academic Publications Japan.

ocean–continent boundary would significantly impact the spatial distribution of crustal velocities associated with glacial isostatic adjustment, especially the horizontal velocities derived from space-geodetic data. Klemann et al. (2009), who defined the base of the elastic lithosphere by the depth to the 1,100°C continental isotherm, showed that the horizontal motions induced by the waxing and waning of ice sheets are globally $\sim\pm1$ mm yr^{-1} and are influenced by spatial variations in the elastic thickness of the lithosphere, especially those at the mid-ocean ridges.

Other possible influences of the elastic thickness are on the growth and decay of an ice sheet. For example, Crucifix et al. (2001) suggest that while isostatic adjustment reduces the growth of an advancing ice sheet by lowering its surface elevation, thereby increasing the area where melting occurs. Yet, it also slows the melting of a retreating ice sheet because the ice surface is raised to higher elevations, thereby reducing the area where melting occurs. van den Berg et al. (2006) used synthetic data to show that bedrock motion data can be used to constrain lithospheric strength; moreover, that elastic thickness variations could cause ice volume change estimates of up to 10 per cent. Subsequently, van den Berg et al. (2008) used a model of an elastic layer that overlies a viscous substrate (Fig. 4.5) to argue that ice sheet structure depends on certain feedbacks between ice sheet growth and decay and the isostatic response time of the lithosphere. If, for example, subsidence during a glacial cycle is slower than the ice sheet growth rate, the ice sheet surface may begin to rise, thereby decreasing the likelihood of melting and increasing the ice sheet growth; while if subsidence is faster, the ice sheet surface sinks, thereby increasing the likelihood of melting and decreasing the ice sheet growth.

Finally, several workers have argued that links might exist between ice sheet growth and decay and faulting and seismicity. Stewart et al. (2000) showed that in the case of the Laurentian ice sheet the greatest fault and seismic instability occurred soon after deglaciation $\sim$7–4 ka in the central part of the ice sheet and $\sim$13 ka in marginal ice areas. The greatest stability, however, occurred during early deglaciation $\sim$9 ka beneath the central part of the ice sheet and in the flanking bulge area. Stewart et al. (2000) argued that the stresses induced into the lithosphere during ice growth and decay reach several tens of MPa and, when superimposed on regional background stresses, may give rise to a mix of tectonic styles in flanking regions, including normal and thrust faulting. And Hampel et al. (2010) showed that the slip rates on the faults depend indirectly on the strength of the lithosphere: they increase on normal faults if the lower crust is strong and increase on thrust faults if the lower crust is weak.

4.2.1 Late-Glacial Shorelines

As the Laurentide and Fennoscandia ice sheets retreated, they left in their wake a succession of lakes, known as glacial lakes. The evidence for such lakes is found in the raised beaches, notches and other features that once surrounded the lake waters and 'islands' within the lake area. Lake Algonquin, which occupied much of what is now south-central Canada, is probably the most well-known example of such a lake. The present-day Great Lakes of

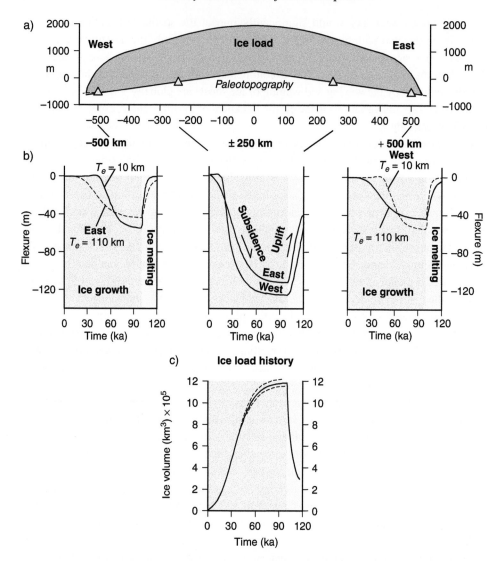

Figure 4.5 Simple model of flexure due to an ice load superimposed on an elastic plate of variable thickness, T_e, that overlies a viscous substrate. The curves in (b) show the flexure for an ice load that progressively increases in size for 100 ka and then melts for 20 ka at distances of 500 and 250 km from the centre of the load. The solid lines show the flexure at $-$500 and $-$250 distance in the west over the weak part of the plate and at $+$250 and $+$500 km distance in the east over the strong part of the plate. The dashed lines compare east and west flexures. Also shown in (c) is the ice load history for a variable T_e. The dashed lines show the case of a uniformly weak (upper curve) and strong (lower curve) plate. Reproduced from figs. 2 and 4 in van den Berg et al. (2006).

southern Canada and northern United States are, in fact, 'relics' of this once more extensive lake.

The ancient shorelines of the glacial lakes reveal a distinct pattern of deformation. Near the ice sheet edge, the shorelines have been uplifted by up to 200 m above present-day sea level. At far distance, the uplift decreases. Typically, the shorelines are at sea level at distances 200–300 km from the ice sheet edge. In between, the curve that connects shorelines of the same age has an exponential form. By carefully dating marine deposits, we know that the age of the Lake Algonquin shorelines is approximately 12 ka (Walcott, 1972b). Other shorelines have now been recognised (Fig. 4.6) that range in age from 14 to 9 ka. In general, the youngest shorelines are found near the ice edge while the oldest ones are located at great distance. The uplift and dating data suggest therefore that the marginal uplifts followed the ice retreat.

Broecker (1966) suggested that the marginal uplift was simply the response of the crust and mantle to the retreat of the ice sheet. He used a model in which the glacial lake depression was a function of the time constant for rebound and the rate of retreat of the ice sheet and showed that the model could explain well the uplift data compiled by Chapman (1954) for Lake Algonquin. The problem is that neither the response time nor the rate of ice sheet retreat is very well known.

A more serious problem is that Broecker used an Airy-type model for the crustal response, and so ignored flexure. As Walcott (1972b) first showed, it is possible to explain the glacial lake depression by a model in which the ice sheet acts as a load on the lithosphere, which deforms by flexure under its weight. According to this model (e.g., Fig. 4.7), the deformation extends from beneath the load into flanking regions because of the

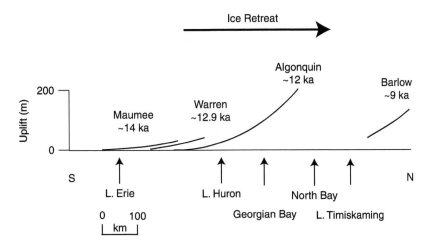

Figure 4.6 The flexure of North America that followed retreat of the last advance of the Laurentide ice sheet. The lines represent the heights of former shorelines of the lakes that once filled the flexural depressions that flanked the ice sheet. Reproduced from fig. 17 of Walcott (1972b), copyright by the American Geophysical Union.

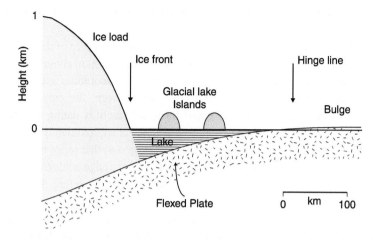

Figure 4.7 Simple model for the flexure of the lithosphere by an ice sheet load. The flexural depression between the edge of the load and the flexural bulge is infilled by water that forms a broad inland lake.

strength of the lithosphere. In regions that flank the load edge there is a wedge-shaped flexural moat, beyond which is a bulge. As the ice melts, the moat would fill with water. The load edge and the flanking flexural bulge act as natural barriers to the water and so a glacial lake would form. If there are 'islands' within the lake, then we would expect the position of the lake shorelines to be recorded in beach deposits, sea caves, notches and other indicators.

The flexure model predicts that once a load is removed, the region that was once depressed by the load would rise, while the region that was uplifted would sink. Islands within the lakes would also rise. The height of any shorelines which had been etched into these islands should therefore reflect the original flexed surface.

The shorelines of Lake Algonquin potentially provide one of the best data sets that we have with which we can evaluate the role of flexure in the continents. Chapman (1954) argued that the lake evolved in at least three stages. The first (~14 ka) corresponds to the maximum advance of what is known as the Laurentide ice sheet when the ice front extended to a line connecting Port Elgin and Alliston (Fig. 4.8). By the end of the second stage (~12 ka), the ice had retreated to Sundridge, a distance of about 160 km NNE of Alliston. As the ice retreated, the depression caused by the ice sheet load filled, and a vast lake, Lake Algonquin, flooded the region between Sundridge in the north and Bayfield in the south. It was at this stage that the shorelines were cut. The final stage (~10–12 ka) is marked by a retreat of the ice to the north of Sundridge. With further retreat, the floor of the pro-glacial lake together with its islands rebounded. The uplift changed the drainage patterns and caused the lake to drain south into the Mississippi River system. There is evidence, however, that the lake did not start to drain until the ice front had crossed the Ottawa River valley near Fossmill. Eventually, the cold melt waters were diverted, possibly east

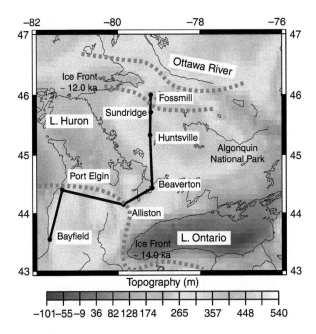

Figure 4.8 Localities in central Canada where the uplift of the former shorelines of Lake Algonquin have been measured. The thick dashed lines show the approximate position of the Laurentide ice front at 14.0, 12.2 and 11.2 ka based on fig. XII-16 of Prest (1970).

into the St. Lawrence River system where they entered the North Atlantic Ocean and disrupted its thermohaline circulation (Broecker et al., 1989).

The main difficulty with isolating the effects of flexure at the Laurentide ice sheet is its large size (radius ~1,660 km), so, as Gunn had shown earlier (e.g., Fig. 2.20), the response immediately beneath the ice sheet load will appear Airy in type at this size ($bm = 6.97$), irrespective of the elastic thickness structure of the lithosphere. However, as a number of workers have pointed out (e.g., van den Berg, 2006, 2008) motions of the bedrock at the edge of an ice sheet load are still sensitive to the elastic thickness, T_e. This is illustrated in Fig. 4.9 which compares the observed height of the ancient shorelines of Lake Algonquin to the predictions of a simple elastic plate model with T_e of 70, 85 and 100 km. The comparison shows a close agreement between the predicted and observed data. We may conclude, therefore, despite the width of the Laurentide ice sheet, that flexure plays an important role in the evolution of the ice sheet, especially in the regions that flank it.

4.2.2 The Holocene (0–12 ka) Sea Level Problem

It was shown in the previous section that viscoelastic models that include both the ice load history and the distribution of meltwater predict changes in sea level that are in general agreement with observations. More detailed comparisons, however, reveal discrepancies.

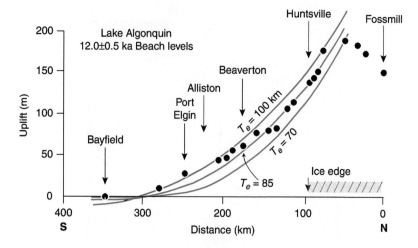

Figure 4.9 Comparison of observed upper beach levels of Lake Algonquin with calculated profiles based on an elastic plate model with T_e of 70, 85 and 100 km. The decrease in uplift at 0–50 km distance is due to a rapid retreat of the ice at Fossmill about 10–12 ka, which allowed Lake Algonquin to drain into the Ottawa-Bonnechere graben (a Mesozoic rift) and eventually into the Champlain Sea, the St. Lawrence River and the North Atlantic Ocean. Reproduced with permission from fig. 11 of Walcott (1970b).

The most notable is that the viscoelastic layered models predict a greater uplift in the near field than is observed.

Figure 4.10 shows, however, that these discrepancies extend into far-field regions also. The Atlantic coast of Brazil, for example, is in Zone 4, which is predicted to be a region of submergence. Observations of Holocene 'beach ridges' between Cananéia-Iguape and Ilha Grande Bay (Suguio et al., 1980), however, reveal that since about 5–6 ka the coast has been emergent. The beach ridges extend for > 200 km along-strike of the coast, suggesting that they are some sort of continental margin 'edge effect' related to water loading during the last major sea level rise.

To test this, we need first to consider the loading effects of a rise in sea level. Figure 4.11 shows a simple model for the shifts in shorelines due to a sea level rise. The model is similar to the one illustrated in Fig. 1.35, except that we now include the possibility that the continent is not stable but may subside at different rates relative to the ocean. Case (a) is when both the continental shelf and slope subside at a similar rate to the ocean following a sea level rise. The vertical height between the original notch (i.e., S_1) and the new notch (i.e., S_2) indicates the amount of sea level rise. Case (b) is when the continental shelf subsides at a slower overall rate than the slope and ocean basin. The vertical height between notches will now be greater than the sea level rise. Finally, case (c) is when the shelf subsides at a faster rate. The vertical height between the notches is now less than the sea level rise. Using the definition that eustatic sea level is the average height of sea level with respect to the Earth's centre, then we can say that case (b) corresponds to higher sea level than case (a), while case (c) corresponds to a lower one.

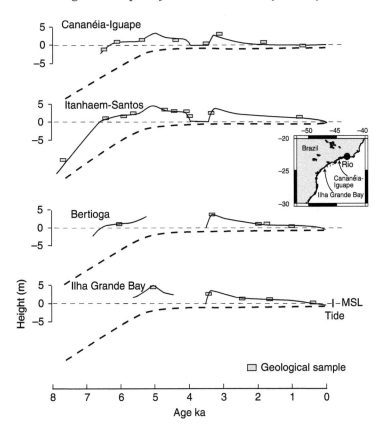

Figure 4.10 Sea level curves from the Cananéia region of the south-east coast of Brazil. The emergence of the region, which began at about 5–6 ka, is attributed to flexure of the continental lithosphere by meltwater loads. The thick dashed lines show the predicted sea level curve for Zone 4 from fig. 9 of Clark (1980). Reproduced from fig. 4 of Suguio et al. (1980) with permission of John Wiley & Sons Limited.

There are several reasons why a continental shelf might subside at different rates than a slope and rise. For example, sediments might be transported from the continent, across the shelf, to the slope. The shift in load would cause the slope to subside at a faster rate than either the shelf or the rise. The shelf would still subside, however, because of flexure, but not by as large an amount as the slope regions.

These considerations suggest that flexure may play an important role in coastline evolution. The case in Fig. 4.12, for example, shows a wedge-shaped meltwater load that has been emplaced on a continental margin. If the continental shelf, slope and adjacent ocean basin were perfectly rigid, there would be no deformation. The distance measured vertically between a notch cut at A and one cut at B would therefore be a 'true' indicator of the sea level change. If, however, the region responds by flexure, then the distance between the two notches will not necessarily reflect the sea level change. This is because the wedge-shaped

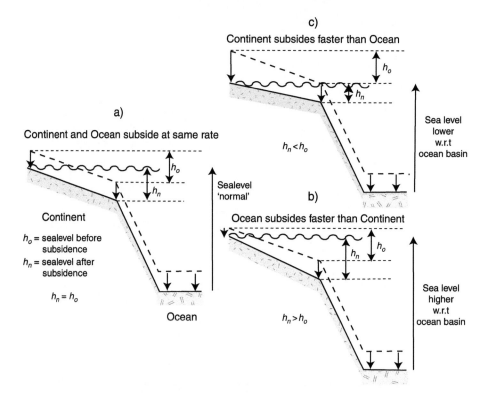

Figure 4.11 Simple models showing the change of a rise in sea level as it would appear to an observer on a continental margin that was subsiding at (a) the same rate, (b) a slower rate and (c) a faster rate than the adjacent ocean basin.

meltwater load is of greater size on the slope and adjacent ocean basin than it is on the shelf. This difference in load size causes the slope and adjacent ocean basin to subside at a faster rate than the shelf, and hence the notch that is cut at A after subsidence (i.e., A') will be deeper than the one that is cut at B (i.e., B'). Because of flexure, the net sea level is lower for the deformable margin than the undeformed one. This is because some of the meltwater infills the flexural depression, thereby leaving less water to flood coastal regions than in the undeformed case.

The effect of flexure on the eustatic sea level curve is illustrated in the inset in Fig. 4.12, which has been modified from Chappell (1974). We see that a consequence of a sea level rise is that the height of sea level with respect to the ocean basin increases. The load of the extra water, however, depresses the ocean basin, which causes it to move downwards with respect to the centre of the Earth. The net result is that sea level with respect to the centre of the Earth (i.e., eustasy) increases, reaches a peak and then decreases. The actual curve will depend, however, on both position and time.

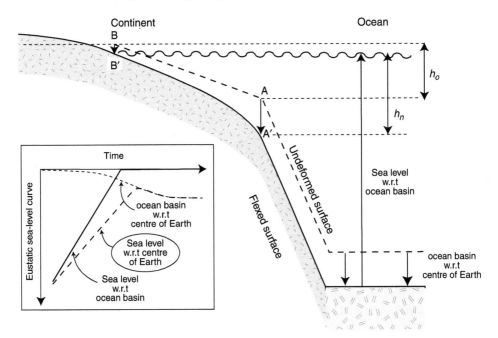

Figure 4.12 The rise in sea level that would appear to an observer on land for a margin that responds to a meltwater-shaped load by flexure. w.r.t = With Respect To. Note that the vertical distance, h_n, between the notches created by flexure exceeds the distance, h_o, and so a sea level rise will appear larger in amplitude than it would on an undeformed margin. The inset, which has been modified from fig. 3 of Chappell (1974), shows, however, that the effect of flexure on a eustatic sea level rise with respect to the centre of the Earth is to decrease the overall height of eustatic sea level, which leads to a less steep rise followed by a fall.

Chappell (1974) constructed a model, which he dubbed a 'hydro-isostatic flexure' model, that combined elastic and viscous effects. He showed that this model, when considered together with the basin-wide deformation of the ocean floor due to a sea level rise or fall, explains many features of the observed coastal sea level curves during the past 16 ka. As in previous models, the deformation of the ocean basin by meltwater is the main cause of the sea level rise between 16 and 7 ka. Since 7 ka, however, other, more local, factors such as the shape of the shelf and slope and the T_e structure of the underlying lithosphere will exert the main control on sea level. As demonstrated by Thom and Chappell (1978), sea level curves for the inner part of a gently sloping shelf show a strong dependence on the flexural rigidity (Fig. 4.13): a high T_e increases the relative fall in sea level, whereas a low one decreases it. This is because the higher the T_e, the greater the distance into the coastal regions that flexure will extend.

This discussion does not imply that all Holocene beach ridges are solely the products of local water loading and flexure. Bittencourt et al. (1999), for example, have argued that

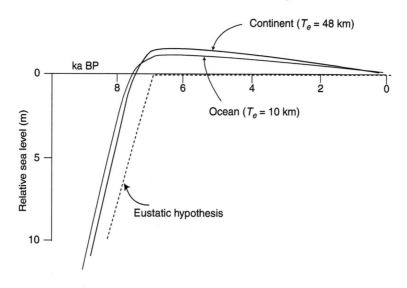

Figure 4.13 Relative change in sea level curves for the inner margin of continental shelves due to a eustatic sea level rise. Reproduced from Thom and Chappell (1978) with permission of the Royal Society.

beach ridges of the Eastern Brazil coastal zone may be caused by sediment loading offshore, while Opdyke et al. (1984) attributed the ridges in northern Florida to erosional unloading due to karstification of limestone. Figure 4.14 shows how submarine canyon formation offshore may contribute to shoreline bulges. Flushing of canyons due, for example, to repeated sediment infilling and removal due to cascading of dense shelf waters (Canals et al., 2006) could induce temporal variations in onshore bulge amplitude and wavelength. Finally, there are several examples in New Zealand and elsewhere where beach ridges are of tectonic origin having been raised during earthquakes (e.g., Little et al., 2009). Irrespective of the different loading and unloading scenarios, flexure is still likely to be involved and therefore is an important control on the landscape of coastal regions.

4.2.3 *Mountain Glacier Erosion and Meltwater Charged River Incision*

While flexure is clearly involved in large-scale processes such as ice sheet and meltwater loading and unloading, it may also be important to consider when modelling smaller-scale climate-induced processes such as those associated with mountain glacier erosion and meltwater-charged river incision. Pelletier (2004), for example, used the mapped extent of glaciation in the Western US, together with a Digital Elevation Model (DEM), to assess the ratio of rock uplift caused by erosion to glacier erosion. They found that the elastic thickness was a major control on the rock uplift and that its spatial variation impacts the topographic relief in a glaciated or deglaciated region. Similar studies have been carried out on the relief created by glacial and inter-glacial climate cycles modulating ice loads in

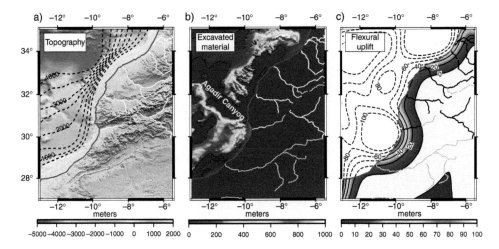

Figure 4.14 Simple model for the flexural rebound caused by submarine canyon formation on a rifted continental margin. The dashed line in (a) shows the average topography of the Moroccan continental margin used to separate out the submarine canyons and calculate the load that has been excavated in (b). Panel (c) shows the uplift due to this unload, which is predicted to reach up to 20–80 m along the Moroccan coastline.

mountains (e.g., Alps – Champagnac et al., 2008) and meltwater-charged river incision in distal plains to advancing or retreating ice sheets (e.g., south-central England – Lane et al., 2007;, Fig. 4.15). Repeated glacial and inter-glacial events, say at 100 kyr intervals or more, could lead, for example, to alternating re-surfacing of the landscape, regional tilting of pre-existing strata, and the creation of a new 'scarp and vale' topography.

4.3 Seamounts and Oceanic Islands

The ocean floor is littered by seamounts. Menard (1964), for example, estimated that there are > 300,000 seamounts in the Pacific Ocean alone. More recent estimates based on extrapolation of shipboard and satellite altimeter data place the number at > 3 million (Hillier and Watts, 2007). Rising upwards of 5 km above the mean depth of the surrounding seafloor, most seamounts are volcanic in origin. Some (e.g., Loihi, south-east of Hawaii) are known to be growing upwards on the seafloor. Others (e.g., Surtsey, south-west of Iceland) have built up to such a height that they have formed ocean islands. Finally, there are examples, especially in the Pacific Ocean, of seamounts that were once islands but have been trimmed by the prevailing fetch of the waves and subsequently subsided below sea level. The proportion of seamounts that are growing up on the seafloor compared to those that are sinking is not currently known.

Seamounts and oceanic islands are found at the present day in a variety of tectonic settings. Of particular significance is their association with divergent plate boundaries, especially the mid-ocean ridges. They are also associated with transform faults and with

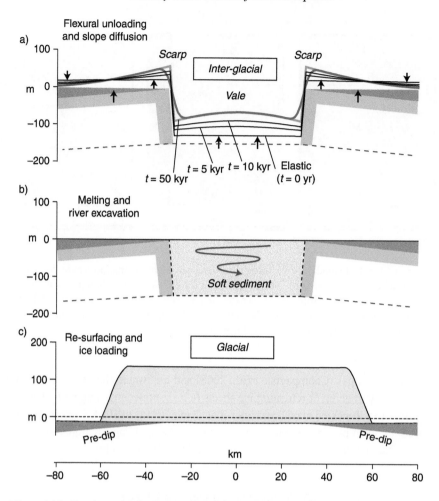

Figure 4.15 Simple model for the development of a 'scarp and vale' topography following glacial loading, melting, excavation and subsequent unloading by meltwater-charged rivers and slope diffusion. The initial flexure due to glacial loading occurs on a thick elastic plate, so it is small and occurs over a broad region. On ice retreat, the plate starts to relax and the flexed surface rebounds. Meltwater-charged rivers induce excavation and further rebound. The relaxation locally weakens the plate, which results in the creation of a 'scarp and vale' topography.

fracture zones, especially so in the central Atlantic Ocean. A large majority of seamounts and oceanic islands, however, are found in the interior of the plates far away from active plate boundaries.

Oceanic volcanoes are made up of successive basalt lava flows that are derived by fractionation from relatively shallow sub-crustal magma chambers. Once hot magma rises and reaches the sea floor, it is rapidly quenched by seawater. The mechanisms by which submarine volcanoes build up on the sea floor are poorly understood. Some have complex ridges and isolated conical-shaped features on their flanks, which suggest that they grow upwards and outwards by flank eruptions. Others, however, have very smooth flanks. These seamounts probably build up by summit eruptions. The flanks of all seamounts are vulnerable to large-scale slope failures (e.g., Moore et al., 1989), and recent swath bathymetry and seismic reflection profile data (e.g., Morgan et al., 2003; Got et al., 2008) suggest that seamounts are spreading outwards on detachment surfaces that separate the base of a volcanic edifice from the top of the underlying oceanic crust. The amount of spreading appears to be governed by the roughness of the top of oceanic crust being greater on smooth than rough crust (McGovern et al., 2014). Some seamounts are associated with large-scale sector collapses which have generated debris flows that have travelled several tens of kilometres across the sea floor.

There is evidence that the time it takes for a seamount to grow upwards on the sea floor is relatively short. One line of evidence is the high eruption rates of sub-aerial lavas. The mean eruption rates of basaltic lavas on Hawaii, for example, are about 0.050 km^3 yr^{-1}. (See summary in Crisp, 1984, table II.) Since the volume of Hawaii (above and below sea level) is about 1.13×10^5 km^3, then at these rates it would have taken only about 2.3 Myr to form the island's edifice.

The basalts that make up submarine volcanoes displace either water or air and so represent a load on the crust and mantle which should subside under their weight. The subsidence will limit the height that an individual oceanic volcano can stand above the surrounding depth of the sea floor. Since the subsidence will at least partly infill the flexural depression created by the load, a considerably greater volume of volcanic material is required to build a submarine volcano than is likely to be visible on the seafloor today. Menard (1986), for example, estimated that a seamount that now stands 3 km above the depth of the surrounding seafloor would have had to grow upwards 9 km because during loading its base would subside by as much as 6 km.

A simple model for the growth of a seamount consistent with current views on load age, eruption rates and flexural isostasy is shown in Fig. 4.16. The model assumes that (a) a water depth of 5 km is available for basaltic lavas to displace, (b) a seamount builds upwards and outwards by summit eruptions and (c) it takes about 1 Myr for a seamount grow up to sea level. The loading model is based on a viscoelastic rather than an elastic plate, since such models consider the effects of a load-induced stress relaxation (Section 6.7). Initially, the plate appears rigid and the seamount load builds up rapidly, reaching 50 per cent of its final height above the surrounding seafloor within 130 kyr. However, as the load widens, the plate weakens, subsidence increases, and it takes a further 870 kyr for the seamount to reach its final height.

The Hawaiian Islands, in the central Pacific Ocean, are considered by many as the type example of oceanic volcano loading. The islands are relatively young, ranging in age from < 0.4 Ma for Hawaii to ~5 Ma for Kauai, and they have been emplaced on 80–100-Ma-old Pacific oceanic crust. Bathymetry maps (e.g., Fig. 4.17a) show that the islands rise from a west-north-west to east-south-east trending 'ridge' (the Hawaiian Ridge), which is made up of a series of en echelon ridges that are separated by narrow channels. The Hawaiian Ridge extends beyond Kauai to latitude 32.5° N and longitude 172° E where it joins the

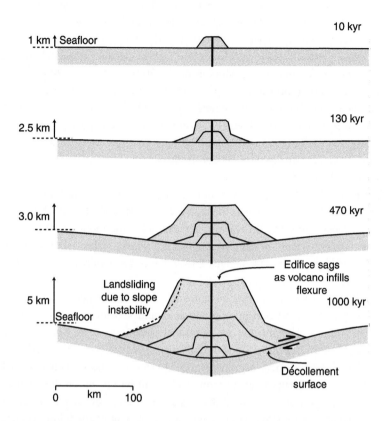

Figure 4.16 Simple model for the growth of a seamount on the seafloor by summit eruptions. The ages are estimates based on consideration of eruption rates and the timescales of stress relaxation in the oceanic lithosphere. The vertical and horizontal growth of the seamount may be accompanied by down-sagging of the central part of the edifice (van Wyk de Vries and Matela, 1998), large-scale slope failures (Moore et al., 1989) and thrusting and seismicity along a décollement plane (Morgan et al., 2003; Got et al., 2008).

Emperor Seamounts, a north-north-west to south-south-east trending chain of flat-topped seamounts that extend to the intersection of the Aleutian and Kuril trenches.

As Clague and Dalrymple (1987), Sharp and Clague (2006) and O'Connor et al. (2013) have shown, radiometric dating of the tholeiitic, alkali and highly alkali lavas on individual islands and seamounts suggests the age of volcanism progressively increases along the Hawaiian-Emperor seamount chain. This has led to the suggestion the chain formed as the Pacific plate migrated over a 'hotspot' in the underlying deep mantle. The trend and the progression in ages therefore reflect the direction and magnitude of the absolute motion of the Pacific plate with respect to the deep mantle. The age of the seamounts in the 'bend' region where the Hawaiian Ridge and Emperor seamounts intersect is about 50 Ma, which suggests (Clague and Jarrard, 1973) that prior to 50 Ma the absolute motion of the Pacific

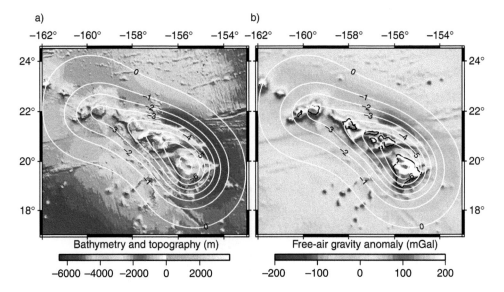

Figure 4.17 Bathymetry, topography and free-air gravity anomaly in the region of the Hawaiian Islands. White lines show the flexure (contour interval is 1 km) assuming the Hawaiian Islands act as a load on the Pacific plate with an elastic thickness (T_e) of 30 km. (a) Bathymetry and topography based on a 1 × 1 minute grid of satellite altimetry, shipboard soundings and onshore data (Sandwell and Smith, 1997, V19.1). (b) Free-air gravity anomaly based on a 1 × 1 minute grid (Sandwell et al., 2014, V29.1).

plate had a more northerly than a westerly direction. However, paleomagnetic measurements from samples acquired during Ocean Drilling Project (ODP) leg 197 indicate that the Emperor seamounts did not form at the same paleolatitude as Hawaii (the latitude difference is as large as about 20°), which raises the possibility that it was the hotspot itself that may have moved southward, rather than the Pacific plate that moved northward (Tarduno et al., 2003).

Irrespective, the continuity of the Hawaiian Ridge and Emperor Seamounts suggest that the Hawaiian hotspot has been active for much of the past approximately 82 Myr. An important question from the flexure viewpoint, then, is the loading history. We know, for example, from bathymetric maps offshore and topographic maps onshore (Bargar and Jackson, 1974) that some 7.4×10^5 km^3 of volcanic rock has been added to the surface of the Pacific oceanic crust at the Hawaiian Ridge. An additional 3.3×10^5 km^3 of volcanic rocks have been added to the crust at the Emperor Seamounts. Unfortunately, these volume estimates only concern the volcanic load that is presently visible above the mean depth of the surrounding seafloor. They therefore ignore the volume of the volcanic material that infills the flexure caused by loading. However, by making assumptions about the infill and its density, we can estimate the amount of volcanic material that has been added to the surface of the oceanic crust at different times during the evolution of the seamount chain and

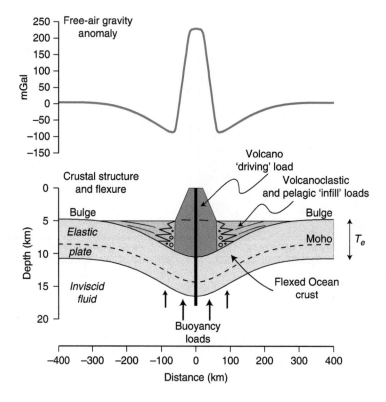

Figure 4.18 Simple model for the gravity anomaly and flexure of the lithosphere at a seamount or oceanic island.

therefore estimate the total volume and hence the 'pulse' of the Hawaiian hotspot as a function of time (e.g., Wessel, 2016).

Figure 4.18 shows a two-dimensional model for the flexure of the lithosphere caused by a submarine volcano, the approximate shape of the Hawaiian Ridge. The flexure calculation is based on an assumed density of seawater, volcanic load, infill material and mantle of 1,030, 2,800, 2,800 and 3,330 kg m^{-3} respectively and $T_e = 25$ km . The volcanic load is assumed to have been emplaced on 6-km-thick crust that, prior to loading, was at a mean depth of 5 km. The figure shows that volcanic loading results in a broad region of subsidence that is flanked by one of uplift. The subsidence extends from immediately beneath the load to the flanking regions. At about 150 km from the load flank, there is a node, where there is neither subsidence nor uplift.

The superposition of loads on the surface of the oceanic crust disturbs the local isostatic balance and, hence, is associated with gravity anomalies. In submarine volcanic loading, there are two main contributors. One, a positive anomaly, arises from the mass excess associated with the displacement of seawater by high-density volcanic rocks. The other, a negative anomaly, arises because the volcanic load flexes relatively low-density oceanic crust downwards into a denser mantle. The combined anomaly due to the load and flexure has a characteristic shape (e.g., Fig. 4.18) with a large-amplitude gravity anomaly high over the load and a low in the flanking regions. The low is flanked, in turn, by a small-amplitude gravity high associated with the flexural bulge.

The gravity calculation, like the flexure, depends on the densities assumed for the load, infill and mantle. In addition, the gravity anomalies are sensitive to the thickness and density of the pre-existing oceanic crust and to the density of the underlying mantle. The main contribution to the gravity anomaly is usually assumed to be from the base of the crust, or Moho. This is not to imply that there will be no contribution from deeper layers. Depending on the density contrast, there may, for example, be a contribution from the base of the elastic layer. Seismic modelling suggests, however, that the main density contrast is likely to be at or near the Moho.

The simple elastic plate model has proved a useful 'first-order' model with which to predict both the flexure and the gravity anomalies caused by volcanic loads. By comparing the predicted flexure and gravity anomalies with observations, it has been possible to estimate the long-term mechanical properties of the oceanic lithosphere. Historically, the first observations to be used were the gravity anomalies.

Vening Meinesz (1941b) first showed from measurements on a submarine that the Hawaiian Ridge was associated with high-amplitude positive free-air gravity anomalies that are flanked by negative anomalies. The amplitude and wavelength of the anomalies, he argued, could not be explained by an Airy model (Fig. 2.14) and required a regional scheme of compensation based on flexure in order to explain them.

The gravity data of Vening Meinesz have now been supplemented by onshore and offshore gravity surveys. Onshore, Woollard (1951) and colleagues mapped the gravity field of Oahu and other Hawaiian Islands. Offshore, surface ship (Watts and Talwani, 1975b) and satellite-derived (Sandwell and Smith, 1997) gravity data have been acquired. A gravity field, based on available land, shipboard and altimetry measurements, is shown in Fig. 4.17. The plot shows that the Hawaiian ridge is associated with a narrow belt of positive free-air gravity anomalies that are flanked by a belt of negative anomalies. The positive–negative anomaly belts are, in turn, flanked by a broad belt of positive anomalies.

Gravity anomalies are a sensitive indicator of the flexural properties of the lithosphere. Figure 4.19 compares, for example, observed and calculated free-air gravity anomalies along a profile that intersects the Hawaiian Ridge between Oahu and Molokai. The calculated gravity anomaly depends strongly on T_e. A high T_e predicts an anomaly that is of large amplitude and long wavelength. This is because the plate is rigid and the compensation for the volcanic load is spread out over a large area. A low T_e, in contrast, predicts an anomaly that has a small amplitude and short wavelength. In this case, the load is compensated more locally. The T_e that best accounts for the amplitude and wavelength of the observed anomalies appears to be 30 km.[1]

One uncertainty in the model calculations concerns the density of the volcanic load. The calculations in Fig. 4.19, for example, are based on a density of 2,730 kg m^{-3}. However, as

---

[1] A MATLAB® file, chap4_Hawaii_grv_crust.m, that compares a profile of the observed free-air gravity anomaly of the Hawaiian Ridge with a calculated anomaly based on the bathymetry and its flexural compensation for different values of T_e is available from the Cambridge University Press server.

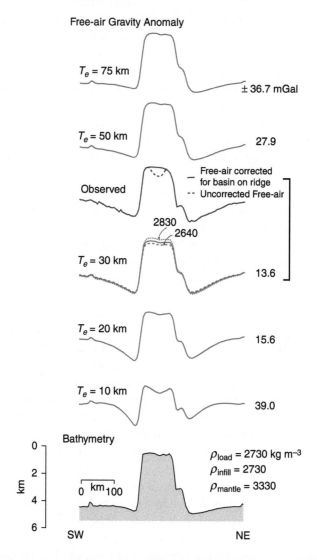

Figure 4.19 Comparison of observed and calculated gravity anomalies along a profile that intersects the Hawaiian Islands between Oahu and Molokai. The observed profile is based on data acquired during cruise KK8213 of *R/V Kana Keoki* in 1982. The calculated profiles are based on an elastic plate model with $T_e = 10, 20, 30, 50$ and 75 km. The numbers on the right-hand side of each profile are the root mean square differences between observed and calculated anomalies.

Hyndman et al. (1979) have shown from drilling in the Azores and Bermuda (Fig. 4.20), the density of a volcano could vary from 2,640 to 2,830 kg m^{-3} depending on whether it is made up of mainly submarine or sub-aerial basalts.

Probably the most detailed density study that has been carried out to date was the one by Hammer et al. (1994) at Jasper, a 3.8-km-high, 60-km-wide seamount in the eastern Pacific

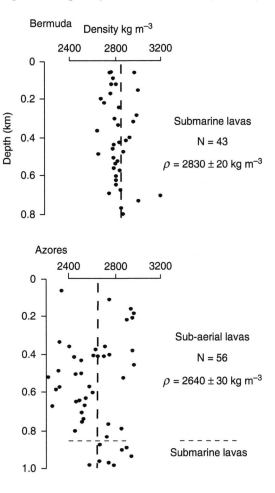

Figure 4.20 Density as a function of depth in the Bermuda and Azores boreholes. Reproduced from fig. 4 in Hyndman et al. (1979), copyright by the American Geophysical Union.

Ocean, offshore California. In an early study, Harrison and Brisbin (1959) used the Nettleton 'slope method' (Nettleton, 1939) to suggest that the seamount was cored by rocks with an average density of 2,300 kg m^{-3}. In 1992, Hammer et al. (1994) selected Jasper as the target for the first seismic tomographic study of a seamount. These authors showed P-wave velocities that increased from 2 km s^{-1} on the surface of the seamount to > 6 km s^{-1} at its base. The average velocity was 3.5 km s^{-1}, which suggests a density of 2,100 kg m^{-3} using the empirical relationship between P-wave velocity and density of Carlson and Raskin (1984) for the oceanic crust.

Ocean islands where densities have been estimated from gravity and seismic studies include Ascension Island in the South Atlantic and the Canary Islands (Tenerife) in the

North Atlantic. At Ascension, Minshull and Brozena (1997) used the gravity anomaly to deduce a density for the volcanic load of 2,500 kg m^{-3}. In 1999, T. Minshull and colleagues selected Ascension as the target of a seismic tomographic study and used the P-wave velocity structure to deduce a density of 2,300 kg m^{-3} for the volcanic core (Evangelidis et al., 2004). At Tenerife, in contrast, a 'fan' profile refraction study (Canales et al., 2000) revealed a high P-wave velocity (6.6–7.3 km s^{-1}) for the volcanic core, which implies a high density (2,902–2,989 kg m^{-3}), assuming the empirical P-wave velocity and density relationship of Carlson and Raskin (1984). Thus, while the density of the volcanic core at Tenerife is at the high end of the measured density values in the Bermuda and Azores deep drill holes (Fig. 4.17), the densities of the cores at Jasper seamount and Ascension Island are at the low end.

One reason for such a large variation in seamount and ocean island densities is that they reflect structural differences as they evolve from submarine through to sub-aerial volcanoes. Typically, seamounts comprise basalt, alkali basalt, pyroclastic, breccias and hyaloclastite deposits (Staudigel and Schmincke, 1984). Sample data from seamounts and oceanic islands in French Polynesia (e.g., Tahiti – Hildenbrand et al., 2004) suggest the edifice is formed of sheet lava flows, usually basalts. As the volcano evolves, conditions (e.g., pressure) in the magma chamber change such that the lavas become more highly evolved and more undersaturated in silica, indicating decreasing degrees of partial melting in the mantle. Finally, as a seamount builds up to sea level, more explosive volcanism occurs, hydroclastic breccias and sediments form and, eventually, sub-aerial lavas may dominate over submarine lavas.

The density of the load will affect both the gravity anomaly and the flexure. Modelling shows, however, that the effect is usually small because load density affects the anomaly in opposing ways. Increasing the load density, for example, increases the amplitude of the 'high' but increases the amount of flexure and hence the amplitude of the flanking 'lows'. The combined effects are therefore not as large as one might at first expect (see, for example, the T_e = 30 km case shown in Fig. 4.19 where the gravity anomaly has been calculated for load densities of 2,640, 2,730 and 2,830 kg m^{-3}). Significantly, the main effects are limited to the region of the high and there is little difference in the region of the flanking lows. This is because it is the amplitude and wavelength of the lows, rather than the amplitude of the high, that are usually used to constrain T_e. While the gravity anomaly may therefore be weakly sensitive to the load density, the effect on the flexure may, of course, be significant (Kim et al., 2010). For example, lower densities imply less flexure and higher densities more flexure. The uniform density load that best fits the gravity anomaly might overestimate the flexure and underestimate T_e at a seamount where the core density exceeds the uniform density, and the density of the flanking material is less than the uniform density.

More problematic for flexure calculations is the infill density, whether the plate is continuous or broken, and spatial variations in T_e (Fig. 4.21). We have assumed in the calculations so far, for example, that the infill density is the same as that of the load. However, we know that the amount of moat infill varies between volcanoes. Some moats are only partially filled by sediment. In this case, the infill density would be expected to be significantly less than the load. Others appear to be over-filled. Since the moat sediments are

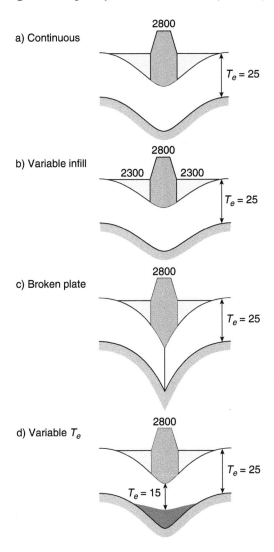

Figure 4.21 Schematic models of flexure at a seamount or oceanic island. (a) The continuous plate model, (b) the variable infill model, (c) the broken plate model and (d) the variable T_e model.

likely to comprise volcanoclastic sediments, mass wasting products and intercalated pelagic muds and oozes, it is likely that that they will be generally lower in density than the load. Indeed, many workers assume a lower value. Others, however, assume the same density as the load (e.g., as in Fig. 4.18). One consequence of assuming an infill density that is lower than that of the load is to decrease the flexure and hence enhance the gravity high. Another is to increase the amplitude of the gravity lows over their flanking flexural moats.

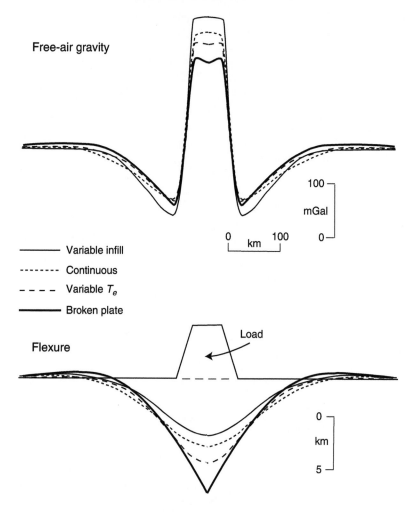

Figure 4.22 Gravity anomalies and flexure associated with each of the different flexure models illustrated in Fig. 4.21.

While we might be able to constrain the load and infill densities of volcanic loads within plausible limits, it is more difficult to determine if the plate is broken or if T_e varies spatially beneath a loaded region (Fig. 4.21). As Fig. 4.22 shows, the main differences in the gravity anomaly associated with the different models occur in the region of the high, where other factors, such as shallow basement structure and density variations within the load, may also contribute to the gravity anomaly. Differences also exist in the wavelength of the lows, but these are more subtle and might easily be interpreted in terms of a variation in T_e structure.

These considerations suggest that it might not be possible, using gravity anomaly data alone, to determine whether the plate is broken or whether T_e varies spatially. However, Fig. 4.22 shows that the flexure associated with the various models is quite different. This suggests that if we can constrain the flexure by more direct methods, such as seismic reflection and refraction, then it should be possible to determine whether the plate is broken or continuous and whether or not T_e varies spatially.

The first seismic experiment to be carried over a submarine volcano which had as its main aim constraining the flexure model was the one carried out in 1982 at the Hawaiian Ridge by Watts et al. (1985). Similar experiments have subsequently been carried out at the Marquesas Islands (Caress et al., 1995), Canary Islands (Watts et al., 1997), Tuamotu Islands (Ito et al., 1995), La Réunion (Charvis et al., 1999), Great Meteor seamount (Weigel and Grevemeyer, 1999) and more recently, Ascension Island (Evangelidis et al., 2004), the Cape Verde Islands (Pim et al., 2008), the Louisville Ridge (Contreras-Reyes et al., 2010a) and the Emperor Seamounts (Watts et al., 2021; Xu et al., 2022).

The 1982 Hawaiian experiment comprised a mix of multi-channel seismic reflection and refraction profiling. Refraction data were acquired using the two-ship Expanding Spread Profile (ESP) technique (Stoffa and Buhl, 1980). Profiles were oriented sub-parallel to the local trend of the Hawaiian Ridge and were spaced sufficiently closely to define the velocity structure of the volcanic load as well as the flanking flexural moat and bulge. The end-points of the ESPs were connected by single ship seismic reflection profiles that intersected the ridge between the islands of Oahu, Molokai and Kauai. In addition, each ESP mid-point was connected by two two-ship seismic reflection profiles.

Seismic refraction data (Fig. 4.23) have revealed the thickness and velocity structure of the flexed oceanic crust in the vicinity of the Hawaiian Ridge. In bulge regions, the upper part of the crust is associated with an increase in *P*-wave velocity from 3.9–4.2 km s^{-1} to about 5.1–6.1 km s^{-1}, while the base of the crust corresponds to an increase in velocity from 6.9–7.1 km s^{-1} to 7.8–8.3 km s^{-1}. Beneath the load and moat regions, the structure of the crust resembles that of the bulge region. The main differences are that the oceanic crust has been flexed downwards by about 3 km beneath the load, and the velocity change at the base of the crust is more gradual beneath the load and flanking moat than in bulge regions.

One of the most interesting results to emerge from the refraction part of the experiment was the possibility of a deep body below the base of the flexed crust. The velocities associated with the body are intermediate between lower oceanic crustal and upper mantle rocks (i.e., in the range of 7.4–7.8 km s^{-1}). This observation led Watts et al. (1985) and, later, ten Brink and Brocher (1987) to speculate that the deep body might represent magmatic material that had ponded at the base of the flexed crust during the later stages of volcano building. Similar-velocity deep-crustal bodies have been reported from rifted margins (e.g., Eldholm et al., 1995; Bauer et al., 2000) where they have been interpreted as magmatic material that has 'underplated' stretched continental crust during the initial stages of continental rifting.

An important question is whether magmatic underplating modifies the flexed oceanic crust. Underplating thickens the crust, which, depending on its bouyancy, would cause uplift. Watts et al. (1985), for example, argued that underplated material beneath the Hawaiian Ridge might act as a sub-surface upward load that, in effect, unbends the oceanic crust flexed downwards by the surface loads. The result would be, depending on its spatial extent, to broaden the flexure and make the lithosphere appear more rigid than it is. This

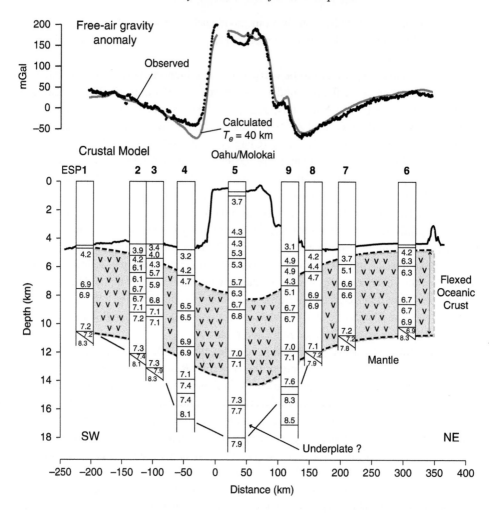

Figure 4.23 Gravity anomalies, bathymetry and crustal structure of the Hawaiian Islands, in the region of Oahu and Molokai. Solid red lines and the blue shaded region show the calculated gravity anomaly and crustal structure based on a simple model of flexure with $T_e = 40$ km. Reproduced from figs. 9 and 18 of Watts and ten Brink (1989), copyright by the American Geophysical Union.

might explain the relatively high value (compared, for example, to that derived in Fig. 4.19) of $T_e = 40$ km deduced by Watts and ten Brink (1989). Indeed, Wessel (1993) has shown that once the long-wavelength topographic swell on which the Hawaiian Islands are superimposed is removed from the flexed surfaces of the oceanic crust prior to modelling, then the value of T_e required beneath the ridge is reduced to about 25 km. While underplating might therefore play a role in modifying the shape of the flexed crust at Hawaii, its role at other oceanic islands is not clear. Some islands (e.g., Marquesas – Caress et al., 1995;

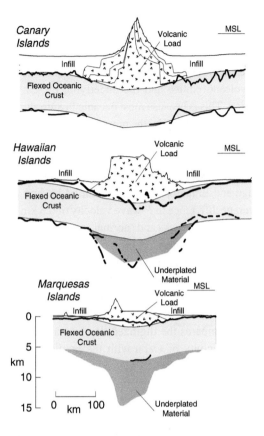

Figure 4.24 Comparison of the crustal structure at the Canary (Watts et al., 1997), Hawaiian (Watts and ten Brink, 1989) and Marquesas (Caress et al., 1995) islands with the predictions of simple elastic plate models. Reproduced from fig. 7 of Watts et al. (1997).

La Réunion – Gallart et al., 1999), for example, appear to be underplated. Several other islands (e.g., Canary – Watts et al., 1997; Cape Verde – Pim et al., 2008) do not (Fig. 4.24).

Recent seismic studies using large-volume air gun arrays and long seismic 'streamers' have found little evidence for crustal underplating, suggesting it may not be as common an occurrence beneath seamounts and ocean islands as previously thought. For example, studies at the Louisville Ridge (Contreras-Reyes et al., 2010a) and the Emperor Seamounts (Watts et al., 2020; Xu et al., 2022) show little evidence of magmatic underplating, although they do reveal lateral changes in the density in the sub-crustal mantle. They do reveal, however, high-velocity, high-density, intrusive material either within the flexed oceanic crust or within the overlying edifice. Interestingly, both the Louisville Ridge and Emperor Seamounts were emplaced on young oceanic lithosphere on or near a mid-ocean ridge crest where the elastic thickness of the lithosphere is small, so, as Contreras-Reyes et al. (2010a) point out, magmatic material may more easily penetrate the lithosphere than at features such as Hawaiian and Marquesas islands, which were formed on relatively old oceanic crust. This is supported by flexure modelling which shows that a T_e of 10 km and 14 km is required at the Louisville Ridge and Emperor seamounts respectively, to best fit the depth to the seismically constrained Moho (Fig. 4.25).

The significance of dense intrusive material of probable mafic or ultramafic composition within the core of the volcano edifice and the underlying pre-existing crust is important regarding flexure modelling. This is because most oceanic flexure studies assume all the loads that have acted on the plate are as defined by the present-day bathymetry, and so they therefore ignore the effect of any sub-surface loading. This is seen in the calculated gravity anomaly, which, while generally fitting the wavelength of the observed gravity anomaly, underpredicts (Contreras-Reyes et al., 2010a; Xu et al., 2022) the amplitude of the gravity over the crest of both the Louisville Ridge and Emperor seamounts. Buried loading might contribute to the flexure, but the small average difference in P-wave velocity of 0.3 km s^{-1} and density contrast of 100 kg m^{-3} between the intrusive rocks and 'normal' oceanic crust suggest that the surface load is significantly larger, and therefore more effective in driving the flexure, than the buried load. Nevertheless, as Watts et al. (2021) have shown, the sub-surface loads are large enough to explain the amplitude of the gravity high over the crest of the volcanoes.

Seismic reflection data when combined with refraction data has the potential to constrain the surfaces of flexure, which can then be directly compared to model predictions. Figure 4.26, for example, compares the digitized depth to a reflector that defines the top of the oceanic crust in the vicinity of Oahu with the predictions of simple elastic plate models. The models assume the volcanic load is given by bathymetry above a mean depth of 4,500 m, a load and infill density of 2,650 kg m^{-3}, and T_e values of 10, 25 and 50 km. There is excellent agreement between the general shape of the observed and predicted depths, with $T_e = 25$ km providing the best fit to the data.

Perhaps the most significant use of seismic reflection profile data, however, has been to delineate the stratigraphic 'architecture' of the deep-water flexural moats that flank oceanic islands and seamounts. At Hawaii, the data show the flexural moats to be infilled by a thick wedge of semi- to well-stratified material with individual reflectors that generally diverge towards the islands (ten Brink and Watts, 1985). The stratigraphic patterns (e.g., Fig. 4.27) comprise onlap of oceanic basement in the lowermost section and offlap that migrates back as downlap towards the islands in the uppermost section. There is evidence for ponding of the youngest sediment in the deepest part of the moat. ten Brink and Watts (1985) attributed the transition from an onlap to a downlap pattern to some form of viscoelastic stress relaxation in the underlying oceanic lithosphere.

Alternatively, the onlap and offlap patterns are the consequence of 'progressive loading' of an elastic plate (Watts and ten Brink 1989). According to this model, illustrated in Fig. 4.28, the load of a new volcano modifies the shape and, hence, the stratigraphic patterns that develop in the moats of pre-existing islands. The result is a pattern of reflectors that is dominated by onlap in the lowermost part of the section and offlap in the uppermost part. Similar reflector patterns characterise river deltas (see Section 4.4). As Fig. 4.28 shows, onlap and offlap patterns would be expected in both 'cross-moat' and 'along-moat' profiles, but they would be best developed in the along-moat profiles because of the alignment of these profiles in the direction of load migration. Although ten Brink and Watts (1985)

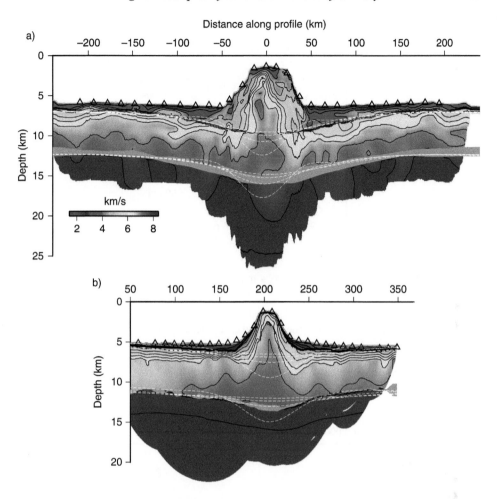

Figure 4.25 Crustal structure at the Emperor seamounts at 46° N (Xu et al., 2022) and Louisville Ridge seamount chain at 26° S (Contreras-Reyes et al., 2010a). Colours show the *P*-wave velocity structure as derived from modelling refracted and reflected arrivals generated by a ~100-litre multi-element air-gun array onboard M/V *SONNE* and R/V *Marcus G. Langseth* and recorded at Ocean Bottom Seismometers/Hydrophones (OBSs/OBHs) with an average spacing of 10 km (Louisville) and 16 km (Emperor). The grey shade shows the uncertainty in the modelled depth to Moho, which has a mean value of 0.1 km (Louisville) and 0.4 km (Emperor) along the profile. The dashed white lines show the predicted flexure based on a three-dimensional elastic plate model with a density of the water, load, infill, crust and mantle of 1,030, 2,800–2,882, 2,800, 2,900 and 3,400 kg m^{-3} respectively and $T_e = 5$ (the lowermost curve beneath the volcano edifice), 10, 15 and 20 km.

presented the evidence for onlap on a cross-moat profile, they lacked the along-moat data that could have substantiated the progressive loading model.

To better define the stratigraphic patterns, Rees et al. (1993) carried out a 'high-resolution' seismic reflection survey of the Hawaiian moat to the north of Oahu, Molokai

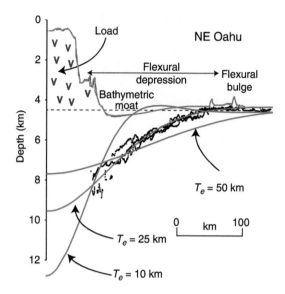

Figure 4.26 Comparison of observed and calculated flexure to the north-east of Oahu and Molokai in the Hawaiian Islands. The observed data (filled black circles) is based on 'picks' of the top of Pacific Ocean crust from depth-converted seismic reflection profile data. The calculated profiles are based on an elastic plate model, surface bathymetric loading and $T_e = 10, 25$ and 50 km.

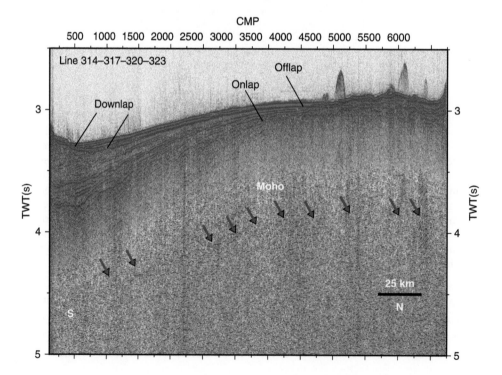

Figure 4.27 Seismic reflection profile of the Hawaiian flexural moat and arch north of Oahu. Original two-ship profile published in ten Brink and Watts (1985). The figure shows a reprocessed profile by P. Cilli.

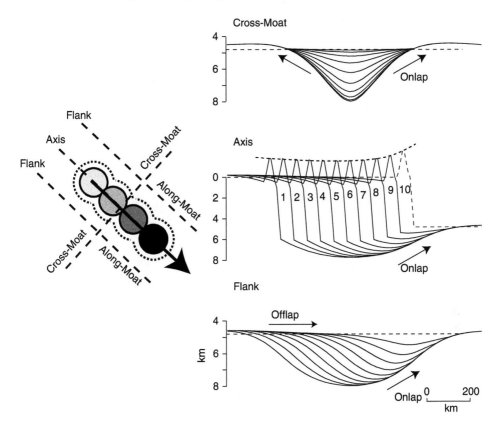

Figure 4.28 The progressive loading model and its relationship to the development of stratigraphic patterns of onlap and offlap in the flexural moats that flank oceanic volcanic island chains. Reproduced from figs. 24 and 25 of Watts and ten Brink (1989).

and Hawaii. They showed that although the moat was infilled by a substantial thickness of semi- to well-stratified material, seismic reflectors were often interrupted by thick 'chaotic' bodies that lacked internal structure. They interpreted these bodies as the products of large-scale slope failures on the flanks of the Hawaiian Islands, especially on their north-facing flanks. Rees et al. (1993) recognised the same patterns of onlap and offlap in cross-moat profiles as were described by ten Brink and Watts (1985). However, they were unable to identify the offlap and onlap patterns predicted on along-moat profiles. The most likely explanation for this is that the stratigraphy is 'disturbed' by the products of slope failures, which, because they are sourced to the flanks of individual islands, would tend to influence reflector sequences more on along-moat than cross-moat profiles.

Seismic reflection profiling of the moats that flank other oceanic islands show stratigraphic patterns similar to those described from the Hawaiian Islands. The Canary Islands, for example, have a well-developed moat that contains > 2 km of semi- to well-stratified material (Fig. 4.29). Within the moat sequences is an angular unconformity. Scientific ocean drilling suggests that the unconformity separates Neogene sediments associated with

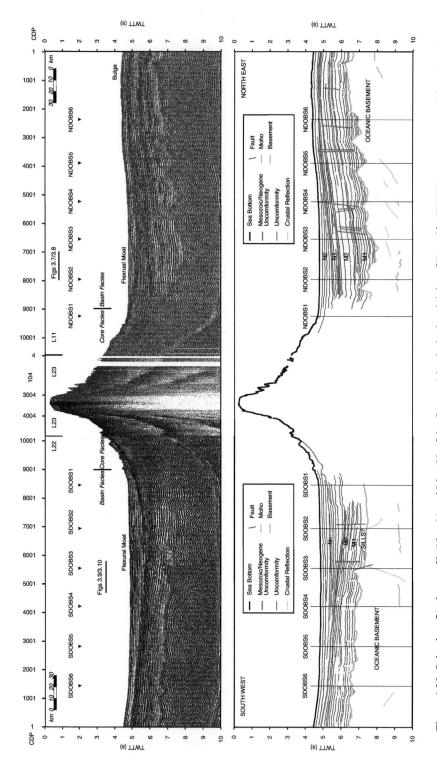

Figure 4.29 Seismic reflection profile Line 11 and Line 22 of the Canary Islands, in the vicinity of Tenerife. TWTT = two-way travel time. Based on data acquired during cruise CD82 of RRS *Charles Darwin*. See Fig. 4.30 for Line location. Reproduced from fig. 3.11 of Dalwood (1996).

volcano loading above from pre-existing Mesozoic sediments and the oceanic crust of the nearby Moroccan continental margin below.

Although the ages of the edifices that make up the Canary Islands are not as well known as those of their Hawaiian counterparts, there is evidence of an age progression, with the westernmost islands of Fuertaventura and Lanzarote being the oldest and the easternmost islands of La Palma and El Hierro being the youngest.

A progressive loading model predicts that the lithosphere that underlies the Canary Islands would have been successively deformed as the main volcanic load centres migrated from east to west. This would have resulted in patterns of stratigraphic onlap and offlap on both cross-moat and along-moat profiles. Cross-moat seismic profiles of the Canary Islands moat (e.g. Fig. 4.30) are dominated by stratigraphic onlap, as they are at Hawaii. Unlike Hawaii, however, where load migration rates are much higher than at the Canary Islands, there is also evidence on along-strike profiles for onlap. Figure 4.30 shows, for example, an along-strike profile of the moat to the north of Gran Canaria. This profile clearly shows evidence of shifts in the patterns of onlap and offlap that are in general agreement with the predictions of the progressive loading model.

At La Réunion (de Voogd et al., 1999) and Cape Verde (Ali et al., 2003) the stratigraphic architecture of the flexural moat differs quite markedly from Hawaii and Tenerife (Fig. 4.31). Rather than thickening down dip towards the volcanic load, as it does at Hawaii (Fig. 4.27) and Tenerife (Fig. 4.29), the moat infill at La Réunion and Cape Verde appears to be uplifted and, as a consequence, dips away from rather than towards the load. We speculate that the uplifted moat infill is the consequence of vertical motions related to the La Réunion and Cape Verde mantle plumes, both of which are associated with mid-plate topographic swells and long-wavelength gravity or geoid anomaly highs (Bonneville et al., 1988; McNutt, 1988).

The flexural loading models discussed thus far have implications for the geomorphic evolution of oceanic islands. For example, they imply that during the process of island construction, volcanic loads 'sink' into the moats of the flexed crust. The result is that a considerable portion of the edifice of a volcano will be submerged below the undeformed depth of the surrounding seafloor. Figure 4.24 shows that as much as 2–4 km of the volcano is now infilling the flexure below the undeformed depth of the sea floor around Hawaii, Tenerife and the Marquesas. An important question, then, is the nature of this subsidence and whether or not it is an ongoing phenomenon.

There is evidence that Hawaii has been subsiding for much of the past 200 years. Apple and Macdonald (1966), for example, describe a series of features on the west coast of the island that include the remains of horse ramps, retaining walls and protective walls surrounding prominent coconut trees. These features were constructed by the indigenous Hawaiians and yet are now in the process of being submerged by the sea.

One of most intriguing lines of evidence for the subsidence of Hawaii is the egg-crate 'papamu'[2] (see inset in Fig. 4.32), which were used by the indigenous Hawaiians to play

---

[2] 'Pa' in Hawaiian means flat and 'papa' means very flat or a big board. The meaning of 'mu' is more obscure. According to Brigham in Hiroa (1957), mu is the name of the official appointed to capture men for sacrifice, and therefore 'papamu' simply means the board or flat surface on which the mu assembles his men.

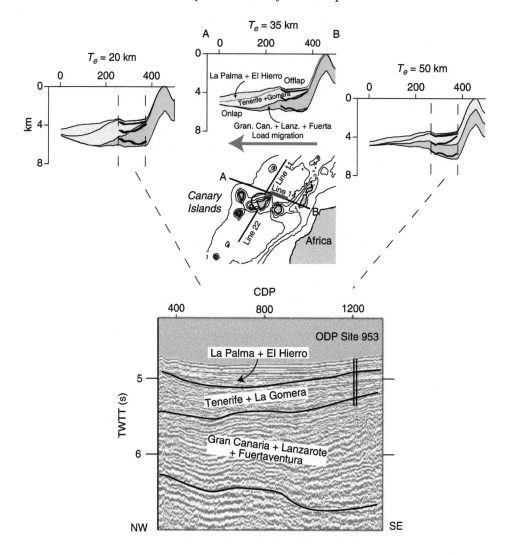

Figure 4.30 Comparison of observed and calculated stratigraphy of the flexural moat to the north of Tenerife and Gran Canaria, Canary Islands. The top three figures show the calculated stratigraphy based on a progressive loading model with T_e = 20, 35 and 50 km. The middle figure shows the location of seismic reflection profiles CD82 Line 22 and 11 (Fig. 4.26) and Line 14 (thick blue line). The lower figure shows Line 14 and the individual contribution to the flexure of the loads of Gran Canaria+Lanzarote+Fuertaventura, Tenerife+La Gomera and La Palma+El Hierro. Modified from figs. 5 and 13 of Collier and Watts (2001).

a checker-like game known as 'konane' (Hiroa, 1957). The konane boards were carved in the basaltic lavas with hammer stones. Today, the papamu, which are found on the west side of the island a short distance from Honaunau Bay, are only visible at low tide. It is difficult to explain why the Hawaiians would have laboured for so long to carve the boards if it was only possible to play the game at low tide. Apparently, the konane was a favourite pastime

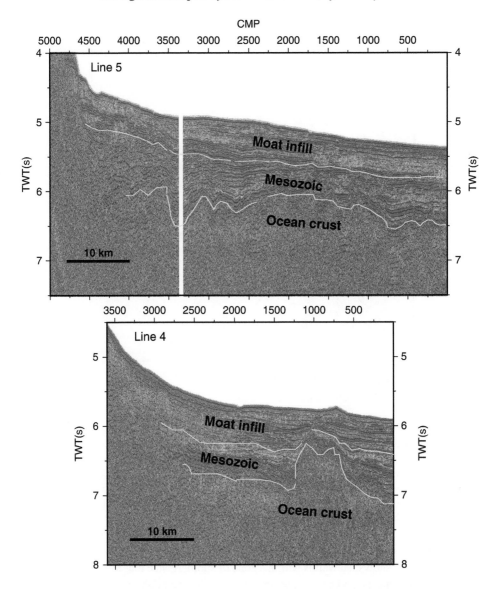

Figure 4.31 Seismic reflection profiles in the vicinity of the Cape Verde islands showing an uplifted flexural moat infill sequence. (a) Line 5 extends east of Boa Vista and Boa Vista seamount. (b) Line 4 extends south of the ridge between Brava and Fogo. Based on data acquired on RRS *Charles Darwin* in Ali et al. (2003).

of the elderly, and Hiroa (1957) cites a game that started in the morning and had barely finished by nightfall. There seems little doubt therefore that the boards would have been carved in lava that was normally dry.

From a consideration of all the historic sites, Apple and Macdonald (1966) estimate that since about 1750 sea level on Hawaii has risen by about 3 mm yr^{-1}. We do not know, of

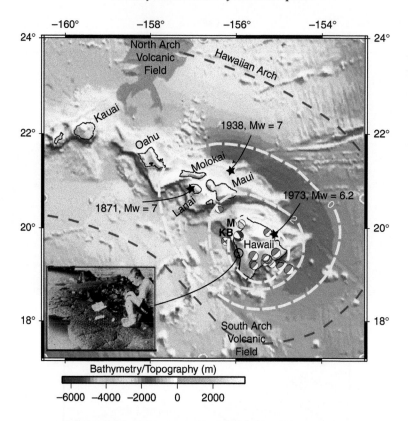

Figure 4.32 Contours of annual subsidence in mm yr^{-1} based on tide-gauge data in the Hawaiian Islands. Reproduced from fig. 1 of Moore (1970). Filled circles show earthquakes with Centroid Moment Tensor (CMT) solutions since 1976. M and KB show the October 15, 2006, events which have been related to volcano loading and flexure (e.g. McGovern, 2007). The inset shows an example of the papamu in the Pu'uhonua o Honaunau National Historic Park on Hawaii's west coast. Photo reproduced with permission of the National Park Service of the United States Department of Interior.

course, from the sites themselves, whether it is the island that is sinking or the seas that are rising. As Fig. 4.2 shows, Hawaii is within a zone of submergence that followed melting of the last major ice sheets, and so it is possible that the sinking might, at least in part, be the result of a sea level rise.

Since Apple and Macdonald's review, tide-gauge data from five stations on Oahu, Maui, Kauai and Hawaii have become available. These data, which span some 90 years, show that while Hawaii has been subsiding at average rates of up to 4 mm yr^{-1}, Oahu has been stable (Moore, 1970). Moreover, Maui is subsiding at an intermediate rate of about 2 mm yr^{-1}. These data cannot therefore be explained by a sea level rise, because this should have affected each island equally. Rather, it requires that Hawaii has been subsiding relative to its neighbours (Fig. 4.32).

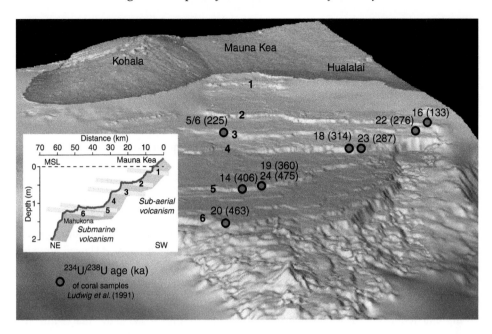

Figure 4.33 Perspective view (from the north-west) to the west of Hawaii showing a 'staircase pattern' of marine terraces offshore the flank of Moana Kea. The inset shows individual terraces are tilted inwards towards the center of the island. The image has been constructed using GeoMapApp. Red filled circles locate the coral sample sites of Ludwig et al. (1991).

Sonar imaging of the seafloor south-west of Hawaii suggests that the island has been subsiding for at least the past 0.5 Myr (Moore and Fornari, 1984; Moore and Campbell, 1987; Ludwig et al., 1991). The images reveal several submarine terraces between Molokai and Hawaii that are believed to have once been wave-cut terraces that have since subsided below sea level (Fig. 4.33). Reefal limestones from the 150-, 360- and 580-m-deep terraces, for example, yield ages of 13, 145 and 255 ka, respectively. Each terrace is tilted inwards towards the centre of Hawaii. By comparing other terraces offshore Maui and Lanai with the dated terraces between Molokai and Oahu, it has been possible to estimate a 650-ka age for the oldest terrace. Below the oldest terrace is a major slope change that Moore and Campbell interpret as the shoreline when the major tholeiitic shield building ceased and the sub-aerial phase of volcanism began.

The historical data, together with the tide-gauge and sonar data, indicate that despite its present-day high relief, subsidence rather than uplift has played the major role in the evolution of Hawaii. The subsidence appears to have occurred concentrically around the island and there is evidence (e.g., Fig. 4.32) that the maximum subsidence, which extends from Hawaii to Molokai, correlates with a number of deep earthquakes beneath the region. One of the largest was a $Mw = 6.2$ earthquake that occurred at a depth of 40 km beneath Honomu on 26 April 1973. An even larger earthquake ($Mw = 6.7$) occurred on

15 October 2006 at a depth of 39 km beneath Kiholo Bay (KB, Fig. 4.32) which was followed 6 minutes later by a $Mw = 6.0$ earthquake at a depth of 19 km beneath Mahukona (M, Fig. 4.32). The 1973 and 2006 earthquakes, both of which caused injury and damage on Hawaii, are clearly in the sub-crustal mantle, and their association with the maximum contours in Fig. 4.17a suggests they may be related to flexure (McGovern, 2007). This is supported by focal mechanism studies that show that the deeper Kiholo Bay event, which is similar in focal mechanism and depth to the Honomu event, is normal with a strike-slip component and below the neutral surface of flexure, while the shallower Mahukona event is compressional and above the neutral surface of flexure.[3]

The most likely explanation of the regional subsidence, and possibly also the deep earthquakes, is that they are caused by flexure due to the load of the big island of Hawaii. We will refer to this load-induced subsidence as the loading subsidence. The fact that the loading subsidence is greatest beneath Hawaii, the most active island in the chain, suggests that isostatic adjustment here is an ongoing process. Therefore, the study of Hawaii provides important clues as to the loading effects of new volcanic islands on pre-existing ones.

The contribution of the island of Hawaii to the subsidence and uplift of its neighbours can be estimated by first isolating its load and then calculating the flexure. The calculations (Watts and ten Brink, 1989), which are illustrated in Fig. 4.34, show that while the subsidence is greatest immediately beneath the island, Hawaii's influence extends as far north-west as the west coast of Molokai. Since the west coast of Molokai is approximately 1.9 Ma, these results suggest that the subsidence due to Hawaii has been ongoing for at least this time. This is in accord with the evidence, from sub-aerial eruption rates, of the time it would take to build the island of Hawaii. The pattern of subsidence, together with its associated deep earthquakes, is of particular interest because it suggests that even a ~2 Ma volcanic load is capable of inducing significant stresses, which have led to earthquakes and faulting in the upper part of the lithosphere.

The most visible evidence of subsidence perhaps is the topography of Maui and Molokai. These islands are in Hawaii's flexural moat and so would have experienced subsidence. Indeed, the topography of Maui and Molokai systematically decreases away from Hawaii. However, the topography of the Hawaiian Islands is also influenced by climatic, erosional and other factors, and so the role of load-induced flexure is not clear. Offshore regions, however, are unaffected by erosion, and so the wave-cut terraces on the submerged flanks of Maui and Molokai (Fig. 4.35) more clearly record the effects of subsidence. We see, for example, that there is an excellent agreement between the locus of the observed and predicted submarine terraces.

The evidence from the Hawaiian Ridge suggests the big island of Hawaii has influenced pre-existing 'upstream' islands for at least the last ~2 Myr. The subsidence and uplift is ongoing. Eventually, however, the influence of Hawaii on its neighbours will decrease as the crust beneath it approaches isostatic equilibrium. By then, however, Hawaii may itself have

---

[3] The depth to the neutral surface can be estimated by adding the depth to the base of the volcanic edifice (~10 km, Fig. 4.32) to one-half of the T_e at Hawaii (~25 km, Fig. 4.25). This yields a depth to the neutral surface of $10 + 12.5 = 22.5$ km.

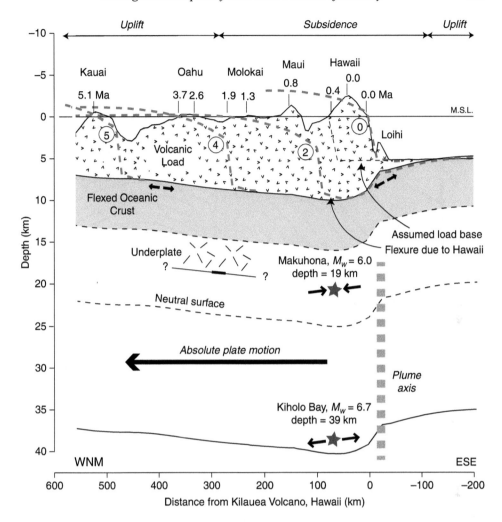

Figure 4.34 Progressive loading of the Hawaiian Ridge by successively younger loads that are emplaced on its south-east end. The thin dashed line delineates the approximate load of Hawaii Island, which has caused subsidence at Maui and Molokai and uplift at Oahu. The thick red dashed lines show the 0, 2, 4 and 5 Ma 'chrons' within the volcanic ridge. Filled purple stars show the projected depth of the Kiholo Bay and Mahukona earthquakes described by McGovern (2007). Reproduced with modification from fig. 15 of Watts and ten Brink (1989).

started to be influenced by the newest load in the chain, Loihi, which is presently forming on its south-east flank.

There is good evidence that as well as causing subsidence, the load of Hawaii has also caused uplift. Grigg and Jones (1997), for example, used sea level indicators such as corals, beach deposits, wave-cut notches and wave-eroded terraces to show that up to 50–100 m of uplift has occurred on Lanai, Molokai and Oahu during the past 500 kyr. Systematic differences in the timing of the uplift occur between the islands. The uplift on Oahu, for

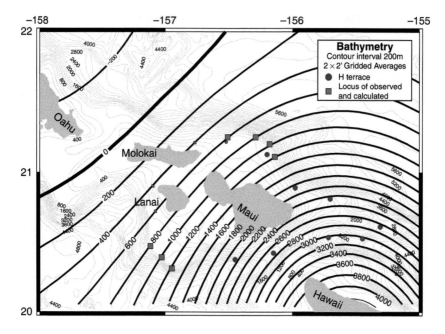

Figure 4.35 Comparison of the calculated flexure due to the load of Hawaii with bathymetry data in the region of the upstream islands of Maui, Molokai and Oahu. Calculated flexure is contoured at the 200 m interval. Thick solid line shows the node, a region of no subsidence or uplift. The red filled squares show the 'locus' points on the outermost edge of wave-cut terraces where the calculated flexure and observed bathymetry contours intersect to form a 'staircase' pattern. The terraces were probably cut during times of rising sea levels, while the 'staircase' pattern is a consequence of a flexural subsidence induced by the load of Hawaii. Reproduced from fig. 22 of Watts and ten Brink (1989).

example, spans 0–500 ka (McMurtry et al., 2010), suggesting that it was in progress quite soon after completion of the main submarine phase of shield building on Hawaii and is continuing today. In contrast, Lanai has experienced little vertical movement in the past 30 ka (Webster et al., 2006), and offshore terraces (e.g., Fig. 4.33) suggest that prior to this it has undergone subsidence. The occurrence of elevated coral reef deposits at heights greater than the 30 ka surface (thought originally by Griggs and Jones to be due to uplift) have been re-interpreted by Webster et al. (2007) as the result of a mega-tsunami. Since Oahu is furthest from Hawaii, while Lanai is nearest and the first node of flexure is between them, then it is likely that the loading of Hawaii could have caused both the uplift at Oahu and the subsidence at Lanai. However, the fact that Lanai has been stable since 30 ka raises an interesting question, especially as Hawaii, together with Loihi on its south-west flank, is continuing to grow. One possibility, suggested by Zhong and Watts (2002), is that the uplift and subsidence patterns along the Hawaiian Islands are modified by the dynamic effects of the Hawaiian plume. These effects would not be evident at Hawaii because load-induced subsidence there will dominate any dynamic effects but would be more apparent

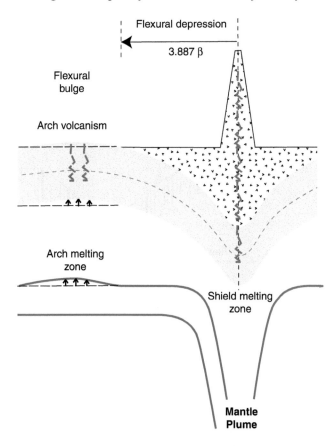

Figure 4.36 Schematic diagram illustrating the relationship between volcano loading, flexure and bulge (arch) volcanism. Modified from Bianco et al. (2005).

'upstream', on islands such as Lanai where the subsidence is smaller and the ascending plume beneath Hawaii has been sheared by the moving Pacific plate.

In addition to causing uplift of pre-existing islands, there is evidence that volcano-induced flexure may have generated secondary volcanism along the Hawaiian Islands and in offshore bulge regions. For example, shield building on Hawaii is coeval with alkali volcanism on Maui, Molokai and Oahu (e.g., Honolulu Volcanic Series) and with seafloor volcanism in the so-called South Arch and North Arch fields (Frey et al., 2000; Lipman et al., 1989) which, as Fig. 4.32 illustrates, are clearly related to the position of the flexural bulge. ten Brink (1991) and Hieronymus and Bercovici (2000) have attributed this secondary volcanism to the interaction of tectonic and flexural bending stresses, while Bianco et al. (2005) have attributed it to decompression melting of the mantle beneath the bulge crest (Fig. 4.36). The arch volcanics cover a broad swathe of relatively deep seafloor, however, and they suggest that not only the 'big island' of Hawaii but the flexural bulges of upstream older islands may also be involved.

A seamount which has experienced both a volcanic load-induced flexural uplift and subsidence is Cross, a flat-topped, 400-m-deep, Cretaceous 'guyot' located about 275 km

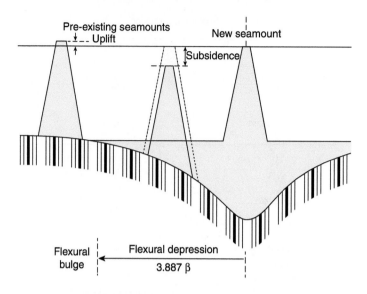

Figure 4.37 Simple model of flexure showing the influence of a new seamount load on pre-existing seamounts that find themselves in either the flexural depression or flexural bulge of a new seamount. Reproduced from fig. 2 of McNutt and Menard (1978), copyright by the American Geophysical Union.

south-west of Hawaii (Wessel and Keating, 1994). A seamount at this distance would be in Hawaii's flexural bulge (Fig. 4.32). Therefore, the subsidence cannot be attributed to the load of Hawaii. Other islands in the chain must have been involved. Wessel and Keating (1994) showed that the subsidence of Cross can be explained by the cumulative effects of being in the flexural moats of Maui, Lanai, Molokai and the south-east flank of Oahu. Since Oahu, the most westerly of these islands, began its constructional phase about 3.2 Ma, Wessel and Keating (1994) suggested that Cross was uplifted and truncated then and has since subsided 400 m.

Some of the best evidence for flexural loading due to new volcanic islands has come from widely scattered islands rather than islands that formed along a single hotspot-generated ridge. This is because flexure will deform the oceanic crust in 'waves' that are centred on where a new load has been applied. Island clusters are therefore more likely to reveal the full effects of flexure than islands that form a single hotspot-generated ridge: islands that are in the flexural moats of a new volcanic load will be subject to subsidence and may submerge, while islands on the flexural bulges will experience uplift and hence a period of emergence (Fig. 4.37).

A good example is seen in the Southern Cook Islands in the south-central Pacific. McNutt and Menard (1978) have shown, for example, that the subsidence and uplift history of three of the older (Dickinson, 1998) islands in the group – Atiu, Mauke and Mangaia – can in large part be attributed to flexure caused by the younger islands of Aitutaki and Rarotonga (Fig. 4.38). A special feature of the flexure is that it shows how the subsidence and uplift history of a single island may be influenced by more than one younger neighbour. Atiu, for example, is located on the flexural bulge of both Aitutaki and Rarotonga. Model calculations based on a $T_e = 14$ km suggest that Atiu was raised 15 m by Aitutaki and

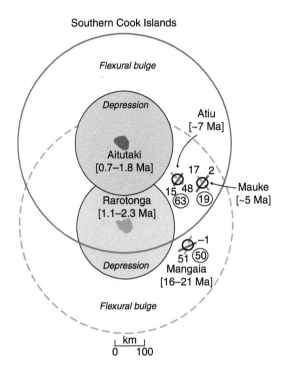

Figure 4.38 Comparison of calculated and observed uplift at Mauke, Mangaia and Atiu in the Southern Cook Islands. The colours show the extent of the flexural moat and bulge associated with Aitutaki (purple) and Rarotonga (light green). The short, coloured lines show the calculated uplift due to flexural loading of Aitutaki and Rarotonga at the makatea islands of Atiu, Mauke and Mangaia. Modified from McNutt and Menard (1978), copyright by the American Geophysical Union.

43 m by Rarotonga. The cumulative effect,[4] therefore, was to raise the island by 63 m. As Fig. 4.39a shows, this is close to the average height of Atiu's central volcanic 'core'. The 'core' is flanked by a raised coral (limestone) reef which forms a continuous fringe around the island 5–20 m above sea level. Between the surface of the raised reef and the volcanic core is a low-lying swampy area. Islands, such as Atiu, Mauke and Mangaia, with both a visible volcanic 'core' and a fringing raised limestone reef form a distinct landscape and are collectively known by the general term of 'makatea' (e.g., Dickinson, 1998, 2004).

Similar evidence of island uplift by younger nearby volcanic loads have been reported by Dickinson and Green (1998) from Samoa and American Samoa (where Tutuila has been raised by Savai'i) and by Pirazzoli (1983) from the Society Islands (where Makatea has been raised by Mehetia).

---

[4] Note that this estimate excludes the effect of Manuae, an island located about 90 km ESE of Aitutaki. According to McNutt and Menard (1978), Atiu should have experienced 45 m of subsidence because of its location in the moat of Manuae. However, Manuae is probably younger than Atiu (Lambeck, 1981b). The loading effects of this island have therefore been omitted in Fig. 4.38.

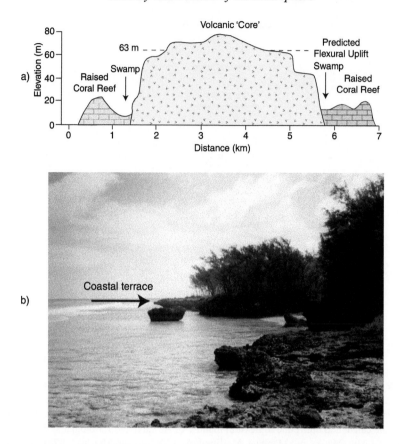

Figure 4.39 Atiu island in the Southern Cook Islands. (a) Schematic geological cross section of the uplifted makatea island of Atiu. Note the plateau surface of the central volcanic core is ~63 m above sea level, which agrees well with the calculated flexural uplift (Fig. 4.38). Most settlements are located on the surface of the raised coral reef. Reproduced in part from fig. 1 of Jarrard and Turner (1979), copyright by the American Geophysical Union. (b) Ora Varu beach (site of Captain Cook's landing on April 4, 1777) showing a coastal wave-cut terrace 1–2 m above sea level.

A feature of several coastlines in the southern Cooks, Samoa and American Samoa and Society islands is a 1- to 2-m-high coastal terrace that has been cut either into the paleoreef (e.g., Fig. 4.39b) or the volcanic core (Dickinson, 2001). The terraces are approximately mid-Holocene in age, and the uplift is what would be expected from the location of these islands in an emergent zone (Zone 5) of the global ice melting models of Clark et al. (1978). In such a zone, seas would have risen in early Holocene times and then have fallen. Therefore, while flexural loading is an important contributor to the uplift of these and other island groups, there may well have also been a contribution from hydro-isostasy, especially to their late Holocene uplift history.

Ocean islands and seamounts, once formed, are vulnerable to large-scale sector collapse. Several different factors may act to destabilize an island or seamount, including sea level changes (Quidelleur et al. 2008), height above the surrounding seafloor, lithology and the formation of a décollement surface on which collapse blocks can slump and slide. Once material has been removed, it will cause a flexural rebound and, in some cases, a tilting of the edifice. Smith and Wessel (2000), for example, used a simple flexure model of unloading and calculated uplifts of 50–100 m, which they showed depended on the elastic thickness T_e, width, and sector collapse depth.

So far, we have only considered the flexure associated with seamounts and ocean islands in the plate interiors. However, flexure may be important in other submarine volcanic settings such as island arcs where flexural bending stresses induced by one arc volcano may control the spacing of subsequent ones (ten Brink, 1991). Historically active volcanoes in the Indonesia, Aleutian, Tonga-Kermadec and Scotia arcs, for example, are spaced about 50–70 km apart which, if we use ten Brinks' volcano point loading model, implies $15 < T_e < 35$ km for arc lithosphere. In addition, Hoffman et al. (2010) have shown that volcano-induced flexure, in combination with sediment loading, of arc lithosphere with $5 < T_e < 20$ km may explain drowned carbonate platforms that have grown up on volcanoes along the Bismarck arc in the south-west Pacific.

4.4 River Deltas

In contrast to oceanic islands and seamounts, which form relatively rapidly (a few Myr), the sediments deposited in river deltas represent loads that have accumulated on the crust over long periods of time. Present-day examples such as the Nile and Ebro in the Mediterranean Sea, the Mississippi in the Gulf of Mexico, the Niger and Amazon in the Atlantic Ocean and

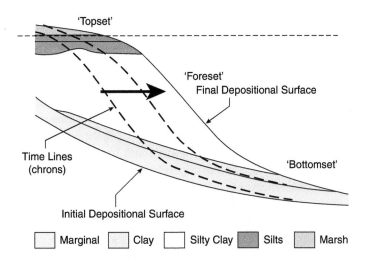

Figure 4.40 Schematic diagram illustrating the progradational phase of river delta growth. Solid lines show the initial and final depositional surface. Dashed lines show intermediate depositional surfaces at different times. Colours show the sedimentary facies. Reproduced from Scruton (1960).

the Indus and Bengal fans in the northern Indian Ocean, for example, span ages that range from a few to several tens of Myr.

River deltas (e.g., Fig. 4.40) typically comprise three distinct stratigraphic units: 'topset', 'foreset' and 'bottomset' beds. The topset beds form a continuous sedimentary sequence from the river's flood plain, across a graded shelf, to a shelf slope break. The deposits are fluviatile and comprise gently dipping, coarse-grained silts and sands that contain fresh rock fragments. Below the topset beds are the foreset beds. These beds dip more steeply than the topset and comprise mainly silty clays that were deposited as turbid river water began to spread out. The foreset beds are underlain by the bottomset beds, which are fine-grained clay deposits that may themselves build out great distances. Usually, however, the bottomset beds are progressively overstepped by the prograding foreset beds.

As a river delta builds and sediments displace water, the load increases, and so the underlying crust will subside, and flanking regions will be raised. These load-induced vertical movements of the crust and mantle have a profound effect on the delta. Subsidence increases the overall water depth and, hence, increases the accommodation space available for sediments to infill. Uplift has two competing effects, depending on which delta flank we are concerned with. On the oceanward side, uplift reduces the accommodation space and may, despite its small size, act as a barrier that limits the lateral extent of a delta. On the landward side, however, uplift increases the height of a source area which, in turn, may increase the sediment supply. It was this feedback between subsidence and uplift in the river delta system that so fascinated the early isostasists, especially workers such as Barrell, Bowie and Lawson.

While the importance of river deltas was recognised by the early isostasists, they disagreed on the role of flexure. Barrell (1914e) argued, for example, that there was no evidence for large amounts of subsidence and therefore that the lithosphere was able to support loads such as the Nile and Niger deltas because of its great strength. Bowie (1922), however, disagreed (p. 82). He argued that the sediment loads were so large that they would sink into and depress the crust such that only about one-third of the height of a sediment deposit would appear above its original surface. The remaining two-thirds would form a downward protruding 'root' in the crust.

Lawson (1942) published a seminal paper on river deltas in which he considered the role of both regional and local schemes of isostatic compensation. However, by focusing on the Mississippi delta, he considered a sediment load that was so wide that there was little difference between the maximum thickness of sediments estimated using the two isostatic schemes. He therefore saw no need to incorporate flexure. It was not until 30 years later (Walcott, 1972a) that the notion of flexure at river deltas, an idea originally put forward by Barrell, re-appeared.

Prograding river delta systems have discrete geometries, depending on whether they are dominated by processes in the river system or by processes in the depositional basin (Elliot, 1986). In plan view, both fluviatile- and basinal-dominated delta systems display a generally lobate, arcuate pattern that reflects the interactions between the discharge of sediments outward from the river mouth and the competing effects of basinal processes such as long-shore drift and ocean currents. There is evidence that during the constructional phase, the lobes of river delta systems maintain a fixed migration path. Once the supply of sediment to the delta is reduced, however, individual lobes become abandoned, and lobes may switch (Coleman, 1988).

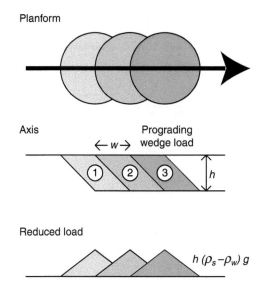

Figure 4.41 Simple model for a prograding sediment wedge. Each wedge is assumed to be circular in planform and to be aligned in a constant direction. The sediment wedge has a density, ρ_s, and displaces water with density, ρ_w. Reproduced from fig. 1 of Watts (1989).

A simple model for the loading of successive wedges of sediment in a prograding river delta system is shown in Fig. 4.41. The model assumes that each load displaces the same amount of water and that they migrate in a constant direction. The shape of the wedge in planform is such that each load reduces to a series of cone-shaped loads.

Figure 4.42 shows how the lithosphere might respond to three successive cone-shaped loads, assuming it to behave as a thin elastic plate that overlies an inviscid fluid. The upper profile shows the response to a single load. The response is flexural in form and consists of a subsidence that reaches its maximum value immediately beneath the load and an uplift in flanking regions. The subsidence extends beyond the limits of the original load because of the lateral strength of the underlying crust and mantle. Surfaces that were once flat before loading will therefore be deformed after loading. If the downward flexures are subsequently infilled by sediment, then thin wedge-shaped basins will form in the 'wings' of the newly formed basin. The net effect is to produce an *S*-shaped sediment wedge, the internal architecture of which is dominated by a downdip thickening and an updip thinning. The middle profile in Fig. 4.42 shows the effect of adding a second load, similar in size, on the seaward side of the first load. The first load is now in the flexural depression of the second load and so will be depressed, and slightly rotated, by the weight of the new load. The rotation makes the depositional surface of the first load steeper than that of the new load. Since the second load is added seaward of the first load, it oversteps the region of maximum flexure of the first load and forms an offlap pattern landward of the slope break and an onlap pattern seaward of the slope break. Finally, the lower profile in Fig. 4.42 shows the effects of adding a third load. As before, the new load depresses and rotates previous loads. The effect on the first load is not as pronounced, however, because this load now finds itself in the

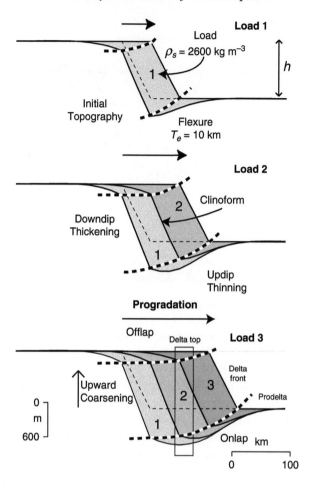

Figure 4.42 Schematic model for the progressive flexure of the lithosphere due to a prograding sediment load. The upper profile shows the initial load and resulting flexure. The middle profile shows the effect of adding a second load. The lower profile shows the effect of adding a third and final load. Reproduced from fig. 2 of Watts (1989).

flexural bulge of the new load. This may cause uplift rather than subsidence and, if the uplift triggers erosion, removal of the featheredge of some of the sedimentary infill material that formed during emplacement of previous loads.

The elastic model can be used to predict the facies distribution in the final sediment wedge by tracking the two slope breaks: one between the shelf and slope (the upper one) and the other between the slope and the basin floor (the lower one). These are shown as thick dashed lines in Fig. 4.42. Since the positions of the slope breaks move with the progradation, they intersect the depositional boundaries at high angles. The region above the upper line represents sediments of the delta-top facies and would therefore be relatively coarse grained. The region below the lower line, in contrast, represents sediments of the bottomset facies and hence would be fine grained. Between the two thick lines, the sediments are

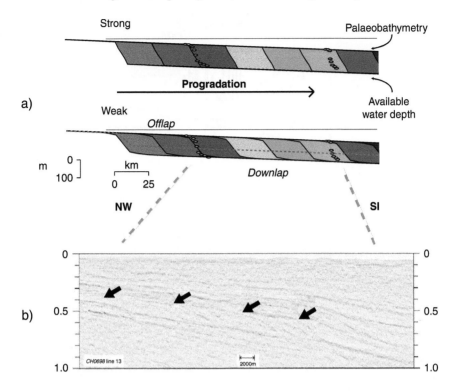

Figure 4.43 Comparison of the calculated stratigraphy due to progressive loading to *S*-shaped 'clinoform' structures observed on seismic reflection profiles. (a) Calculated profiles based on a strong ($T_e = 50$ km) and a weak ($T_e = 2$ km) margin. Dashed line indicates the water depth that is required to produce the same sediment thickness in the strong and weak cases. Reproduced from fig. 3 of Watts (1989). (b) Seismic reflection profile showing the Miocene clinoforms imaged on R/V *Ewing* Line 1003 (Monteverde et al., 2008).

entirely in slope facies and so we would expect them to be intermediate in coarseness. The net result is a sedimentary wedge that is dominated by upward coarsening.

The predictions of the prograding flexure model are compared with observations in Figs. 4.43 and 4.44. Two cases of sediment progradation during the Tertiary are shown: one at relatively small scale at the East Coast, United States, and the other at large scale at the Texas Gulf Coast margins.

Seismic reflection profiles offshore of New Jersey suggest that during the Miocene a series of clastic wedges prograded across the shelf almost to the slope break. The evidence for progradation is seen in seismic reflection profiles as a series of *S*-shaped reflectors. Known as clinoforms, each reflector is characterised by downlap in a basinward direction and offlap (i.e., a downward shift in the pattern of coastal onlap) in a landward direction. Greenlee and Moore (1988) interpret the offlap as the result of erosion of an otherwise onlapping surface during a major sea level fall.

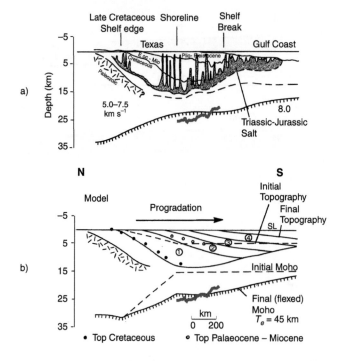

Figure 4.44 Comparison of the calculated and observed crustal structure along a transect of the Texas Gulf Coast region. Solid purple line shows the seismically constrained Moho depth along nearby GUMBO Line 2 (Eddy et al. 2018). (a) The observed profile is based on the work of Antoine et al. (1974). (b) The calculated profile assumes that sediments have prograded about 120 km into a 5-km-deep water-filled basin. The sediments represent a load on the crust that responds by a downward flexure of its upper and lower surface. Reproduced from fig. 4 of Watts (1989).

The flexural effects of a series of prograding sediment loads, similar in geometry to the US East Coast, Miocene loads, is illustrated in Fig. 4.43a for $T_e = 2$ and $T_e = 50$ km, which correspond to a weak and a strong margin, respectively. The width of each sequence was assumed to be similar to the spacing between clinoforms observed on seismic data. In the case of high T_e, there is negligible flexural subsidence (or uplift), and so sediments would simply fill the available water depth up to the palaeobathymetry surface. Because the flexure is small, there is no rotation of previous depositional surfaces. In the case of low T_e, however, there is flexure, and the sediments are able to deform the sea floor downwards beneath the loads and upwards in flanking regions. There are two consequences of flexure. First, the available water depth does not need to be as large, otherwise a larger volume of sediment would be required in order for sediments to build up to the same palaeobathymetry surface. Second, successive loads will deform and rotate previous depositional surfaces into a distinct S-shape. Unfortunately, it is difficult to constrain T_e accurately from the observed clinoforms, although the data are generally suggestive of low rather than high values.

There is evidence that the progressive loading model can also account for larger-scale features of river delta systems, such as their stratigraphic 'architecture' and deep crust and mantle structure. One of the best examples is the Mississippi delta offshore the Texas Gulf Coast. Here, thick sequences of Tertiary–Recent clastic sediments were deposited in off-lapping wedges over mainly carbonate beds of Cretaceous age (Fig. 4.44a). The offlapping wedges formed following the culmination of the Laramide orogeny during the Late Cretaceous to Early Tertiary, when there was a rapid influx of sediment into the Gulf of Mexico region that caused the depositional shelf break to migrate from a point 150 km inland of the present-day coastline to about 100 km seaward of the coast.

Figure 4.44b shows that a prograding flexure model can explain both the stratigraphic 'architecture' and the deep structure of the Mississippi delta. In particular, a model in which sediments prograde into a 4-km-deep, water-filled basin that is underlain by thinned crust accounts for the presence of offlap and onlap patterns. It also accounts for the deep crustal structure. The sediment load, for example, flexes the thinned crust downward by up to 9 km, which results in a gradual decrease in the depth to the Moho from ~35 km beneath northern Texas, through ~24 km beneath the shoreline to ~17 km beneath the Gulf of Mexico. The modelled Moho is in reasonable agreement with the seismically constrained Moho of Antoine et al. (1974) and the recent GUMBO experiment of Eddy et al. (2014), Van Avendonk et al. (2015) and Eddy et al. (2018) which demonstrates that much of the 12- to 15-km-thick sediments that underlie the Gulf of Mexico overly continental crust that has been stretched to < 10 km. The main feature of the flexure model not clearly visible is the downdip thickening and onlap of the prograding wedges onto basement. The most likely explanation for this is post-depositional salt movement during the late stages of rifting in the Triassic and Early Jurassic.

The models so far assume that sediments prograde into water-filled basins of constant depth. But what about the case where the water depth varies with time, as would be expected, for example, if the sediment wedges had prograded onto oceanic crust that was subsiding with age? Figure 4.45 shows, for example, the expected stratigraphic patterns and facies if a succession of sediment wedges prograde into a basin that was subsiding with time. The figure shows that flexure due to a succession of three wedges, comprising three, four and three sediment loads, respectively, can create quite complex patterns of strati-graphic onlap and offlap and facies.

We have shown that flexure, due to sediment loading, can explain several features of the stratigraphic 'architecture' and sedimentary facies of river delta systems. A key question, then, is what T_e should be used to model the flexure? Unfortunately, it is difficult to constrain T_e at river deltas because, unlike seamount and ocean island loads, which form during relatively short periods of time, river delta loads are progressively added to the crust over long periods of time.

One parameter that may be used to constrain the T_e at river deltas is the free-air gravity anomaly (Walcott, 1972a). River deltas disturb the local isostatic balance and so would be expected to cause gravity anomalies.

Figure 4.46 illustrates the predicted free-air gravity anomaly for a simple model of a sediment wedge that progrades into a 5-km-deep, water-filled basin. The anomaly can be considered as comprising two parts: one due to the initial crustal structure, and the other due to the prograding sediments. The gravity anomaly of the initial crustal structure is given by the combined effect of the mass deficiency of a water-filled basin and the mass excess of its compensation, which, in the case of Airy isostasy, takes the form of crustal thinning. The

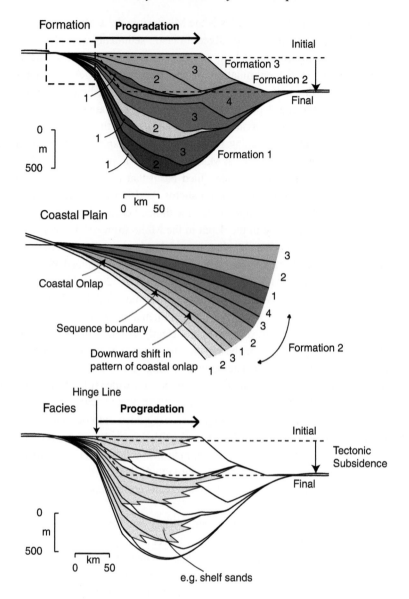

Figure 4.45 Simple model for the stratigraphy and facies due to progressive sediment loading on a subsiding margin. Three 'pulses' of sediment are considered. The first (Formation 1) involves three wedges that prograde 150 km towards the ocean basin. The second (Formation 2) involves four wedges, and the third (Formation 3) three wedges. Each formation is associated with offlap in the landward direction and onlap in the seaward direction, which reflects the progradation. The tectonic subsidence is assumed to increase with time about a fixed 'hinge line'. The net effect is a stratigraphic sequence where the formations are bounded by unconformities and the facies are repeated in a vertically sectioned sequence. Reproduced from fig. 6 of Watts (1989).

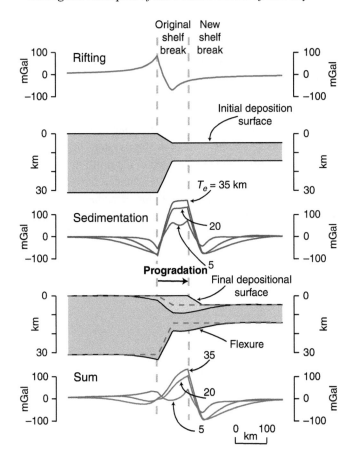

Figure 4.46 Simple model showing the contributions to the gravity anomaly of rifting which thins the crust and a prograding sediment wedge which loads the rifted margin. The amplitude and wavelength of the sum anomaly depend strongly on T_e with the largest amplitude anomalies associated with a high T_e or strong margin and the smallest with a low T_e or weak margin.

resulting anomaly comprises an 'edge effect' with a high and a flanking low over the basin margin. Because crustal thinning is usually associated with rifting, this anomaly is referred to in Fig. 4.46 as the rifting anomaly. The gravity anomaly of the sediments comprises two parts: a high over the sediment load, and a low over flanking regions. The high arises because sediments are denser than the water they displace, while the lows are the consequence of the downward flexure of relatively low-density sediment into the crust and relatively low-density crust into the mantle. The total anomaly, referred to in Fig. 4.46 as the sedimentation anomaly, is a strong function of T_e, being larger in amplitude and longer in wavelength for loads on strong lithosphere than for loads on weak lithosphere. If crustal thinning and sediment progradation have been the only processes to have modified the basin margin, then the sum of the rifting and sedimentation anomalies should reflect the total gravity anomaly at the margin. As Fig. 4.46 shows, sedimentation modifies the gravity anomaly associated with rifting, causing the original edge-effect high and low to migrate seaward from the shelf break at the time of rifting to the new shelf break after sedimentation. The modification depends, however, on T_e. At weak basin margins (i.e., low T_e), the high is distributed at the initial and final shelf break, causing a 'double' high. In contrast, at strong basin margins (i.e., high T_e), the initial high is entirely removed and replaced by a single large-amplitude high at the new shelf break.

One river delta where constraints exist on both the magnitude of sediment loads and the gravity anomaly is the Amazon Cone, a deep-sea fan system in the north-east Brazilian continental margin (Fig. 4.47c). The fan is located to the north-north-east of the mouth of the Amazon River, the drainage basin of which is bounded by the Bolivian Andes in the west, the Guinea Highlands in the north, and the Brazilian Highlands in the south. Sedimentation rates derived from ODP Site 354 (Fig. 4.47c) and piston cores suggest that the fan is mid- to late Miocene in age (i.e., 8–12 Ma) (Damuth, 1975). This correlates with

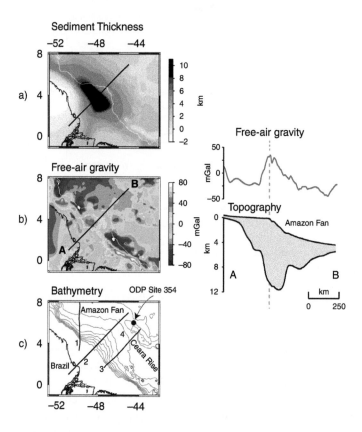

Figure 4.47 Sediment thickness, gravity anomaly and bathymetry at the Amazon fan, north-east Brazilian continental margin. (a) Sediment thickness based on Kumar et al. (1979). (b) Shipboard and satellite-derived free-air gravity anomaly. (c) Bathymetry. Contour interval = 500 m.

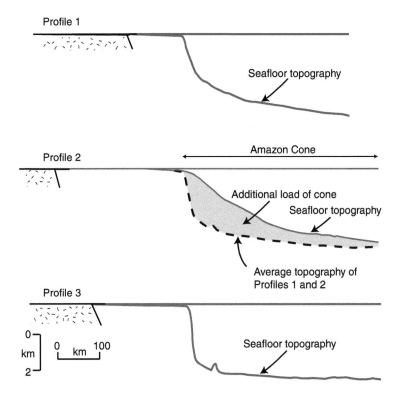

Figure 4.48 Topography profiles 1–3 of the north-east Brazilian continental margin. Based on data acquired during cruise C1509 of R/V *Robert D. Conrad*. See Fig. 4.47 for profile location.

a period of rapid uplift and erosion in the Bolivian Andes (Benjamin et al., 1987) suggesting that, despite its great distance, the region was a significant source for the fan sediments.

The Amazon fan is recognised in bathymetry maps by a 'bulge' in individual seafloor depth contours (Fig. 4.47c). The volume of sediment in the fan can therefore be estimated by comparing profiles of the Brazilian margin in the region of the fan with profiles to the north and south (e.g., Fig. 4.48). Profiles show it to have a characteristic shape: a concave-up upper part, a convex-up middle part and a concave-up lower part. In contrast, profiles to the north and south of the fan reveal a sea-floor morphology that is more typical of a continental slope and rise.

It is estimated by considering profiles, as well as three-dimensional bathymetric grids, that the additional volume of sediments present in the cone, and which are not present elsewhere along the north-east Brazilian continental margin, is about $1.4 \times 10^5 \, \text{km}^3$.

The fan sediments represent a load on the crust that should flex under their weight. The volume of sediments estimated from the bathymetry data may therefore not indicate the true volume, because sediment infills the flexure. Figure 4.49 shows the flexure that would be expected from the additional fan load for different values of T_e. The flexure comprises a circular-

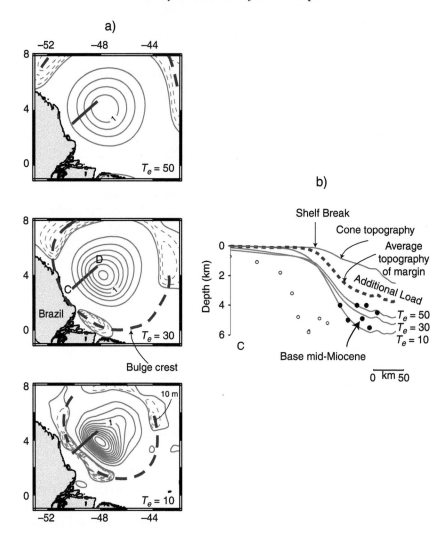

Figure 4.49 Comparison of the flexure of the lithosphere caused by the additional load of the Amazon fan with observations. (a) Predicted flexure. Solid lines show subsidence, dashed lines show uplift. The thick dashed lines show the crest of the flexural bulge. (b) Topography, depth to the base mid-Miocene reflector and flexure along the profile CD. Circles are based on the data of Braga (1991). Filled circles = data fit by the flexure model. Open circles = unexplained data.

shaped region of subsidence that is flanked by uplift. The subsidence is at a maximum beneath the fan load, where it reaches values of 1.5–4 km, depending on the value of T_e. The fan load thickness reaches a maximum of about 2 km, so the flexure calculations suggest that a total thickness (i.e., load + flexure) of 3.5–6 km of sediments is present in the fan.

The base of the fan sediments is recognised on seismic reflection profile data as a major angular unconformity. The depth to the unconformity on Profile CD, which is based on Braga (1991), is shown as filled and open circles in Fig. 4.49b, together with the predicted depths based on flexure modelling. The figure shows reasonable agreement between observed and calculated surfaces beneath the cone load, with $T_e \sim 30$ km providing the best overall fit to the data.

We see from Fig. 4.49b that flexure due to the additional load of the fan cannot explain all the details of the angular unconformity. The main problem is in explaining the large thickness of mid-Miocene and younger sediments that underlie the outer part of the shelf, on the landward side of the fan. These sediments require an additional load, and we speculate that this is caused by a 'background' thermal subsidence that continued to modify the margin while the deep-sea fan was being constructed.

The flexure calculations allow an estimate of the total mass of the sediments in the cone to be made that, in turn, can be compared with the yield in the Amazon River. The volumes of sediment in the load and the flexure are 1.4×10^5 km^3 and 3.2×10^5 km^3, respectively. The combined volume is therefore 4.6×10^5 km^3. Assuming the cone sediments are 10 Ma and that the sediments have a mean density of 2,400 kg m^{-3}, then this volume implies an average sediment accumulation rate of 1×10^{11} kg yr^{-1}. This is nearly a factor of four smaller than the estimated yield of 3.6×10^{11} kg yr^{-1} in the Amazon River, which according to Holeman (1968) is the seventh largest of any river in the world. This suggests, if the measured yield is representative, that only about a quarter of the sediment supplied by the Amazon River is deposited in the fan. A large amount of sediment therefore must be bypassing the fan and being transported along-strike rather than across-strike the northeast Brazilian continental margin. This appears also to be the case for the Nile (Table 4.1). The Mississippi and Ganges, however, appear to be cases where the yield in the river is of the order of that which is deposited in the delta.

We see from the calculations in Fig. 4.49 that flexural loading of the north-east Brazilian margin by the Amazon fan has caused not only subsidence, but also a bulge in flanking regions. The amplitude of the bulge and its distance from the load centre depend on T_e. For example, for $T_e = 10$ km, the bulge reaches its maximum value of 25 m over the inner continental shelf, while for $T_e = 50$ km, the bulge is less than 5 m and is located onshore. For $T_e = 30$ km, the

Table 4.1 *Comparison of river yield and accumulation rate in large river deltas*

Delta	River Yield (kg yr^{-1} × 10^{11})	Volume (km^3 × 10^5)	Mass (kg × 10^{18})	Age (Ma)	Accumulation Rate (kg yr^{-1} × 10^{11})	References for Volume and Mass
Amazon	4.0	4.6	1.0	10	1.0	This study
Mississippi	3.5	110.0	26.2	65	4.0	Lawson (1942)
Nile	1.2	0.9	0.2	30	0.07	Barrell (1914e)
Ganges	16.0	101.0	20.2	9	22.4	Curray and Moore (1971)
Niger	0.02	5.0	1.2	45?	0.26	Hospers (1965)

* Assumes density = 2,400 kg m^{-3}.
Source: Holeman (1968).

crest of the bulge coincides with the coastline. Unfortunately, the lack of seismic reflection data, and the fact that other factors may have influenced the subsidence and history of the north-east Brazilian margin, make it difficult to distinguish between the different T_e models.

The Amazon fan is associated with a distinct 'edge-effect' gravity anomaly with a high and a flanking low (Fig. 4.47b). As Fig. 4.46 illustrates, the edge-effect anomaly can be considered as the sum of the gravity effect of the initial crustal structure (the 'rifting' anomaly) and the sedimentation. Indeed, Cochran (1973) used the anomaly, together with seismic refraction constraints on the total sediment thickness, to determine the combination of the load (which determines the initial crustal structure) and flexure that best explains the observations in the region of the Amazon fan (Fig. 4.50). He showed that $T_e = 31.1$ km is consistent with both the amplitude and wavelength of the edge-effect anomaly and the total thickness of sediment in the cone region.

Once the T_e structure of the margin has been constrained by gravity modelling, then its implications for the topography onshore can be explored. Driscoll and Karner (1994), for example, in their study of the Amazon fan, chose a T_e value of 38 km, a somewhat higher

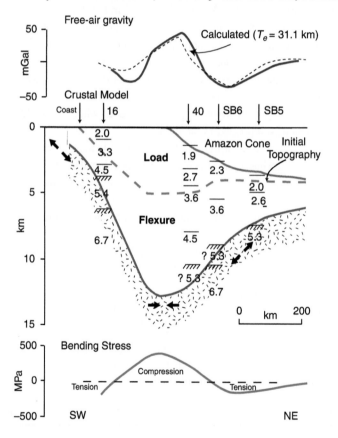

Figure 4.50 Gravity anomaly, topography, flexure and bending stress at the Amazon fan based on the calculations in fig. 7 of Cochran (1973). The seismic *P*-wave velocities are based on the refraction data in Houtz et al. (1978).

value than the one estimated by Cochran (1973), and showed that the bulge would be located onshore, in the region of the Gurupá arch which defines a change in the morphology of the Amazon flood-plain. The bulge crest also correlates with a drainage divide that separates east-flowing rivers east of the divide from west- and north-flowing rivers to the west. Based on these observations, Driscoll and Karner (1994) speculate that the stratigraphic development of the fan has been dominated by an isostatically driven 'feedback' that links sediment loading offshore with bulge uplift in the hinterland and modulates the spatial and temporal delivery of sediment to the fan.

In 2003, the first deep seismic experiment was attempted on the Lower, Middle and part of the Upper Amazon fan (Watts et al., 2009). Results from the experiment have enabled a better definition of the sediment thickness, T_e, and the flexural depression and flanking bulge associated with the fan load. Figure 4.51 shows, for example, a composite 240-km-long seismic reflection profile of the fan which extends from the north-east Brazilian continental shelf, across the slope to the Ceara Rise. The profile shows a thick succession of Cretaceous and younger sediments which comprise an early Cretaceous to mid-Miocene sequence associated with the initial development of the margin which is overlain by a later Mid-Miocene to Recent sequence associated with the Amazon fan. Seismic and gravity modelling (Fig. 4.52) reveals the maximum thickness of sediments (~12 km), the crustal structure, both at the present day and immediately following margin formation, and the T_e associated with sediment loading.

The flexure associated with the youngest, Mid-Miocene to Recent layer (Layer 2) is shown in Fig. 4.53. The figure shows that the flexure, based on the best-fit T_e structure, reaches 1–2 km and is a maximum beneath the middle fan. Flanking the flexure is a bulge, which onshore correlates with the crystalline basement rocks of the Gurupá Arch region, where it acts as a drainage divide separating generally east- and west-flowing rivers. Flexure predicts, of course, 'waves' of deformation within a plate so that the first bulge will be flanked by a second flexural depression (referred to by DeCelles and Giles (1996) as the 'back-bulge') which, in turn, will be flanked by a second bulge. Detecting these secondary flexures in the continental rock record is difficult, because of the modifying effects of erosion and sedimentation; but in the case of the north-east Brazil margin they are manifest in the depositional and non-depositional history of the Amazon River. The second flexural depression, for example, corresponds to a region of 'white sand' deposition which forms the unique seasonally flooded savannah ecosystem known as the Gurupá Várzea (e.g., Vogt et al., 2016), while the second bulge corresponds to a region of non-deposition and crystalline basement rock outcrop which Roddaz et al. (2005) referred to as the Monte Alegre Arch.

While bulges similar to those onshore north-east Brazil have been recognised from other river deltas (e.g., Mississippi – Jurkowski et al., 1984), the Holocene to Recent history of large river deltas, including the Mississippi, are dominated by coastal subsidence, not uplift. The cause of the subsidence at the Mississippi delta has been debated. Tornqvist et al. (2008) argued it is caused by sediment compaction, while Blum and Roberts (2009) proposed insufficient sediment supply and loading combined with sea level rise as the primary control. Goodbred and Kuehl (2000) argued that the Late Quaternary subsidence of the Ganges-Brahmaputra delta, with its active tectonic hinterland and high-yield sediment supply, was caused by the relative rate difference between sediment input and sea level

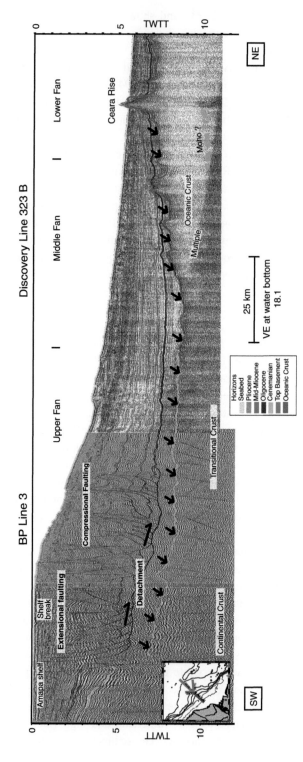

Figure 4.51 Seismic reflection profile showing the flexure of the Amazon deep-sea fan system. TWTT = Two-Way Travel Time. Based on a merge of data acquired by BP and during cruise D275 of RRS *Discovery*. The insets show a legend with the approximate age of prominent reflections and a location map with the ship tracks. The detachment surface (half-arrows) is part of a gravitational collapse structure that links extensional faults beneath the outer shelf with compressional faults beneath the slope. Full arrows delineate the top of the flexed continental (green reflector) and oceanic basement (brown reflector). Reproduced from fig. 4 in Watts et al. (2009).

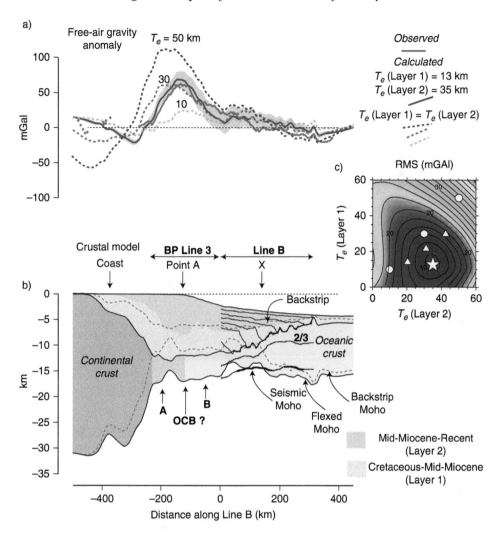

Figure 4.52 Gravity anomalies and crustal structure of the north-east Brazil continental margin in the vicinity of the Amazon deep-sea fan system. (a) Comparison of the observed free-air gravity anomaly (offshore) and Bouguer anomaly (onshore) to the calculated anomaly based on one sediment load and uniform values of T_e of 10, 30 and 50 km (dashed green, red and purple lines respectively) and two sediment loads, one of Mid-Miocene to Recent age and the other of Cretaceous to Mid-Miocene age, and a T_e that varies with age since the initiation of the margin (solid blue line). (b) Crustal structure based on seismic and gravity modelling. Red dashed lines show the initial topography and Moho, as inferred from flexural backstripping of the sediments and crustal restoration assuming Airy isostasy. (c) RMS difference between observed and calculated gravity anomalies. The white filled star shows the best-fit two-sediment load model with $T_e = 13$ km for Layer 1 and $T_e = 35$ km for Layer 2. The other white filled symbols show the case of a single-sediment load model and either a uniform T_e (circles) or T_e as a function of age (triangles).

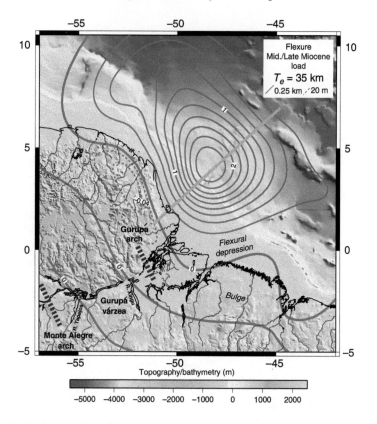

Figure 4.53 Flexure of the lithosphere due to the Mid-Miocene to Recent sediment load. Contour interval = 250 m. The zero onshore delineates the flexural bulge, flexural depression and flexural bulge associated with the Gurupá arch, Gurupá várzea and the Monte Alegre arch, respectively. The Gurupá várzea, between the Xingu and Tapajos rivers, is a unique ecosystem that is mainly savannah rather than dense rain forest of the seasonally and tidally flooded arch regions to the west and east. Flexure due to offshore sediment loading is therefore an important control on landscape evolution onshore, influencing the flora and fauna of a large region extending from the coastline to >400 km into the continental interior and beyond.

change, not compaction. However, Higgins et al. (2014) using InSAR data have attributed recent subsidence of the delta to anthropogenic effects associated with groundwater and hydrocarbon extraction in combination with reduced sediment aggradation. Other deltas where InSAR data have documented recent subsidence include Pearl River and Nile where Wang et al. (2012) and Becker and Sultan (2009) have both attributed it to sediment compaction.

Bulges would be expected seaward as well as landward of prograding delta systems, as have been described and modelled by Kjeldstad et al. (2003) at the Voring Plateau margin offshore Norway and the Niger delta. The role of these submarine bulges is not as clear as that of their sub-aerial counterparts, but at margins such as the Amazon fan where gravitational collapse dominates slope processes, they may serve as ramps on which deep-water fold and thrust structures develop.

These considerations show a prograding flexure model to be a major contributor to our understanding of river deltas which prograde into deep water, especially to their internal and external stratigraphic 'architecture'. As Cisne (1984) has shown, such a model may also be a be a significant contributor to the stratigraphic 'architecture' of near-shore regions of shallow epicontinental seas, although the associated onlap and downlap patterns will be more subtle in this case and difficult to recognise.

More difficult to explain by flexure are the sedimentary facies. While many fluvial-dominated modern deltas and deltas in the rock record (e.g., the Carboniferous of the British Isles) show coarsening-upward sequences from say mudstone to sandstone, as predicted by a prograding flexure model (e.g., Fig. 4.45), others do not (Stanley, 2001). Foremost among them are fining-upward sequences. Reading and Collinson (1996) have argued that such sequences can occur in response to a sea level rise as source areas are flooded and a fluvial-dominated delta switches from sand-rich braided streams to a mixed mud-and-sand meandering river system. In addition, coarsening-upward sequences may form in response to a sea level fall as a source area is exposed and a sand-rich shelf is reworked and redeposited as beach ridges. Therefore, flexure may play a role together with sea level change in controlling the facies of river deltas.

4.5 Deep-Sea Trench and Outer-Rise Systems

It has been known since Gunn's pioneering work that flexure of the lithosphere probably plays an important role in the development of island arc–deep-sea trench systems. We showed in Section 2.7 how Gunn (1943a) used the width of the gravity anomaly associated with the Java Trench to estimate the flexural properties of the lithosphere. Later, he proposed (Gunn, 1947) that the trench and adjoining arc are part of a coupled system where compressive forces applied to the edge of a plate subject to shear failure cause the continental 'side' to overthrust the oceanic 'side'. The shape of the overthrust continental side and the underthrust oceanic side was controlled by flexure, although Gunn recognised that it might subsequently be modified by erosion and sediment deposition.

Gunn's ideas were later to play a role in the development of plate tectonics.[5] It was several years after its development, however, before the role of flexure at island arc deep-sea trench systems was re-examined. The first papers focused on the Kuril Island arc–deep-sea trench system in the north-west Pacific Ocean (Fig. 4.54), which, together with the Japan arc to the south and the Aleutian arc to the east, is the most active region in the world for shallow- and intermediate-depth earthquakes.

Walcott (1970a) pointed out that seaward of the Japan, Kuril and Aleutian trenches, the depth of the sea floor was unusually shallow for its age. A shallow region seaward of the

---

[5] See, for example, the acknowledgement to Ross Gunn in the introduction of *Plate Tectonics* by X. Le Pichon, J. Francheteau and J. Bonnin (1973).

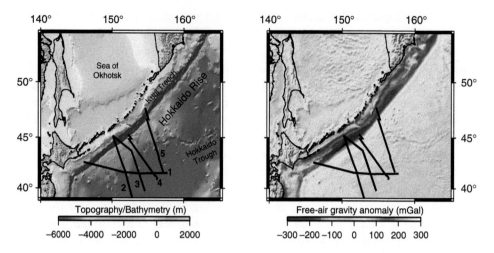

Figure 4.54 Topography/bathymetry and free-air gravity anomaly maps showing the Hokkaido Rise, seaward of the Kuril Trench. The topography/bathymetry is based on GEBCO2020, and the gravity is based on EGM2008. The solid black lines show the R/V *Robert D. Conrad* and R/V *Vema* (Cruises C1405, V2006, C1207) ship tracks along which the data in Fig. 4.55 were acquired.

Kuril Trench had, in fact, been noticed before, and had been dubbed by Japanese marine geologists the 'Hokkaido Rise'. Cooling plate models suggest that the 110-Ma-old oceanic crust approaching the Kuril trench should be at a depth of about 6,000 m. The depth to the crest of the Hokkaido Rise, however, is about 5,300 m, suggesting that it is some 700 m shallower than would be expected for its age (Fig. 4.55). Walcott (1970a) attributed the rise to a flexural bulge that was caused by loading of the oceanic lithosphere at or near the trench axis.

Hanks (1971) noted that large earthquakes (i.e., earthquakes with $M_b > 6.2$) in the region of the Kuril Trench were distinct from those of other arc systems in the western Pacific. In particular, the earthquakes had a high stress drop (~100 MPa) and there was a low incidence of events with tensional focal mechanisms. He showed that these observations were in accord with a flexure model that was subject to vertical and horizontal loading at the trench axis. Horizontal loads were required to explain the absence of tensional mechanisms (e.g., horizontal compression would negate the tension that develops in the upper part of the flexed plate) as well as the high stress drops. Hanks (1971) showed that regional compressive stresses of the order of 400–800 MPa could also explain the steepness of the seaward wall of the trench on some bathymetric profiles.

Hanks realised that some of the ideas in his paper were speculative – especially those concerned with the magnitude of the stresses that exist across convergent plate boundaries. The question was whether his conclusions were biased in any way by his choice of an elastic model. In such a model, stresses could be stored indefinitely. In the final sentence of his paper, he stated that the elastic model should be verified using gravity data.

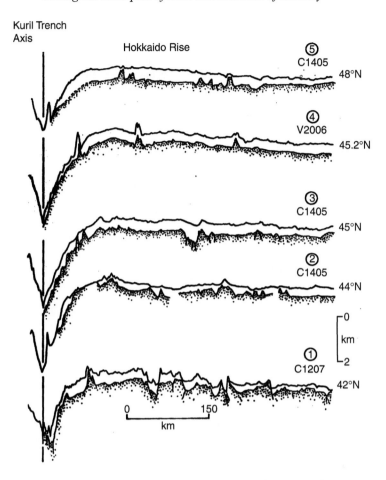

Figure 4.55 Bathymetry and line drawings of seismic reflection profile data over the Hokkaido Rise seaward of the Kuril trench axis. Solid black line shows bathymetry. Stippling shows oceanic basement. Based on the ship track data shown in Fig. 4.54.

Watts and Talwani (1974) used gravity anomaly and bathymetry data to demonstrate that an 'outer rise' seaward of deep-sea trenches was, in fact, a common feature of the Pacific Ocean seafloor. They demonstrated that the correlation between the outer rise and the free-air gravity anomaly was positive and ~70 mGal km^{-1}. Watts and Talwani (1974) pointed out (Fig. 4.56) that such a correlation would not be expected if oceanic crust thickens beneath the rise, as an Airy model would predict. They also demonstrated that if the Moho was flat beneath the rise, then the correlation would exceed the observed one. The observed correlation of about 70 mGal km^{-1} was similar to what would be expected (Figs. 4.57 and 4.58) if oceanic crust was simply warped upward beneath the rise, as predicted by the elastic plate (flexure) model.

Watts and Talwani used a similar elastic plate model as Hanks to model the topography and gravity anomaly seaward of deep-sea trenches. They showed that a compressive stress of a few hundred MPa was required to explain the steep seaward wall of the eastern

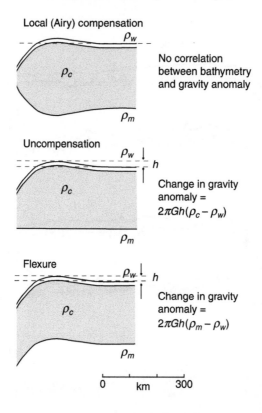

Figure 4.56 The relationship between gravity anomaly and bathymetry for different models of crustal structure at the Hokkaido Rise. Reproduced from fig. 6 of Watts and Talwani (1974).

Aleutian, Kuril, northern Bonin, Ryukyu and Philippine trenches, but that no compressive stress was required for the southern Bonin and Mariana trenches. They attributed these differences in the stress regime to variations in the degree of coupling across convergent plate boundaries, with the Aleutian arc being the most strongly coupled and the Mariana arc being poorly coupled. Such a model was consistent with the distribution of large thrust-fault earthquakes (Kanamori, 1971) and the presence or absence of active extensional marginal basins on the landward side of the island arc.

The work of Hanks, and Watts and Talwani, was followed by much discussion, particularly regarding the magnitude of the stresses required across the plate boundary. Parsons and Molnar (1976), for example, showed that the free-air gravity anomaly high and bathymetric rise seaward of deep-sea trenches could, in many cases, be explained either by a bending moment or by vertical shear forces that were transmitted to the trench axis by the subducting slab. In their view, the bathymetric rise could be explained by the appropriate end-loads, and they did not require a large external horizontal compressive stress in order to produce them.

Caldwell et al. (1976) re-examined the bathymetry data seaward of the Kuril, Aleutian, Bonin and Mariana trenches. They agreed with Watts and Talwani that significant differences

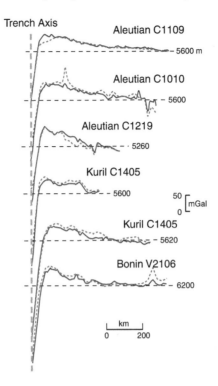

Figure 4.57 Gravity anomaly profiles of the outer rise seaward of the Aleutian, Kuril and Bonin deep-sea trenches in the western Pacific. Solid line is observed free-air gravity anomaly. Dashed line shows the calculated gravity anomaly based on the bathymetry and the flexure model (Fig. 4.56) with $\rho_m = 3,330$ kg m^{-3} and $\rho_w = 1,030$ kg m^{-3}. The observed gravity anomaly and bathymetry data was acquired during cruises V2106 of *R/V Vema* and C1109, C1010, C1219 and C1405 of *R/V Robert D. Conrad*. Reproduced from fig. 2 of Watts and Talwani (1975a).

occurred in the shape of the bulge seaward of trenches. However, rather than interpreting these differences in terms of variations in horizontal compressive stress, they argued that the main influence on the bathymetry was the T_e structure of the oceanic plate approaching the trench. To demonstrate this, they used bathymetry profiles to estimate the height of the bulge above the mean seafloor, w_b, and the distance to the crest of the bulge, x_b, from the first point of zero flexure for several topographic profiles of island arc–deep-sea trench systems. They showed that x_b was least at the Kuril Trench (42 km) and greatest at the Mariana Trench (55 km). They interpreted this as indicating that T_e varied from a low of 20 km at the Kuril Trench to a high of 29 km at the Mariana Trench. Caldwell et al. (1976) showed that when bathymetry profiles from different island arc–deep-sea trench systems are normalised with respect to x_b and w_b, there was little difference between them. They dubbed this profile the 'universal trench profile' (Fig. 4.59). This profile, in their view, describes the flexure of all elastic plates that are acted on only by a vertical end-load. They had therefore used a model in which there were no

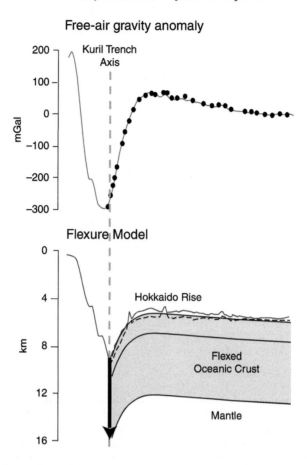

Figure 4.58 Comparison of the observed free-air gravity anomaly, bathymetry and a line drawing of a seismic profile of the Hokkaido Rise to calculated profiles based on a broken elastic plate model. The observed data was acquired during cruise V2006 of *R/V Vema*. The calculated gravity (red solid line), bathymetry and crustal structure (blue shaded region) is based on $T_e = 27$ km, the best-fit value estimated by Hanks (1971) using profiles derived from Scripps bathymetry maps.

horizontal compressive stresses, in accord with the results of Parsons and Molnar. If a horizontal load had been included, they noted that there would have been no discernible difference in the universal trench profile, unless the stress exceeded 1,000 MPa.

Other problems were raised concerning the magnitude of the stresses generated in the flexed oceanic plate. The curvature of the bathymetry profiles seaward of trenches was such that the magnitude of the bending stresses that develop in the seaward wall was high and on the order of several hundreds of MPa. Several workers believed that these stresses were unreasonably large and were unlikely to be supported by the strength of the lithosphere. In fact, there was evidence in some of the earliest seismic reflection profile data (e.g., Fig. 4.60) that stresses were being relieved by normal faulting in the high-curvature regions of the seaward wall of the trench.

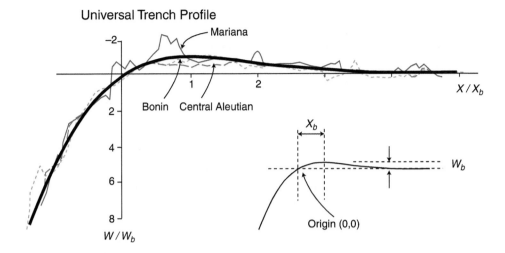

Figure 4.59 The Universal Trench Profile (thick solid line). Thin lines show selected corrected and normalised bathymetry profiles over the Aleutian, Bonin and Mariana trenches. Reproduced from fig. 6 of Caldwell et al. (1976) with permission of Elsevier Science.

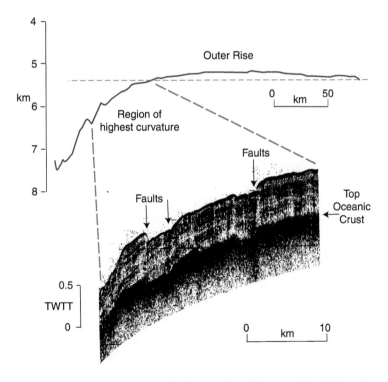

Figure 4.60 Bathymetry and seismic reflection profile of the Japan Trench. Based on data acquired during cruise V3213 of *R/V Vema*. Vertical arrows highlight faults which offset both the seafloor and the underlying top of oceanic crust. Reproduced from fig. 3 of Watts et al. (1980).

To address the bending stress problem, De Bremaecker (1977) used a viscous rather than an elastic plate model. In this model, a viscous plate that overlies an inviscid substratum converges on a trench at a constant rate. The outer rise is maintained more by viscous flow than elastic plate flexure, and so the stresses that develop in the deformed plate are smaller. The problem with De Bremaecker's model was his assumption of a 120-km-thick, 10^{23} Pa s viscosity lithosphere. This viscosity implies a relaxation time of 10^5 yr, which explains the support of young loads such as the Hawaiian Ridge, but it is unable to account for the support of old loads such as the Emperor Seamounts. More importantly, there is no evidence that the amplitude of the outer bathymetric rise seaward of deep-sea trenches depends on plate convergence rates, as would be expected for a viscous model.

One advantage of a viscous model is that the bending stresses are much smaller than would be predicted by an elastic model. Melosh (1978), however, pointed out that due to a mathematical error the stresses computed by De Bremaecker were a factor of nine too small. Melosh (1978) used a three-layered model in which the lithosphere comprised of a 30-km-thick, high-viscosity ($> 10^{25}$ Pa s) upper layer and a 60-km-thick, low-viscosity (10^{22} Pa s) lower layer. Underlying the lithosphere was a low-viscosity ($\sim 10^{20}$ Pa s) asthenosphere. This viscosity structure implies that the lower layer of the lithosphere would behave as a viscous fluid on timescales greater than a few million years. Since it takes about 2 Myr for oceanic lithosphere to move through a bathymetric rise seaward of deep-sea trenches, Melosh (1978) argued that during convergence, flow in the lower layer generates pressure anomalies that are transmitted to the elastic layer, so the lithosphere – as a whole – acts as a thick mechanical unit and the stresses are low.

McAdoo et al. (1978) pointed out that previous flexure models at deep-sea trenches were limited because they were based on rheologically homogeneous elastic or viscous layers. They argued that a better model to use in the high-curvature faulted regions of trenches would be one in which the lithosphere behaves as an elastic–plastic material. Such a model supports loads elastically until a certain yield stress criterion is met. Therefore, regions of the plate where the bending stresses exceed the yield stress would deform plastically. McAdoo et al. (1978) showed that an elastic–plastic model explains the seaward wall of the trench and bathymetric rise, yet it does not produce such large bending stresses as elastic models. They also showed that the bathymetry of the Kuril Trench–Hokkaido Rise system could be explained by a combination of T_e in the range of 25–47 km, axial compressive stress of 25–50 MPa and yield stresses of 50–72 MPa. Bodine and Watts (1979) showed that profiles of the Bonin and Mariana trenches could be explained by a T_e that depended on the age of the approaching plate, equally modest changes in the axial compressive stress and a constant yield stress of 400 MPa.

Rather than use a constant yield stress model, McNutt and Menard (1982) used a model in which the yield stress is constrained by data from experimental rock mechanics. These workers showed that because of yielding, it is important to take curvature into account when estimating T_e. This was especially the case, they found, at deep-sea trench systems where curvatures of up to $1 - 8 \times 10^{-7}$ m^{-1} are observed. They showed that in the high-curvature regions seaward of some trenches (e.g., the northern Bonin Trench), yielding would cause the deformation of a $T_e = 40$ km plate to resemble more closely that of a $T_e = 29$ km plate.

In contrast, the T_e that is associated with a low-curvature plate (e.g., southern Bonin Trench) would be the same.

McNutt (1984) examined all the T_e estimates from deep-sea trench–outer rise systems and showed that much of the scatter in these data can be explained by inelastic yielding in the plate. By correcting for yielding using the measured curvature of the flexed profile, she was able to correct the values of T_e that had been previously determined at trenches using elastic plate models.

By the late 1980s, bathymetric measurements acquired with a shipboard echo sounder data were being supplemented by multibeam 'swath' bathymetry and backscatter data. These data have significantly improved the resolution of outer-rise bathymetry. Among the first such studies was one by Kobayashi et al. (1998) who used SEA BEAM (Farr, 1980) swath data to show that the seaward wall of the Kuril trench was characterised by arcuate fault-bounded horst and graben structures. Later, Hirano et al. (2006) found evidence from submersible dive samples for highly vesicular volcanic lavas and sheet flows and pillow lavas within the seaward wall of the trench in the extensional part and, interestingly, on the 'back bulge' in the compressional part of the flexed plate.

One of the best swath surveyed regions that shows arcuate scarps a majority of which are normal faults is at the Tonga-Kermadec trench (Fig. 4.61). Unlike at the Kuril trench, the tectonic fabric of the subducting Pacific plate[6] at the Tonga-Kermadec trench is almost orthogonal to the trench axis, so faults mapped there are not due to reactivation but can be directly attributed to flexure. Individual faults generally show a flat peak with little or no offset at the fault tips and a maximum offset in the centre. Figure 4.62 shows fault lengths that are in the range of ~15–51 km (mean ~33 km) and offsets that are in the range of ~194–1,041 m (mean ~600 m) which are significantly greater than have been derived from extensional fault data at the present-day fast-spreading East Pacific Rise (e.g., Bohnenstiehl and Kleinrock, 2000).

Offshore Nicaragua and Central Chile, Ranero et al. (2003), Grevemeyer et al. (2005) and Grevemeyer et al. (2018) used swath bathymetry, together with seismic reflection and refraction and heat flow data, to argue that the bend faults imaged in the seaward wall of these trenches extend in depth, offset oceanic crust and Moho and facilitate the hydration and serpentinisation of the sub-crustal mantle. However, data from other outer-rise systems such as offshore Alaska (Bécal et al., 2015; Shillington et al. 2015) show faulting is limited to the uppermost part of the oceanic crust and that Moho is not offset and is a relatively continuous feature (Fig. 4.63).

Subsequent flexure studies have confirmed the existence of a weak zone in the seaward wall of a trench but have differed on whether the subducting oceanic crust has a T_e that is dependent on plate age or not. Contreras-Reyes et al. (2010b) used swath data at the Chile Trench to show a reduction in T_e of 10–50 per cent in the seaward wall, and Zhang et al. (2014) used swath data at the Mariana Trench and found a statistically significant similar reduction of T_e of 21–61 per cent in its seaward wall. Both sets of authors interpret the weak zone as a region of pervasive faulting and high fluid pore pressure in the flexed subducting lithosphere. Contreras-Reyes et al. (2010b) estimated $15 < T_e < 41$ km for the 5–50 Ma

---

[6] As revealed, for example, by the magnetic anomaly lineations.

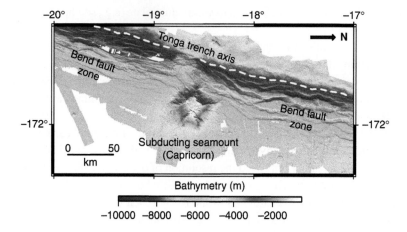

Figure 4.61 Swath bathymetry grid of the Tonga trench in the vicinity of Capricorn seamount showing the arcuate faults associated with flexure. The 100 × 100 m grid was constructed from all available data, including several recent cruises of R/V *SONNE*. Note that some faults appear to continue into the Capricorn seamount, where they are assisting in the deformation of the seamount as Pacific plate motion carries it into the trench.

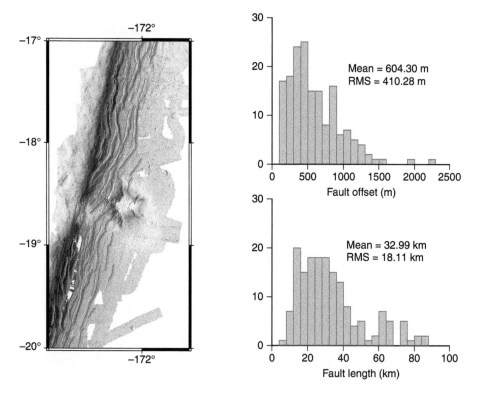

Figure 4.62 The distribution of faults in a ~550-km-long segment of the Tonga trench between latitude 16° S and 21° S. The faults are recognised as offsets in the seafloor. The offset is calculated from the difference in depth between the top (red line) and bottom (green line) of the fault. The mean length and offset of the faults are ~33 km and ~604 m, respectively.

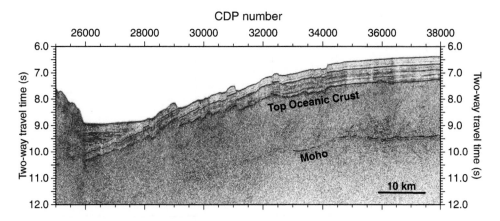

Figure 4.63 Seismic reflection profile showing offsets in the seafloor seaward of the trench offshore Alaska, which extend to depth and offset the top of oceanic crust. The offsets do not appear to extend to the Moho. Reproduced from fig. 4 of Bécal et al. (2015) and Supplementary of Shillington et al. (2015).

oceanic crust seaward of the weak zone with the lowest values associated with relatively young (6 Ma) seafloor near the Chile Rise spreading centre and the highest values associated with the relatively old (48 Ma) seafloor between the Nazca and Perdida Ridges. Nevertheless, these authors conclude that the dependence of T_e on crustal age was weak and they supported the earlier conclusion of Bry and White (2007) that there is little or no correlation between T_e and age of a subducting oceanic plate. We note, however, that the Bry and White (2007) best-fit T_e estimate of 11.6 km for the 40–50 Ma seafloor between the Nazca and Perdida Ridges is significantly less than the ~41 km estimated by Contreras-Reyes et al. (2010b) for the same region and that the Bry and White (2007) best-fit T_e estimate of 16.4 km for the 140–160 Ma seafloor seaward of the Mariana trench is significantly less than the Zhang et al. (2014) estimate of $45 < T_e < 52$ km.

In 2017, Hunter and Watts re-assessed the relationship between T_e and the age of the oceanic crust and lithosphere along circum-Pacific trenches, finding a dependence between T_e and age which they argued was consistent with the results from experimental rock mechanics. They found that oceanic crust and lithosphere was stronger than expected seaward of a point ~100 km seaward of the trench axis and weaker than expected landward of that point (Fig. 4.64). In the strong zone, they described T_e as given by the depth to the ~671°–714°C oceanic isotherm based on the cooling plate model of Parsons and Sclater (1977), while in the weak zone the controlling isotherm was significantly smaller and in the range ~342°–349°C. A dependence of T_e on age at trenches has been recently confirmed by Zhang et al. (2018) who concluded that plate age was the dominant factor controlling the flexure profile at trenches, irrespective of a weak zone within, and boundary forces applied to a subducting oceanic plate.

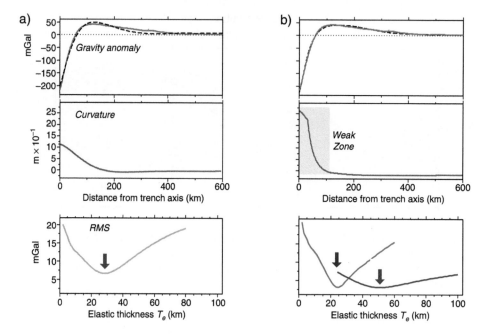

Figure 4.64 Comparison of observed (red line) and calculated (black dashed line) gravity anomaly profile of the Kuril trench – outer-rise system. The curvature (purple line), the RMS between observed and calculated gravity and the best-fit T_e are shown for the case of (a) inversion for a uniform T_e and (b) inversion for a spatially varying T_e. Note how a spatially varying T_e improves the fit to both the amplitude and wavelength of the observed gravity. Modified from figs. 6 and 10 in Hunter and Watts (2016).

The flexure profiles used to calculate the gravity anomaly in Fig. 4.64 assume two- rather than three-dimensionality. Manriquez et al. (2014), however, has argued that T_e may be overestimated if three-dimensionality is not considered. They used a finite element model to examine the outer rise in the Arica bend region, between the Nazca and Perdida Ridges, where the strike of the Peru-Chile trench changes from ~185° offshore Chile to ~315° offshore Peru. In the bend region, Manriquez et al. (2014) found average values of T_e in the weak zone of 16–23 km and of 25 km in the strong zone. Hunter and Watts (2016), who assumed two-dimensionality, examined only one segment in the bend region, finding a T_e of 17 km for both weak and strong zones. However, offshore Peru and Chile, to the north and south of the bend, they found 12.5–15 km in the weak zone and 26–32.5 km in the strong zone; results that are compatible with those of Manriquez et al. (2014), suggesting only a small effect from three-dimensionality.

So far, we have mainly been concerned with the interpretation of the outer bathymetric rise in terms of the flexure of thin elastic, viscous, and elastic–plastic plates. Several workers have suggested, however, that deep-sea trench outer-rise systems are, at least in part, dynamically supported features associated with plate convergence.

A convincing case for a dynamic role was put forward by McAdoo (1981, 1982), who examined geoid rather than gravity anomalies. Geoid data, which are more sensitive to deep structure than are gravity data, show that many convergent plate boundaries are characterised by a long-wavelength geoid anomaly 'high'. The high can be explained if the mass excess associated with the subduction of a cold, dense, downgoing slab beneath island arcs is in some way compensated at depth. Davies (1981), Hager (1984) and Zhong and Gurnis (1994), for example, constructed dynamical models for mantle flow in the region of subducting slabs that are consistent with the geoid observations. All models predict a geoid high that extends beyond the island arc–deep-sea trench system into the flanking regions of the rise. In addition, some of the models predict both a trench (e.g., Zhong and Gurnis, 1994) and a bathymetric rise seaward of trenches. Therefore, it is possible that part of the gravity and topography anomalies seaward of trenches that have been attributed here to oceanic plate flexure might, in fact, be due to pressure anomalies generated by some form of viscous flow in the underlying mantle.

The possibility of a viscous flow has been considered by Billen and Gurnis (2005), who pointed out that the existence of a weak zone with loss of strength seaward of the trench axis was consistent with viscous models, in the sense that it requires coupling between the sinking slab and the subducting plate by viscous rather than elastic stresses. Bai et al. (2018), however, showed that the dynamic contribution of a downgoing slab and subducting plate to the bathymetry and gravity anomaly in the outer-rise region is long wavelength (> 1,600 km) and relatively small in amplitude, resulting in a bathymetric low of up to 200 m and a gravity anomaly high of up to 3 mGal. Nevertheless, these authors found that by considering dynamic effects, T_e estimates at trench–outer-rise systems derived using the elastic model were somewhat reduced.

Regardless of the dynamical contribution, the existence of what appears from surface observations to be a flexural bulge seaward of deep-sea trenches has important geological consequences. Islands, for example, that formed on oceanic crust and are approaching a deep-sea trench will experience uplift as they 'ride' the bulge seaward of the trench (Fig. 4.65).

One of the best examples, first described by Dubois et al. (1975), is Niue Island in the south-west Pacific (Fig. 4.66). Niue is a former atoll which has had its fringing reef and lagoonal floor raised above sea level (Schofield and Nelson, 1978). The uplift caused seawater in the lagoon to evaporate, which, in turn, generated magnesium-rich ground waters that dolomitised the central limestone 'core' of the island. Dubois et al. (1975) suggested the uplift is the consequence of Niue 'riding' a flexural bulge as plate motions carried the island towards the Kermadec Trench.[7] The limestones forming the fringing reefs of the former atoll are now ~70 m above sea level, and Dubois et al. (1975) used bathymetric profiles to estimate that Niue must have travelled ~64 km on the bulge to produce this amount of uplift. Radiometric age dating of soils suggests an age for the emergence of the

---

[7] J. C. Schofield (personal communication) has questioned the flexural bulge interpretation. He points out that a number of islands in the south-central Pacific have raised coral terraces, which he attributes to a fall in sea level from > 70 m above the present (Schofield, 1967). However, the examples that he cites – Mangaia, Atiu and Mauke – are themselves located on a flexural bulge due to recent volcanic loading. Therefore, flexure rather than changes in sea level is the most likely explanation.

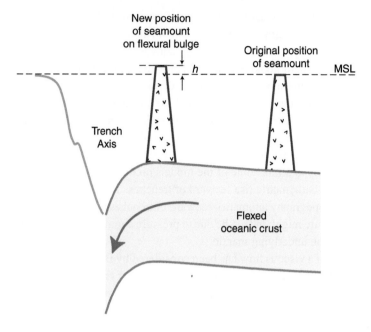

Figure 4.65 Simple model for the uplift of a seamount as it approaches a deep-sea trench–outer-rise system.

central lagoon of ~ 700 ka, and Dubois et al. (1975) used these age estimates to calculate an average convergence rate between the Pacific and Indo-Australian plates of ~91 mm a^{-1}.

Other examples of islands that appear to be riding flexural bulges seaward of deep-sea trenches are Christmas Island (Woodroffe, 1988) and the Cocos (Keeling) Islands (Woodroffe et al., 1990) in the eastern Indian Ocean. These islands, which are located near the crest of the flexural bulge seaward of the Java–Sumatra deep-sea trench, have undergone at least 0.5 m of emergence since about 3 ka. Christmas and Cocos (Keeling) islands are volcanic in origin and comprised of Eocene to Miocene shield basalts (Hoernle et al., 2011).

While these islands probably did not form on an outer rise, some seamounts may well have formed in such a setting. Hirano et al. (2006) propose, for example, that magmas can migrate upwards around the compressional lower part of the flexed plate and into the extensional upper part in the high-curvature fractured region seaward of the trench where they form deep-water basaltic sheet flows and pillow lavas. In addition, they propose magmas can migrate upwards into the extensional lower part of the flexed plate seaward of the outer rise (the so-called 'back-bulge' area of DeCelles and Giles, 1996) and then into the compressional upper part along fractures, where they form small seamounts on the seafloor, dubbing them 'petit spots'. Fractures have indeed been imaged in swath bathymetry surveys of subducting seamounts, notably at Osbourne seamount in the Louisville Ridge (Robinson et al., 2018) and Daiichi Kashima seamount (Lallemand et al., 1989) in the Japan seamounts.

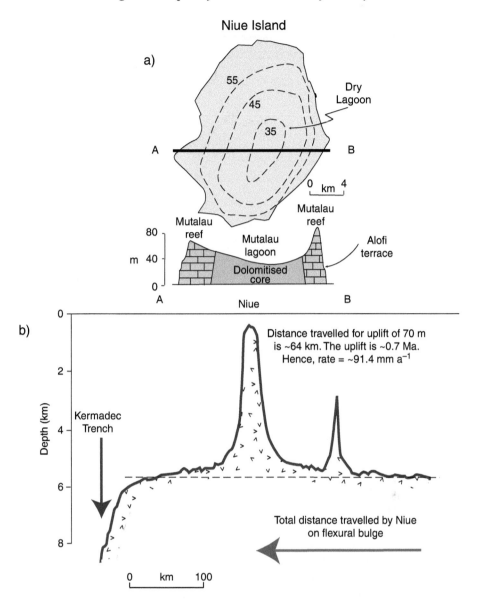

Figure 4.66 Geological cross section and topography of Niue Island in the south-west Pacific Ocean. (a) Topography (contour interval = 10 m) showing the saucer-shaped depression of the uplifted lagoon. Reproduced from fig. 1 of Schofield and Nelson (1978). The 17–18-m-high Alofi terrace is younger than the last inter-glacial (Dickinson, 2001), suggesting a vertical uplift of about 0.14 mm/yr. (b) Bathymetry profile acquired during cruise C1713 of *R/V Robert D. Conrad* showing the location of Niue Island on the crest of the outer bathymetric rise seaward of the Kermadec Trench. An unnamed > 2-km-high seamount located ~37 km south-west of Niue Island is starting to ride the bulge. Both Niue Island and the unnamed seamount will eventually be subducted at the Kermadec Trench.

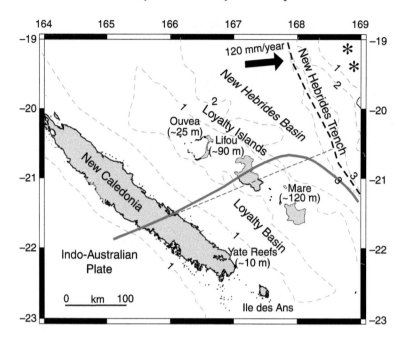

Figure 4.67 The flexure of the Loyalty Islands and New Caledonia as the Indo-Australian plate approaches the New Hebrides Trench. Uplift in brackets in m. Arrow illustrates the motion of the Indo-Australian plate relative to the western North Fiji Basin. Reproduced from fig. 2 of Dubois et al. (1975) with permission of Elsevier Science.

By analogy with these oceanic cases, we would expect flexure to also influence continental crust if it was brought closer to a trench by subduction. Dubois et al. (1974, 1975), for example, suggest that uplift in the Loyalty Islands in the south-west Pacific results from a flexural bulge caused by subduction of oceanic crust in the New Hebrides Basin at the New Hebrides Trench (Fig. 4.67). These karstified limestone islands, which are located 100 km to the north-east of the continental land mass of New Caledonia, comprise uplifted atolls. The island of Mare, nearest to the trench axis, has been raised 120 m, while Lifou has been raised ~90 m and Ouvea ~25 m. There is evidence from Yate reefs that the southern tip of New Caledonia, which are underlain by late Palaeozoic basement rocks, is itself beginning to ride a bulge. And satellite, aerial photographs and field data indicate that the mean azimuth of fractures on Lifou is ~290±35° (Bogdanov et al., 2007), consistent with the Loyalty Islands 'riding' a bulge.

A similar tectonic setting occurs along the Banda arc in the north-east Indian Ocean. Here, the subduction of oceanic and continental crust has brought the north-west continental shelf of Australia into close proximity with the Banda volcanic arc, creating the deep 'underfilled' basin of the Timor Trough. In addition, subduction has caused the continental crust that underlies the north-west shelf to bulge upwards, forming an outer rise (Price and Audley-Charles, 1983; Lorenzo et al., 1998; Londono and Lorenzo, 2004). The flexed crust has been involved in one or more previous extensional rifting events, and Lorenzo et al.

(1998) suggest, on the basis of seismic reflection profile data, that the bulge is associated with a number of shallow normal faults that in this case were created by the reactivation of deeper Jurassic rift-bounding faults.

The existence of flexure seaward of deep-sea trenches has implications for the rock record. Robertson (1998), for example, attributed a hiatus during the Middle Eocene and Oligocene on the Eratosthenes Seamount in the eastern Mediterranean to flexure of Neotethyan oceanic crust that was approaching a north-dipping subduction zone beneath Cyprus during the Early Miocene. The bulge involved a carbonate bank that had been developing on the seamount since the Early Cretaceous.

In an even earlier example, Jacobi (1981) and Mussman and Read (1986) attributed the Early/Middle Ordovician erosional unconformity observed along much of the western margin of the Appalachians to flexure seaward of a trench. According to their model, the flexure involved proto-Atlantic (Iapetus) oceanic crust that was approaching an east-dipping subduction zone during the Taconic orogeny. The bulge first involved oceanic crust and then, as subduction proceeded, an Ordovician carbonate bank that was underlain by continental crust. The emergence of what was once a shallow-water carbonate sequence by the flexural bulge resulted in erosion, karstification (St. George formation) and dolimitisation (Knox-Beekmantown formation). Following uplift, the continental crust was rapidly subducted. The unconformity is therefore a 'precursor' to the culmination of the Taconic orogeny, which peaked, according to Jacobi (1981), once continental crust had 'jammed' the subduction zone. By Upper Devonian, continued east-dipping subduction brought a new terrane (Avalonia) into contact with the newly formed Taconic terrane, and Lash and Engelder (2007) describe a set of north–south joints in a shale of the Lake Erie district which they interpret in terms of extension on a flexural bulge that developed during the resulting Acadian orogeny (Northfieldian).

4.6 Summary

It has been shown in this chapter that lithospheric flexure is a phenomenon that has played an important role in the evolution of the Earth's surface features. Flexure, through the deformation it causes, contributes directly to the crustal structure, the subsidence and uplift history, and the gravity anomaly at several different geological features.

The waxing and waning of ice sheets during the Late Pleistocene flexed the continents on relatively short (i.e., 10^4–10^5 yr) timescales and over large areas. One of the most striking examples is the uplift of the land surface that followed the retreat of the last major ice advance. Spectral studies of the rate of uplift of North America, for example, help constrain the viscosity structure of the asthenosphere. Other, more local features provide information on the lithosphere rather than the asthenosphere. The uplift of shorelines of lakes that once filled the glacial depressions in front of retreating ice loads, for example, provided constraints on T_e, the large size of these lakes being explained by the fact that the ice loads had once encroached onto the old, cold, cratonic 'core' of North America.

There is evidence that water released from melting ice sheets also flexed the continental lithosphere. In the continents, significant volumes of meltwater accumulated in narrow topographic depressions within mountainous regions. The oceans, however, are the receptacle of most meltwater, where it can flood onto the shelves at the margins of the continents. As a consequence, the continental lithosphere is flexed downwards in offshore shelf regions and upwards in onshore regions. Along the Atlantic coast of Brazil and the west coast of Australia, flexure has formed a series of beach ridges, up to a few metres high.

One of the best examples that we have of flexure on long ($> 10^6$ yr) timescales is at oceanic islands and seamounts. These concentrated loads are volcanic in origin and have been on the lithosphere for much longer times than have ice sheet loads. At the Hawaiian Ridge, gravity and seismic modelling reveals that volcanic loading has flexed the underlying Pacific oceanic crust downwards by as much as 3–4 km. There is evidence that the flexed crust beneath Oahu, as well as other islands such as Marquesas, may be underplated by magmatic material. Underplating would thicken the crust and cause uplift. However, underplating is not sufficiently significant to prevent large amounts of subsidence at volcanoes. Hawaii, for example, is presently subsiding at rates of up to 4 mm a^{-1} with respect to Oahu. There is evidence that flexure due to a new volcanic island modifies the subsidence and uplift history of pre-existing islands. Hawaii, for example, has caused subsidence to Maui, which is in its flexural moat, and uplift to Oahu, which is in its bulge. The flexural effect of new volcanic loads is best illustrated in the case of seamount clusters. In the south-west Pacific, for example, subsidence and uplift due to the relatively recent loads of Rarotonga and Atuitaki have raised flanking islands by upwards of 60 m.

The accumulation of sediments in river delta systems represents loads of even longer duration than oceanic islands and seamounts. At small scale, flexure explains many of the details of the stratigraphic 'architecture' of deltas, including offlap, downdip thickening and updip thinning. In addition, flexural subsidence due to prograding sediment loads helps to explain why many ancient depositional surfaces have steeper dips than do modern ones. At large scale, flexure accounts for the large thickness of sediments, the gravity field and, in some cases, the existence of persistent regions of uplift onshore. The Amazon River, for example, drains large areas of South America. A large portion of these sediments are deposited in the Amazon Cone, a deep-sea fan system in the north-east Brazilian continental margin. There is evidence that the additional load of sediments in the Amazon Cone has depressed the underlying Brazilian margin by upwards of 1–2 km. The flexural effects extend onshore, and there is evidence that flexure has raised onshore regions by up to a few tens of meters.

Finally, flexure is an important feature of deep-sea trench–outer-rise systems. In particular, the bathymetric rises seaward of deep-sea trenches can be explained by vertical and horizontal loads that act on the approaching oceanic lithosphere. While the source of these loads is not always clear, they combine to flex the lithosphere downwards by a few kilometres at the trench and upwards by several hundreds of meters in the region of the bathymetric rise. Many of the differences in the bathymetry of deep-sea trench–outer-rise systems can be explained by variations in T_e of the approaching plate. Others may be due to variations in the state of stress that exist across the plate boundary. There is evidence that the flexure is large enough at some trenches to locally weaken the lithosphere and cause it to fail through normal faulting. Seamounts and oceanic islands that 'ride' the topographic rise will experience uplift and then subsidence as they are carried by plate motions into the trench. This will also be the case for continents that arrive at a trench because of the subduction of oceanic crust. Uplift is in the process of affecting New Caledonia and other neighbouring islands, and flexure seaward of trenches may explain several erosional unconformities that are observed in the ancient rock record.

5

Isostatic Response Functions

For the large blocks – for example, a whole continent or whole ocean floor – the theory of isostasy must be accepted without question; but where smaller ones are concerned, such as individual mountains, the law loses its validity. Such small blocks can be supported by the elasticity of the whole block.

(Wegener, 1966, p. 43)

5.1 Introduction

One of the aims of isostasy is to determine the temporal and spatial scales over which the Earth's crust and upper mantle adjust to geological loads. We saw from a consideration of flexure at glacial lakes, seamounts and oceanic islands, river deltas, and deep-sea trench–island arc systems that isostasy is operative over timescales that range from a few thousand to a few million years. Since the early part of the twentieth century, however, much of the debate about isostasy has been concerned with the degree to which a particular surface feature is compensated.

It has been known for some time that one of the main factors that determines the degree to which a particular surface feature is compensated is its size. Surface features such as a narrow hill, for example, will not be compensated, whereas a wide mountain belt will be. This is generally true regardless of the actual strength of the lithosphere. However, the strength of the lithosphere determines the amount of bending and, hence, the degree to which the compensation approaches the predictions of local models, such as those of Airy and Pratt.

In the early part of the twentieth century, views differed on the extent of isostatic compensation. Barrell argued that geological features would have to be of the order of a few hundred kilometres wide before they were compensated. Bowie, however, felt that features as small as 100 km across would be compensated. The reason for these differences is that Barrell and Bowie held fundamentally different views about the long-term strength of the crust and upper mantle.

The works of Vening Meinesz, Gunn and Walcott greatly improved our understanding of the physical properties of Earth's crust and upper mantle. However, it was not until the development of spectral techniques, first in the continents (Dorman and Lewis, 1970, 1972; Lewis and Dorman, 1970) and then in the oceans (McKenzie and Bowin, 1976), that it really became possible to quantify the degree of isostatic compensation of the Earth's main surface features. These techniques, which were a significant improvement on previous statistical techniques (e.g., Neidell, 1963), utilise the fact that the relationship between gravity and topography as a function of wavelength varies with the isostatic model. By analysing the frequency content of observed gravity and topography data over a geological feature and comparing the spectra with the predictions of both local and regional models of isostasy, it has been possible to determine the compensation scheme with much greater precision than is possible by forward modelling alone.

5.2 The Lithosphere as a Filter

It was demonstrated in Chapter 4 that the lithosphere responds to long-term geological loads not locally, as the Airy and Pratt models would predict, but regionally by flexure over a broad region. In a sense then, the lithosphere is behaving as a 'filter' that suppresses the large-amplitude, short-wavelength deformation associated with local models of isostasy and passes the small-amplitude, long-wavelength deformation associated with flexure (Fig. 5.1). We use the term filter here in its usual meaning: that is, to describe a system with an input and output. In the case of a loaded lithospheric plate, the input is the load and the output is the flexure.

Let us assume that the lithosphere is a particular type of filter such that if a load $h_1(x)$ produces a deflection $y_1(x)$ and a load $h_2(x)$ produces a flexure $y_2(x)$, then the load $h_1(x) + h_2(x)$ produces a flexure of $y_1(x) + y_2(x)$. A filter of this type is called a 'linear space-invariant' filter, and its behaviour is a well-known problem in engineering. One of the properties of filters of this type is that when they are subject to a periodic load, their output also is periodic.

The general equation that describes the response of an elastic plate that overlies a weak fluid substratum to a periodic load is given by

$$D\frac{\partial^4 y}{\partial x^4} + \left(\rho_m - \rho_{infill}\right)yg = (\rho_c - \rho_w)gh\,\cos(kx), \tag{5.1}$$

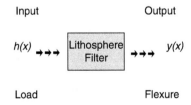

Figure 5.1 Schematic diagram illustrating the behaviour of the lithosphere as a filter.

where D is the flexural rigidity, y is the flexure, x is the horizontal distance, ρ_m and ρ_{infill} are the densities of the mantle and the material that infills the flexural depression, respectively, h is the peak-to-trough amplitude of the load, k is the absolute value of the wavenumber ($k = 2\pi\lambda_w$, λ_w is wavelength), g is average gravity and ρ_c and ρ_w are the densities of the load and the medium the load displaces (i.e. either water or air), respectively.

It is easy to show by substitution that the solution of Eq. (5.1) is also periodic and of the form:

$$y = \frac{(\rho_c - \rho_w)h\,\cos(kx)}{(\rho_m - \rho_{infill})}\left[\frac{Dk^4}{(\rho_m - \rho_{infill})g} + 1\right]^{-1}. \tag{5.2}$$

As $D \to 0$ (i.e., the plate becomes thin or weak), Eq. (5.2) becomes

$$y \to \frac{(\rho_c - \rho_w)h\,\cos(kx)}{(\rho_m - \rho_{infill})}$$

which is the 'Airy response' to a periodic load if $\rho_{infill} \to \rho_c$. Also, we see that if $D \to \infty$ (i.e., the plate becomes infinitely rigid), then

$$y \to 0$$

which is the 'Bouguer response'.

It follows that if a plate is thin or weak, its response to loads is local and approximates that of an Airy model. If, on the other hand, the plate is rigid, there is no response, which is the Bouguer case. When the plate has a finite thickness or strength, it responds by flexure.

Let us now determine the parameter, $\phi_e\,(k)$, which modifies the Airy response to produce the flexure. Consider the input and output to the system. We can write

$$\phi_e(k) = \frac{\text{Output}}{\text{Input}}$$

$$\text{Input} = \frac{(\rho_c - \rho_w)h\,\cos(kx)}{(\rho_m - \rho_{infill})}$$

$$\text{Output} = \frac{(\rho_c - \rho_w)h\,\cos(kx)}{(\rho_m - \rho_{infill})}\left[\frac{Dk^4}{(\rho_m - \rho_{infill})g} + 1\right]^{-1}$$

$$\therefore\ \phi_e(k) = \left[\frac{Dk^4}{(\rho_m - \rho_{infill})g} + 1\right]^{-1}. \tag{5.3}$$

The dependence of $\phi_e\,(k)$ on k for a range of infill densities is illustrated in Fig. 5.2. The figure shows two sets of curves: one for a relatively thin or weak plate ($T_e = 5\,\text{km}$) and the other for a thick or strong one ($T_e = 25\,\text{km}$). We see that for low-wavenumber, long-wavelength loads, the plate behaves as a weak, Airy-type structure. In contrast, for

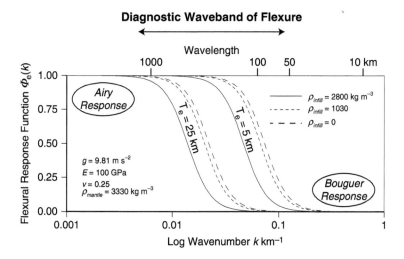

Figure 5.2 Flexural response function, ϕ_e (k) showing the diagnostic waveband of flexure. At long wavelengths, $k \to 0$ and ϕ_e $(k) \to 1$ which is the Airy response, while at short wavelengths, $k \to \infty$ and ϕ_e $(k) \to 0$ which is the Bouguer response. The flexural response function describes the mechanical behaviour of the lithosphere in response to loading.

high-wavenumber, short-wavelength loads, the plate appears rigid. This behaviour can be verified by considering the limiting values of k. As $k \to 0$, for example,

$$y \to \frac{(\rho_c - \rho_w)h}{(\rho_m - \rho_{infill})} \tag{5.4}$$

which is the Airy response if $\rho_{infill} \to \rho_c$. Similarly, if $k \to \infty$, $y \to 0$, which is the Bouguer response. At intermediate wavenumbers (i.e., $0.001 < k < 0.100$), the response is flexural. We call this 'the diagnostic waveband of flexure'.

Figure 5.2 shows that ϕ_e (k) determines the wavenumber, and hence wavelengths, for which flexure is important. We follow Walcott (1976) and will refer to this parameter here as the 'flexural response function'.

As expressed here, Eq. (5.2) only gives the flexural response to a load at a particular wavenumber. If, however, a load can be broken down into its individual spectral components, then the equation may be used to compute the response to any two-dimensional, arbitrary-shaped load.

A widely used technique to decompose a spatial data set into its spectral components is Fourier analysis (e.g., Brigham, 1974). Replacing $h \cos(kx)$ in Eq. (5.2) by $H(k)$, the wavenumber domain representation of the bathymetry, we get

$$Y(k) \to \frac{(\rho_c - \rho_w)}{(\rho_m - \rho_{infill})} H(k)\phi_e(k), \tag{5.5}$$

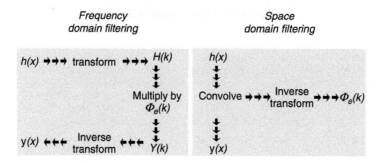

Figure 5.3 Schematic showing principal steps in frequency and space domain filtering of a topographic or bathymetric data set.

where the bold uppercase variable indicates the Fourier transform of a lower-case variable. $Y(k)$ is the wavenumber domain representation of the flexure $y(x)$, and $H(k)$ is the wavenumber domain representation of the topography $h(x)$. The spatial variation in flexure due to two-dimensional loads of arbitrary shape is obtained by first taking the Fourier transform of the load, multiplying it by a wavenumber parameter and a density term, and then inverse transforming the result.

We see from Eq. (5.5) that when

$$H(k) = 1,$$

$$Y(k) \rightarrow \frac{(\rho_c - \rho_w)}{(\rho_m - \rho_{infill})} \phi_e(k) \, .$$

This is the flexural response to a unit load of wavenumber k. The condition that $H(k) = 1$ is that $h(x) = \delta_d(x)$, where $\delta_d(x)$ is called the 'impulse function'. Normally (Brigham, 1974), the impulse function is defined as

$$\delta_d(x) = 0, \; x \neq 0$$

$$\int_{-\infty}^{\infty} \delta_d(x) \, dx = 1.$$

We see from these considerations that Fourier analysis is the equivalent in the space domain to a convolution of an impulse function with discrete 'samples' of a load. As Fig. 5.3 illustrates, either technique can be used to evaluate the flexure due to an arbitrary-shaped load.

5.3 The Gravitational Admittance

As was seen in Chapter 4, it is not always possible to observe the surfaces of flexure in the geological record directly. One parameter that is sensitive to both the load size and its flexural response is the gravity anomaly. The gravity field, which can be measured with relative ease both on land and at sea, has been the principal means with which to estimate the long-term thermal and mechanical properties of the lithosphere.

Although the gravity effect of surface loads and their compensation can be computed using line-integral techniques (e.g., Talwani et al. 1959; Bott, 1960), it is computationally faster to use Fourier methods. Another advantage is that the evaluation of the gravity anomaly can be combined into the same 'algorithm' as the flexure, since calculations of both gravity and flexure use the Fourier transform of the load.

To understand how gravity anomalies can be calculated in the wavenumber domain, we need first to consider the gravity effect of an undulating density interface. Parker (1972) has shown from potential theory that to first order:

$$\Delta \mathbf{g}_P(k) = e^{-kp}\Delta \mathbf{g}_Q(k),$$

where $\Delta \mathbf{g}_P(k)$ is the Fourier transform of the gravity anomaly $\Delta g_P(x)$ at a point P, $\Delta \mathbf{g}_Q(k)$ is the Fourier transform of the gravity anomaly $\Delta g_Q(x)$ at a point Q, and p is the height of P above Q. According to Green's equivalent layer theorem, the gravity anomaly $\Delta \mathbf{g}_Q(k)$ can be replaced by an equivalent surface mass distribution on the same plane, $M_Q(k)$, that is given by

$$M_Q(k) = \frac{\Delta \mathbf{g}_Q(k)}{2\pi G}$$

Since $M_Q(k)$ is a surface mass, it follows that

$$\Delta \mathbf{g}_Q(k) = 2\pi G\rho \mathbf{H}(k),$$

where $\mathbf{H}(k)$ is the Fourier transform of the surface of an undulating interface and ρ is its uniform density contrast with the surroundings. It follows that

$$\Delta \mathbf{g}_P(k) = e^{-kp}2\pi G\rho \mathbf{H}(k). \tag{5.6}$$

We are now ready to compute the gravity anomaly due to different schemes of isostatic compensation. Before doing so, however, it is useful to first define the 'gravitational admittance', $\mathbf{Z}(k)$, which is the wavenumber parameter that modifies the topography to produce the gravity anomaly. This parameter, as we will see later, contains information on the state of isostasy at a surface feature and is given by

$$\mathbf{Z}(k) = \frac{\text{output}}{\text{input}} = \frac{\Delta \mathbf{g}(k)}{\mathbf{H}(k)}. \tag{5.7}$$

5.3.1 Uncompensated Topography

We begin by considering the gravity effect of the seafloor bathymetry that is isostatically uncompensated (e.g., Fig. 5.4). The gravity anomaly at sea level is given by

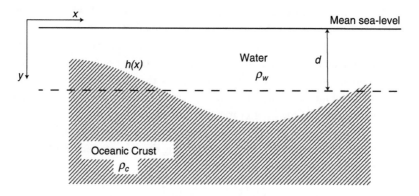

Figure 5.4 The case of bathymetry that is isostatically uncompensated.

$$\Delta g(k)_{topo} = 2\pi G(\rho_c - \rho_w)H(k)e^{-kd},$$

where ρ_c is the density of the crust, ρ_w is the density of seawater, $H(k)$ is the Fourier transform of the bathymetry and d is the mean water depth.

$$\therefore \ Z(k)_{uncomp} = 2\pi G(\rho_c - \rho_w)e^{-kd} \qquad (5.8)$$

The form of Eq. (5.8) is illustrated in Fig. 5.5 for different values of ρ_c and d. We see that $Z(k)_{uncomp}$ decreases with decreasing wavelength.[1] The decrease is most apparent for greater water depths and larger density contrasts between the ocean crust and seawater. In the limits that $k \to 0$, $Z(k)_{uncomp} \to 2\pi G(\rho_c - \rho_w)$ and $k \to \infty$, $Z(k)_{uncomp} \to 0$.

5.3.2 Airy Model

In the Airy model, bathymetry is compensated by lateral changes in the thickness of a uniform-density oceanic crust (Fig. 5.6). Beneath a bathymetric 'high', the compensation takes the form of a crustal 'root' which projects downwards into the mantle. In contrast, beneath a bathymetric 'low', the compensation involves an 'anti-root', a thinning of the crust and a mantle upwarp. The gravity anomaly of an Airy model is therefore the sum of two effects: one from the bathymetry and the other from the compensation.

The gravity effect of the bathymetry is given by

$$\Delta g(k)_{topo} = 2\pi G(\rho_c - \rho_w)H(k)e^{-kd}. \qquad (5.9)$$

The gravity anomaly of the compensation (i.e., the root or anti-root in the Airy model) is given by

---

[1] A MATLAB file, chap5_faa_admit_ocean.m, to compute $Z(k)_{uncomp}$, as well as other selected free-air admittances, are available from the Cambridge University Press server.

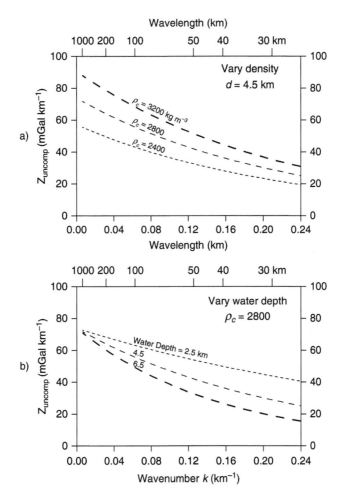

Figure 5.5 Gravitational admittance for uncompensated bathymetry assuming (a) a fixed mean water depth and variable density and (b) a fixed density and variable water depth.

$$\Delta g(k)_{comp} = 2\pi G(\rho_m - \rho_c)R(k)e^{-k(d+t)}, \qquad (5.10)$$

where ρ_m is the density of the mantle, ρ_c is the density of the crust, t is the mean thickness of the crust, $r(x)$ is the compensating root and $R(k)$ is its Fourier transform. In the case of the Airy model, it follows from the discussion in Section 1.11 that the topography and compensation are related by

$$R(k) = -H(k)\frac{(\rho_c - \rho_w)}{(\rho_m - \rho_c)}. \qquad (5.11)$$

The minus sign in Eq. (5.11) indicates that a positive load on the surface is associated with a negative compensating root at depth. It also ensures that the gravity anomalies associated

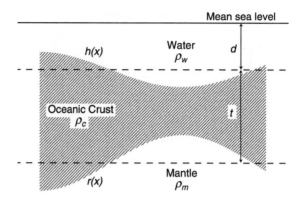

Figure 5.6 The Airy model of isostatic compensation.

with the mass excess of the topography and the mass deficiency of the compensating root are of the correct sign.

The total gravity anomaly is obtained by combining the gravity effect of the topography with that of the root; that is,

$$\Delta g(k)_{total} = \Delta g(k)_{topo} + \Delta g(k)_{comp}.$$

Combining Eqs. (5.9) and (5.11) gives

$$\Delta g(k)_{total} = 2\pi G(\rho_c - \rho_w)H(k)e^{-kd}\left(1 - e^{-kt}\right)$$

$$\therefore \; Z(k)_{airy} = 2\pi G(\rho_c - \rho_w)e^{-kd}\left(1 - e^{-kt}\right). \tag{5.12}$$

In contrast to the uncompensated case considered earlier, Fig. 5.7 shows that $Z(k)_{airy} \to 0$ for both short and long wavelengths. The decrease at long wavelengths is due to isostasy, whereas the decrease at short wavelengths is due to the increased attenuation of the gravity anomalies by the water depth.

We note that Eq. (5.12) does not include a term that depends on the 'normal' thickness of zero-elevation crust. This enables consideration of the general case in which the oceanic crust prior to loading may have been unusually thin or thick compared with the predictions of an Airy model of isostasy. If, on the other hand, we wish to maintain Airy isostasy, then we need to ensure that columns of the undeformed ocean basins and their margins are in isostatic balance. This will be the case when t is given by

$$t = T_c - d\left[\frac{(\rho_m - \rho_w)}{(\rho_m - \rho_c)}\right],$$

where T_c is the 'normal' thickness of the crust in regions of zero elevation.

In many geological applications, it is useful to consider a modification of the Airy model that allows for the possibility that the upper and the lower crust may have different densities.

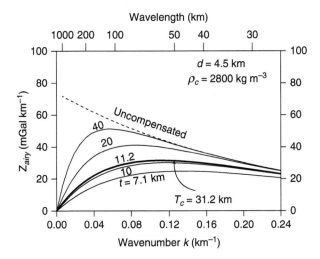

Figure 5.7 Gravitational admittance for the Airy model of isostasy. Heavy solid line shows the admittance for $t = 11.2$ km, which is the oceanic crustal thickness that is in isostatic balance with a zero-elevation crustal thickness of 31.2 km.

Such a model (Fig. 5.8) implies a 'hidden layer' (e.g., Karner and Watts, 1982) that separates the two crustal layers. By itself, the uniform thickness hidden layer has no effect on the gravity anomaly. This is because gravity anomalies are caused by lateral rather than vertical changes in density. However, the density contrast values at the upper crust/water and lower crust/mantle interfaces in the hidden-layer model do change and this, in turn, will alter the gravity anomaly.

It follows from Eqs. (5.9) to (5.11) that the admittance, $Z(k)_{airy\ hidden}$, for the hidden-layer model is given by

$$Z(k)_{airy\ hidden} = 2\pi G(\rho_2 - \rho_w)e^{-kd}\left(1 - e^{-kt}\right), \tag{5.13}$$

where

$$t = T_c - d\left[\frac{(\rho_2 - \rho_w)}{(\rho_m - \rho_3)}\right]$$

and ρ_2 and ρ_3 are the densities of the upper crust and the lower crust, respectively.

5.3.3 Pratt Model

In the Pratt model, seafloor topography is compensated for by lateral density changes in the crust and mantle. The gravity anomaly is now the sum of two effects: the bathymetry and the lateral density changes associated with the compensation.

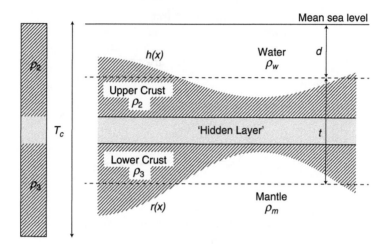

Figure 5.8 The 'hidden layer' Airy model of isostatic compensation.

Consider the model in Fig. 5.9, which shows a column of crust and mantle in a region of flat-lying seafloor compared with one in a region of undulating bathymetry. Equating pressures at the base of the two columns gives

$$D'_c \rho_o + d\rho_w = (d - h(x))\rho_w + \left(D'_c + h(x)\right)\rho(x),$$

where ρ_0 is the sub-seafloor density of the flat-lying column, $\rho(x)$ is the sub-seafloor density of a column of the undulating topography, D'_c is the depth of compensation, d is the mean water depth and $h(x)$ is the topography. The density of the column of the undulating topography is given by

$$\rho(x) = \frac{D'_c \rho_o + h(x)\rho_w}{D'_c + h(x)}$$

which can be re-arranged as

$$\rho(x) - \rho_w = \frac{D'_c(\rho_o - \rho_w)}{D'_c + h(x)}, \quad \rho(x) - \rho_o = \frac{h(x)(\rho_w - \rho_o)}{D'_c + h(x)}. \tag{5.14}$$

Like Airy model, the Pratt admittance is obtained by combining the gravity effect of the topography and its compensation. The gravity effect of the topography is

$$\Delta g(k)_{topo} = 2\pi G e^{-kd} \mathcal{J}\left\{(\rho(x) - \rho_w)h(x)\right\},$$

where $\mathcal{J}$ indicates the Fourier transform. Substituting for $\rho_{(x)} - \rho_w$ from Eq. (5.14) gives

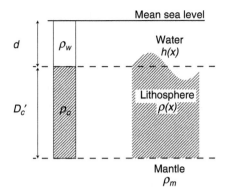

Figure 5.9 The Pratt model of isostatic compensation.

$$\therefore\ \Delta g(k)_{topo} = 2\pi Ge^{-kd}(\rho_o - \rho_w)\mathcal{J}\left\{\frac{D'_c h(x)}{D'_c + h(x)}\right\}.$$

The gravity effect of the compensation is

$$\Delta g(k)_{comp} = 2\pi Ge^{-k(d+D'_c)}\mathcal{J}\left\{(\rho(x) - \rho_o)D'_c\right\}.$$

Substituting for $\rho(x) - \rho_o$ from Eq. (5.14) gives

$$\Delta g(k)_{ccomp} = -2\pi Ge^{-k}(d + D'_c)(\rho_o - \rho_w)\mathcal{J}\left\{\frac{D'_c h(x)}{D'_c + h(x)}\right\},$$

$$\Delta g(k)_{total} = \Delta g(k)_{topo} + \Delta g(k)_{comp},$$

$$\therefore\ \Delta g(k)_{total} = -2\pi Ge^{-kd}\left(1 - e^{-kD'_c}\right)(\rho_o - \rho_w)\mathcal{J}\left\{\frac{h(x)}{1 + h(x)/D'_c}\right\}.$$

We note that $h(x) << D'_c$, and so $h(x)/D'_c \to 0$. Therefore, it follows that

$$\Delta g(k)_{total} = 2\pi Ge^{-kd}\left(1 - e^{-kD'_c}\right)(\rho_o - \rho_w)H(k)$$

$$\therefore\ Z(k)_{pratt} = 2\pi G(\rho_o - \rho_w)e^{-kd}\left(1 - e^{-kD'_c}\right). \qquad (5.15)$$

which is the same as the Airy model if $\rho_0 \to \rho_c$ and $D'_c \to t$ (Eq. 5.12).

In geological applications (e.g., to a mid-oceanic ridge), it is useful to modify the classical Pratt model to include a crust of constant thickness (Fig. 5.10). The addition of a crust means that the gravity anomaly at the crust/water interface changes and that there is an additional contribution to the gravity anomaly from the base of the crust. It can be shown that

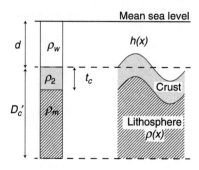

Figure 5.10 The Pratt model of isostatic compensation including a crust of uniform density and thickness.

$$Z(k)_{pratt\ crust} = 2\pi Ge^{-kd}\left[(\rho_2 - \rho_w) + (\rho_m - \rho_2)e^{-kt_c} - (\rho_m - \rho_w)e^{-k\left(D'_c - t_c\right)}\right]. \quad (5.16)$$

We see that with $t_c \to 0$, $\rho_2 \to \rho_o$ and $\rho_m \to \rho_o$, Eq. (5.16) approaches the classical Pratt model.

5.3.4 Elastic Plate (Flexure) Model

In its simplest form, the flexure model of isostasy resembles the Airy model. Both models predict a crustal thickening in the vicinity of a seafloor bathymetric 'high' and a crustal thinning in the vicinity of a seafloor bathymetric 'low'. The main difference between them is the form of the thickening and thinning: in the Airy model the compensation is local, but in the flexure model the compensation is spread out over a broad region (Fig. 5.11).

The elastic plate model involves a 'load' and a 'flexural response' to that load. Referring to Fig. 5.6, the load is then the bathymetry, $h(x)$, with the mean water depth subtracted from it, and the flexural response is now the root, $r(x)$, with the mean water depth and crustal thickness subtracted from it. It follows then from Eqs. (5.9) and (5.10) that the gravity effect of the load and the flexure is, respectively,

$$\Delta g(k)_{load} = 2\pi G(\rho_c - \rho_w)H(k)e^{-kd},$$

$$\Delta g(k)_{flexure} = 2\pi G(\rho_m - \rho_c)Y(k)e^{-k(d+t)},$$

where $H(k)$ is the Fourier transform of the load and $Y(k)$ is the Fourier transform of the flexure due to that load. We saw earlier in the derivation of Eq. (5.3) that the flexure is determined by a certain wavenumber parameter, $\phi'_e(k)$, which modifies the Airy response to produce the flexure. We can then write

$$Y(k) = -H(k)\frac{(\rho_c - \rho_w)}{(\rho_m - \rho_c)}\phi'_e(k), \quad (5.17)$$

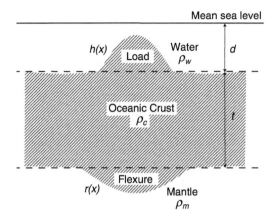

Figure 5.11 The elastic plate or flexure model of isostatic compensation. The model assumes that bathymetry, $h(x)$, acts as a load on the surface of a uniform thickness plate which is compensated for by the downward flexure of the base of the crust into a denser mantle.

where

$$\phi'_e(k) = \left[\frac{Dk^4}{(\rho_m - \rho_c)g} + 1 \right]^{-1}.$$ (5.18)

The minus sign in Eq. (5.17) indicates that the flexure is negative for a positive load.[2]

$$\Delta g(k)_{total} = 2\pi G(\rho_c - \rho_w)_o H(k)e^{-kd}\left(1 - \phi'_e(k)e^{-kt}\right)$$

$$\therefore Z(k)_{flexl} = 2\pi G(\rho_c - \rho_w)e^{-kd}\left(1 - \phi'_e(k)e^{-kt}\right)$$ (5.19)

The flexure model discussed so far assumes the following:

- The crust is the same density as the load.
- The material that infills the flexure is the same density as the crust.
- The crust is of uniform density.

In geological applications, these assumptions often break down. For example, the average density of a volcanic load is not necessarily the same as that of the underlying oceanic crust. Neither is the density of the infill material in the moat that flanks a volcanic load necessarily the same as the load. Finally, the oceanic crust is generally layered according to its *P*-wave velocity structure and may also be vertically stratified in terms of its density.

Let us consider a more general case (e.g., Fig. 5.12) of a load that is emplaced on the surface of a layered crust. As before, we have for the gravity effect of the load

$$\Delta g(k)_{load} = 2\pi G(\rho_l - \rho_w)H(k)e^{-kd},$$ (5.20)

---

[2] The load will be positive when the bathymetry is shallower than the mean water depth and negative when the bathymetry is deeper.

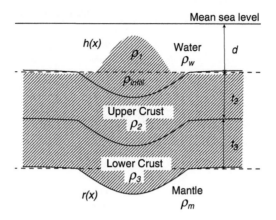

Figure 5.12 The elastic plate or flexure model of isostasy. The figure illustrates the case of a positive (i.e., downward) load acting on the surface of a layered oceanic crust.

where ρ_l is the density of the load. We can write the gravity anomaly contributions from the surfaces flexed by the load as follows:

$$
\begin{aligned}
\Delta g(k)_{infill/layer\,2} &= 2\pi G(\rho_2 - \rho_{infill})Y(k)e^{-kd} \\
\Delta g(k)_{layer\,2/layer\,3} &= 2\pi G(\rho_3 - \rho_2)Y(k)e^{-k(d+t_2)} \\
\Delta g(k)_{layer\,3/moho} &= 2\pi G(\rho_m - \rho_3)Y(k)e^{-k(d+t_2+t_3)}
\end{aligned}
\qquad (5.21)
$$

where ρ_{infill}, ρ_2, ρ_3 and ρ_m are the densities of the infill, oceanic layer 2, oceanic layer 3 and mantle, respectively, and t_2 and t_3 are the thicknesses of layer 2 and layer 3, respectively. In the general case, $Y(k)$ is related to $H(k)$ by

$$
Y(k) = -H(k)\frac{(\rho_l - \rho_w)}{\left(\rho_m - \rho_{infill}\right)}\phi_e(k),
$$

where $\varphi_e(k)$ is defined in Eq. (5.3). The total gravity anomaly is given by

$$
\Delta g(k)_{total} = \Delta g(k)_{load} + \Delta g(k)_{inf\,ill/layer\,2} + \Delta g(k)_{layer\,2/layer\,3} + \Delta g(k)_{layer\,3/moho}.
$$

It follows that the gravitational admittance for an elastic plate that loads a vertically stratified crust is given by

$$
Z(k)_{flex\,2} = 2\pi G(\rho_l - \rho_w)e^{-kd}
$$
$$
\left\{ 1 - \varphi_e(k)\frac{\left((\rho_2 - \rho_{infill}) + (\rho_3 - \rho_2)e^{-kt_2} + (\rho_m - \rho_3)e^{-k(t_2+t_3)}\right)}{(\rho_m - \rho_{infill})} \right\}.
\qquad (5.22)
$$

The admittance curve, $Z(k)_{flex1}$, is illustrated in Fig. 5.13 where it is compared with $Z(k)_{uncompensated}$, $Z(k)_{airy}$ and $Z(k)_{pratt}$. All the isostatic models show $Z(k) \to 0$ as $k \to 0$ and

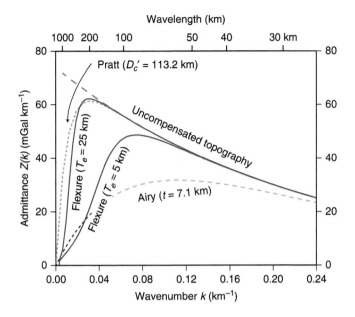

Figure 5.13 Gravitational admittance for the uncompensated model (red long-dashed line) and the Airy (green intermediate-dashed line), Pratt (green short-dashed line) and elastic plate flexure (blue solid lines) models of isostatic compensation.

$k \rightarrow \infty$. The main differences between the models are in the shape of the admittance. At long wavelengths, the Pratt model is associated with the highest gravity anomalies, while the Airy model is associated with the smallest anomalies. This is because the compensation of bathymetry extends to much greater depths in the Pratt model than in the Airy model. The flexure model is generally intermediate in admittance, with its high T_e case (i.e., $T_e = 25$ km) approaching the Pratt value, and its low T_e case (i.e., $T_e = 5$ km) approaching the Airy value. At short wavelengths, the admittance in all models approaches 0 because of the increasing effects of attenuation of the gravity effect of short-wavelength bathymetry by the water depth.

Figure 5.14 compares the spatial domain equivalent of the admittance, or filter coefficient, of the different isostatic models.

5.3.5 'Buried' Loads

So far, we have only been considering surface loads that are emplaced on the surface of oceanic crust. Not all loads act on the surface, however. Some, the 'buried loads', are in the sub-surface. Buried loads occur in a wide range of geological settings in oceanic as well as continental crust. In compressional and extensional settings, they include loads caused by intra-crustal thrusting, ophiolite obduction, sinking slabs, magmatic underplating and the forcible intrusion of dykes, sills and central igneous complexes. Each of these loads, which may or may not have surface expressions, can cause large-scale flexures of the lithosphere. Finally, buried loads include processes, such as density-driven convection in the sub-crust

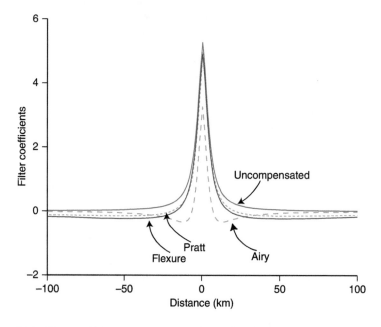

Figure 5.14 Filter coefficients for the uncompensated, Airy, Pratt and flexure models of isostasy. The uncompensated model is associated with a larger amplitude peak, but an absence of flanking troughs compared to the Airy, Pratt and elastic plate flexure models.

and lithosphere mantle, that induce normal stresses on the base of the tectonic plates which deform them below.

Let us consider first the general case (Fig. 5.15) of an upward force that acts on the base of the lithosphere. The gravity anomaly is made up of two contributions: one from the upward acting force and the other from the surface deformation that it causes. If we express the upward-acting force in terms of an equivalent load of height, $W_b(k)$, and density contrast $(\rho_m - \rho_L)$, then it follows that the gravity anomaly is

$$\Delta g(k)_{\,upward\ force} = 2\pi G(\rho_m - \rho_c)W_b(k)e^{-k(d+Z_L)}, \qquad (5.23)$$

where ρ_L is the density of the buried load, ρ_m is the density of the surrounding mantle, d is the mean water depth and Z_L is the mean depth of the upward-acting force below the mean water depth and, in this case, below the Moho.

In surface loading, the deformation is driven by the density difference between the bathymetry and its surroundings. In contrast, buried loading is driven by density differences in the sub-surface. In both types of loading, the deformation is limited by the strength of the lithosphere and is modulated by restoring forces that act on the top and base of the flexed plate. In surface loading, these forces arise from the difference between the downward forces of the material that infills the flexure and the upward buoyancy of the underlying mantle. In buried loading, restoring forces also result from material which infills the flexure and the buoyancy of the underlying mantle. In surface loading, the infill is usually of the same density as the load. In contrast, in buried loading the infill refers to the material that is

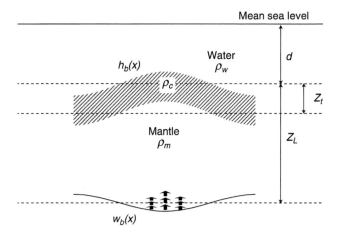

Figure 5.15 The buried loading or 'hotspot' model for isostatic compensation.

displaced by or infills the flexure, and so in the case considered here has the density of water or air.

In surface loading, the deformation, $Y(k)$, is related to the bathymetry, $H(k)$, by

$$Y(k) = -H(k)\frac{(\rho_c - \rho_w)}{(\rho_m - \rho_c)}\phi_e'(k), \qquad (5.24)$$

where $(\rho_c - \rho_w)$ is the density contrast between the load and water, and $(\rho_m - \rho_c)$ is the density contrast between the material that underlies the flexed crust and the material that infills its surface. This is the case if the density of the load is the same as the density of the infill, which, in turn, is the same as the density of the flexed crust.

In buried loading, $Y(k)$ is replaced by $H_b(k)$, the contribution to the bathymetry of buried loading, and $H(k)$, the surface load, is replaced by $W_b(k)$, the buried load. In addition, there is now a new load, so $(\rho_c - \rho_w) \rightarrow (\rho_m - \rho_L)$. Also, water rather than crust is displaced by or infills the flexure, so $(\rho_m - \rho_c) \rightarrow (\rho_m - \rho_w)$. Equation (5.24) therefore becomes

$$H_b(k) = -W_b(k)\frac{(\rho_m - \rho_L)}{(\rho_m - \rho_w)}\phi_e''(k),$$

where

$$\phi_e''(k) = \left[\frac{Dk^4}{(\rho_m - \rho_w)g} + 1\right]^{-1} \qquad (5.25)$$

$$\therefore W_b(k) = -H_b(k)\frac{(\rho_m - \rho_w)}{(\rho_m - \rho_L)\phi_e''(k)}. \qquad (5.26)$$

The gravity anomalies due to buried loading and the deformation that it causes can now be calculated. The gravity anomaly associated with the buried load is obtained by substitution of Eq. (5.26) in Eq. (5.23). This gives

$$\Delta \mathbf{g}(k)_{upward\ force} = -2\pi G(\rho_m - \rho_w)\mathbf{H}_b(k)\frac{e^{-k(d+Z_L)}}{\phi_e''(k)}.$$

The gravity anomaly due to the deformation is the sum of two effects: one from the surface and the other from the crust/mantle boundary. Combining the two gives

$$\Delta \mathbf{g}(k)_{deformation} = 2\pi G(\rho_c - \rho_w)\mathbf{H}_b(k)e^{-kd} + 2\pi G(\rho_m - \rho_c)\mathbf{H}_b(k)e^{-k(d+Z_t)},$$

where Z_t is the mean thickness of the crust. Hence,

$$\Delta \mathbf{g}(k)_{deformation} = 2\pi G\mathbf{H}_b(k)e^{-kd}[(\rho_c - \rho_w) + (\rho_m - \rho_c)e^{-kZ_t}],$$

$$\Delta \mathbf{g}(k)_{buried} = \Delta \mathbf{g}(k)_{upward\ force} + \Delta \mathbf{g}(k)_{deformation},$$

$$\mathbf{Z}(k)_{buried} = 2\pi Ge^{-kd}\left\{(\rho_c - \rho_w) + (\rho_m - \rho_c)e^{-kZ_t} - \frac{(\rho_m - \rho_c)e^{-kZ_L}}{\phi_e''(k)}\right\}.$$

$$(5.27)$$

This is the gravitational admittance for the case of buried loading that acts at a mean depth Z_L below sea level to flex the overlying oceanic crust.

The form of $\mathbf{Z}(k)_{buried}$ is illustrated in Fig. 5.16. The admittance curves for buried loading, unlike those of surface loading, show a distinct peak and trough. The peak indicates the wavelength range for which the gravity contribution from the bathymetry is dominant over that of buried loading. The amplitude of the peak depends strongly on Z_L. A shallow buried load, for example, reduces the gravity contribution of the bathymetry and, hence, the admittance. The trough, however, indicates the wavelength range for which the gravity contribution from buried loading is dominant over that of surface loading. In this wavelength range, the lithosphere is strong enough that the response to buried loading is dampened. Hence, the gravity contribution of the surface deformation is relatively subdued, while the gravity contribution of the buried load is enhanced. The trough is seen in all cases but is best developed for relatively low values of Z_L.

We have considered so far a simple model of an elastic plate that overlies a weak fluid substrate and is deformed by a surface or sub-surface (i.e., buried) load. But what about convection which acts below the plates and causes a surface deformation?

With density-driven convection it is necessary to also consider the temperature and viscosity of the mantle. McKenzie (2010), for example, gives the gravitational admittance for convection that incorporates the equations for temperature and viscous flow in a half-space. Figure 5.17 shows that the admittance for such a flow is a strong function of thickness, T_e, of the elastic plate that overlies the viscous half-space. The admittance is positive for $0 < T_e < 28$ km and negative for $T_e > 28$ km. The negative values can be attributed to elastic plate strength, which minimises the gravity effect of the surface

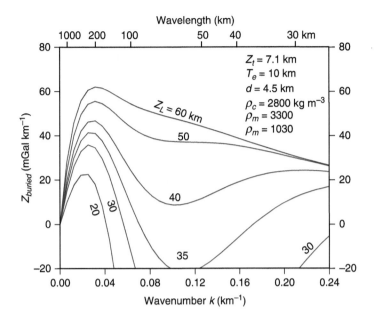

Figure 5.16 Gravitational admittance for the buried loading or 'hotspot' model. The curves represent different depths for the buried loading.

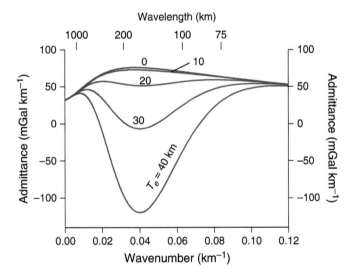

Figure 5.17 Gravitational admittance for the case of a thin elastic plate that overlies a viscous half-space. The admittance has been computed using the formation of McKenzie (2010) with a temperature anomaly at the top of the half-space of 100°C, a thermal thickness of the lithosphere of 120 km, a coefficient of volume expansion of the lithosphere of $4 \times 10^{-5\circ}C^{-1}$, a density of sub-crustal mantle of 3,330 kg m^{-3} and values of the elastic thickness of the lithosphere, T_e, in the range of 0–40 km.

response to convective motions and maximises the gravity effect of the sub-plate structure. Unlike the surface and buried loading cases considered previously, the admittance for convection in oceanic regions at long wavelengths is non-zero and approaches instead a constant value of ~30 mGal km^{-1}.

5.4 High-Order Terms

It follows from this discussion that once it has been defined, a theoretical admittance may be used to model the gravity anomaly caused by any bathymetric feature and its compensation. The procedure is a three-step process. First, $Z(k)$ is calculated for a particular isostatic model. Then, $Z(k)$ is multiplied by the Fourier transform of the bathymetry, $H(k)$. The final step is to inverse Fourier transform the product of $Z(k)$ and $H(k)$ to obtain $\Delta g(x)$, the calculated gravity anomaly in the space domain (see Fig. 5.3). By comparing the calculated anomaly with observations, constraints may be placed on the various isostatic parameters that were assumed when constructing $Z(k)$, such as t (Airy), D_c (Pratt) and T_e (flexure).

We have assumed so far, however, that in the wavenumber domain the gravity anomaly is linearly related to the bathymetry. As Parker (1972) has shown, this 'first-order' approach is only an approximation. The gravity anomaly due to an undulating interface of uniform density contrast, such as the seafloor, needs to be written as a series that involves successively higher orders in the bathymetry.

The complete expression for the gravity effect of uncompensated bathymetry is given (Parker 1972) by

$$\Delta g(k) = 2\pi G(\rho_c - \rho_w)e^{-kd} \sum_{n=1}^{\infty} \frac{k^{n-1}}{n!} \mathcal{J}\{h^n(x)\}, \tag{5.28}$$

where d is mean seafloor depth below the sea surface. When $n = 1$ in Eq. (5.28), we get the expression for the gravity effect of bathymetry that was used in generating the admittance in Section 5.3. Equation (5.28) shows, however, that there are additional terms in the series expansion for the gravity effect that depend on the bathymetry. While the series is infinite, it converges rapidly. For example, the maximum values of the first-, second- and third-order terms over an uncompensated volcanic bathymetric ridge that is 5 km high, 240 km wide at its base and 100 km wide at its surface are 370, 15 and 1 mGal, respectively. These calculations assume the plane of observation (i.e., the sea surface) is always above the seafloor. Otherwise, the series will not converge.

Figure 5.18 compares the gravity anomaly due to an uncompensated volcanic ridge computed by a Fourier technique with $n = 1$ with the anomaly calculated using a line-integral[3] technique (e.g., Talwani et al., 1959, Bott, 1960). The first point of difference between the two curves concerns the levels. The line-integral technique is an exact theory so that the calculated anomaly is zero over the flat seafloor and maximum over the ridge. The

---

[3] A MATLAB script, chap5_gbott.m, to calculate the gravity effect of an arbitrary-shaped body based on the method of Bott (1960) is available from the Cambridge University Press server.

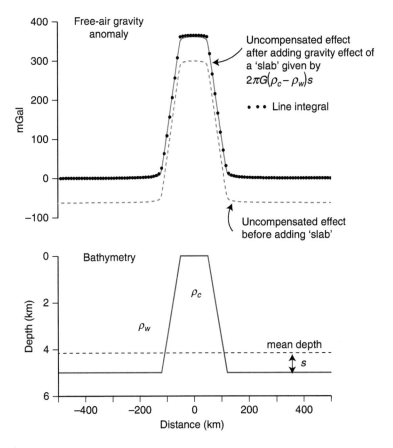

Figure 5.18 Comparison between the gravity effect of a seamount using the Fourier transform with $n = 4$ and one based on the line-integral method. Note that a mean depth has been removed from the bathymetry in the Fourier transform and the gravity effect of the resulting slab added to fit the calculated gravity based on the line-integral method.

Fourier technique, however, involves the subtraction of a mean and the separation of the load into a positive and a negative part. When account is taken of the mean, there is general agreement between computed anomalies.

In detail there are differences, however, due to the limitations of first-order theory (Fig. 5.19). However, the difference is < 10 per cent of the total amplitude of anomaly. Also, the difference is limited to the region of the slopes of the load, which are unusually steep in the case considered. By $n = 3$, the differences are negligible, and so the usual approach in forward modelling is to evaluate Eq. (5.26) up to and including $n = 4$.

5.5 Isostatic Response Functions

When discussing the gravity anomaly, we found it useful to consider a certain wavenumber parameter, known as the 'gravitational admittance', that described the function that modified

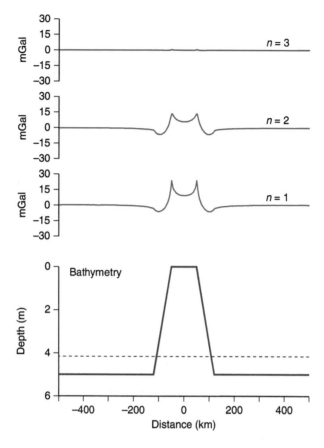

Figure 5.19 Differences between the gravity effect calculated using the Fourier transform and line-integral methods for different values of n.

the bathymetry to produce the gravity anomaly. While the gravitational admittance depends on the flexural properties of the lithosphere, it also depends on the mean water depth and the density contrast that is assumed between the crust and seawater. To compare the state of isostasy at different features, we therefore seek a wavenumber parameter that modifies the gravity effect of the bathymetry to produce the gravity anomaly and depends on the main isostatic parameters, for example, the zero-elevation crustal thickness, the depth of compensation and T_e.

The wavenumber parameter, $\varphi_e(k)$, that describes the behaviour of such a system is given by

$$\varphi_e(k) = \frac{\text{output}}{\text{input}},\tag{5.29}$$

$$\text{input} = 2\pi G e^{-kd}(\rho_c - \rho_w)\,H(k).$$

For the Airy model, we have that

$$\text{output} = 2\pi G(\rho_c - \rho_w)H(k)e^{-kd}\left(1 - e^{-kt}\right),$$

$$\therefore \boldsymbol{\varphi}_e(k)_{airy} = \left(1 - e^{-kt}\right). \tag{5.30}$$

In the case of the flexure model (in its simplest form), we have

$$\text{output} = 2\pi G(\rho_c - \rho_w)\boldsymbol{H}(k)e^{-kd}\left(1 - \boldsymbol{\phi}'_e(k)e^{-kt}\right),$$

$$\therefore \boldsymbol{\varphi}_e(k)_{flex1} = \left(1 - \boldsymbol{\phi}'_e e^{-kt}\right), \tag{5.31}$$

where ϕ'_e is defined in Eq. 5.18. We see from Eqs. 5.30 and 5.31 that $\boldsymbol{\varphi}_e(k)$ is independent of the mean water depth and the density difference between the crust and seawater. The parameter depends only on the thickness of the crust in the Airy model and the properties of the lithosphere in the flexure model. Hence, it provides 'direct' information on the state of isostasy at a geological feature and for this reason is referred to here as the 'isostatic response function'.

5.6 Estimating Admittance, Coherence, and Isostatic Response Functions from Observations

An advantage of the spectral approach is that statistical methods can be used to estimate $\boldsymbol{Z}(k)$ and $\boldsymbol{\varphi}_e(k)$ directly from observations. By comparing the 'observed' values for these parameters with calculations based on different isostatic models, it should therefore be possible to constrain the state of isostasy at different features.

We saw from Eqs. (5.7) and (5.29) that

$$\boldsymbol{Z}(k) = \frac{\Delta \boldsymbol{g}(k)}{\boldsymbol{H}(k)} \tag{5.32}$$

and

$$\boldsymbol{\varphi}_e(k) = \frac{\Delta \boldsymbol{g}(k)}{2\pi G e^{-kd}(\rho_c - \rho_w)\boldsymbol{H}(k)}. \tag{5.33}$$

In principle, then, it should be a relatively simple matter to determine $\boldsymbol{Z}(k)$ by dividing the Fourier transform of the observed free-air gravity anomaly by the Fourier transform of the bathymetry. $\boldsymbol{\varphi}_e(k)$ can then be estimated from the mean water depth, d, and the density difference between the bathymetry and seawater, $(\rho_c - \rho_w)$. In practice, this is difficult to do because both data sets need to be transformed, and because of, for example, instrument and navigation errors they could produce spectral estimates with considerable scatter.

Several methods have been proposed to reduce the noise in spectral estimates. In oceanic regions, gravity anomaly and bathymetry data are acquired along individual ship tracks. Therefore, methods to smooth spectra have largely focused on two-dimensional averaging techniques. In continental regions, however, gravity anomaly and bathymetry data are usually available as large regional 'grids'. Therefore, methods to smooth spectra have mainly been based on three-dimensional averaging techniques. We now consider these methods in more detail.

5.6.1 Oceans

The first applications of spectral techniques to free-air gravity and bathymetry data in the oceans were by McKenzie and Bowin (1976) in the Atlantic and Watts (1978) in the Pacific.

A common approach to estimate the admittance, $Z(k)$, between two time-series is to divide the Fourier transform of the output function by the transform of the input function. Munk and Cartwright (1966), however, pointed out that this method ignores the effect of noise that will inevitably be present in the output function. They argued that in the presence of noise, a better estimate of $Z(k)$ is obtained by dividing the cross-spectrum of the input and output functions by the power spectrum of the input function.

McKenzie and Bowin (1976) used 'cross-spectral techniques' to estimate $Z(k)$ from free-air gravity anomaly and topography measured along segments of an individual ship track. The complex admittance for a particular wavenumber is given by

$$Z(k) = \frac{C_c(k)}{E_t(k)}, \tag{5.34}$$

where C_c is the cross-spectrum of the gravity anomaly and topography and E_t is the power spectrum of the topography. C_c and E_t are given by

$$C_c(k) = \frac{1}{N} \sum_{m=1}^{N} \Delta g_m(k) H_m^*(k),$$

$$E_t(k) = \frac{1}{N} \sum_{m=1}^{N} H_m(k) H_m^*(k) \tag{5.35}$$

The asterisk denotes the complex conjugate, and $\Delta g(k)$ and $H(k)$ are now the discrete Fourier transforms of the observed free-air gravity anomaly and bathymetry, respectively.

McKenzie and Bowin (1976) smoothed the spectral estimates by summing them along equal-length segments of a ship track and dividing by the total number of segments. They justified their approach by pointing out that the data along segments of a ship track could be regarded as an independent estimate of the relationship between gravity anomaly and bathymetry over a particular portion of the seafloor.

A difficulty with along-track smoothing, however, is that the spectral estimates may be 'contaminated' by the isostatic response from different tectonic provinces. This problem did not affect the conclusions of McKenzie and Bowin (1976) because their ship track mainly crossed features (e.g., Walvis Ridge) that formed at or near a mid-oceanic ridge crest. Hence, isostatic parameters such as T_e would, in any event, be expected to be similar.

To determine T_e at different geological features, Watts (1978) modified the McKenzie and Bowin (1976) smoothing technique. Rather than dividing a single ship track into several segments, he smoothed the spectra along different profiles of the same geological feature.

The methodology of McKenzie and Bowin (1976) and Watts (1978) was similar. Shipboard free-air gravity anomaly and bathymetry data along equal-length profiles were linearly interpolated to obtain evenly spaced values. After removal of the trend, the ends of the profiles were tapered using a cosine taper to avoid edge effects. Then, the discrete Fourier transform of each data set was used to calculate the cross-spectrum between the gravity and bathymetry and the power spectrum of the bathymetry and, hence, the gravitational admittance, $Z(k)$.

In addition to determining $Z(k)$, McKenzie and Bowin (1976) and Watts (1978) found it useful to compute two additional spectral parameters: the 'coherence' and the 'phase of the admittance'. The coherence, $\gamma^2(k)$, is given by

$$\gamma^2(k) = \frac{C_c(k)C_c^*(k)}{E_{\Delta g}(k)E_t(k)}, \qquad (5.36)$$

where

$$E_{\Delta g}(k) = \frac{1}{N}\sum_{m=1}^{N} \Delta g_m(k)\Delta g_m^*(k)$$

In the presence of noise, Munk and Cartwright (1966) indicate that a better measure of coherence is given by

$$\gamma_o^2(k) = \frac{\left(N\gamma^2(k) - 1\right)}{N - 1},$$

where N is the 'degree of freedom' and is either the number of along-track segments or the number of profiles of the same geological feature.

The usefulness of $\gamma^2(k)$ is that it provides a measure of the portion of the observed free-air gravity anomaly that is caused by the bathymetry. At short wavelengths, the coherence is zero because the water depth attenuates the gravity effect of the bathymetry. The gravity anomaly will therefore also be small, even though the bathymetry may be significant. The coherence will also be small at long wavelengths because the gravity effect of the bathymetry competes with the effect of isostatic compensation. This is the case irrespective of which model of isostasy is used. At intermediate wavelengths, the coherence approaches 1, and it is within these wavelength ranges that most of the free-air gravity anomaly is caused by bathymetry.

The phase of the admittance, $\phi_z(k)$, which is defined by

$$e^{-i2\phi_z(k)} = \frac{Z(k)}{Z^*(k)}, \qquad (5.37)$$

is also important since, although the admittance is complex, we would expect it to be real. The phase in the wavelength range of high coherence should therefore be close to zero.

An example of the admittance, coherence and phase of the admittance determined by Watts (1978) using bathymetry and free-air gravity anomaly profile data intersecting the Hawaiian–Emperor seamount chain is shown in Fig. 5.20. Since the methods were subsequently applied to data from other geological features (e.g., Cochran, 1979; Detrick and

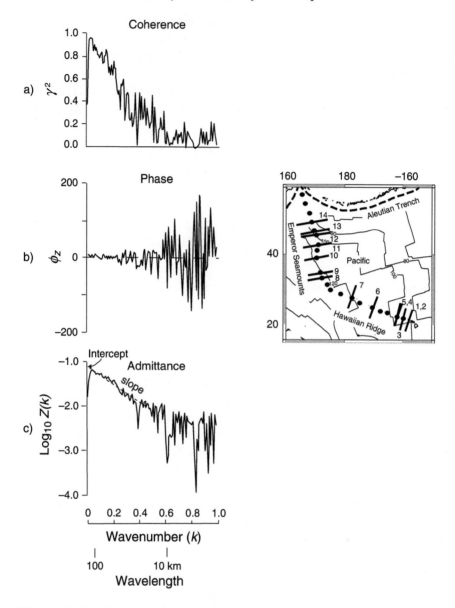

Figure 5.20 The coherence (a), phase (b) and gravitational admittance (c) determined from spectral analysis of bathymetry and free-air gravity anomaly data along the Hawaiian–Emperor seamount chain. The inset shows the location of the projected profile data used to obtain smooth spectral estimates. Isochrons (Ma) are based on Müller et al. (1997). Modified from fig. 4 of Watts (1978).

Watts, 1979; Karner and Watts, 1982) it is useful to outline them here. Firstly, the bathymetry and gravity data along individual ship tracks (there were 14 in the Hawaiian–Emperor study) were projected onto profiles normal to the local trend of the seamount chain. Each profile was selected to be 800 km in length. A total of 2^8 (256) samples of the gravity anomaly and bathymetry data were then extracted from each profile, yielding a regularly spaced data set at intervals of 3.137 km. The spatial wavelengths represented in the data therefore potentially range from 6.274 km to 1,600 km, equivalent to wavenumbers of $0.0039 < k < 1.0016$. Secondly, the trend (including the mean) was removed, and the ends of the profiles were tapered using a cosine window on 5 per cent of the total length of the profile. The final step was to take the discrete Fourier transform of the sampled, trend-removed and tapered gravity anomaly and bathymetry data sets and use Eqs. (5.34), (5.36) and (5.37) to calculate the spectral estimates for each pair of profiles.

The averaging of data acquired along different ship tracks of the same geological feature yields smooth spectral estimates. In particular, the coherence is high (> 0.5) for $0.01 < k < 0.22$ (Fig. 5.20a), indicating that this is the 'critical' wavenumber band within which most of the energy in the free-air gravity anomaly is caused by the bathymetry. The phase (Fig. 5.20b) and admittance (Fig. 5.20c) are also smooth within this waveband, the latter reaching its peak value at $k = 0.037$. Figure 5.21 compares admittance estimates in the waveband $0 < k < 0.16$ with the predictions of both the Airy and simple flexure model. It shows that the decrease at low wavenumbers is caused by isostasy, while the decrease at high wavenumbers is due to the attenuation of the gravity anomaly by the water depth. Figure 5.20c shows that at $k > 0.57$, the admittance flattens, probably because of instrument noise in the shipboard gravity data.

The part of the admittance not caused by isostasy and instrument noise reflects the uncompensated bathymetry. Therefore, this part of the spectrum contains information on the mean water depth and the density of the seafloor. The uncompensated admittance is given by

$$Z(k) = 2\pi G(\rho_c - \rho_w)e^{-kd}.$$

Taking logs of both sides yields

$$\log_{10}Z(k) = -kd \log_{10}e + \log_{10}\left(2\pi G(\rho_c - \rho_w)\right).$$

A log–linear plot of $Z(k)$ against k therefore has a slope d of 0.4335 and an intercept on the $Z(k)$ axis given by $\log_{10}(2\pi G(\rho_c-\rho_w))$. Taking these values from Fig. 5.20c yields a mean depth and density of seafloor along the Hawaiian–Emperor seamount chain of 4,500 m and 2,800 kg m^{-3}, respectively.

The significance of the spectral approach is that it allows us to distinguish between the different models of isostasy at a particular geological feature. At the Hawaiian–Emperor seamount chain, for example, Fig. 5.21a shows that an Airy model cannot explain the observed admittance unless the mean oceanic crustal thickness is in the range of 35–80 km. These estimates are significantly higher than the generally accepted thickness of oceanic crust in the Pacific region, which is ~5.5–7.5 km (Mutter and Mutter, 1993; Grevemeyer et al., 2018). Therefore, the Airy model can be ruled out as the isostatic mechanism along the seamount chain.

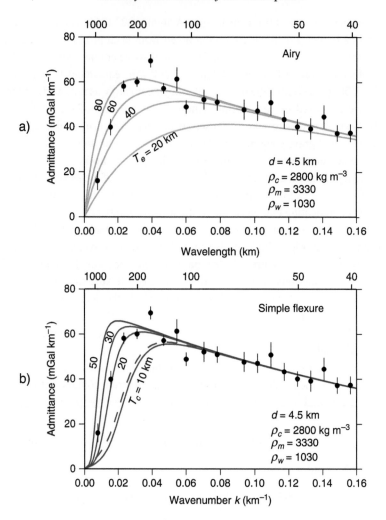

Figure 5.21 Comparison of the observed gravitational admittance along the Hawaiian–Emperor seamount chain with the predictions of (a) Airy and (b) simple flexure models of isostasy. The curves show sensitivity of the admittance to (a) zero-elevation crustal thickness, T_c, and (b) elastic thickness, T_e.

Figure 5.21b shows that the admittance along the Hawaiian–Emperor seamount chain can be best explained by a flexure model with T_e in the range of 20–30 km. This range of estimates was verified by forward modelling of individual topographic profiles of the chain (Table 5.1).

The success of these early studies led to a rapid increase in applications of the spectral technique to the mid-oceanic ridge crest (Cochran, 1979), aseismic ridges (Detrick and Watts, 1979), fracture zones (Louden and Forsyth, 1982) and, particularly, passive margins (Karner and Watts, 1982; Diament et al., 1986; Mello and Bender, 1988; Verhoef and Jackson, 1991). Figure 5.22 compares, for example, the isostatic response function, $\varphi_e(k)$,

Table 5.1 *Comparison of estimates of the elastic thickness from forward modelling and from admittance studies along the Hawaiian–Emperor seamount chain*

Profile number in inset of Fig. 5.20	Island/Seamount	Elastic thickness, T_e, (km)
1	Oahu	30.5
2	Oahu	32.0
3	Kauai/Oahu	30.0
4	Kaula	30.0
5	Kaula	20.0
6	Raita bank	17.1
7	Midway	37.0
8	Koko	30.5
9	Koko	36.0
10	Ojin	–
11	Nintoku	20.0
12	Suiko	–
13	Jimmu	10.5
14	–	13.0

Average elastic thickness from forward modelling = 25.5±8.5 km
Range of elastic thickness from admittance = 20–30 km

determined from free-air gravity anomaly and bathymetry data at the Hawaiian–Emperor seamount chain with values at the mid-Atlantic Ridge and the East Pacific Rise. The figure illustrates the significant differences that exist between the state of isostasy of features that form in off-ridge (e.g., Hawaiian–Emperor seamount chain) and on-ridge settings (e.g. the mid-Atlantic ridge crest and the East Pacific rise).

The difference between the isostatic response functions of features that formed on-ridge and off-ridge was used by Detrick and Watts (1979) to determine the tectonic setting of features of unknown origin on the sea floor. Figure 5.23 shows, for example, the application of the technique to a gravity anomaly and bathymetry profile of the Walvis Ridge in the South Atlantic Ocean. The calculated profiles show the expected gravity anomaly over the Walvis Ridge, if it had formed in a similar tectonic setting and had the same isostatic response as the Hawaiian–Emperor seamount chain and the Mid-Atlantic Ridge. The best agreement between the observed and calculated gravity is for a model in which the Walvis Ridge formed in a mid-oceanic ridge setting. This model explains well the amplitude and wavelength of the observed anomaly. In contrast, there is a poor agreement between the observed and calculated gravity for a model in which the Walvis Ridge formed off-ridge, in a similar setting to the Hawaiian–Emperor seamount chain.

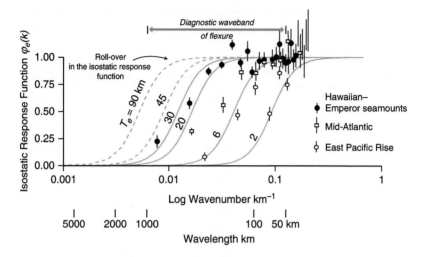

Figure 5.22 Comparison of observed and calculated isostatic response functions. The observed data are based on Watts (1978) for the Hawaiian–Emperor seamount chain and Cochran (1979) for the Mid-Atlantic Ridge and East Pacific Rise. The calculated curves are based on the isostatic response function defined in Eq. 5.31 and assume a thickness of oceanic crust, t, of 5 km, $\rho_{infill} = 2{,}700$ and $\rho_m = 3{,}330$. The 'diagnostic waveband of flexure' refers to the wavelength band for which loads at the mid-ocean ridges and the plate interior appear to be flexurally isostatically compensated.

We have so far only been concerned with the application of spectral methods to gravity anomaly and bathymetry data acquired along profiles. These methods can, of course, be extended to three dimensions.[4] The admittance, for example, is then given by

$$Z(k) = \frac{\Delta g(k)}{H(k)},$$

and since $Z(k)$ is isotropic, the wavenumber vector, k, can be replaced by

$$\sqrt{k_x^2 + k_y^2}.$$

The problem with three-dimensional spectral studies is that they require closely spaced gravity and bathymetry data over a broad area, and these data are difficult to acquire using shipboard surveys. Some surveys, however, have been carried out along closely spaced ship tracks. McNutt (1979), for example, analysed a grid of shipboard bathymetry and gravity anomaly data over the crest of the East Pacific Rise and showed that the small values of T_e that had been obtained by modelling two-dimensional profiles extended over broad regions of the seafloor.

The advent of satellite altimetry data, together with shipboard 'swath' bathymetry data, have significantly improved grid resolution and, hence, estimates of the three-dimensional admittance.

---

[4] A MATLAB script, chap5_flex3d_FFT_GMT.m, to calculate the flexure due to an arbitrary-shaped three-dimensional load is available from the Cambridge University Press server.

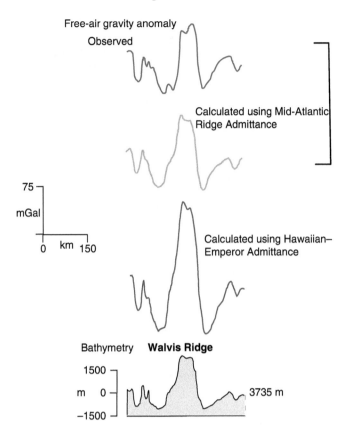

Figure 5.23 Comparison of the observed gravity anomaly over the Walvis Ridge compared with the anomaly that would be expected if the Walvis Ridge had formed in a similar off-ridge setting to the Hawaiian–Emperor seamount chain and a similar on-ridge setting to the Mid-Atlantic ridge. The predicted anomalies were obtained by multiplication (in the frequency domain) of the Walvis Ridge topography with the Hawaiian–Emperor seamount chain and Mid-Atlantic Ridge admittance. Reproduced from fig. 15 of Detrick and Watts (1979).

Figure 5.24 compares, for example, the observed isostatic response function for different window sizes of the Central Pacific Ocean based on the admittance derived from a 2 × 2 minute satellite-derived gravity anomaly (Sandwell and Smith, 1997) and a 1 × 1 minute GEBCO grid to the calculated isostatic response function for $5 < T_e < 50$ km. The analysis, which has been carried out in 'windows' centred on the Hawaiian Islands with sizes that range from 9 × 9° (~1,000 km × 1,000 km) to 49 × 49° (~5,400 km × 5,400 km), reveals that the highest T_e correspond to the smallest windows and the lowest T_e to the largest windows. We can explain this geologically when consideration is given to the inclusion in the smallest windows of mostly seamounts and oceanic islands (e.g., the Hawaiian Ridge) emplaced on old oceanic lithosphere, while the larger windows include mostly features (e.g., Line Islands, Musician Seamounts, Necker Ridge, Mid-Pacific Mountains) emplaced on young lithosphere on or near a mid-ocean ridge crest.

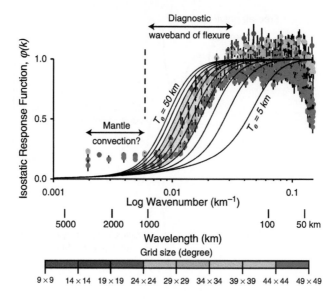

Figure 5.24 The isostatic response function for a range of window sizes from 9° × 9° (~1,000 km × 1,000 km) to 49° × 49° (~5,400 km × 5,400 km) centred on the Hawaiian Islands in the Central Pacific Ocean. The solid black lines show theoretical isostatic response functions for 5 < T_e < 50 km. The best fit is for 20 < T_e < 35 km with the higher values corresponding to the smaller windows and the lower values to the larger windows. Based on fig. 4.6 in Wilson (2004).

The agreement between observed and calculated isostatic response functions in Fig. 5.24 is close, the rollover being particularly well explained for 20 < T_e < 35 km. The main departure is at long wavelengths (λ > 1,000 km), where the observed isostatic response function approaches an approximately constant value, $\varphi_e \rightarrow 0.2$, while the calculated functions approach zero. This value for φ_e implies using Eq. (5.33) an admittance of ~23 mGal km^{-1}, assuming a density of water and the bathymetry of 1,030 and 3,330 kg m^{-3} respectively, a mean water depth of 4,800 m and a 'warped' model for oceanic crust (see, for example, Fig. 4.56). This admittance is close to the gravity/bathymetry ratio of 22 mGal km^{-1} determined by Watts (1976) from 5 × 5° averages of bathymetry and free-air gravity anomaly data over the Hawaiian swell and attributed by him to some form of convection in the sub-crustal mantle.

Other applications of the three-dimensional admittance technique have been by Ito and Taira (2000) to the Ontong Java Plateau, Ali et al. (2003) to the Cape Verde Islands, Luis and Neves (2006) to the Azores Plateau and Daly et al. (2004) to the continental margin offshore Ireland. Each of these studies consider the possibility of both surface and sub-surface loading. Ito and Taira, Ali et al. and Luis and Neves used single cosine tapers and the Fourier transform, while Daly et al. used a multitaper and a wavelet transform technique. Ito and Taira argued for two stages of loading with surface loading of a low-T_e young lithosphere followed by sub-surface loading of a higher-T_e older lithosphere. Ali et al. (2003) suggested a higher T_e (~29 km) and a relatively small, buried load (20 per cent of the surface load) for the old (~140 Ma) Cape Verde lithosphere, while Luis and Neves (2006) suggested a low T_e (~4 km) and a relatively large, buried load (33 per cent of the surface load) for the young (~0–10 Ma) Azores lithosphere.

Synthetic tests show that the window size plays a critical role in the resolution of a three-dimensional admittance (Crosby, 2007). In the flexural wavelength band of 200–1,000 km, low T_e (20 km) is well resolved (within ± 5 km) with a window size of 1,000 km. Higher T_e, however, requires larger windows. Kalnins and Watts (2009), for example, used moving windows with a range of sizes from 400×400 km to $1,400 \times 1,400$ km. They found that when a 'bias correction' was applied for windowing, T_e could be recovered from a weighted average of the windows to an accuracy of ± 5 km for plates with a T_e up to 30 km. They used the technique to map spatial variations in T_e in the western Pacific Ocean, showing that while T_e in oceanic lithosphere could vary substantially, it was generally highest in the vicinity of subduction zones.

5.6.2 Continents

Spectral techniques have also been applied with some success to continent-wide gravity and topography-gridded data sets over North America (Dorman and Lewis, 1970, 1972; Lewis and Dorman, 1970; Bechtel et al., 1990; Armstrong and Watts, 2001; Fluck et al., 2003; Audet and Mareschal, 2007; and McKenzie, 2010), South America (Ussami et al., 1993; Ojeda and Whitman, 2002; Tassara et al., 2006; Pérez-Gussinyé et al., 2007; Sacek and Ussami, 2009), Australia (McNutt and Parker, 1978; Zuber et al., 1989; Simons et al., 2000; Swain and Kirby, 2003, 2006; Kirby and Swain, 2009), Europe (McNutt and Kogan, 1987; Armstrong, 1997; Poudjom Djomani et al., 1999; Pérez-Gussinyé and Watts, 2005), the Former Soviet Union (FSU) (Kogan and McNutt, 1987; Kogan et al., 1994) and Africa (Banks and Swain, 1978; Hartley et al., 1996; Poudjom Djomani et al., 1995; Doucoure et al., 1996; Ebinger and Hayward, 1996; Upcott et al., 1996; Stark et al., 2003; Tessema and Antoine, 2003; Pérez-Gussinyé et al., 2009).

Unlike oceanic regions, it has been traditional in the continents to use the Bouguer rather than the free-air gravity anomaly. There are two main reasons for this. Firstly, the Bouguer anomaly incorporates a correction (the Bouguer correction) that accounts for the mass of topography above or below sea level. The resulting Bouguer anomaly therefore represents mass anomalies that are within the crust or the sub-crustal mantle. Secondly, the Bouguer correction reduces the free-air gravity anomaly to the value that it should have on the geoid, which at the coast corresponds to mean sea level. This facilitates the comparison of observed and calculated anomalies because the latter are usually evaluated on a level surface such as sea level. In principle, however, there should be no difference in the interpretation of spectral data whether free-air or Bouguer gravity anomalies are used.

We will begin by deriving the Bouguer gravitational admittance for two simple models of flexure: one where loading is from the surface, and the other where the loading is buried. Since we are dealing with Bouguer rather than free-air gravity anomalies, we do not need to consider the gravity contribution of the topography, as this should already have been accounted for in the Bouguer reduction.

Surface Loading

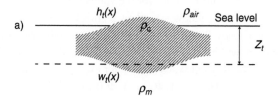

Buried Loading

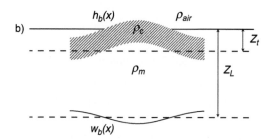

Figure 5.25 The surface loading and buried loading models as they apply to the continents.

In the surface loading case (e.g., Fig. 5.25a), the only contribution to the Bouguer anomaly is from the gravity effect of the compensation. Hence,

$$\Delta g(k)_{flexure} = 2\pi G(\rho_m - \rho_c)W_t(k)e^{-kZ_t},$$

where $W_t(k)$ is the Fourier transform of the compensation. If the compensation is the consequence of flexure due to a surface load, $H_t(k)$, then it follows from Eq. (5.17) that

$$W_t(k) = -\frac{H_t(k)\rho_c\phi'_e(k)}{(\rho_m - \rho_c)}, \qquad (5.38)$$

where ϕ'_e is defined in Eq. 5.18

$$\therefore Z(k)_{surface} = -2\pi G\rho_c e^{-kZ_t}\phi'_e(k) \qquad (5.39)$$

For buried loading (Fig. 5.25b), the only contribution to the Bouguer anomaly is from the deformation on the base of the crust that is caused by upward or downward motions in the underlying mantle. The gravity anomaly due to the buried loads is given by

$$\Delta g(k)_{load} = 2\pi G(\rho_m - \rho_L)W_b(k)e^{-kZ_L},$$

where $W_b(k)$ is the equivalent topography of the buried load. If the buried loads cause surface deformation, and if the deformation displaces air rather than water, then it follows from Eq. (5.24) that

$$W_b(k) = \frac{H_b(k)\rho_m}{\phi_e'''(k)(\rho_m - \rho_L)},$$ (5.40)

where

$$\phi_e'''(k) = \left[\frac{Dk^4}{\rho_m g} + 1\right]^{-1},$$ (5.41)

$$\therefore \Delta g(k)_{load} = -2\pi G H_b(k)\rho_m \frac{e^{-kZ_L}}{\phi_e'''}.$$ (5.42)

The gravity anomaly due to the deformation on the base of the crust by the buried load is given by

$$\Delta g(k)_{deformation} = 2\pi G(\rho_m - \rho_c)H_b(k)e^{-kZ_t}.$$ (5.43)

Combining Eqs. (5.42) and (5.43) gives

$$Z(k)_{buried} = 2\pi G\left\{(\rho_m - \rho_c)e^{-kZ_t} - \rho_m \frac{e^{-kZ_L}}{\phi_e'''}\right\}$$

$$\therefore Z(k)_{buried} = 2\pi G(\rho_m - \rho_c)\left\{e^{-kZ_t} - \frac{e^{-kZ_L}(Dk^4 + \rho_m g)}{(\rho_m - \rho_c)g}\right\}.$$ (5.44)

Equation (5.44) is the same as eq. (5) of Forsyth (1985) and gives the Bouguer admittance for the case of buried loading at some depth Z_L. If we now let $Z_t = Z_L$, we get the special case of buried loading at the base of the crust. Equation (5.44) then becomes

$$Z(k)_{buried} = 2\pi G(\rho_m - \rho_c)\left\{e^{-kZ_1} - \frac{e^{-kZ_t}Dk^4}{(\rho_m - \rho_c)g} - \frac{e^{-kZ_t}\rho_m g}{(\rho_m - \rho_c)g}\right\},$$

$$Z(k)_{buried} = -2\pi G\left\{e^{-kZ_t}(\rho_m - \rho_c) - \frac{e^{-kZ_t}Dk^4}{g} - e^{-kZ_t\rho_m}\right\},$$

$$Z(k)_{buried} = -2\pi G\rho_c e^{-kZ_t}\left[1 + \frac{Dk^4}{\rho_c g}\right],$$

$$\therefore Z(k)_{buried} = -2\pi G\rho_c \frac{e^{-kZ_t}}{\phi_e''''(k)},$$ (5.45)

where

$$\phi_e''''(k) = \left[\frac{Dk^4}{\rho_c g} + 1\right]^{-1}.$$ (5.46)

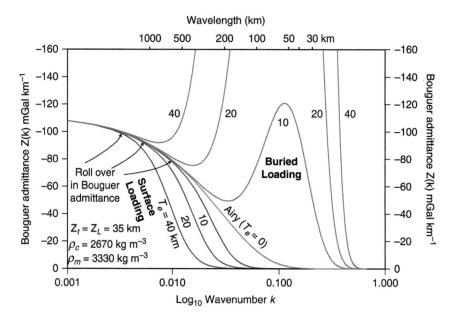

Figure 5.26 Bouguer gravitational admittance for the surface (blue solid lines) and buried (purple solid lines) loading models and different values of T_e. Note that for $T_e = 0$ km (i.e., the Airy model) the two admittances are the same.

The Bouguer admittances for surface and buried loading[5] are compared in Fig. 5.26. At long wavelengths (i.e., $k \to 0$), both types of loading are compensated and $\mathbf{Z}(k) \to -2\pi G \rho_c$. This is the familiar 'Bouguer slab formula' in which ρ_c is the density used in the Bouguer reduction. At short wavelengths (i.e., $k \to \infty$), $\mathbf{Z}(k) \to 0$. This is because there is little deformation of the crust by either surface or buried loads at short wavelengths, regardless of the strength of the lithosphere. The admittance for surface loading shows a rollover, the position of which is controlled by the strength of the lithosphere. For buried loading, the admittance has a more complex shape. It is difficult to recognise a rollover, except for cases of very low T_e. The reason for this is that the admittance becomes strongly negative at wavelengths when the strength of the lithosphere is sufficiently high to dampen the response to buried loading. The result is a large change in Bouguer gravity anomaly for a small change in topography. Figure 5.26 shows that in the case of an Airy model (i.e., $T_e = 0$ km), the admittance is the same for surface and buried loading.

The observed admittances for North America (Dorman and Lewis, 1970, 1972; Lewis and Dorman, 1970), Australia (McNutt and Parker, 1978) and Africa (Hartley et al., 1996) are compared with the predictions of the Airy model in Fig. 5.27. The observed data show a well-defined rollover that resembles that of the predicted curves. The wavelength of the rollover varies between the continents, being about 500 km for North America and 1,000 km for Africa and Australia. The main variable in the Airy model is the mean crustal thickness, Z_t. As the figure

---

[5] A MATLAB script, chap5_ba_admit_continent.m, to calculate the Bouguer admittance $\mathbf{Z}(k)_{surface}$ and $\mathbf{Z}(k)_{buried}$ is available from the Cambridge University Press server.

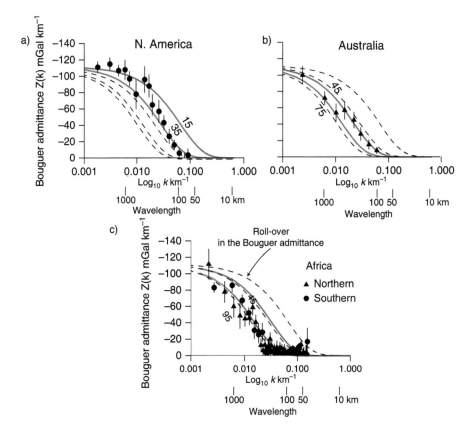

Figure 5.27 Comparison of observed and calculated Bouguer gravitational admittance for (a) North America, (b) Australia and (c) Africa. The observed admittances are based on analyses of Dorman and Lewis (1970), Lewis and Dorman (1970), McNutt and Parker (1978) and Hartley et al. (1996). The calculated admittance is based on an Airy model (i.e., $T_e = 0$ km), a crust and mantle density of 2,670 and 3,330 kg m^{-3} respectively, and a crustal thickness, Z_t, of 15, 35, 45, 75 and 95 km.

shows, the Z_t that best explains the observed admittance is in the range of 15–95 km, with the low end of the range being preferred for North America, and the high end for Northern Africa. This is a possible range for the thickness of the continental crust according to seismic refraction data.

The problem with the Airy model, of course, is that it ignores the strength of the lithosphere. We know that the strength of the lithosphere is a major cause of departures from the Airy model in the oceans, and there is no reason to suppose that this is not also the case in the continents. If strength, whether it lies mainly in the crust or mantle, is important, then, as Fig. 5.26 shows, we need to know whether surface or buried loads are dominant. Unfortunately, the proportion of these loads on a continent-wide scale is unknown.

There is an indication, however, from the shape of the observed admittance that a strong lithosphere may be required. The admittance for North America, for example, is steep in the waveband $0.012 < k < 0.003$ and, as Fig. 5.27a shows, it is difficult to fully explain with an

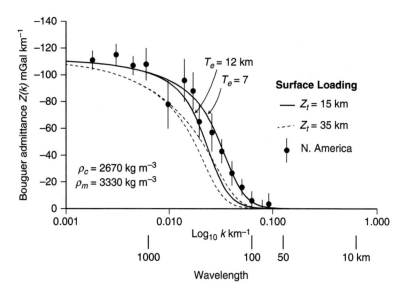

Figure 5.28 Comparison of observed and calculated Bouguer gravitational admittance for North America. The calculated admittance is based on surface loading with $T_e = 7$ and 12 km and a crustal thickness of $Z_t = 15$ km (solid lines) and 35 km (dashed lines).

Airy model without invoking a wide range of crustal thickness values. The steep admittance, together with the lack of a peak at $k = 0.100$, is suggestive of surface rather than buried loading. Indeed, Walcott (1976), Banks et al. (1977) and Cochran (1980) have all shown that surface loading with T_e in the range of 7–22 km provides a good overall fit to the admittance data of North America. The problem, as Fig. 5.28 shows, is that this model requires the main source of the gravity effect of the compensation to be at relatively shallow depths in the crust (e.g., 15 km) rather than at the Moho (e.g., 35 km).

By itself, surface loading does not consider buried loads. While buried loads are not necessarily associated with topography, they are reflected in the gravity field, especially in orogenic belts. We therefore need a method that combines the admittance due to surface and buried loading. Unfortunately, it is not possible simply to combine the admittances of surface and buried loading. Although the individual admittances are real, the combined admittance may be complex if the surface and buried loading are out of phase. In the Appalachians, for example, there is evidence (Karner and Watts, 1983) that the maximum depth to the Moho is laterally offset by up to 50 km from the peak in the surface topography because of buried loading.

Forsyth (1985) derived an expression for the Bouguer admittance for combined surface and buried loading that considers the possibility the two types of loads may be out of phase. He showed that

$$Z(k)_{surface+buried} = -2\pi G \rho_c e^{-kZ_t} \left\{ \frac{\frac{H_b^2}{\phi_e'''(k)} + H_t^2 \phi_e'}{H_b^2 + H_t^2} \right\}, \tag{5.47}$$

where H_t is the contribution of surface loading to the topography, and H_b the contribution of buried loading and ϕ'_e and ϕ'''_e are defined in Eqs. 5.18 and 5.41, respectively. H_t and H_b are related to the observed topography, H, by

$$H = H_t + H_b \tag{5.48}$$

To determine H_t and H_b from H, however, we need an equation that relates them: a process that has been referred to as 'load deconvolution'.

Forsyth (1985) introduced a 'load ratio', f_i, which he defined as the ratio between the weight of the applied load at the base of the crust to the weight of the applied load on the surface. The applied loads in this case are the initial loads that were applied to the surface and base of the crust – before deformation.

Let the initial height of the surface topography before deformation be H_i. We then have (e.g. from Fig. 5.29) that

$$H_i = H_t - W_t,$$

$$H_i = H_t \left[1 + \frac{\rho_c \phi'_e(k)}{(\rho_m - \rho_c)}\right],$$

$$H_i = H_t \left[1 + \frac{\rho_c g}{\left(Dk^4 + (\rho_m - \rho_c)g\right)}\right],$$

$$H_i = H_t \left[\frac{\left(Dk^4 + (\rho_m - \rho_c)g + \rho_c g\right)}{\left(Dk^4 + (\rho_m - \rho_c)g\right)}\right],$$

The applied surface load, $S_L(k)$, is $\rho_c g H_i$. Therefore,

$$S_L = \rho_c g H_t \left[\frac{\left(Dk^4 + (\rho_m - \rho_c)g + \rho_c g\right)}{\left(Dk^4 + (\rho_m - \rho_c)g\right)}\right]. \tag{5.49}$$

Similarly (Fig. 5.29), the height of the buried load before deformation, W_i, is given by

$$W_i = W_b - H_b. \tag{5.50}$$

When the buried loading is at the base of the crust rather than in the mantle below it, we have that

$$W_b(k) = -\frac{H_b(k)\rho_c}{\phi'''_e(k)(\rho_m - \rho_c)}, \tag{5.51}$$

$$-W_i = H_b \left[\frac{\rho_c}{(\rho_m - \rho_c)\phi'''_e(k)} + 1\right],$$

Surface Loading

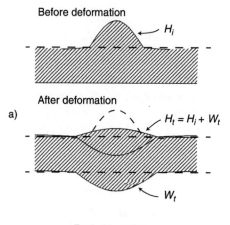

Buried Loading

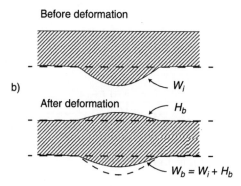

Figure 5.29 Schematic diagram illustrating the concepts of H_i, the initial surface topography due to buried loading, and W_i, the corresponding initial buried load topography. H_i and W_i are subsequently modified by the downward flexure, W_t, due to H_i and the upward flexure, H_b, due to W_i.

$$-W_i = H_b \left[\frac{(Dk^4 + \rho_c g)}{(\rho_m - \rho_c)g} + 1 \right]. \tag{5.52}$$

The applied load at the Moho, $S_M(k)$, is $W_i(\rho_m - \rho_c)g$. It then follows from Eq. (5.48), that

$$S_M = -H_b \left(Dk^4 + \rho_c g + (\rho_m - \rho_c)g \right). \tag{5.53}$$

Therefore, we have for f_l that

$$f_l = \frac{S_M}{S_L} = \frac{|H_b|\left(Dk^4 + (\rho_m - \rho_c)g\right)}{\rho_c g H_t},$$

$$f_l = \frac{|H_b|(\rho_m - \rho_c)}{\phi'_e \rho_c H_t}. \tag{5.54}$$

So, once f_l is assumed, we can derive H_b from H_t and therefore complete the process of deconvolution.

Figure 5.30a shows the admittance for combined surface and buried loading for different values of f_l. The admittance for $f_l = 0$ corresponds to the case of surface loading only. As the role of buried loading increases, however, the curves shift towards lower wavenumbers. This suggests that if an observed admittance had been explained in terms of surface loading with a low T_e and Z_t, it should be possible to explain the same data by a combined surface- and buried-loading model with a high T_e and the same Z_t.

The admittance for North America, which Banks et al. (1977) and Cochran (1980) interpreted in terms of surface loading with T_e in the range of 7–12 km (Fig. 5.28), can be explained by a higher T_e, provided that buried loading is included (Fig. 5.30a). Model curves based on $f_l = 2$ (i.e., the buried load is twice the size of the surface load) and $T_e = 20$ km explain the observed admittance data well. Therefore, in orogenic belts where surface topography and buried loads are significant, the Bouguer admittance may be biased towards the surface-loading case and, as a result, may significantly underestimate T_e.

Forsyth (1985) found, however, that unlike the admittance, the coherence based on the Bouguer anomaly is not very sensitive to f_l. He therefore suggested the use of the coherence rather than the admittance to estimate T_e. While coherence has been used to estimate T_e in oceanic regions (e.g., Ito and Taira, 2000), the technique has most widely been applied to gravity and topography data in the continents.

The coherence can be most easily understood by consideration of 'end-member' cases. At short wavelengths, the coherence approaches zero. This is because the lithosphere appears strong at these wavelengths, regardless of its actual T_e structure. Surface loads therefore cause a small flexure, and so there will be little contribution from these loads to the Bouguer anomaly. Buried loads will also cause a small flexure, and so there will be little contribution from these loads to the topography. The net result is a poor correlation between the Bouguer anomaly and the topography. At long wavelengths, however, the coherence approaches 1. This is because the lithosphere appears weak at these wavelengths. Surface and buried loads therefore cause a large flexure, and so these loads contribute significantly to the Bouguer anomaly. The net result is that the Bouguer anomaly will be strongly correlated with the topography. The rollover in the coherence from low to high values therefore reflects the wavelength range for which loads (surface and buried) are supported by the finite strength of the lithosphere.

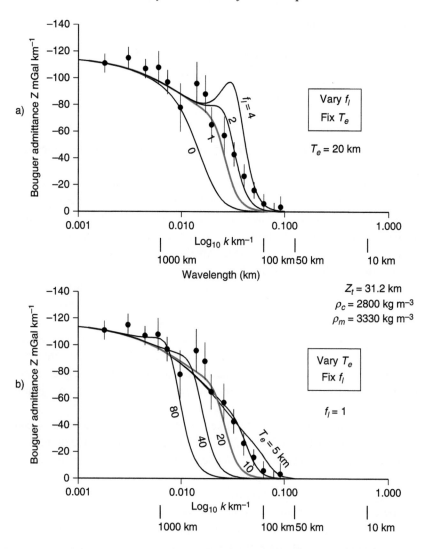

Figure 5.30 Combined surface- and buried-loading model showing the sensitivity of the Bouguer admittance of North America to (a) variations in f_l and (b) variations in T_e. The red solid line shows the case of $f_l = 1$ (i.e., equal surface and buried loading and $T_e = 20$ km). Note that $Z_t = 31.2$ km, the assumed depth of the Moho in both loading cases.

Forsyth (1985) showed that the coherence, considering both surface and buried loads, can be written as follows:

$$\gamma^2 = \frac{\langle H_t W_t + H_b W_b \rangle^2}{\langle H_t^2 + H_b^2 \rangle \ \langle W_t^2 + W_b^2 \rangle},$$

(5.55)

where < > indicate spectral averaging. This form of the coherence assumes that the surface and buried loadings are statistically independent. The coherence calculation therefore requires an estimate of the parameter pairs H_t, H_b and W_t, W_b. These parameters are related (e.g., Eqs. 5.34 and 5.47) by the flexural response functions for surface and buried loading, respectively. Also,

$$\begin{aligned} H &= H_t + H_b \\ W &= W_t + W_b \end{aligned}$$ (5.56)

where H is the combined surface topography due to surface and buried loading, and W is the combined flexure at the base of the crust due to surface and buried loading. Since H and W are known, we have all the information needed to evaluate H_t, W_t, H_b and W_b. H is simply the observed topography, H_t and H_b can be found from H if we assume f_l and T_e, W is unknown. An estimate of W can be obtained by downward continuation of the observed Bouguer gravity anomaly and replacing it by a single, undulating density interface at depth. W_t and W_b can be determined directly from H_t and H_b; however, if we assume f_l and T_e.

The observed and calculated coherences for North America, Africa, Brazil and Australia are compared in Fig. 5.31. The observed coherence, which is based on Eq. 5.36, is shown for both the entire continent-wide region (dashed line) and smaller sub-regions (filled circles and triangles). The calculated coherences have been computed using Eq. 5.51 by evaluating H_t, W_t, H_b and W_b, assuming T_e values that are in the range of 5 to 120 km. The figure shows that the coherence for the continent-wide regions is generally flatter than the calculated coherence. In contrast, the coherence for the sub-regions fits the calculated coherence well. One explanation is that the continent-wide coherence is a 'composite' of the coherence from smaller, tectonically distinct sub-regions that have their own T_e. The continent-wide coherence for North America, for example, shows no obvious rollover and cannot be explained by a single calculated coherence curve. In contrast, the coherences for its sub-regions (Bechtel et al. 1990) of the northern Basin and Range (NBAR), an archetypical region of continental extension, and the cratonic lithosphere of the eastern Canadian shield (ECNS) fit the calculated curves well based on a T_e of 12 and 82 km, respectively. A similar relationship occurs between the continent-wide coherence for Africa (Hartley et al. 1996) and those of its sub-regions of the White Nile rift (WNR), an extensional system, and the cratonic lithosphere of the Leo uplift (LU) and Kalahari craton (KC).

The calculated coherence curves in Fig. 5.31 assume an equal amount of surface and buried loading (i.e., $f_l = 1$). The coherence plot for North America shows, for example, that an increase in f_l (due say to an increase in the amount of buried loading) shifts the rollover in the calculated coherence curves to shorter wavelengths. Hence, a T_e that is determined by comparing the observed coherence with calculated curves based on $f_l = 1$ may be an overestimate. The plot also shows, however, that an increase in f_l does not change the steepness of the calculated coherence curves. Moreover, the shift in the rollover is relatively small (at least in comparison to the influence on the rollover of a change in T_e). Therefore, a change in f_l does not significantly alter the conclusion that North America is associated with a wide range of T_e values.

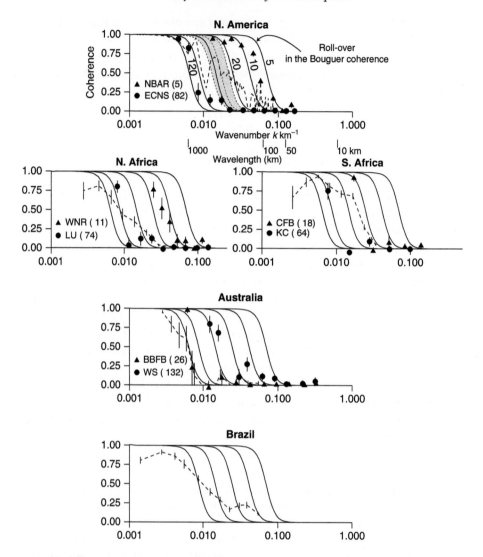

Figure 5.31 Comparison of observed and predicted coherences for North America, Africa, Australia and Brazil. The observed coherences are from Bechtel et al. (1990), Hartley et al. (1996), Zuber et al. (1989) and Ussami et al. (1993). CFB = Cape fold belt, BBFB = Bowen basin/northern Lachlan fold belt, WS = western shield. Other sub-regions are defined in the text. The numbers in parentheses are the best-fit estimates of T_e. The calculated coherences are (from right to left) based on T_e = 5, 10, 20, 40, 80 and 120 km and the same density and crustal thickness parameters as those used in Fig. 5.30. The thin dashed lines on the North America plot show the calculated coherence for f_l = 0.5, 2 and 5 (from left to right) and T_e = 40 km (the region 0.5 <f_l<5 is shaded grey).

The results of these early continental admittance and coherence studies are summarised in Table 5.2.[6] The studies have, for the most part, been based on the classical Fourier transform technique, the periodogram method, in which smooth spectral estimates are obtained by repetitive mirroring of windowed tapered data (e.g., Lewis and Dorman, 1970, 1972; Forsyth, 1985). The application of these techniques to North and South America, Africa, Eurasia and Australia show that the continental lithosphere is characterised by a wide range of T_e values and, hence, is capable of both great weakness and great strength.

Results from coherence studies are displayed in a histogram in Fig. 5.32, where they are compared with the results of 'forward modelling' from a wide range of tectonic settings including foreland basins, intracratonic basins and continental rifts. Forward modelling includes those cases where simple models of flexure are used to calculate the gravity anomaly and topography (and, in some cases, crustal structure, stratigraphy and crustal motions), compared to observations and the best fitting T_e determined by 'trial and error'. Some workers (e.g., Kogan and McNutt, 1987) have already noted that the results from spectral studies agree reasonably well with the results of forward modelling in certain tectonic areas such as Eurasia. Figure 5.32 suggests that there is general agreement in the case of the entire data set, at least regarding the range of values. There is also a hint that the peak in low values at shallow depths (15–25 km) in the forward modelling data is repeated in the spectral data.

Despite this success, the application of spectral techniques to the continents has not been without controversy. The main problem is whether it is valid to use the Bouguer coherence to estimate T_e when the present-day topography may not only be the product of surface or buried loads. Other factors, such as erosion and sedimentation, may modify continental topography, especially at short wavelengths.

During the past two decades, new methodologies have been proposed which have advanced our understanding of continental T_e. Most notable have been the maximum entropy (e.g., Lowry and Smith, 1994; Wang and Mareschal, 1999), multitaper (e.g., McKenzie and Fairhead, 1997; Ojeda and Whitman, 2002) and wavelet (e.g., Stark et al., 2003; Audet and Mareschal, 2007) methods. The main advance over the earlier periodogram techniques has been to reduce the variance and improve the spatial resolution of spectral estimates. The application of the maximum entropy technique to North America, for example (Lowry and Smith, 1995; Wang and Mareschal, 1999), resulted in smoother spectral estimates and showed that continental T_e varies on horizontal length scales that are as small as a few hundreds of kilometres.

In the multitaper technique, the Cosine taper is replaced by an orthogonal set of tapers designed to minimise spectral leakage. McKenzie and Fairhead (1997), for example, used the technique to evaluate the effect of erosion: finding it significant enough that they questioned the validity of T_e estimates based only on the Bouguer coherence. These workers point out that, irrespective of T_e, the removal, by erosion, of the topographic expression of surface and buried loads will always reduce the coherence between gravity anomaly and

---

[6] Table 5.22 is available from the Cambridge University Press server.

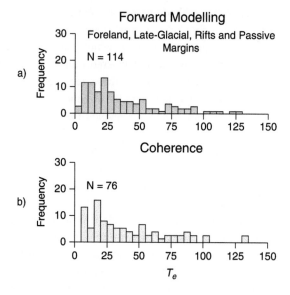

Figure 5.32 Comparison of T_e results based on (a) forward modelling and (b) coherence. The coherence results are based on spectral studies which utilise the Bouguer anomaly and either the method of Forsyth or Lowry and Smith. The results are based on data in Tables 6.2 and 6.3.

topography data. If the effects of erosion occur at short wavelengths, then the effect on the Bouguer coherence would be to shift the rollover to longer wavelengths. The T_e based on the Bouguer coherence may not, therefore, be a true estimate, but an upper bound.

In order to address the problem, McKenzie and Fairhead (1997) recommend that spectral studies in the continents be based on the free-air rather than the Bouguer gravity anomaly. They point out that the surface topography, unlike buried loading, is a known load that acts on the crust and lithosphere. Moreover, the free-air anomaly (unlike the Bouguer anomaly) has not had the contribution to the gravity anomaly of the surface topography removed from it. Therefore, in their view the free-air anomaly provides the best measure of the coherence between gravity and topography that exists, to some degree, in all models of isostatic compensation.

Most previous coherence studies have been based on the Bouguer anomaly. McKenzie and Fairhead (1997) therefore suggest two criteria need to be satisfied before the T_e determined from these studies can be considered reliable. The criteria both concern the behaviour of the free-air anomaly at short wavelengths. First, the free-air coherence should approach one (i.e., the coherence between free-air gravity anomaly and topography should be high). Second, the power in the gravity effect of the uncompensated topography should be comparable to the power in the observed free-air gravity anomaly.

McKenzie and Fairhead (1997) applied their criteria to gravity and topography data over Western Australia. They showed that the Bouguer coherence could be explained by Forsyth's model with $f_l = 1$ (i.e., equal surface and buried loading) and $T_e = 130$ km, in agreement with earlier results (Zuber et al., 1989). They also showed, however, that Western Australia fails their two criteria. The coherence between the free-air anomaly and the topography is not close to 1 at short wavelengths. Moreover, the power in the gravity effect of the uncompensated topography at short wavelengths is significantly less than the power in the observed free-air anomaly.

As regards future spectral studies, McKenzie and Fairhead (1997) recommended the free-air admittance and surface loading be used rather than the Bouguer coherence to constrain T_e. They based their argument on the fact that (a) the surface topography is a known load such that part of the free-air gravity field will always be coherent with the topography, and (b) the extent of the correlation between gravity and topography will be more obvious in the admittance than the coherence. Unfortunately, McKenzie and Fairhead (1997) were unable to obtain a good fit between the calculated free-air admittance based only on surface loading and the observed admittance of Western Australia – probably because of the subdued topography in this region. They did, however, obtain good fits to the admittance of East Africa, Peninsula India and for different sub-regions of the United States and Siberia. Their resulting T_e values were in the range of 6–24 km, which are significantly lower than values determined by previous workers that were based on the Bouguer coherence.

The criteria used by McKenzie and Fairhead (1997) certainly apply in the oceans. The fact that the free-air coherence does not approach one at short wavelengths or that the power in the gravity effect of the uncompensated topography is less than the observed power in the free-air anomaly can be attributed to the attenuating effects on the gravity of the water depth and instrument noise. Otherwise, the coherence in the oceans would probably approach one and the power in the uncompensated topography would be equal to that of the observed free-air anomaly.

The question is whether it is realistic to expect that the coherence between free-air gravity and topography would be high at short wavelengths in the continents. The free-air coherence approaches one in the oceans (e.g., Fig. 5.20a), but these regions are dominated by surface loads. Moreover, erosion is not as important in the oceans as it is in the continents. In the continents, forward modelling studies suggest that buried loads occur in addition to surface loads. Furthermore, there is evidence that these loads modify both the free-air and the Bouguer coherence at short wavelengths, as has been clearly shown by Forsyth (1985).

The fact that Western Australia is a region where both the free-air coherence and the power in the gravity effect of the uncompensated topography are low might therefore be due more to buried loading than erosion. The main reason for this is that buried loads (especially those within the crust) have the capacity to leave their signature in the free-air gravity anomaly, but not necessarily the surface topography, especially in the case of high T_e.

Other regions where McKenzie and Fairhead (1997) documented that the power in the uncompensated topography is low include the eastern US, central US, Peninsula India and the Siberian shield. These regions might therefore also be characterised by a significant contribution of buried loading.

The last point is illustrated for the Siberian Shield in Fig. 5.33. The upper profile shows a comparison of the observed and calculated free-air admittances based on a model of surface

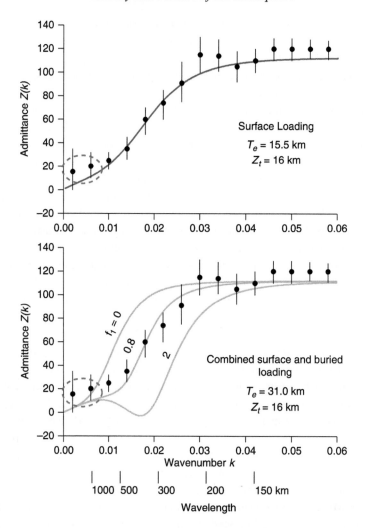

Figure 5.33 Comparison of the observed free-air admittance for the Siberian Shield with calculated curves based on surface loading and the combined surface and buried loading. Note the sensitivity of the admittance to the type of loading, with surface loading suggesting $T_e = 15.5$ km and combined surface and buried loading suggesting $T_e = 31$ km. The red dashed lines highlight high admittance values ($\sim$20 mGal km^{-1}) at long wavelengths which are difficult to explain by any loading model. The observed values (filled circles) are based on fig. 7 of McKenzie and Fairhead (1997), copyright by the American Geophysical Union.

loading only (i.e., $f_l = 0$), $T_e = 15.5$ km and a mean crustal thickness of $Z_t = 16$ km. This is the preferred model of McKenzie and Fairhead (1997). The lower profile compares the observed admittance to calculations based on a combined surface and buried loading model with the same crustal thickness as used in the upper profile, but with $T_e = 31$ km and $f_l = 0$, 0.8 and 2. The comparisons show that it is possible to explain the Siberian Shield admittance with a value of T_e that is twice as high as that determined by McKenzie and Fairhead (1997), provided that it is assumed that buried loading also contributes to the gravity and topography field.

Those continents where McKenzie and Fairhead (1997) found that the uncompensated and observed power was comparable at short wavelengths might therefore be regions with little or no buried loads. In these cases, the T_e deduced from the Bouguer coherence will not be an upper bound, but a true estimate. The continents in this category are East Africa, western US and eastern Siberia. Indeed, the Bouguer coherence and the free-air admittance results for these regions agree to within a factor of 2 and $6 < T_e < 34$ km, which supports the view of McKenzie and Fairhead (1997) that continental T_e can be low. T_e values also appear low ($10 < T_e < 24$ km) for eastern US, Peninsula India, Siberian shield and central US. However, as demonstrated in Fig. 5.33 for the Siberian Shield, the free-air admittance in these regions might be explained equally well by higher T_e values, provided that buried loading is considered.

McKenzie (2003) subsequently considered buried loading, but he disagreed with the conclusions that would have been drawn had the Bouguer anomaly and Forsyth's load deconvolution technique been used. In topographically subdued areas such as cratons, he argued, Forsyth's technique would overestimate T_e because the topographic response to buried loading would be small and so T_e would appear high even if it was not. Coherence would also be low, and so the 'rollover' he argued would likely shift to longer wavelengths.

The issue of low coherence is an important one that has triggered much debate. Swain and Kirby (2003), for example, used a modified sub-surface loading model for 'density anomalies' in the crust, together with synthetic tests, to argue that Forsyth's method would underestimate, not overestimate T_e. They agreed with McKenzie (2003) that the topography of many cratons is subdued and has low coherence. However, Bouguer gravity anomalies over cratons, especially Precambrian sutures, are significant (e.g., Gibb and Thomas, 1976), and they imply loads in the crust and mantle that have been supported by the strength of the crust and lithosphere for long periods of geological time. It is precisely because topography is subdued that a strong elastic lithosphere, not a weak one, is required.

Pérez-Gussinyé et al. (2004) have shown that T_e estimates obtained by comparing observed multitaper admittance or coherence functions to theoretical functions could be biased downwards. However, when observed and theoretical functions are windowed in the same way, then the resulting T_e is not biased. They found that when a self-consistent windowing scheme was applied to Europe, for example, that the Bouguer coherence and free-air admittance recover essentially similar results. T_e in Europe correlates well with geological terranes, being generally high over Precambrian cratons, intermediate over the Caledonian, Variscan (Hercynian) and Alpine orogenic belts and low over continental rifts.

In 2003, a technique based on a wavelet rather than a Fourier transform was introduced by Stark et al. (2003). Wavelets have the advantage that properties at a particular point can be examined across a range of scales. They obtain, in a sense, the local spectrum of a signal. Stark et al. (2003) used the technique to show that T_e in South Africa is in the range of 25–50 km with the highest values over the Kaapvaal craton.

Since 2003, a mix of inverse techniques have been used. Tassara et al. (2006), Audet and Maraschal (2007), Audet and Burgmann (2011) and Audet (2011) have pursued wavelets, while Swain and Kirby (2006), Pérez-Gussinyé et al (2007), Pérez-Gussinyé et al (2009) and Kirby and Swain (2009) have focussed on multitaper techniques. Sacek and Ussami (2009) compared the observed depth to the base of the sub-Andean foreland basin sequence to the predicted depth implied by the T_e structure results from wavelets, multitaper and forward modelling in the bend region of the Central Andes. They found that overall, the multitaper technique performed better than the wavelet. However, they concluded that the best agreement to observations was for the results based on forward modelling.

Nevertheless, the results from spectral methods have provided important new insights into plate properties. In the oceans, a close agreement exists between the results from spectral and forward modelling. In the continents, the agreement is not as good. A problem that has plagued all methods has been the strong asymmetry in plots of the Root Mean Square (RMS) difference between the observed and calculated gravity anomaly, which results in it being relatively easy to determine how low T_e can be, but more difficult to determine how high it might be. However, a consensus is now emerging that continental T_e has a rich structure and shows both low and high values. The highest values exceed 70 km (e.g., Tassara et al, 2006; Kirby and Swain, 2009; Audet, 2011) and may exceed 100 km and correlate with cratons, while the lowest values are < 5 km and generally correlate with rifts (e.g., Lowry and Smith, 1995; Pérez-Gussinyé and Watts, 2005). Cratons, however, are not uniformly high (e.g., Jordan and Watts, 2005; Swain and Kirby, 2006; Audet and Mareschal, 2007). Neither are all rifts low.

We need to keep in mind, of course, limitations in spectral methods. Unlike forward modeling, which focuses on the flexure that is associated with a particular loading event, a spectral estimate of T_e reflects all the loads in a window that have perturbed the lithosphere through time. This includes loads associated with a particular geological event (e.g., orogeny or rifting) as well as subsequent loads and unloads such as those associated with sedimentation and erosion. While erosion generally reduces the power in the topography, it can also enhance it through excavation, large-scale slumping and sliding and regional uplift.

Regions of high coherence do exist in the continents. Paxman et al. (2016), for example, have shown that the coherence between free-air gravity anomaly and topography in the sub-glacial Gamburtsev Mountains of Eastern Antarctica is > 0.8 for $0.05 < k < 0.28$ (i.e., wavelengths in the range 22–125 km), which is as high as that achieved in the oceans. The Gamburtsev have been ice covered since ~34 Ma, and so the high coherence can be explained by ice which has essentially 'entombed' the topography and prevented the normal processes of sedimentation and erosion following orogeny from reducing the power in the topography.

Irrespective of this special case, sedimentation and erosion generally work to modify topography after the 'true' T_e due to surface and buried loading have left their signatures in the gravity and topography fields (Simons et al, 2000). The timescales of these processes, which involve both loads and unloads of the crust, may differ, and as we will see in a later chapter result in a significantly different T_e for each process. Spectral studies may therefore reflect only an average of the combined effects of all the loads and unloads that have acted on a plate through time from the initial topography-forming events up to the present day.

5.7 The Long-Wavelength Admittance

It has been recognised for some time that the long-wavelength gravitational admittance does not always follow the predictions of simple isostatic models, which imply that the admittance should approach zero at long wavelengths. In many instances the admittance is too high. This is well seen in the oceanic case in Fig. 5.24 which shows that the long-wavelength admittance departs significantly from what would be expected for flexural isostatic models. Other departures are seen in the case of the continents, for example, in Fig. 5.33.

When considering the long-wavelength admittance, it is necessary to carefully consider window size (Ojeda and Whitman, 2002; Pérez-Gussinyé et al, 2004; Crosby, 2007). If, as suggested by the spectral data in Figs. 5.24 and 5.33, the wavelength of convection is >1,000 km, then the window size needs to be 2,000 km or higher (Crosby, 2007), which is significantly higher than usually used in spectral studies. The Siberian analysis window used to construct the data in Fig. 5.33, for example, is 1,000 × 1,000 km, and so admittances are likely poorly resolved at the wavelengths associated with convection. However, in the Central Pacific analysis windows in Fig. 5.24, all but the first two windows are greater than 2,000 km, which should yield a reliable admittance. Indeed, the long-wavelength admittance of 18 mGal km^{-1} that has been deduced compares well with the slope of 22 mGal km^{-1} found by Watts (1978) and the 15–20 mGal km^{-1} found by McKenzie (1994) from free-air gravity anomaly and topography profiles of the Hawaiian swell region. The swell has been interpreted as an upwelling region in the mantle that has perturbed the overlying plate and may be one of a number of such focus sites for convection in the central Pacific Ocean.

That long-wavelength admittance data might reflect processes in the convecting mantle beneath the continents has been recognised by a number of workers (e.g., Lago and Rabinowicz, 1984; D'Agostino and McKenzie, 1999), including McKenzie (2010) and Kirby and Swain (2014) who analysed gravity and topography data over the same 2,000 × 2,000km-wide region of the Canadian shield. McKenzie (2010) used a multitaper Fourier method to show that the admittance does not approach zero and is ~60–75 mGal km^{-1} at wavelengths > 1,000 km. He interpreted the high values at long wavelengths as the result of a combination of the flexural response to surface loads, glacial isostatic rebound and mantle convection. Kirby and Swain (2014) also used the multitaper Fourier transform method (with the same number of tapers) and found a similar long-wavelength high in the admittance (Fig. 5.34d).

The main difference between the two studies was in the spatial resolution of the admittance. The multitaper admittance of McKenzie (2010) had large error bars and showed no clear isostatic 'rollover', while the multitaper admittance of Kirby and

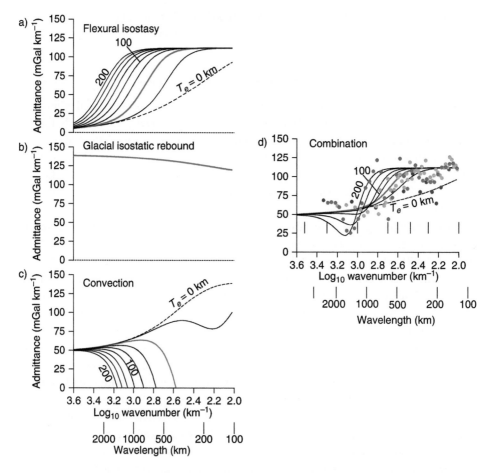

Figure 5.34 Comparison of observed and calculated free-air admittances over North America. (a) Calculated admittance for surface (topographic) flexural loading assuming crustal thickness and crust and mantle densities of 30 km, 2,670 and 3,330 kg m^{-3} respectively and different values of T_e. Solid red line shows the case for $T_e = 50$ km. (b) Calculated admittance for glacial isostatic rebound assuming densities of the upper crust and the lower crust of 2,700 and 2,900, respectively. (c) Calculated admittance for the convection model of McKenzie (2010) which assumes an elastic layer overlying a viscous half-space, convection driven by a temperature variation of 100°C, a coefficient of volume expansion of 4×10^{-5}°C^{-1}, a lithosphere thickness of 120 km, a ratio of glacial to convective topography of 0.5 and different values of the elastic thickness, T_e. (d) Comparison of observed admittances derived by McKenzie (2010, blue filled circles) and Kirby and Swain (2014, green filled circles) from multitaper methods with a calculated admittance based on a combination of the flexural, glacial isostatic and convection admittances derived by Kirby and Swain (2014, red filled circles) from wavelet methods. The dashed line shows the case $T_e = 0$ km (i.e., an Airy model).

Swain (2014) had small error bars and a clear rollover at wavelengths of ~300–400 km (Fig. 5.34d). The existence of an isostatic rollover was confirmed by Kirby and Swain (2014) using the wavelet transform method. They found the rollover had shifted to a wavelength of ~500 km and, interestingly, that there was a dip at wavelengths ~800–1,200 km between the long- and short-wavelength observed admittances. These observations are consistent, they claimed, with a significantly higher T_e than that derived with the multitaper Fourier transform method (the best fit obtained by McKenzie, for example, was 29 km) of > 80 km over the shield.

5.8 Summary

Spectral techniques have provided a powerful new tool with which to quantitatively evaluate the state of isostasy at Earth's bathymetric or topographic features. These techniques can be used to determine information on isostasy directly from observed gravity anomaly and bathymetry or topography data (inverse modelling), as well as to calculate the gravity anomaly from bathymetry or topography data for different isostatic models (forward modelling).

In the oceans, the free-air admittance technique has been successful in characterising the state of isostasy at a wide range of features on the seafloor including seamounts and oceanic islands, aseismic ridges and mid-oceanic ridges. The reasons for this success are that surface loads in oceanic regions are generally well preserved, show a high coherence between their bathymetry and free-air gravity anomaly, and generally dominate over sub-surface or buried loads in the diagnostic waveband of flexure.

In the continents, the Bouguer coherence technique, which considers both surface and sub-surface loading, shows that the continents are characterised by a wide range of T_e values. Recent studies, which are based on the multitaper Fourier and wavelet transform techniques, have significantly improved the spatial resolution of continental T_e estimates. They have shown that in North America T_e varies on length scales that are as small as a few hundreds of kilometres.

The application of spectral techniques to isostatic problems has not been without controversy. In the continents, for example, the usefulness of the Bouguer coherence technique has been questioned. Early controversies focused on the admittance and whether it underestimates T_e in regions of high topography where there may be buried as well as surface loads. More recent controversies have focused on window size and the role that other factors, such as sedimentation and erosion, play in modifying the topography that is produced by surface and sub-surface loading and whether these loads will resolve the actual T_e structure in regions where topography is subdued.

The results of spectral studies compare well with those of forward modelling, especially in oceanic regions at seamounts and oceanic islands.[7] They also agree reasonably well in the continents. Spectral studies, for example, support the results from forward modelling which suggest that the continental lithosphere is capable of both weakness (i.e., low T_e) and

---

[7] More examples of forward modelling are given in Chapters 4 and 7.

strength (i.e., high T_e). However, some workers disagree, suggesting instead that continental T_e does not exceed 25 km.

Irrespective of the relative role of surface and sub-surface loading and the actual value of the elastic thickness, however, there is general agreement that both oceans and the continents are characterised by a long-wavelength admittance that is higher than expected from crustal isostatic models. The most likely explanation for this admittance is that it reflects some form of thermally driven convection in Earth's sub-crustal mantle.

6

Isostasy and the Physical Nature of the Lithosphere

I have usually treated the Earth's crust as behaving like a perfectly elastic solid until a certain stress-difference is attained, and then breaking or bending according to the temperature and pressure.

(Jeffreys, 1954, p. 358)

6.1 Introduction

Isostasy provides information on the way that geological loads are supported by the Earth's crust and mantle. It should therefore tell us about the materials that go to make up the lithosphere. It was realised early on, however, that the lithosphere has had a complex evolution, and it was not immediately clear what, if anything, the phenomenon of isostasy could tell us about its composition.

The early isostasists found it helpful to describe the phenomenon of isostasy by analogy with the behaviour of commonly occurring materials. Airy, for example, concluded that the Earth's crust and upper mantle behaved like 'rafts of timber floating on water'. He regarded the crust as a fundamentally weak material that had been broken up in such a way as to behave as isolated blocks. The underlying mantle, in his view, entered the support of loads only through the buoyancy it provided.

Barrell, as we saw in Section 2.3, held a different view. He likened the behaviour of the Earth's outer layers more to a 'bending plate' that overlies a weak fluid. The term 'plate' implied a material with a finite strength that was able to support surface loads by flexing over large distances. His use of the term 'fluid', rather than liquid, was also a judicious one. It conveyed the all-important idea that the sub-crustal mantle could also be involved in the support of these loads.

In his book *The Physics of the Earth's Crust*, Fisher (1881) argued that isostatic equilibrium occurred when the pressures on the crust due to tangential loads, its own weight, and buoyancy from the weak fluid underlying layer were in balance. In his view, movements of the crust involved changes in all three of these stress states. In considering its response to tangential compression, however, he thought that the crust was not completely flexible. It could bend to produce folding, but at a certain stress its materials would yield in such a way that the crust would thicken. The ideal material that showed these properties, in his view, was a plastic body.

We saw from Chapter 4 that it is possible, in many cases, to explain the observations of flexure in terms of a model of an elastic plate that overlies an inviscid fluid (i.e., a fluid with zero viscosity). There are several problems, however, with this model. Firstly, it is time-invariant and does not consider the possibility that flexure may change because of loading or unloading. Thirdly, it predicts bending stresses that are relatively high compared with the stresses induced by glacial loading and the stress drops observed in earthquakes. Finally, the elastic plate model does not always account for the observations of flexure.

Despite these problems, the elastic model has been widely applied. As a result, there are now a number of estimates of the long-term mechanical properties (e.g., flexural rigidity, elastic thickness) of the oceanic and continental lithosphere in different tectonic settings. The question that now needs to be addressed is, can these estimates be used to show how the lithosphere is actually behaving and, once this has been carried out, can we speculate on its physical nature?

6.2 The Behaviour of Earth Materials

The response of the lithosphere to surface loads depends not only on size, but also the timescale of loading (e.g., Fig. 6.1). For example, at the timescales of seismic surface waves (i.e., a few seconds to few hundred seconds), the sub-crustal mantle behaves elastically down to very great depths. We also know that at the relatively short geological timescales of ice loading (i.e., a few thousand to a few tens of thousands of years), the upper mantle flows like a viscous fluid. Finally, on long geological timescales (i.e., > 1 Myr), the crust and upper mantle behaves in a similar way as a thin elastic plate that overlies an inviscid substratum. Therefore, both elastic and viscous behaviour are likely ingredients in any model for the lithosphere's actual mechanical behaviour.

Elastic and viscous materials deform in fundamentally different ways. In a perfectly elastic model, the applied stress, σ, is related to the deformation, ε, by Hooke's law; namely,

$$\sigma = E\varepsilon, \tag{6.1}$$

where E is Young's modulus. We can schematically represent such a model as a spring that is fixed to a support (Fig. 6.2a). An elastic-perfectly plastic material responds initially elastically and then, when a certain yield stress is exceeded, responds plastically at constant load and can be represented by a spring in series with a friction plane (Fig. 6b). In a viscous model, the applied stress is related to the rate of change of the deformation with time. We then have

$$\sigma = \eta\dot{\varepsilon}, \tag{6.2}$$

where η is the viscosity and the superscript dot indicates the change in strain with time or strain rate. A viscous material can be represented by a dashpot (Fig. 6.2c). Unlike an elastic

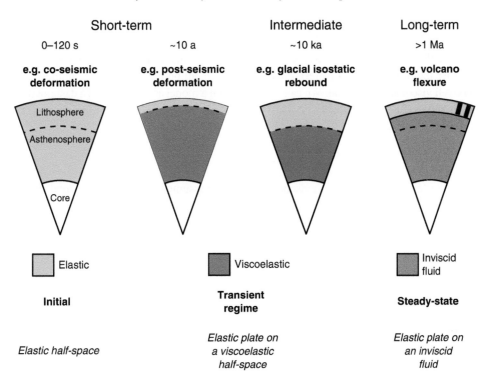

Figure 6.1 Schematic diagram illustrating how the outermost layers of the Earth respond to surface loads of different duration.

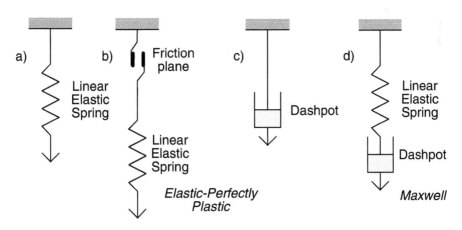

Figure 6.2 The behaviour of elastic and viscoelastic materials by analogy with springs, friction planes and dashpots.

model, which predicts a deflection that does not change, a viscous model will continue to deform with time.

Since the lithosphere may have both elastic and viscous properties, it is useful to first consider the properties of a material that combines them both. Probably the simplest is a Maxwell viscoelastic material. We can illustrate such a model as a spring and a dashpot that are in series (Fig. 6.2d). Since the same stress, σ, is applied to both the spring and the dashpot, we can write that

$$\sigma = E\varepsilon_e, \qquad \sigma = \eta\dot{\varepsilon}_v,$$

where ε_e is elastic strain and ε_v is viscous strain. If we assume linearity, then the total strain, ε, is given by

$$\varepsilon = \varepsilon_e + \varepsilon_v.$$

If the system is initially unstrained (i.e. $\varepsilon = 0$ at $t = 0$), then

$$\varepsilon_e = \frac{\sigma}{E}, \qquad (6.3)$$

and since

$$\dot{\varepsilon}_v = \frac{d\varepsilon_v}{dt} = \frac{\varepsilon_v(t) - \varepsilon_v(0)}{t - t_0} = \frac{\varepsilon_v}{t}$$

$$\varepsilon_v = \frac{t\sigma}{\eta} \qquad . \qquad (6.4)$$

$$\therefore \varepsilon = \frac{\sigma}{E} + \frac{\sigma t}{\eta}$$

We see from Eqs. (6.3) and (6.4) that the strain in a viscoelastic material is made up of both an elastic and a viscous part. There is a time, the Maxwell relaxation time, when the elastic strain that has accumulated is equal to that of the viscous strain. If this time is τ, then it follows that

$$\tau = \frac{\eta}{E}. \qquad (6.5)$$

6.3 Flexure of a Viscoelastic Plate

Walcott (1970a) first applied a viscoelastic plate model to the lithosphere. Using the solution of the general equation for the flexure of a viscoelastic plate obtained by Nadai (1963), Walcott derived the viscoelastic response functions for a plate overlying an inviscid fluid.

In the late 1970s and early 1980s, the viscoelastic plate model enjoyed some success. It was used, for example, to account for the flexure of the lithosphere at seamounts and oceanic islands (Lambeck and Nakiboglu, 1981), deep-sea trench-outer rise systems (De Bremaecker, 1977), the stratigraphic patterns that develop at the edge of sedimentary basins

(Sleep and Snell, 1976; Beaumont, 1978; Lambeck, 1984) and continent–wide Bouguer anomaly admittance data (McNutt and Parker, 1978).

However, as is shown later, the viscoelastic plate model fails to fully explain observations of the flexural rigidity and the equivalent elastic thickness of the lithosphere, T_e, which tends to stabilise on geologically long timescales. Nevertheless, it is instructive to review the theory and explain where it fails.

The general equation for the flexure of a viscoelastic plate overlying an inviscid substratum is given (e.g., Nadai, 1963) by

$$D_o \frac{\partial^4 \dot{y}}{\partial x^4} + (\rho_m - \rho_{infill})g(\tau \dot{y} + y) = 0, \tag{6.6}$$

where D_o is the instantaneous flexural rigidity of the plate, y is the flexure and $\dot{y}$ is the derivative of y with respect to time.

We can show by substitution in Equation (6.6) that the flexure of a viscoelastic plate due to a periodic load is also periodic. For example, if the applied load is of the form

$$(\rho_l - \rho_w)gh\cos(kx),$$

where ρ_l is the density of the load, ρ_w is the density of the material displaced by the load and h is the height of the load, then

$$y = \frac{(\rho_l - \rho_w)h\cos(kx)}{(\rho_m - \rho_{infill})\left[1 + \frac{D_0 k^4}{(\rho_m - \rho_{infill})g}\right]} \left[1 + \frac{D_0 k^4}{(\rho_m - \rho_{infill})g}\left[1 - e^{-\frac{t}{\tau}\left[1 + \frac{D_0 k^4}{(\rho_m - \rho_{infill})g}\right]}\right]\right]. \tag{6.7}$$

The behaviour of a viscoelastic plate can be illustrated by considering the limiting cases that the loading time, t, is small in comparison to the Maxwell relaxation time, τ. We see from Eq. (6.7) that

$$t \to 0, \quad y \to \frac{(\rho_l - \rho_w)h\cos(kx)}{(\rho_m - \rho_{infill})\left[1 + \frac{D_o^{} k^4}{(\rho_m - \rho_{infill})g}\right]},$$

which is the response of an elastic plate with an initial flexural rigidity D_o. Therefore, a viscoelastic plate behaves in a similar way as an elastic plate when the loading time is short. Also, it follows that

$$t \to \infty, \quad y \to \frac{(\rho_l - \rho_w)h\cos(kx)}{(\rho_m - \rho_{infill})},$$

which is the Airy response if $\rho_l \to \rho_c$ and $\rho_{infill} \to \rho_c$. Therefore, the response of a viscoelastic plate approaches that of an Airy model at long loading times.

It is useful, as it was in the case of the elastic plate model, to define the wavenumber parameter, ϕ_v, which modifies the Airy response to produce the viscoelastic flexure. To distinguish it from ϕ_e, the flexural response function (see Section 5.2), we will call it the 'viscoelastic flexural response function'. The function is obtained by considering the Airy response of a viscoelastic plate as the input and the viscoelastic flexural response as the output. We then have

$$\phi_v(k,t) = \frac{\text{Output}}{\text{Input}}$$

$$\therefore \ \phi_v(k,t) = \frac{\left[1+\dfrac{D_o k^4}{(\rho_m - \rho_{infill})g}\left[1 - e^{-t/\tau\left[1+\dfrac{D_o k^4}{(\rho_m - \rho_{infill})g}\right]}\right]\right]}{\left[1 + \dfrac{D_o k^4}{(\rho_m - \rho_{infill})g}\right]}. \qquad (6.8)$$

The flexure for the general case of an arbitrary-shaped load on a viscoelastic plate can now be computed. The flexure is given, for example, by Eq. (5.5):

$$Y(k,t) = \phi_v(k,t)H(k)\frac{(\rho_l - \rho_w)}{(\rho_m - \rho_{infill})}, \qquad (6.9)$$

where $H(k)$ is the Fourier transform of the topography and $Y(k)$ is the Fourier transform of the flexure. The density terms have already been defined. The flexure is obtained by first taking the Fourier transform of the load, multiplying it by the response function and a density term, and then inverse Fourier transforming the result[1].

Figure 6.3 shows the flexure of a viscoelastic plate for cases of a 'wide' and a 'narrow' load on the sea floor and values of $t/\tau = 1$, 10 and 50. The green dashed line shows the Airy response, while the red dashed line shows the initial (i.e., $t/\tau = 0$) elastic response. We see from the figure that as t/τ increases, the flexure immediately beneath where the load is applied increases, approaching that of the Airy case. How rapidly it approaches the Airy case, however, depends on the load width. For a narrow load, the flexure does not reach the Airy case even after a time has elapsed that is long when compared with the relaxation time. For wide loads, the Airy response is approached more rapidly. Isostatic adjustment in the viscoelastic model therefore depends not only on load age, but also on the size of the load.

As shown in Chapter 4, it is the flexural rigidity, D, that is usually estimated from observations. The flexure equations suggest, however, that for each profile that is calculated using a viscoelastic plate model, there should be an equivalent profile that describes the flexure of a purely elastic plate. The problem is that the viscoelastic plate equations are

---

[1] A MATLAB file, chap6_maxwell_viscoelastic.m, to calculate the flexure of a viscoelastic plate that overlies a weak (inviscid) fluid by an arbitrary-shaped two-dimensional load is available from the Cambridge University Press server.

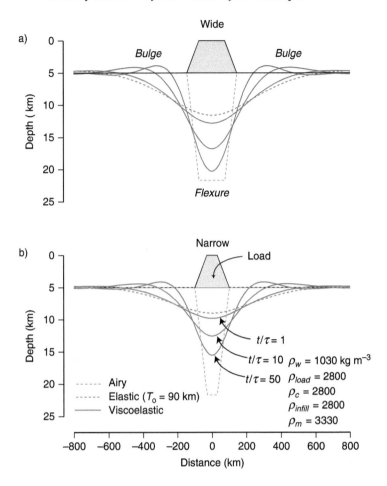

Figure 6.3 Flexure of a viscoelastic plate that overlies a weak (inviscid) fluid by (a) a 'wide' and (b) a 'narrow' surface load for different values of the ratio of time to Maxwell relaxation time, t/τ.

highly non-linear and so it is difficult (De Rito et al., 1986) to determine the equivalent elastic thickness of a viscoelastic plate exactly.

One way to approach the problem is to consider the response functions for the elastic and viscoelastic plates, since both functions give the response of the plate to a unit load. The flexural response function for an elastic plate (Eq. 5.3), for example, is given by

$$\phi_e(k) = \left[\frac{D_e k^4}{(\rho_m - \rho_{infill})} + 1 \right]^{-1}, \tag{6.10}$$

where D_e is the flexural rigidity. Equating Eqs. (6.8) and (6.10) then gives

$$D_e = \frac{(\rho_m - \rho_{infill})g}{k^4} \left[\frac{\left[1 + \frac{D_o k^4}{(\rho_m - \rho_{infill})g}\right]}{\left[\left[1 + \frac{D_o k^4}{(\rho_m - \rho_{infill})g}\right]\left[1 - e^{-t/\tau\left[1 + \frac{D_o k^4}{(\rho_m - \rho_{infill})}\right]}\right]\right]} - 1 \right], \quad (6.11)$$

where D_e is the equivalent flexural rigidity of an elastic plate that describes the flexure of a viscoelastic plate and D_o is the flexural rigidity of a viscoelastic plate at $t = 0$.

A log–linear plot of D_e/D_o against t/τ for a 'narrow' and a 'wide' load is shown in Fig. 6.4. The terms 'narrow' and 'wide' are used in this case to describe a load with a wavelength, λ_w, of 100 and 1,000 km, respectively. Other parameters assumed in the calculations are the same as were used to construct Fig. 6.3. We see from the figure that if τ is constant, then the flexural rigidity that best accounts for the flexure of a viscoelastic plate decreases with time. The decrease depends on load size, being more rapid for the long-wavelength load than for the short-wavelength one. It also depends on load age. When $t/\tau = 1$, D_e/D_o is 0.50 for 'narrow' loads and 0.45 for 'wide' loads. The flexural rigidity of a viscoelastic plate that has an initial value of 6.48×10^{24} N m will therefore be reduced to 3.44×10^{24} N m and 2.92×10^{24} N m for a narrow and wide load, respectively, after 1 Myr, assuming $\tau = 1$ Myr. The corresponding reduction in the equivalent elastic thickness, T_e, is from 90.0 km to 62.0–72.9 km, assuming $E = 100$ GPa and $\nu = 0.25$. After 10 Myr, the elastic thickness is further reduced to 39.0–41.8 km. Therefore, for a particular relaxation time, T_e, of a viscoelastic plate depends strongly on the age of the load, being large soon after a load is emplaced and smaller at subsequent times.

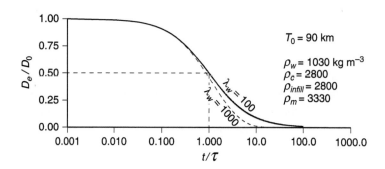

Figure 6.4 Normalised plot of D_e/D_0 against t/τ for a 'narrow' (solid line) and 'wide' (dashed line) load.

6.4 Relationship of Elastic Parameters to Load and Plate Age

Walcott (1970a) first suggested that the flexural rigidity of Earth's lithosphere may depend on load age. He based this suggestion on a comparison of values at young loads, such as at glacial lakes, and older loads, such as seamounts and oceanic islands. In particular, he noted that young glacial lakes tend to be associated with high values of flexural rigidity while old seamounts and oceanic islands have lower ones. The decrease in rigidity with increase in load age could be quite well described by a Maxwell viscoelastic plate model with a relaxation time, τ, of ~ 0.1 Myr, which implies an effective viscosity for the lithosphere of $\sim 1.05 \times 10^{23}$ Pa s.

However, Walcott's model raised several questions. Firstly, by including values from the oceans and continents on the same plot of flexural rigidity against load age, was he comparing 'like with like'? Several authors thought not. Secondly, why was it that, when individual features were examined, a single value of the parameter τ could not explain all the observations? Watts (1978), for example, compared the observed admittance along a number of profiles of the Hawaiian–Emperor seamount chain with calculations based on a viscoelastic model. He showed that a different relaxation time would be required along the chain on either side of the bend in the Hawaiian–Emperor seamount chain. The admittance data for the younger Hawaiian Ridge, for example, could be explained by τ of 0.10–1.00 Myr, while the data for the older Emperor seamounts required τ of 1.00–10.0 Myr. Such differences in τ could not be explained by differences in the initial elastic thickness, because the age of the oceanic lithosphere was similar on either side of the bend in the seamount chain and, hence, the initial flexural rigidity would be expected to be similar. When the continents were considered separately, even higher values of the relaxation time were found. McNutt and Parker (1978), for example, concluded that differences in continent-wide admittances for Australia and North America could be explained by the difference in age between the two continents, with the lower admittance for the older Australian continent (e.g., Fig. 5.27) suggesting a relaxation time of some 45 Myr.

A final question, and perhaps the most critical one, was how could a viscoelastic lithosphere explain why old loads were apparently supported for such long periods of time? For example, there is evidence that the oceanic lithosphere can support loads such as the Emperor Seamounts and the Louisville Ridge, which are at least >50 Ma in age, essentially elastically. In the continents, plate and load age are difficult to resolve. However, the presence of large-amplitude free-air gravity anomalies over Palaeozoic orogenic belts and granites in the Precambrian Shield of North America is a strong argument that there has been no significant viscous relaxation to the stresses associated with orogeny and pluton emplacement.

The possibility that the flexural rigidity of the lithosphere may change with load age is of much significance, especially to geodynamics. The relaxation times deduced by Walcott (1970a), for example, from observations of continental and oceanic flexure have been widely used in flexural loading studies. We therefore need to examine the relation between elastic thickness and load age more closely. Since Walcott's study, there have been more estimates of the flexural rigidity published from a wider range of geological settings. We will now examine this new data and the information that it provides about the relationship between flexural rigidity and load and plate age.

6.4.1 Oceans

The results of some 67 studies of oceanic T_e published during 1970 to 2019 are summarised in Table 6.1.[2] The table summarises the type of load as well as information on the age of the underlying lithosphere, t_{sf}, the age of the load, t, and the elastic thickness, T_e. Also tabulated are the likely ranges and the main references to each estimate of T_e.

The first step in determining whether the oceanic lithosphere is viscoelastic or not would be to plot the ratio t/τ against D_e/D_o for each of the estimates listed in Table 6.1.

The main problem is in estimating D_o, the initial (i.e., $t = 0$) flexural rigidity. One approach is to assume that the initial elastic thickness, and hence flexural rigidity, corresponds to the short-term seismic thickness of the lithosphere.

Early surface wave studies (e.g., Leeds et al., 1974) showed a simple dependence between the seismic thickness and the age of the oceanic lithosphere. However, recent studies have questioned this dependency because of the effects of seismic anisotropy. Nishimura and Forsyth (1989) have considered anisotropy and estimate that the thickness of 0–4-Ma-old Pacific lithosphere is 15–35 km, increasing to 70–110 km for the oldest seafloor (i.e., 80–160 Ma). As Fig. 2.23 shows, these data can be explained well by the depth to the 1,100°C oceanic isotherm based on plate cooling models. More recent studies based on surface wave tomography, long-range seismic refraction using ocean bottom and borehole seismometer, and petrology (e. g., Ritzwoller et al. 2004; Kawakatsu et al. 2009; Priestley et al. 2018) confirm this to be a reasonable assumption. This isotherm may therefore be used with some confidence to estimate the initial elastic thickness and, hence, D_o.

The other unknown is the Maxwell relaxation time, τ, which is a proxy for the viscosity of the plate. This parameter could be estimated by comparing the predicted curves in Fig. 6.4 with observed values, the flexural rigidity and age of which have been normalised with respect to D_o and τ, respectively. Since the predicted curves are based on load wavelengths that are within the range of most of the features tabulated in Table 6.1, the value of τ that brings most of the observations into the range of these predicted curves can be determined.

It is apparent, however, that D_e/D_o is small. In general, it is < 0.05. We therefore cannot meaningfully superimpose the observations onto the calculated curves in Fig. 6.4. A more expanded plot, such as that shown in Fig. 6.5, is required instead. This

---

[2] Table 6.1 of oceanic T_e estimates is available from the Cambridge University Press server.

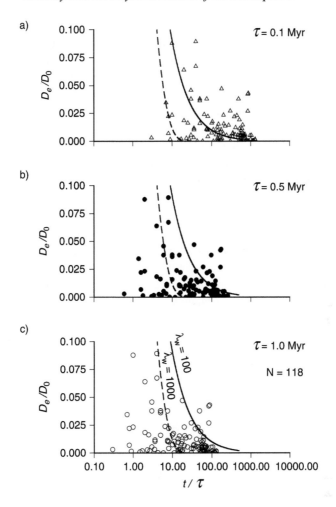

Figure 6.5 Comparisons of observed and predicted D_e/D_o against t/τ for seamounts and oceanic islands. D_e and t are based on the observations in Table 6.1, and D_o is estimated from the age of the underlying seafloor, t_{sf}. The predicted curves are based on load wavelengths of 100 (solid black line) and 1,000 (dashed black line) km. The best overall fit between observed and predicted values is for $\tau = 0.5$ Myr which corresponds (Eq. 6.5) to a viscosity of 1.6×10^{24} Pa s, assuming $E = 100$ GPa.

figure shows that a τ of the order of 0.5 Myr is required to bring most of the observed flexural rigidity estimates into the range of the predicted values. A smaller τ fails to explain the high t/τ values and a larger τ the low t/τ values. We note that a Young's Modulus, E, of 100 GPa and $\tau = 0.5$ implies (Eq. 6.5) a viscosity of the plate of $\sim 1.6 \times 10^{24}$ Pa s, which is three to four orders of magnitude higher than that deduced for the sub-plate asthenosphere from glacial isostatic rebound.

Before concluding that a viscoelastic plate with $\tau = 0.5$ Myr is an adequate description of the time-varying behaviour of the oceanic lithosphere, it is necessary to examine in more detail the relationship between the flexural rigidity, D_e, and the age of the load, t. This is ideally carried out by examining D_e at different age loads that were emplaced on oceanic lithosphere of the same age. Unfortunately, there are only a few examples of such cases. We will therefore examine instead those seamounts and oceanic islands that formed on > 80-Ma seafloor. This 'cut-off' age was chosen because seismic studies (e.g., Nishimura and Forsyth, 1989) indicate that the thickness of the lithosphere sphere changes only slowly for seafloor ages > 80 Ma. Hence, the effects of load age should dominate those of seafloor age in these tectonic settings.

Figure 6.6 shows a plot of D_e against t for seamount and oceanic island loads that formed on old (> 80 Ma) seafloor. The figure shows the data together with predicted relaxation curves for a viscoelastic plate with τ of 0.01, 0.1, 1 and 10 Myr. We see that the bulk of the data can be explained by $0.1 < \tau < 1$ Myr. These include estimates from such widely separated features as the Hawaiian Ridge in the central Pacific, Great Meteor and Canary Islands in the Atlantic, and Crozet Island in the

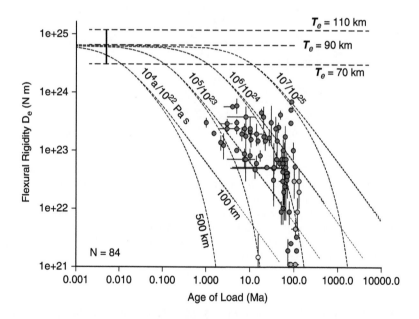

Figure 6.6 Comparison of the flexural rigidity, D_e, against age of load, t, for seamounts and oceanic islands that formed on seafloor older than 80 Ma (i.e., $t_{sf} > 80$ Ma) with predictions based on a viscoelastic plate model. The short, dashed lines show the rigidity for load wavelengths of 100 and 500 km. Labels (e.g., $10^6/10^{24}$) show the Maxwell relaxation time and equivalent viscosity of the plate. Orange filled circles are seamounts in the French Polynesia region.

Indian Ocean. However, some of the data plot outside of this range. The data for Koku (#22 in Table 6.1) and the New England seamounts (#200–202), for example, require $\tau > 1.0$ Myr, while Madeira (#111), Corner (#110), Maria (#148) and north-east Indian Ocean seamounts (#242, 245, 247) require $\tau < 0.1$ Myr. Therefore, when examined in detail, oceanic flexure data requires a large range of τ values. The τ values cannot be explained by local differences in D_o and therefore some other cause is required.

An important observation in Fig. 6.6 is that there are few low values of D_e for loads older than 1 Myr. In fact, if we exclude the Maria and Corner seamounts and the north-east Indian Ocean seamounts, there are no values $< 8 \times 10^{21}$ N m, which is equivalent to $T_e < 20$ km. This seems to suggest that T_e does not continue to decrease with increase in load age, as the Maxwell viscoelastic plate model would predict, but flattens out instead.

However, as De Rito et al. (1986) have demonstrated, it is possible to construct a viscoelastic model that is restrained enough to have significant long-term strength. These workers define a 'strength index' that is a ratio of the stress in a viscoelastic fibre to the stress that would be present in an elastic fibre that undergoes the same strain. Thus, a strength index of one is where elastic strains dominate and the lithosphere is strong, while an index of zero is where the viscous strains dominate and the lithosphere is weak. Like the viscoelastic plate, the behaviour of their model can be described in terms of a linear spring that is in series with a dashpot. The main difference is that the effectiveness of the dashpot (De Rito et al., 1986) is limited and depends inversely on a viscosity 'coefficient' (Fig. 6.7a). Such a model predicts a flexure that is initially elastic, subsides rapidly, and then – depending on a strength index – flattens with time.

Unfortunately, because the response was non-linear, De Rito et al. (1986) were unable to express their time-varying deflection as a response function and therefore it is difficult to apply their model to the general problem of the flexure of the lithosphere caused by arbitrary-shaped topographic or bathymetric loads.

A simpler modification of the viscoelastic plate model that incorporates a long-term strength has been given by Karner (1982). He used a model of a spring that is coupled with a dashpot in parallel with a second spring (Fig. 6.7c). Such a material is well known in engineering (e.g., solids in which vibrations are damped by internal friction), where it is referred to as a 'general linear substance'.

The response function of a general linear substance[3] is given (Karner, 1982) as follows:

$$\phi_{vl}(k,t) = \left[E_f + \left(E_i - E_f \right) e^{-\left(\frac{t}{\tau}\right)\frac{D_i E_i}{D_f E_f}} \right] e^{-kt}, \qquad (6.11\text{a})$$

where D_i and D_f are the initial and final flexural rigidity of the plate, respectively, τ is the Maxwell relaxation time, and the parameters E_i and E_f are given by

---

[3] A MATLAB file, chap6_hawaii_general_linear.m, to calculate the flexure of a viscoelastic plate that overlies a weak (inviscid) fluid by an arbitrary-shaped two-dimensional load is available from the Cambridge University Press server.

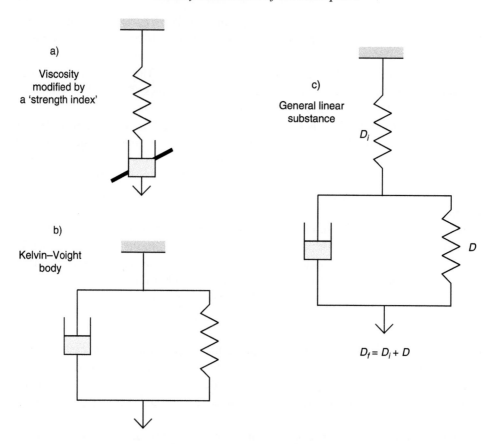

Figure 6.7 The behaviour of viscoelastic materials in which the viscous element is limited in some way. (a) The strength-limited model of De Rito et al. (1986) of a spring in series with a dashpot. (b) The Kelvin–Voight body of a spring in parallel with a dashpot. (c) The general linear substance model of Karner (1982) in which a spring is coupled with a dashpot that is in parallel with a second spring.

$$E_i = \left[1 + \frac{D_i k^4}{(\rho_m - \rho_{infill})g}\right]^{-1}$$

and

$$E_f = \left[1 + \frac{D_f k^4}{(\rho_m - \rho_{infill})g}\right]^{-1}$$

The flexure can then be obtained from

$$Y(k,t) = \phi_{vl}(k,t)H(k)\frac{(\rho_l - \rho_w)}{(\rho_m - \rho_{infill})}. \tag{6.12}$$

We see from Eq. 6.11a that when $t \to 0$, $\phi_{vl} \to E_i$, which is the response function for an elastic plate that has a flexural rigidity D_i. As $t \to \infty$, $\phi_{vl} \to E_i$, which is the response function for a plate with a flexural rigidity D_f. The linear viscoelastic plate model is therefore similar to a viscoelastic plate model for short times of loading, but, unlike this model, it approaches that of an elastic plate with a finite flexural rigidity at long times.

The behaviour of the general linear plate model is illustrated in Figs. 6.8 and 6.9. The figures show that, like the Maxwell viscoelastic plate, the subsidence of a general linear viscoelastic plate is initially rapid. At long times, however, the subsidence differs and approaches a constant value.

Figure 6.10 compares the predicted flexural rigidity of the linear viscoelastic plate model with the T_e data. The figure shows that a model in which the final T_e is in the range of 20–30 km can explain the bulk of the data. A recurring problem is that the model cannot account for the general trend of the data (indicated by the purple filled arrows) which intersects the calculated curves. Moreover, a wide range of final T_e values is required to explain the data.

The question that remains is the long-term behaviour of T_e. Does T_e data approach the prediction of the Airy model as the Maxwell viscoelastic plate model would predict, or a constant value as the general linear viscoelastic plate would predict? There is a suggestion that neither model can account for the T_e data. The flexural rigidity and, hence, elastic thickness appear to decrease more slowly than the Maxwell viscoelastic model would predict, yet not as slowly as the general linear plate. If this is the case, then we might wish to re-examine the dependency of T_e on load age to see whether other factors such as plate age may in some way be involved.

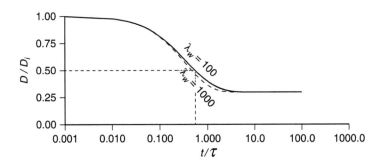

Figure 6.8 Plot of the ratio of the flexural rigidity to the initial flexural rigidity, D/D_i, against the ratio of the time elapsed and the relaxation time, t/τ, for the general linear viscoelastic model. The solid and dashed lines are based on a load wavelength of 100 and 1,000 km, respectively. The dotted lines show that when $t/\tau = 0.55$, the flexural rigidity has been reduced by a factor of 2.

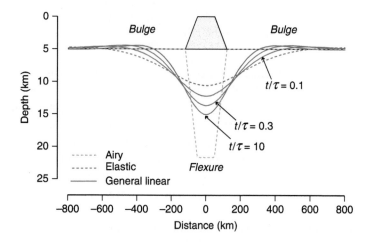

Figure 6.9 Flexure of a general linear viscoelastic plate for a 'narrow' load (Fig. 6.3) and different values of the ratio t/τ. The green-dashed line shows the Airy response and the red-dashed line elastic response for $t/\tau = 0$. Note that the flexure (red-solid line) approaches the Airy response as the subsidence increases and the bulge migrates in towards the load as t/τ increases but does not reach it.

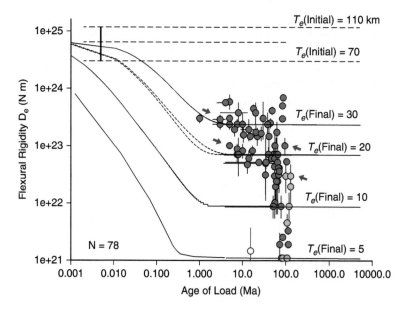

Figure 6.10 Comparison of the D against t for seamounts and oceanic islands that form on seafloor older than 80 Ma (i.e., t_{sf}, the age of the seafloor, > 80 Ma) with predictions based on a general linear viscoelastic model. The predicted curves are based on T_e (Final) of 5, 10, 20 and 30 km and a $D_i = 6.48 \times 10^{24}$ N m, which corresponds to T_e (Initial) = 90 km. The two dashed lines correspond to load wavelengths of 100 and 500 km. Purple filled arrows highlight possible trends in the data that are not well reproduced in the predicted curves.

Watts (1978) carried out the first detailed study of the change in T_e along the length of a single geological feature. Using gravity and bathymetry profile data from along the Hawaiian–Emperor seamount chain, he found that the highest T_e values were at the young end of the chain while the lowest were at the old end. The seamount chain formed on oceanic crust that varied in age only from about 80 to 100 Ma. Therefore, if the oceanic lithosphere is viscoelastic, then different relaxation times would be required to explain any significant differences in T_e along the chain.

An alternative view is that the lithosphere is elastic rather than viscoelastic. Watts (1978) noted that while the volcanoes that make up the seamount chain formed on oceanic lithosphere of similar age, the age of the lithosphere at the time of loading along the Hawaiian Ridge differs significantly from that of the Emperor Seamounts. In particular, the 3–18-Ma-old Hawaiian Ridge formed on 77–105-Ma seafloor, whereas the 44–58-Ma-old Emperor Seamounts formed on 22–71-Ma seafloor. Watts (1978) plotted the Hawaiian-Emperor seamount chain values, together with two other values that had been recently published from the Mid-Atlantic Ridge and Kuril Trench and he showed there to be a general trend, with T_e increasing with the age of the lithosphere at the time of loading.

The dependence of T_e on the age of the oceanic lithosphere at the time of loading has been examined in several subsequent studies (e.g., Caldwell and Turcotte, 1979; Lago and Cazenave, 1981; Calmant et al., 1990; Watts and Zhong, 2000; Watts, 2007) and has been found to be a robust result. Oceanic T_e is given approximately by the depth to the 300–600°C oceanic isotherm based on a cooling plate model with more than 75 per cent of all the estimates plotting within this depth range. The main exceptions are some estimates in French Polynesia, which are lower than expected, and some fracture zones and deep-sea trench–outer rise systems where some of the estimates are higher.

Figure 6.11 shows a plot of T_e against age of the lithosphere at the time of loading based on a compilation of published elastic thickness estimates (Table 6.1). We have not plotted all 262 estimates listed in the table because some are revisions of earlier estimates based on better data. Others are corrections to earlier data sets for improvements in data sets, plate curvature and the bathymetry of mid-plate swells. The reduced data set amounts to some 201 estimates.

There are two main conclusions that can be drawn from Fig. 6.11. Firstly, as the oceanic lithosphere cools and ages, it becomes more rigid in the way it responds to loads. Therefore, loads that form on young seafloor on or near a mid-oceanic ridge have a lower T_e, and hence flexural rigidity, than the same-sized load that is emplaced on the flanks of a ridge on old seafloor.

Secondly, T_e is significantly less than the seismic thickness of the lithosphere, as inferred from surface wave and ocean bottom seismometer studies. If we use, for example, the two estimates of Nishimura and Forsyth (1989) that have been corrected for anisotropy – one for young (0–5 Ma) oceanic crust and the other for old (80–160 Ma) crust – then T_e is approximately 25–50 per cent of the seismic thickness, with most values about 40 per cent. This has been confirmed by the more recent studies of

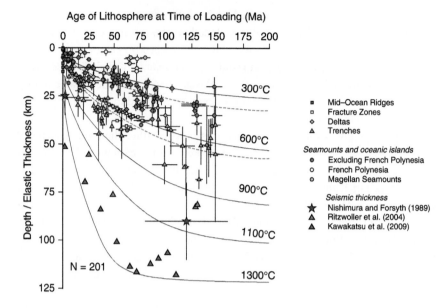

Figure 6.11 Plot of T_e, the elastic thickness of oceanic lithosphere, against age of the lithosphere at the time of loading (i.e., $t_{sf} - t$ where t = the age of loading). Data based on Table 6.1. The symbols and colour coding are according to the type of geological feature. The triangles show the seismic thickness of the lithosphere deduced by Nishimura and Forsyth (1989) and Ritzwoller et al. (2004) from surface wave measurements and Kawakatsu et al. (2009) from borehole broadband ocean bottom seismometers. The solid black lines show the depth to the oceanic isotherms based on the plate cooling model (Parsons and Sclater, 1977) and dashed red lines the depth of the 300°C and 600°C oceanic isotherms with a correction for the temperature dependence of the thermal conductivity (McKenzie et al. 2005).

Ritzwoller et al. (2004) and Kawakatsu et al. (2009). Allowing for the fact that it may take upwards of 1–2 Myr to construct the edifice of an oceanic volcano, then most of the T_e estimates plotted in Fig. 6.11 represent loads that are 1–2 Myr or older. In contrast, the seismic estimates can be considered as loads of the duration of a few seconds to a few hundred seconds. We can conclude, therefore, that the thickness of the lithosphere that mechanically supports a load must decrease from its short-term seismic thickness to its long-term elastic thickness. The timescale over which this relaxation occurs is apparently in the range 0 to 1–2 Myr.

An important question is whether there is any evidence that the relaxation continues on longer timescales. If it does, then we would expect a dependency of T_e on load age, with a lower T_e for old loads than for young ones. The evidence should be apparent in Fig. 6.11. However, the clearest evidence would be in a plot of T_e against the age of the oceanic lithosphere, because such a plot does not a priori assume a dependency of T_e on the age of the lithosphere at the time of loading and would test for a dependence of T_e on load age for a particular age of oceanic lithosphere.

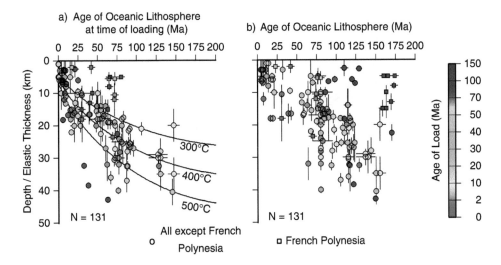

Figure 6.12 Plot of (a) T_e against age of oceanic lithosphere at time of loading (i.e., $t_{sf} - t$) and (b) age of oceanic lithosphere (i.e., t_{sf}) for seamounts and oceanic islands. Symbol colour coding is according to load age. Note that while there is little relationship between T_e and load age in (a) where T_e generally increases with age of the oceanic lithosphere at the time of loading, T_e generally decreases with load age in (b) for a particular age of oceanic lithosphere. G = Gambier and C = Corner, S = Suiko and B = Bermuda, K = Koku and A = Atlantis.

Figure 6.12 is a plot of T_e against age of the oceanic lithosphere at the time of loading and age of the oceanic lithosphere for seamounts and oceanic islands. The estimates are colour coded according to load age, with the oldest features blue and the youngest, red. Figure 6.12a shows an increase in T_e with age of the lithosphere at the time of loading being the most dominant trend in the data. There is no obvious dependence of T_e on load age in the data. In contrast, Fig. 6.12b shows that when T_e is plotted against age of the lithosphere, T_e appears to depend on load age: old loads generally have a lower T_e than young loads. Some old loads have a low T_e, and some young loads have a high T_e because they formed on young and old oceanic lithosphere, respectively. However, seamounts such as Koku (48 Ma) and Suiko (63 Ma) formed on similar age oceanic lithosphere at the time of loading (52–67 Ma) yet are associated with T_e that ranges from 36 km at Koku (Watts, 1978) to 14 at Suiko (Watts et al. 2021), suggesting that factors in addition to the age of oceanic lithosphere at the time of loading might be involved. One possibility is a plate weakening due, for example, to stress relaxation following loading. Stress relaxation may also account for T_e differences between other pairs of seamounts, such as Bermuda (30 Ma) and Atlantis (75 Ma), which formed on older oceanic lithosphere at the time of loading (87–95 Ma) yet are associated with T_e that ranges from 32.5 km at Bermuda (Calmant et al. 1990) to 22 km at Atlantis (Calmant et al. 1990), and Gambier (0.7) and Corner (73) which formed on younger oceanic

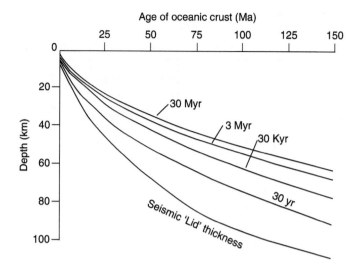

Figure 6.13 The relationship between the seismic and rheological properties of oceanic lithosphere and plate and load age. Reproduced from fig. 2 of Anderson and Minster (1980).

lithosphere at the time of loading (25–30 Ma) yet are associated with T_e that ranges from 18 km at Gambier (Goodwillie and Watts, 1993) to 6 km at Corner (Calmant et al. 1990). These considerations suggest then that oceanic lithosphere may relax not only on short timescales but on long timescales of 1–2 to ~100 Myr.

The observations in Figs. 6.11 and 6.12 of a load-induced stress relaxation in the lithosphere are in general accord with previous ideas, notably those of Beaumont (1979), Anderson and Minster (1980) and Willet et al. (1985). Figure 6.13, for example, shows the thickness of the oceanic rheological lithosphere as a function of plate and load age according to Anderson and Minster (1980). These calculations assume a surface load of 100 MPa, an oceanic isotherm and a dislocation climb creep law. The authors argue that transient creep, seismic attenuation and elastic plate flexure data are consistent with the ideas that the oceanic lithosphere relaxes and that it would appear as a strong, thick plate at short seismic timescales and a weak, thin plate at longer times.

But is there any more direct evidence the oceanic lithosphere relaxes on long timescales? The most obvious expression of a load-induced stress relaxation would be the vertical motion history of the crust and mantle. A viscoelastic plate model predicts, for example, subsidence beneath a volcanic load and uplift of flanking regions, which, depending on the relaxation time and hence viscosity assumed for the underlying lithosphere, may continue for some time after loading. The vertical motion history of ocean islands is recorded in borehole (e.g., Ascension Island, Minshull et al., 2010) and Interferometric Synthetic Aperture Radar (InSAR) data (e. g., Tenerife Island, Fernandez et al., 2009).

Arguably the best example of a subsiding oceanic island is the 'big island' of Hawaii. Although its two largest volcanoes (Kilauea and Mauna Loa) are active today, there is evidence the island is in the final stages of volcano growth (Staudigel and Schmincke, 1984). The reason for this is that absolute Pacific plate motions have caused the island to migrate off the main source of the Hawaiian plume. The main centre of volcanism is currently located beneath Loihi, an active volcano with lava flow lobes and magma supply rates of up to 8 m^3 per minute, which is building upwards on the submarine extension of the south-east flank of Kilauea.

The evidence for subsidence on Hawaii is based on tide gauge data, archaeological artefacts and submarine terraces (Moore and Fornari, 1984). Tide gauge data indicate that the island has been subsiding at rates of about 4 mm yr^{-1} relative to Oahu during the past 100 years (Fig. 4.32). Archaeological artefacts include evidence from the papamu (see Section 4.3), drowned forests and submerged horse trails, which show that since Captain James Cook's fatal visit to Hawaii in February 1779 the island has subsided some 0.25 m. There is evidence, however, that the subsidence has continued over much longer timescales. The dating of submarine terraces south-west of Hawaii (Ludwig et al., 1991), together with the age-dating results of a scientific 3.1-km-deep drill hole in Hilo harbour (Sharp and Renne, 2005; Garcia et al., 2007; Stolper et al., 2009), suggest, for example, that Hawaii has subsided at rates of up to 2.5–3.0 mm yr^{-1} over the past 0.5 Myr (Fig. 6.14).

While Hawaii has been subsiding, there is evidence from 'upstream' islands for contemporaneous uplift (Grigg and Jones, 1997). In particular, Oahu has experienced uplift during the past 0.5 Myr, while Lanai and Molokai emerged about 0.35 Ma.

Unfortunately, the Maxwell viscoelastic plate model considered earlier is limited in its ability to explain both the short-term subsidence that is predicted as the lithosphere thins from the seismic thickness to the elastic thickness (e.g., Fig. 6.11) and the long-term subsidence associated with the construction of seamounts and oceanic islands (e. g., Fig. 6.12) such as Hawaii. The reason for this is that the viscoelastic model is a single-layer model with a single relaxation time. This is illustrated in Fig. 6.15, which shows that while a viscoelastic plate model may be able to explain subsidence rates of up to 1 mm yr^{-1} 0.1–1.0 Myr following loading, the model approaches the predictions of an Airy rather than a flexure model at longer times.

A better approach might therefore be to use a layered viscoelastic model, which would be associated with a range of possible relaxation times. Such models have, for example, been constructed by Zhong (1997) and used by him to examine the temporal evolution of the topography and Moho depth in orogenic belts.

Figure 6.16 shows the flexure that would be expected beneath the centre of a load[4] on a two-layer viscoelastic plate. Two models are illustrated in the figure: Model 1 in which the upper layer has a relatively low viscosity (i.e., 10^{24} Pas) and Model 2 which has a relatively high viscosity upper layer (i.e., 10^{27} Pas). In both models, the thickness of the upper layer is set to 30 km, the viscosity of the lower

---

[4] The load is approximated by a frustrum with a base radius, top radius and height of 40, 20 and 4 km, respectively.

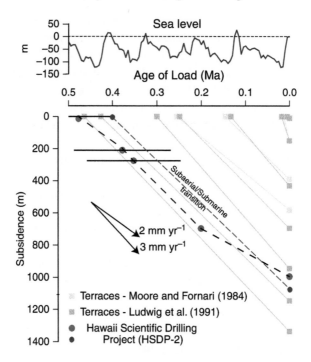

Figure 6.14 Summary of subsidence data from Hawaii Island. The squares show data from submerged terraces, the outermost edge of which have been dated using uranium decay (Ludwig et al. 1991). The filled circles show results from the Hawaiian Scientific Drilling Project in Hilo harbour (red filled circles – Stolper et al. 2009; blue filled circles – Sharp and Renne, 2005). The uppermost profile shows sea level change for the past 500 ka, based on Miller et al. (2005). Note the terraces appear to have formed at times of relatively low sea level compared to that of the present day.

layer is 10^{21} Pa s and the thickness of the lower layer is of the order of the mantle thickness (i.e., 2,900 km). Because the entire mantle is elastic initially, the flexure at $t = 0$ is small. As time elapses, however, viscous effects become more important, and the flexure increases. The flexure occurs in two phases: the earliest, fast, phase, is determined mainly by the viscosity of the lower layer, whereas the latest, slow, phase, is controlled mainly by the viscosity of the upper layer.

Before comparing the predictions of the two-layer viscoelastic model with observations, it is of interest to first compare the two-layer models to the predictions of the thin single-layer elastic and viscoelastic plate models. We will consider two models: an elastic model with $T_e = 30$ km and a viscoelastic model with $\tau = 1$ Myr, $T_o = 30$ km and $\eta = 10^{24}$ Pa s.

Figure 6.16 shows that the flexure of the two-layer viscoelastic model with a high-viscosity upper layer (i.e., Model 2) agrees well with the predictions of the elastic model. However, the degree of fit depends on the viscosity that is assumed for the lower layer.

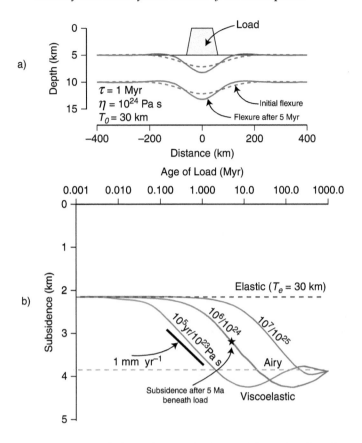

Figure 6.15 Flexure of a single-layer viscoelastic plate model as a function of age of load. (a) Flexure profile 0 Myr (red dashed line) and 5 Myr (red solid line) after loading. (b) Flexure beneath the centre of the load 1,000 kyr to 100 Myr after loading for a plate with an effective viscosity, η, in the range of 10^{23}–10^{25} Pa s, corresponding to a relaxation time, τ, in the range 0.1–10 Myr. The elastic plate ($T_e = 30$ km) and Airy models are time invariant.

The 10^{21} Pa s lower layer model, for example, agrees with the elastic model for a wider range of load ages than does the 10^{22} Pa s model. This is because the relaxation in the 10^{21} Pa s model is more rapid than in the case of the 10^{22} Pa s model and so approaches the predictions of an elastic plate sooner. Also shown in Fig. 6.16 is that the flexure of the two-layer model with a low-viscosity upper layer (i.e., Model 1) agrees well with the predictions of the viscoelastic plate model. However, as was the case for Model 2, the extent of the agreement depends on the viscosity assumed for the lower layer. The 10^{21} Pa s lower layer model more closely follows the viscoelastic plate model than does the 10^{22} Pa s model. This is because the relaxation in the 10^{21} Pa s model is again more rapid than in the case of the 10^{22} Pa s model and so approaches the predictions of a viscoelastic plate sooner.

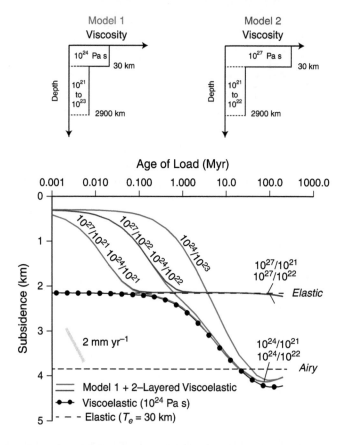

Figure 6.16 Flexure beneath the centre of a load emplaced on a two-layer viscoelastic plate. The red solid lines show the flexure of Model 1 with a low-viscosity upper layer, and the blue solid lines show the flexure of Model 2 with a high-viscosity upper layer. The thickness of the upper layer (30 km) is fixed in both models. The viscosity of the lower layer varies from 10^{21} to 10^{23} Pa s in Model 1 and from 10^{21} to 10^{22} Pa s in Model 2. The inset shows the slope of a 2 mm yr^{-1} subsidence for an age of load between 0.1 and 1.0 Myr. Note that all models with a low-viscosity upper layer approach the Airy case, while all models with a high-viscosity upper layer approach an elastic plate at long load ages.

The usefulness of the comparisons in Fig. 6.16 is they indicate the widely used thin elastic and viscoelastic plate models can be considered as 'end-member' cases of a multilayered viscoelastic model. A two-layer model, for example, with a high-viscosity upper layer (in this case > 6 orders of magnitude greater than that of the lower layer) is essentially behaving as an elastic plate for loading ages in the range of 0.1–100 Myr. At shorter and longer loading ages, the flexure differs from the elastic model by an amount that depends on the viscosities assumed for the lower and upper layers, respectively. Moreover, a two-layer model with an upper layer which has a low viscosity is essentially behaving as a

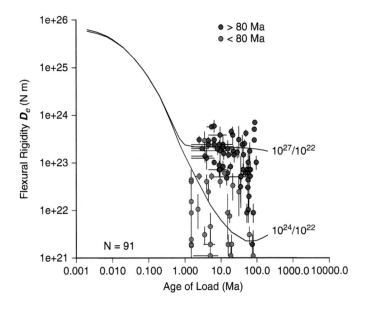

Figure 6.17 Comparison of observed and predicted flexural rigidity as a function of plate and load age for seamounts and oceanic islands, except those in French Polynesia. Data based on Table 6.1. The blue filled circles are features formed on > 80 Ma seafloor. The red filled circles are features formed on < 80 Ma seafloor. The predicted curves are based on the two-layer viscoelastic model in Fig. 6.16.

viscoelastic plate for the same loading age range, provided the viscosity assumed for the lower layer is 10^{21} Pa s or less.

A test of the two-layer viscoelastic model is to compare it with observations. The decrease in flexural rigidity with increase in load in the two-layer model with a viscosity of the lower layer of 10^{22} Pa s 0.1–1.0 Myr after loading, for example, is steeper than in the case of the single-layer model and predicts a subsidence rate of up to 2 mm yr^{-1} which approaches that observed in the Hawaiian region. Moreover, Fig. 6.17 shows that a two-layer model can explain the T_e data at seamounts and oceanic islands (except those in French Polynesia which 'backtrack' to the Pacific super swell region, an anomalous region of stable isotopes and thermal structure), depending on the viscosity assumed for the upper layer. The high-viscosity upper layer model, for example, explains the flexural rigidity estimates at seamounts and ocean islands on old (i.e., > 80 Ma) lithosphere, while the low-viscosity upper layer model is a better fit to loads on young seafloor (i.e., < 80 Ma). This suggests a link between the age, and hence thermal structure, and the viscosity structure of oceanic lithosphere.

The implications of the two-layer model for the evolution of a submarine volcano that builds up on the seafloor to sea level is illustrated in Fig. 6.18. Initially, the model predicts a small subsidence, as the thickness of the material that supports the load involves the crust and much of the mantle. Then, as time elapses, viscous effects

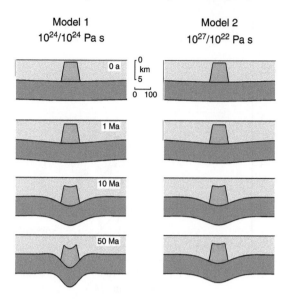

Figure 6.18 Predicted subsidence history of a seamount that builds up to sea level based on the two-layer viscoelastic model. Initially (i.e., $t = 0$), there is a subsidence as the seamount is supported elastically by a thick mantle. Then, as t increases, viscous effects become more important than elastic ones. The viscous effects are similar for load ages up to about 1 Myr but differ later depending on whether the upper layer has a low (Model 1, Fig. 6.16) or high (Model 2, Fig. 6.16) viscosity.

become more important than elastic ones and the seamount subsides. The early part of this subsidence is similar for cases of both a high-viscosity and a low-viscosity upper layer. This is because the short-term relaxation mainly reflects the viscosity of the lower layer, which is the same in both cases. After about 1 Myr, however, the subsidence differs as the relaxation in the different-viscosity upper layers becomes more important. At very long times, there are significant differences in the amount of subsidence on the crest of the seamount and in its flanking moats between the high and low upper-layer viscosity cases.

6.4.2 Continents

The continents are characterised by loads of much longer duration than those in the ocean. They therefore have the potential to provide the best information on how the lithosphere responds to loads on long geological timescales. The problem is that the loads that drive flexure are not as easily defined in the continents as they are in the oceans. Although we know that the continental lithosphere is impacted by both surface and buried loads, their geometry is not well known. Surface topographic loads, due, for example, to thrusting and folding above shallow crustal décollement surfaces, are modified by erosion. Moreover, buried loads, due, for example, to mafic intrusions in the lower crust, have proved difficult to image using seismic techniques.

Another, potentially more difficult problem is that the continental crust has had a complex tectonic evolution and is structurally more heterogeneous than oceanic crust. The Wilson cycle, for example, shows that oceans open and close, and in the process active and passive continental margins may become incorporated into orogenic belts (Dewey and Bird, 1970). In some cases, orogenic belts themselves may become involved in compressional and extensional events. There will always be some uncertainty therefore on whether a particular continental T_e estimate reflects the fundamental mechanical properties of the underlying lithosphere or was inherited from a previous rifting or orogenic event.

Despite such difficulties, continental T_e estimates have shed much light on the long-term mechanical behaviour of the lithosphere. Of particular interest have been the estimates from the flexural depressions and bulges that form in front of migrating thrust and fold belts and fluctuating continent-wide ice sheet loads. The loads of thrust and fold belts form a depocentre into which sediments derived from the orogenic belt are deposited. The depression that results is called a foreland basin. The loads of ice sheets, in contrast, form water-filled depressions known as glacial lakes. Both types of depression are associated with a flanking peripheral bulge and may involve shield areas or cratons as well as other types of continental lithosphere.

Unfortunately, the loads that cause the flexures that flank thrust and fold belts and ice sheet loads may not be visible. This is because orogenic loads are represented not only by the surface topography, but also by buried loads, and ice sheet loads wax and wane because of global climate variations. There is evidence, however, from other data sets for the geometry of these loads. For example, Bouguer gravity anomalies which are corrected for the gravity effect of topography are usually a good indicator of buried loads, while the main centres of ice sheet loading can usually be inferred from the erosional (e.g., rock gouge) and depositional (e.g., moraines) features that they leave in their wake.

Table 6.2[5] summarises the results of 62 T_e studies at foreland basins and glacial lakes. The table also summarises the plate age, t_{co}, and the load age, t, for each feature. Unfortunately, these ages are in most cases uncertain: for example, the Alpine foreland basins developed on crust and mantle that was stabilised during the Hercynian orogeny. We do not know the age of the plate very precisely, because rock ages only reflect the age of the last thermal event. In addition, the duration of a loading event is often long. It is therefore difficult to define the precise plate and load age.

The relationship between continental T_e and plate and load age has been examined in the same way as it was for the oceans. Figure 6.19 compares, for example, the flexural rigidity, D_e, from foreland basins and glacial lakes to the predictions of a Maxwell viscoelastic plate model. The figure shows that the flexural properties of the lithosphere at foreland basins and glacial lakes are highly scattered and, as was the

---

[5] Table 6.2 is available from the Cambridge University Press server.

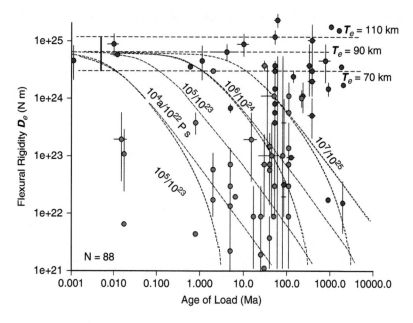

Figure 6.19 Comparison of observed and predicted flexural rigidity as a function of load age for the loads associated with the late glacial lakes and foreland basins. The observed data is based on Table 6.2. Red filled circles shows loads on young lithosphere (<500 Ma) while blue filled circles indicate loads on old lithosphere (>1,000 Ma). The calculated curves are based on the predictions of a viscoelastic plate model with viscosity, η, in the range of 10^{22}–10^{25} Pa s, corresponding to a relaxation time, τ, in the range of 0.01–10 Myr. The first number on the calculated curves is the relaxation time and the second the viscosity. The assumed initial flexural rigidity, D_o, is 6.48×10^{24} N m, which is equivalent to a $T_o(\text{Initial}) = 90$ km.

case in the oceans (Fig. 6.6), there is no single relaxation curve that can explain all the observations. The main difference between the ocean and continent plots, however, is the relatively large number of high rigidity values that characterise the continental lithosphere. The continents are characterised also by many low values. The colour coding in Fig. 6.19 shows that many of the high values are associated with old (>1,000 Ma) lithosphere, while many of the low values are associated with young (<500 Ma) lithosphere. It is apparent, therefore, that the young loads correlate with a lower viscosity and hence shorter relaxation times, while the old loads correlate with a higher viscosity and hence longer relaxation times, a suggestion made earlier by Beaumont (1979) and Willet et al. (1985) for the continental lithosphere.

Figure 6.20 shows a plot of T_e against age of the continental lithosphere at the time of loading. Again, the data are highly scattered. The main difference with the oceanic results is the wide range of values that characterise continental T_e. In the case of foreland basins, for example, they range from small values (i.e., < 5 km) at the Appenine, Po, East Papua New

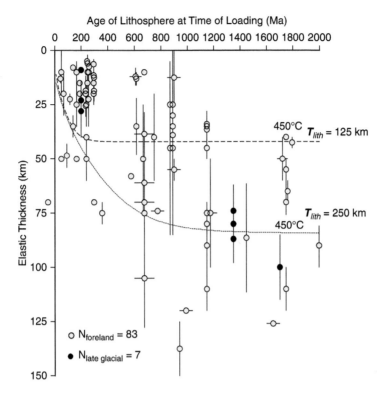

Figure 6.20 Plot of T_e against $t_{co} - t$, the age of the continental lithosphere at the time of loading. Grey filled circles are foreland basin estimates. Black filled circles are late glacial rebound estimates.

Guinea and East Carpathians to high values (i.e., > 100 km) for the Appalachian, Karakorum and Hellenides.

Karner et al. (1983) suggested, using an older data set than used in Fig. 6.20, that, like the oceans, continental T_e is given approximately by the depth to the 450°C isotherm based on a cooling plate model. Figure 6.20 shows, however, that such a model is unable to explain the new data set. While the model explains some values, it does not account for both the very low values in some foreland basins and the high values from some foreland basins and glacial lakes. The fit is not improved if we use a 250-km-thick rather than a 125-km-thick thermal lithosphere.

The plot in Fig. 6.20 therefore suggests there is no simple relationship between continental T_e and the thermal age of the continental lithosphere. Neither does continental T_e appear to depend on load age. There is some evidence that young ice sheet loads generally have a higher T_e than older foreland basin loads (Fig. 6.21) but continental T_e shows considerable scatter, with many young loads having low values of T_e and many old loads having high ones.

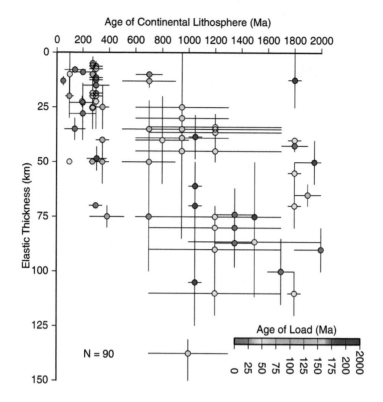

Figure 6.21 Plot of T_e against t_{co}. The colour coding is according to load age, t.

It was shown in Fig. 6.17 that flexural rigidity estimates from seamounts and oceanic islands could be explained quite well by a two-layer viscoelastic model with a viscosity of 10^{24}–10^{27} Pa s for the upper layer and a viscosity of 10^{22} Pa s for the lower layer. Figure 6.22 shows an age of load versus rigidity plot for foreland basins which form on continental lithosphere. When the estimates are colour coded according to the thermal age of continental lithosphere, a pattern emerges that was not apparent in either Fig. 6.20 or Fig. 6.21. Loads associated with foreland basin formation on < 500 Ma lithosphere (red filled circles) plot within the same viscoelastic model parameters as describe the estimates associated with seamounts and oceanic islands, suggesting that continental lithosphere, like oceanic lithosphere, also undergoes a load-induced stress relaxation. The loads associated with foreland basins on > 1,000 Ma lithosphere (blue filled circles) are associated with much higher rigidity estimates that cannot be explained by these parameters. However, they could still be consistent with a stress relaxation, provided the viscosity for the upper layer is at least an order of magnitude greater than 10^{27} Pa s.

Although T_e estimates from forelands and glacial lakes are generally regarded as among the most reliable, there have been several estimates from forward modelling (i.e., non-spectral) of other geological features. The majority have been from extensional basin

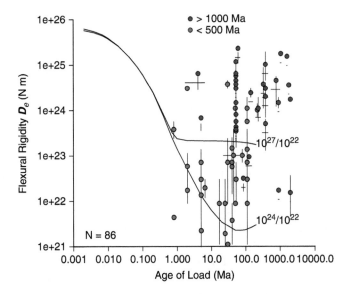

Figure 6.22 Comparison of observed and predicted flexural rigidity as a function of both plate and load age for foreland basin and late glacial rebound estimates. Data based on Table 6.2. The blue filled circles are estimates from > 1,000 Ma continental lithosphere. The red filled circles are estimates from < 500 Ma continental lithosphere. The predicted curves are based on the same two-layer viscoelastic models as those used in Fig. 6.17 for the oceanic lithosphere. We note, though, that the oceanic estimates are enveloped within the predicted curves, but the estimates from > 1,000 Ma continental lithosphere plot above these curves and so would require a significantly higher viscosity ($\gg 10^{27}$ Pa s) for the upper layer in order to explain them.

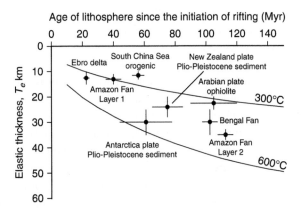

Figure 6.23 Plot of T_e against age of lithosphere since initiation of rifting for selected continental margins on which discrete loads such as river delta, deep-sea fan, Plio-Pleistocene sediment, orogenic and ophiolite loads have been emplaced.

settings (e.g., continental rift valleys, rifted continental margins). These results, which are summarised in Table 6.3,[6] reveal T_e values of 19.0 ± 11.8 and 21.6 ± 15.1 km for continental rift valleys and rifted continental margins, respectively. The most striking difference with the results from glacial lakes and foreland basins is the greater number of low values and the absence of very high values.

The compilations in Tables 6.2 and 6.3 suggest that the continental lithosphere is associated with a wide range of T_e values and hence is capable of showing both great weakness and great strength. Although this is also true in some respects for the oceanic lithosphere, it is the width of range and the dominance of low and high values that distinguish the two lithospheres. In contrast to the oceans, plots of continental T_e against age of the plate at the time of loading show considerable scatter with no clear range of controlling isotherms. However, plots of T_e against age since the initiation of rifting for discrete loads (e.g., orogenic, ophiolite or river delta loads) emplaced on rifted continental margins suggest a behaviour that resembles that of oceanic lithosphere (Fig. 6.23). In addition, there is evidence that both oceanic and continental T_e depend broadly on plate and load age (e.g., Figs. 6.17 and 6.22). We will therefore defer a discussion of these differences and similarities in the long-term behaviour of the oceanic and continental lithosphere until we have considered the rheology of the lithosphere, as constrained from the results of experimental rock mechanics.

6.5 Rheology of the Lithosphere

The phenomenon of flexure has mainly been described thus far in terms of the behaviour of simple elastic and viscoelastic plate models. A model that incorporates a significant degree of elasticity has been favoured because it explains the fact that loads such as seamounts and oceanic islands appear to be supported by the lithosphere for long periods of geological time. The problem with the models, however, is that they are elastic and so there is no limit to the stress that they can withstand.

Since large loads on the crust are more often associated with bending rather than breaking, it could be argued that this is not a problem. In other words, the lithosphere really is strong and is more than capable of supporting the stresses and strains associated with large loads, such as volcanoes, for long periods of time.

We know from extrapolations of experimental rock mechanics data, however, that the lithosphere is unlikely to support large loads without undergoing some form of yielding. The problem is that these data are acquired over much shorter time-periods (usually a few days to months) than the long time-periods associated with flexural loading. However, sufficient evidence exists from the mechanical behaviour of the materials that make up the lithosphere to be able to make inferences about their behaviour at longer times. As a result, experimental rock mechanics data have greatly improved our understanding of flexure and the physical meaning of some of its parameters, such as the elastic thickness.

---

[6] Table 6.3 is available from the Cambridge University Press server.

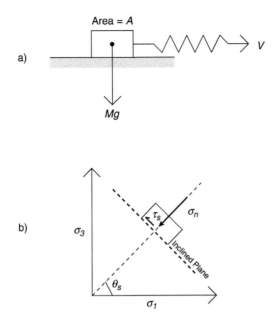

Figure 6.24 Schematic diagrams to illustrate Byerlee's rock friction experiment.

In general terms, experimental rock mechanics data suggest that the lithosphere is dominated by two main types of deformation: brittle deformation in its cool, uppermost part and ductile flow in its hot, lowermost part. Both types of deformation have been extensively studied in the laboratory using a wide range of commonly occurring mineral and rock assemblages. As a result, several empirical laws have been derived that describe the mechanical behaviour of the minerals and rocks that occur in the crust and mantle, and so we will now consider some of them briefly here.

6.5.1 Brittle

It is known from geological field mapping that most of the Earth's surface rocks are brittle and fail by movements along localised fractures. The conditions that lead to failure have been extensively studied in the laboratory. Among the most widely quoted are the rock friction studies carried out by Byerlee and colleagues (Byerlee, 1978) during the late 1970s and early 1980s on a variety of commonly occurring rock types.

A schematic diagram of Byerlee's rock friction experiments is shown in Fig. 6.24. The figure shows a mass M that is free to slide across a flat horizontal surface. The force, V, required to initiate sliding is related to the shear stress, τ_s, and is given by

$$\tau_s = \frac{V}{A},$$

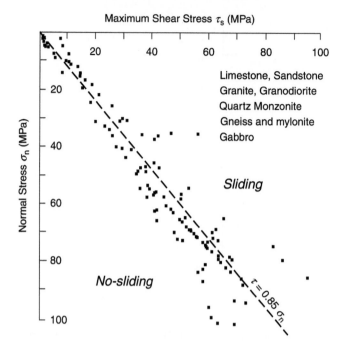

Figure 6.25 Plot of normal stress against maximum shear stress needed to cause sliding for a variety of rock types, including limestone, sandstone, granite, granodiorite, quartz monzanite, gneiss, mylonite and gabbro. Reproduced from fig. 5 of Byerlee (1978).

where A is the area of the surface of contact. The mass exerts a downward force, Mg, that acts on the flat surface. The normal stress, σ_n, that acts across the surface is therefore

$$\sigma_n = M\,g/A,$$

Byerlee showed that

$$\tau_s = S_o + \mu\sigma_n, \tag{6.13}$$

where S_o is an inherent shear strength and μ is the coefficient of friction. This relationship, which determines the stress required to overcome friction and cause sliding, was described by Handin (1969) and has subsequently been referred to as the Coulomb–Navier criterion. One of the most significant results to emerge from Byerlee's experiments to test the criterion was the virtual independence of friction on rock type and temperature. Byerlee (1978) found, for example, that the shear stress required to initiate sliding in rocks as diverse as limestone, sandstone, granite, gneiss and gabbro (Fig. 6.25) is given by

$$\begin{aligned} \tau_s &= 0.85\,\sigma_n, \quad \sigma_n < 200\,\text{MPa} \\ \tau_s &= 0.5 + 0.6\,\sigma_n, \quad \sigma_n > 200\,\text{MPa} \end{aligned} \tag{6.14}$$

The application of Byerlee's law to the lithosphere requires that it first be expressed in terms of the principal stresses. Carslaw and Jaeger (1959) have shown, for example, that σ_n and τ are related to the maximum (σ_1) and minimum (σ_3) principal stresses (Fig. 6.24b) by

$$\sigma_n = \sigma_1 \cos^2\theta_s + \sigma_3 \sin^2\theta_s \qquad (6.15)$$

or

$$\sigma_n = \frac{\sigma_1 + \sigma_3}{2} - \cos2\theta_s \frac{\sigma_3 - \sigma_1}{2}$$

$$\sin 2\theta_s = \frac{2\tau_s}{(\sigma_1 - \sigma_3)} \qquad (6.16)$$

where θ_s is the angle between the maximum principal stress and the normal to the shear plane.

Byerlee's empirical law in terms of the principal stresses (e.g., Brace and Kohlstedt, 1980) is given by

$$\sigma_1 \sim 5\,\sigma_3, \ \ \sigma_3 < 110 \ \text{MPa} \qquad (6.17)$$

$$\sigma_1 \sim 3.1\,\sigma_3 + 210, \ \ \sigma_3 > 110 \ \text{MPa}. \qquad (6.18)$$

The condition for failure is given by the maximum value of $(\sigma_v - \sigma_h)$, where σ_v and σ_h are the vertical and horizontal stresses, respectively. In the Earth, σ_v is usually the vertical stress due to the weight of the overburden. We therefore can write that

$$\sigma_1 = \sigma_v = \rho g h.$$

Byerlee's empirical law for low stress suggests that

$$\sigma_3 = \sigma_h = \frac{\sigma_1}{5} = \frac{1}{5}\rho g h.$$

The condition for failure in tension is therefore given by

$$(\sigma_v - \sigma_h) = \frac{4}{5}\rho g h.$$

Substituting $g = 9.81 \ \text{ms}^{-2}$ and $\rho = 2,800 \ \text{kg m}^{-3}$ gives a slope for tension of $21.97 \ \text{MPa km}^{-1}$. In the case of compression, failure occurs when the compressive stress has built up to a sufficient value that it has effectively nulled the horizontal stress. This occurs when $\sigma_h = 5\rho g h$. The condition for failure is therefore given by

$$(\sigma_v - \sigma_h) = -4\rho g h.$$

Substituting for g and ρ gives a slope for compression of $-109.87 \ \text{MPa km}^{-1}$. For high stress, the corresponding slopes for tension and compression are $18.6 \ \text{MPa km}^{-1}$ and $-57.7 \ \text{MPa km}^{-1}$ respectively.

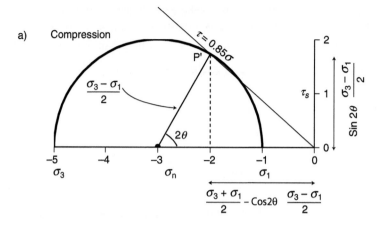

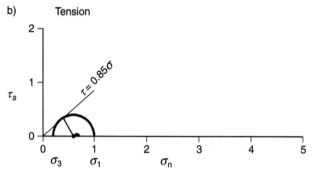

Figure 6.26 The verification of Byerlee's empirical law for low stress in (a) compression and (b) tension using a Mohr diagram.

Byerlee's empirical law can be verified with a Mohr diagram. In this diagram, a pair of orthogonal co-ordinates is laid out and graduated in stress units. Points are located on the horizontal axis, σ_n, which have values that correspond to the principal stresses σ_1 and σ_3. According to Byerlee's law for low stress and compression, failure occurs when $\sigma_3 = 5\sigma_1$. If we draw a circle centred on the σ_n axis and passing through $\sigma_1 = -1$ and $\sigma_3 = -5$ at opposite ends of its horizontal diameter, then it follows that we can find the co-ordinates of any point on the circle. One such point is P, which subtends an angle $2\theta_s$ with the horizontal axis. This point, as Fig. 6.26 shows, has the co-ordinates σ_n and τ_s. Substituting $\sigma_1 = -1$ and $\sigma_3 = -5$ in Eqs. (6.14 and 6.15) gives the co-ordinates of P as $\sigma_n = -2$ and $\tau_3 = 1.73$. The line that connects P to the origin has a slope of 0.85, which verifies Byerlee's law.

A direct application of Byerlee's law implies (a) that the lithosphere has already failed, and (b) that fractures within the lithosphere have a favourable orientation for frictional sliding. Byerlee's law will therefore tend to give a lower bound on the yield stress. The manner in which seismic waves propagate suggests that initially the rocks that comprise the lithosphere are dominated by elastic behaviour. Therefore, when constructing a yield stress envelope for the Earth, it might be more appropriate to assume that the background stress state of the lithosphere is lithostatic such that it only depends on the weight of the overlying rocks (i.e., $\sigma_n = \rho g h$). In other words, the conditions for failure should be computed assuming a stress state that would be expected for a perfectly elastic material.

In an elastic solid, if the principal stress is vertical and given by

$$\sigma_1 = \sigma_v = \rho g h,$$

then the least principal stress is given, for example, by

$$\sigma_3 = \sigma_h = \frac{1}{3}\rho g h.$$

Failure will occur at the maximum value of $(\sigma_v - \sigma_h)$. In the case of tension, this occurs when $\sigma_h = \rho g h / 3$. We then have that

$$(\sigma_v - \sigma_h) = \rho g h - \frac{1}{3}\rho g h = \frac{2}{3}\rho g h. \tag{6.19}$$

Substituting $g = 9.81 \text{ ms}^{-2}$ and $\rho = 2,800 \text{ kg m}^{-3}$ gives a slope of 18.3 MPa km^{-1}. In the case of *compression*, failure occurs when the compressive stress has built up to a sufficient value that it has effectively nulled the horizontal stress. This occurs when $\sigma_h = 3\rho g h$. We then have that

$$(\sigma_v - \sigma_h) = \rho g h - 3\rho g h = -2\rho g h \tag{6.20}$$

which yields a slope of -54.9 MPa km^{-1}.

Since Byerlee's pioneering studies, considerable work has been carried out, especially in the determination of the role of fluid-pore pressure in modifying the friction laws. Brace and Kohlstedt (1980), for example, showed that the influence of pore pressure can be described by a parameter, p_o, which is the ratio of pore pressure to the lithostatic stress. $p_o = 0$ corresponds to 'dry' conditions and $p_o = 0.7$ to 'wet' conditions; $p_o = 0.37$ corresponds to an intermediate condition when the stress state is hydrostatic (i.e., the principal stresses are equal). The failure envelopes for 'dry', 'wet' and hydrostatic conditions are illustrated in Fig. 6.27.

Measurements of the horizontal stress from overcoring in deep mines can be used to quantify the role of pore pressure in the continental crust. Mines in Canada, for example, suggest that Byerlee's law with $p_o = 0.37$ encloses most of the data. However, mines in South Africa are better explained by Byerlee's law with $p_o = 0$. Therefore, the role of pore pressure is likely to vary significantly in the crust. This problem, together with the fact that mines only sample the stress state down to depths of 2–5 km, suggest caution in the application of Byerlee's law to the crust and mantle.

Nevertheless, a simple approach to calculate the failure envelope as a function of depth, h, would be to assume hydrostatic conditions. We can then write

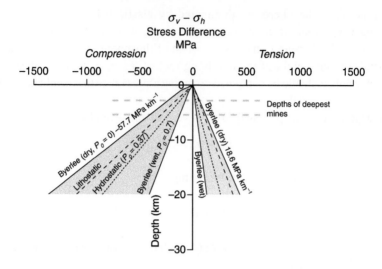

Figure 6.27 Failure envelopes for 'wet' and 'dry' conditions in the crust.

$$\tau_s = 0.85(\sigma_n - P_f), \qquad \sigma_n < 200 \text{ MPa}$$
$$\tau_s = 0.5 + 0.6(\sigma_n - P_f), \qquad \sigma_n > 200 \text{ MPa}$$

where P_f is the fluid-pore pressure which is given by $\rho_w g h$ where ρ_w is the density of water.

6.5.2 Ductile

As depth increases, pressures increase, and most rock materials become ductile and start to flow. The rate of flow depends on the magnitude of the stress applied. For a Newtonian substance, the strain rate $\dot{\varepsilon}$ is related to the applied stress σ by

$$\dot{\varepsilon} = \frac{\sigma}{\eta}, \tag{6.21}$$

where η is the viscosity. For a non-Newtonian substance, the strain rate is given by

$$\dot{\varepsilon} = A_p \frac{\sigma^n}{\eta}, \tag{6.22}$$

where the exponent n is a positive integer and A_p is the power law stress constant.

Equations (6.21) and (6.22) ignore the effects of temperature and pressure. We know, for example, that as depth in the Earth increases, so do temperature and pressure, and this has a profound effect on the flow rate of rocks.

Goetze (1978) was among the first to quantify the mechanical behaviour of olivine, an important constituent of the sub-crustal mantle. He subjected 'dry' polycrystalline olivine aggregates to stresses of up 5,000 MPa and temperatures of up to 1,600°C. By conducting the experiments over periods of the order of 1 to 100 days, he was able to observe the behaviour of olivine at strain rates of 10^{-6} to 10^{-4} s^{-1}. At low stress, Goetze found that the behaviour was described by a power law:

$$\dot{\varepsilon} = A_p(\sigma_1 - \sigma_3)^n e^{-Q_p/R_g T}, \qquad (\sigma_1 - \sigma_3) < 200 \text{ MPa}, \qquad (6.23)$$

where $n = 3$, T is the applied temperature, $(\sigma_1 - \sigma_3)$ is the applied stress difference, R_g is the universal gas constant and Q_p is the power law olivine activation energy. At high stress, Goetze found that the Dorn law better described the behaviour:

$$\dot{\varepsilon} = B_d e^{-Q_d/R_g T} \left[1 - \left(\frac{(\sigma_1 - \sigma_3)}{A_d} \right)^2 \right], \qquad (\sigma_1 - \sigma_3) > 200 \text{ MPa}, \qquad (6.24)$$

where Q_d is the Dorn law olivine activation energy, A_d is the stress constant and B_d is the strain rate constant.

Although Goetz's experiments were carried out at high strain rates, he was able to extrapolate the behaviour of olivine to the much lower strain rates associated with geological processes. Observations at strain rates 10^{-4} to 10^{-6} s^{-1} were sufficient to show that the onset of ductility would occur at shallower depths as the strain rate was decreased. Therefore, the behaviour of olivine at high strain rates may be extrapolated to low strain rates.

The stress-difference at different strain rates can be computed by re-arrangement of the power law and the Dorn law. We have, for example, from Eq. (6.23), that

$$(\sigma_1 - \sigma_3) = \sqrt[n]{\frac{\dot{\varepsilon}}{A_p e^{-Q_p/R_g T}}}. \qquad (6.25)$$

Substituting values of T in kelvins and $\dot{\varepsilon}$ in s^{-1} together with R_g in J mol^{-1} K^{-1}, Q_p in J mol^{-1} and A_p in s^{-1} Pa^{-n} gives $(\sigma_1 - \sigma_3)$ in pascals. For the Dorn law, we need first to take the logarithm of both sides to get

$$\ln \dot{\varepsilon} = \ln B_d - \frac{Q_d}{R_g T} \left[1 - \frac{(\sigma_1 - \sigma_3)}{A_d} \right]^2.$$

Re-arranging then gives

$$(\sigma_1 - \sigma_3) = A_d \left[1 - \left[\ln \left(\frac{B_d}{\dot{\varepsilon}} \right) \frac{R_g T}{Q_d} \right]^{1/2} \right]. \qquad (6.26)$$

Substituting R_g in J mol^{-1} K^{-1}, Q_d in J mol^{-1}, A_d in pascals, B_d in s^{-1}, T in kelvins and $\dot{\varepsilon}$ in s^{-1} gives $(\sigma_1 - \sigma_3)$ in pascals.

Goetze (1978) used the experimental creep data on olivine to estimate the parameters n, A_p, A_d and B_d. They found a cubic dependence (i.e., $n = 3$) of strain rate on the applied stress. In addition, by setting Q_p and Q_d using values based on different physical processes

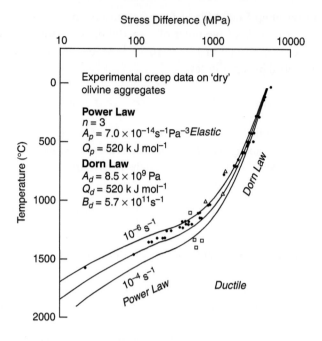

Figure 6.28 Comparison of experimental creep data on 'dry' olivine aggregates in the ductile regime with the predictions of the power and Dorn law. Data reproduced from fig. 4 of Goetze and Evans (1979).

(e.g. hardness data), they estimated A_p from the power law and A_d and Q_d from plots of $(\sigma_1 - \sigma_3)$ versus $T^{1/2}$. We summarise the best-fit parameters in Fig. 6.28, where the power law and Dorn law, based on strain rates of 10^{-4}–10^{-6} s^{-1}, are compared directly with experimental creep data.

Figure 6.29 shows a plot of stress difference, $(\sigma_1 - \sigma_3)$, against temperature, T, for strain rates in the range 10^{-5}–10^{-17} s^{-1}. The high end of the range corresponds to those that were achieved in the laboratory experiments. The low end is based on the power law and Dorn law. Taking into account all the formal errors, Goetze (1978) estimate the accuracy of these strain rate curves to be better than 50°C.

As pointed out earlier, geological strain rates are significantly lower than those that can be achieved in the laboratory. Estimates range from 3×10^{-15} s^{-1} for compressional deformation (Cloetingh and Burov, 1996), to 10^{-15} to 10^{-17} s^{-1} for extensional deformation (Newman and White, 1997). The motions of the plates themselves (Gordon, 1998) suggest strain rates of 6×10^{-20} to 10^{-17} s^{-1} for the plate interiors.

Since Goetze's pioneering work, there have been several subsequent studies of the ductile deformation field. It is now known, for example, that olivine shows either diffusion or dislocation creep and that the role of these creep mechanisms varies according to tectonic setting. Dislocation creep, for example, is applicable to hot, shallow regions in the crust, whereas diffusion is favoured in cold, shallow regions. In addition, it has been

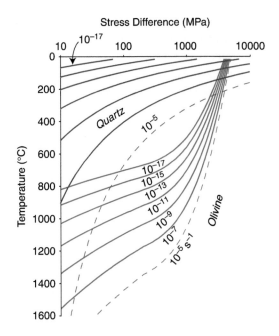

Figure 6.29 Plot of stress differences against temperature for olivine and quartz for strain rates in the range 10^{-5}–10^{-17} s^{-1}. Parameters defining the flow laws of olivine and quartz are as given in Table 6.4.[7]

found (e.g., Karato and Wu, 1993) that diffusion creep is grain size dependent, whereas dislocation creep is not. Water, however, is known to enhance the creep rate in both diffusion and dislocation creep.

Unlike olivine, which retains significant strength up to temperatures as high as 1,000–1,200°C, other minerals weaken at significantly lower temperatures. The onset of ductility for quartz, for example, occurs at temperatures that are upwards of 400°C lower than olivine.

The parameters A_p, n and Q_p for quartz, as well as other commonly occurring minerals in the crust and mantle (Carter and Tsenn, 1987; Kirby and Kroenberg, 1987) are summarised in Table 6.4.

In recent years, new technologies have enabled a focus on ductile deformation at the conditions that prevail in the lithosphere. At these conditions we now know that dislocation creep can occur by two main mechanisms. At lower temperatures (<600–800°C) and higher stress (>100–300 MPa) dislocation creep is controlled by the intrinsic resistance of the lattice, known as the Peierls stress (Goetze, 1978). This mechanism is commonly referred to as low-temperature plasticity, and its constitutive equation is of the form:

$$\dot{\varepsilon} = A\sigma^n e^{-\left(\frac{H_0^\star}{RT}(1-(\frac{\sigma}{\sigma_p})^p)^q\right)}$$

[7] Table 6.4 is available from the Cambridge University Press server.

where $\dot{\varepsilon}, \sigma, T$ and R are the strain rate, differential stress, temperature and gas constant, respectively; A is a pre-exponential term that depends on pressure, grain size and volatile content; n is a non-dimensional parameter that arises from the stress-dependence of dislocation density; H_0^* is the zero-stress activation enthalpy; σ_P is the Peierls stress or glide resistance and p and q are non-dimensional parameters that depend on the geometry of kinks (Kocks 1975). At higher temperatures (>600–800°C) and lower stress (<100–300 MPa) dislocation creep is controlled by the climb of dislocations (Goetze, 1978). This mechanism is commonly referred to as power-law creep and its constitutive equation is of the form

$$\dot{\varepsilon} = B\sigma^n e^{\left(-\frac{E^* + PV^*}{RT}\right)},$$

where B is a material-dependent parameter that can include the influence of water content, n is the stress exponent, E^* is the activation energy, P is pressure and V^* is the activation volume. The experimentally determined parameters required to construct the olivine flow laws applicable to lithospheric conditions are given in Raterron et al. (2004), Katayama and Karato (2008) and Mei et al. (2010) for low-temperature plasticity, and by Hirth and Kohlstedt (2003) for power-law creep. The ones used in illustrations within this book are summarised in Table 6.4.

6.6 The Yield Strength Envelope

The Yield Strength Envelope (YSE) attempts to combine the brittle and ductile deformation laws into a single strength profile for the lithosphere. In theory, the strength profile can be used to determine how a surface load might be supported at depth. In practice, it only provides a qualitative guide to the actual strength. This is because the YSE is based on data from experimental rock mechanics that constrains how rocks and minerals behave at strain rates of 10^{-4}–10^{-6} s^{-1}. These strain rates are up to 10 orders of magnitude higher than those associated with geological processes such as extension and compression. Furthermore, the influence of pore pressure and other factors, such as dynamic recrystallisation and preferred orientation, are still not well enough understood in how they modify the brittle and ductile flow laws (Rutter and Brodie, 1991). Nevertheless, the YSE has proved useful in our understanding of the physical meaning of T_e and, hence, how flexural isostasy is actually achieved in the lithosphere.

Figure 6.30a shows examples of the YSE for different thermal ages of the oceanic lithosphere. The YSE has the shape of a 'sailboat'. The strength profile first increases linearly with depth and then decreases. The increase is given by Byerlee's law, which varies for tension and compression. The decrease is given by the olivine power law, which depends on the magnitude of the applied stress and exponentially on the temperature. The two portions of the strength profile intersect at the brittle–ductile transition, but the transition is unlikely to be as sharp as shown in Fig. 6.30a. The curves are based on a strain rate of 10^{-14} s^{-1} and a geotherm defined by the cooling half-space (dashed lines) and cooling plate models (solid lines).

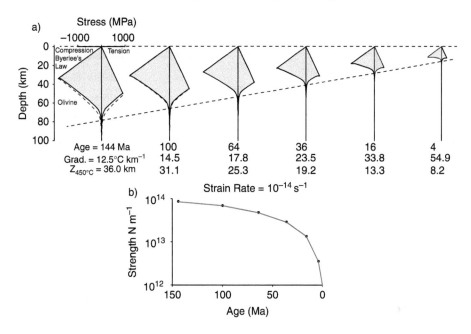

Figure 6.30 Yield Strength Envelope (YSE) profile and integrated strength of oceanic lithosphere. (a) YSE for age of oceanic lithosphere 4–144 Ma. (b) Integrated strength of oceanic lithosphere.

The numbers beneath each YSE in Fig. 6.30a indicate (from top to bottom) the thermal age, the geothermal gradient, and the depth to the 450°C oceanic isotherm. Since stress has units of pascals and depth is metres, the area under the YSE has units of Pa m or, equivalently, $N\ m^{-1}$. The YSE provides an estimate of the stress differences that can be supported by the lithosphere before yielding. Furthermore, because the maximum stress difference that the lithosphere can support is indicative of its strength, the area under the YSE provides a measure of the integrated strength of the lithosphere. Figure 6.30b shows that the integrated strength of the oceanic lithosphere increases exponentially with age, which is expected since the base of the YSE depends on the temperature and, hence, thermal age.

The first to apply the YSE to oceanic T_e data were Bodine et al. (1981), Lago and Cazenave (1981), McNutt and Menard (1982) and McAdoo et al. (1985). These workers showed that the YSE could explain the fact that the elastic thickness was significantly less than the seismic thickness, and that T_e increased with the age of the oceanic lithosphere at the time of loading. As we will examine in a later section, the YSE also helps to explain the relationship between T_e and curvature.

The YSEs in Fig. 6.30 are based on the power law of Goetze and Evans which was derived from experiments on 'dry' olivine samples at stresses of < 200 MPa. Figure 6.31 shows a selection of more recent laws which are, as pointed out earlier, based on

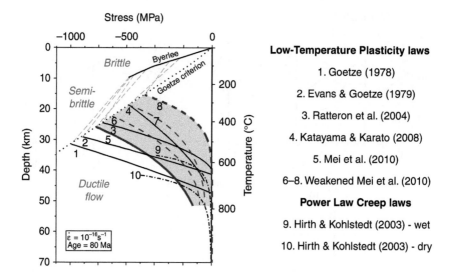

Figure 6.31 The oceanic YSE (for compression) calculated by selected brittle and ductile flow laws. The ductile flow laws are derived assuming a thermal structure given by 80 Ma oceanic lithosphere, a strain rate of 10^{-16} s^{-1} and olivine Low-Temperature Plasticity (LTP) and Power Law Creep (PLC). Note the differences between the LTP laws and the wet and dry samples in the PLC law. Modified from Hunter and Watts (2016).

pressure (and temperature) conditions more applicable to the sub-crustal oceanic mantle. The YSEs shown in the figure are for compressive applied stresses only, the parameters given in Table 6.4, and have been applied in two recent oceanic T_e studies by Zhong and Watts (2013) and Hunter and Watts (2016). Zhong and Watts showed that one of these flow laws (Mei et al. 2010) is too strong at the Hawaiian Ridge in the Pacific plate interior and needs to be weakened significantly. However, Hunter and Watts found that the Mei et al. (2010) flow law explains well the 'background' T_e of the subducting Pacific plate at circum-Pacific deep-sea trenches. It is not immediately clear why the same flow law does not apply to the Pacific plate in both tectonic settings. However, some mechanically induced weakening may be required at deep-sea trenches, especially in the region of bend faults seaward wall of the trench axis and some thermally-induced weakening may be required at the hotspot-generated Hawaiian Ridge. Alternatively, the parameters defining the flow laws (e.g., creep activation energy) may vary between these tectonic settings.

It is clear from the oceanic results, however, that the YSE will be a more difficult concept to apply to the continents. The main reasons are that continental crust is generally thicker than oceanic crust and so crustal composition will play a greater role than it does in the oceans. Moreover, the continental crust is heterogeneous in its structure and therefore it is not clear what compositions would be best to assume.

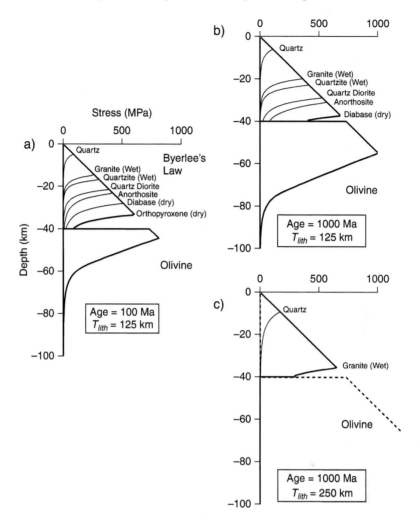

Figure 6.32 Plot of Yield Strength Envelope (for extension) profiles for the continental lithosphere. (a) 'Standard' YSE based on a strain rate of 10^{-14} s^{-1}, the cooling plate model (thermal thickness of 125 and age of 100 Ma), a crustal thickness of 40 km, an olivine mantle rheology and a crustal rheology that depends on composition. Parameters defining the flow laws of olivine and the various crustal rocks/minerals are given in Table 6.4. (b) Same as 'standard' but age of 1,000 Ma. (c) Same as 'standard' but thermal thickness of 250 km and age of 1,000 Ma.

Figure 6.32 shows some examples of continental YSEs for a range of crustal compositions. The main difference with the oceanic case is that there is now a 'notch' in the YSE at deep crustal depths. This is because the onset of ductility in crustal-type rocks (e.g., quartzite, granite) occurs at much shallower depths than in olivine. This means that the brittle–ductile transition in the continental crust occurs at shallower depths than in oceanic crust of the same thermal age. It also means that, given the same geothermal gradient, the integrated strength of the continent will usually be less than that of the ocean. The amount that it is less depends on the composition. The more quartz-rich its composition, then the greater the difference.

The YSE for the continents depends not only on crustal composition, but also on the crustal thickness, the geothermal gradient, and the strain rate. Fig. 6.33a, for example, compares a 'standard' YSE based on a 30-km-thick plagioclase-rich crust that overlies an olivine-rich mantle and a geotherm given by 80-Ma-old oceanic lithosphere to calculated profiles. The calculated profiles are based on thin and thick crust (Fig. 6.33b), a hot and cold lithosphere (Fig. 6.33c) and a high and low strain rate (Fig. 6.33d). The figure shows a wide variation in the area under each YSE profile. The largest area, and hence the greatest integrated strength, corresponds to the case of a thin crust, cold lithosphere and high strain rate. The smallest, and hence the least integrated strength, is for a thick crust, hot lithosphere and low strain rate. Meissner and Wever (1986) referred to those cases in which the lower continental crust was a zone of enhanced ductility, or weakness, between stronger layers in the upper crust and upper mantle as like 'jelly in a sandwich', hence the term jelly sandwich has been used to describe it.

Among the first to apply the YSE to continental T_e data were Karner et al. (1983), McNutt et al. (1988) and Burov and Diament (1992, 1995). Karner et al. (1983) suggested that the high T_e estimates at some orogenic belts could be modelled as the simple response to loading of a YSE based on a quartz rheology for the crust and a dry olivine rheology for the underlying mantle. McNutt et al. (1988) showed how a weak zone in the crust, as implied by some YSE models, might reduce the T_e of an otherwise strong plate if it is flexed to a high enough curvature. Finally, Burov and Diament (1992, 1995) developed the first analytical models that related the integrated strength of the YSE to continental T_e estimates. They argued that T_e could be calculated from

$$T_e^{(n)} = \left(\sum_{i=1}^{n} \Delta h_i^3 \right)^{\frac{1}{3}}, \qquad (6.26a)$$

where n is the number of competent layers and Δh_i is the depth of the ith layer at which the yield strength is 5 per cent of the lithostatic pressure (i.e., the dashed line in Fig. 6.33). For example, for the standard, hot and cold two-layer cases in Fig. 6.32b it follows from Eq. 6.26a that

$$T_e = T_e^{(2)} = \left(h_1^3 + (h_2 - h_c)^3 \right)^{\frac{1}{3}},$$

which with $n = 2$, $h_1 = 40$ km and $(h_2 - h_c) = 62 - 40 = 22$ km gives $T_e = 42$ km for the 'standard' case with a weak lower crust compared to $T_e = 22$ km and 72 km for the single-layer hot and cold cases, respectively. The thermal structures and crust and mantle compositions assumed in Fig. 6.33, therefore, imply that continental T_e could vary significantly,

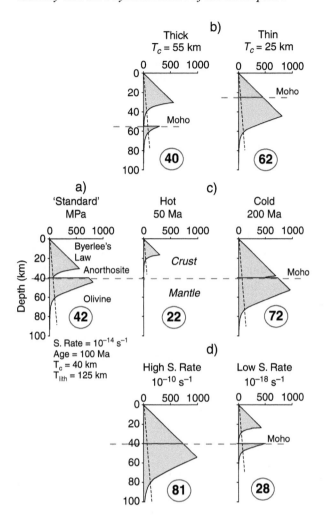

Figure 6.33 Plots of Yield Strength Envelope profiles for the continental lithosphere showing the effects on a 'standard model' of (a) an anorthosite crust and olivine mantle, (b) thickening and thinning the crust, (c) decreasing and increasing the thermal age and (d) increasing and decreasing the strain rate. Parameters defining the flow laws of anorthosite and olivine are given in Table 6.4. Numbers in bold circles show the equivalent T_e derived using Eq. 6.26 and the 'cubic rule' of Burov and Diament.

from 22 to 72 km, which supports the notion that the continental lithosphere is capable of both great weakness and great strength. However, caution should be applied when applying Burov and Diament's so-called 'cubic rule'. This is because it is not immediately obvious that a weak zone within the crust would have no effect on T_e. Bellas et al. (2022), for example, have shown using a two-dimensional viscoelastic model that the applicability of the cubic rule is limited, may only be true for certain conditions, and that the strength of a multilayer viscoelastic plate with a weak zone is mainly determined by the thickness of its uppermost competent layer.

We demonstrated earlier that, unlike oceanic regions, no simple relationship appears to exist between T_e and the thermal age of the continental lithosphere. One possibility then is that the composition of the continental crust plays a much greater role than it does in oceanic crust in determining the geothermal gradient. Radiogenic heat-producing elements, for example, are known to contribute in a major way to the present-day surface heat flow. If this is correct, then it suggests that continental T_e may be controlled more by the present-day geothermal gradient than by secular cooling of the lithosphere.

A relationship between continental T_e and the present-day geothermal gradient has been suggested by Pinet et al. (1991). These workers measured the heat flow along a 'transect' of the Appalachians, across Logan's line to the Canadian Shield, where the crustal structure had already been determined from seismic refraction data. Pinet et al. (1991) used the seismic structure to estimate the composition and, hence, the contribution to the heat flow of the heat-producing elements in the crust. By subtracting the calculated crustal contribution from the observed heat flow, they were able to estimate the mantle contribution and, hence, the geothermal gradient along the transect. As Fig. 6.34 shows, the calculations yield significant differences in the geothermal gradient, with the Appalachians having a relatively 'hot' geotherm and the Canadian Shield a relatively 'cold' one. If, like the oceans, continental T_e is determined by the depth to the 450°C isotherm, then T_e would be expected to range from about 45 km in the hotter, thicker crust of the Appalachians to 75 km in the colder, thinner crust of the shield regions. These estimates are in good general agreement with the values of T_e along the transect based on admittance and coherence studies (Betchel et al., 1990).

These considerations suggest that crustal composition, through its control on the rheology and geothermal gradient, is a major factor in explaining the wide range of continental T_e estimates that are observed. Other factors concern the state of coupling between the crust and mantle, the curvature, and the sediment cover, especially in regions when the lithosphere is subject to extension and compression.

Burov and Diament (1995) have argued, for example, that the strength profile in the continental lithosphere is so extreme that when the lithosphere is subject to side-driven loads, the weak lower crustal layer may, in effect, decouple the strong uppermost crust from the strong underlying mantle. This results in a 'sudden' switch from a strong to a weak lithosphere that might help to explain the preponderance of low values that characterise some extensional and compressional regions. Therefore, according to this model, high values represent those portions of the continental lithosphere, such as cratons, that have remained intact and unchanged when subjected to stress.

Lavier and Steckler (1997) used a parameterisation of the YSE that includes the effects of crust–mantle decoupling to explain the T_e data at foreland basins. Their best-fitting model was one that considered the thermal effects of sediment blanketing. They showed, assuming

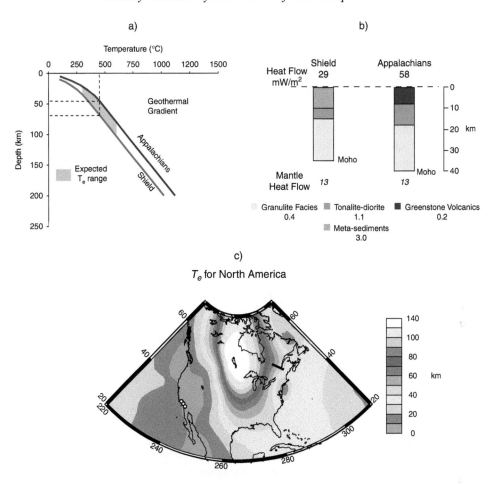

Figure 6.34 Comparison of observed and predicted T_e in North America. (a) The present-day geotherm for the Appalachians and Canadian Shield regions based on Pinet et al. (1991). The orange shading shows the expected T_e, assuming that T_e is defined by the depth to the 300–600°C isotherm. (b) The composition of the crust used by Pinet et al. (1991) to compute the contribution of radiogenic heat-producing elements to the surface heat flow. (c) The observed T_e based on Sheffels and McNutt (1986), Watts (1988), Watts (1988) and Betchel et al. (1990). Reproduced in part from fig. 4 in Watts (1992).

a plagioclase-rich crust, that where the lithosphere is young and hot (e.g., Apennines, Alps) it decouples, and this explains the low T_e here. Where the lithosphere is old and cold (e.g., Ganges and Himalayas) and there is no weakening due to crustal thickening, high curvature or thick sediments, the lithosphere does not decouple, and this explains the high T_e here. Finally, in some forelands (e.g., Zagros, Tarim, Kunlun, southern Alps and Andes), sediments play a critical role by weakening what would otherwise have been a strong lithosphere.

This dichotomy in the response of the continental lithosphere to loads on long geological timescales has led to a lively debate on its strength and where within the plate this strength resides. The debate was triggered by Jackson (2002) when referring to previous work such as that by Meissner and Wever (1986) who had likened the case of a weak, ductile, lower continental crust to jelly in a sandwich between stronger layers in the upper crust and upper mantle. Despite evidence from T_e estimates (e.g., Fig. 6.22) and YSEs (e.g., Fig. 6.33) that continental lithosphere shows both great weaknesses, for example when decoupled, and great strength, for example when coupled, Jackson (2002) argued that the strength of continental lithosphere resides mainly in the brittle upper crust which is characterised by faulting and earthquakes and that the so-called jelly sandwich model should be abandoned. 'Is it all in the crust?', queried Lamb (2002). Burov and Watts (2006), however, while agreeing the upper crust could be strong, argued that like the oceans, the continental mantle must be involved in supporting loads, for example, in cratonic areas away from plate boundaries and other actively deforming regions.

6.7 Time-Dependent Flexure

If yielding is important in the lithosphere, as the data from experimental rock mechanics suggest, then the flexure associated with a particular geological load will vary with time. In the brittle field, yielding is determined by the friction laws and is independent

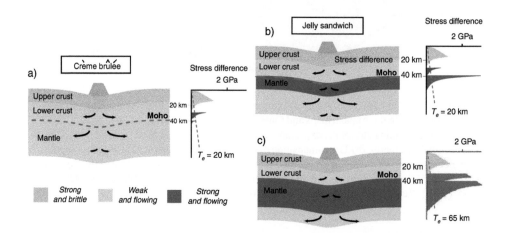

Figure 6.35 Schematic diagram illustrating (a) the 'crème-brûlée' model and (b) and (c) the 'jelly sandwich' model for the long-term strength of continental lithosphere. In the crème-brûlée model the strength is confined to the brittle upper crust, and compensation for a surface load is achieved mainly by flow in the weak upper mantle, while in the jelly sandwich model the upper mantle is strong and compensation is achieved mainly in the underlying asthenosphere. The YSEs have been calculated assuming a wet quartz upper crust, a dry diabase lower crust and a wet (crème-brûlée) and dry (jelly sandwich) olivine upper mantle. Other parameters are tabulated in Burov and Watts (2006).

of strain rate. However, in the ductile field, yielding is a strong function of the strain rate. As time increases, the strain rate decreases, and the depth to the onset of ductility (e.g., Fig. 6.29) shallows. This results in a thinning of the elastic 'core' of the lithosphere that supports a load and, hence, a flexure that changes with time. We can calculate the flexure at a particular time, provided some assumption is made concerning the ductile flow laws and, because of their dependence on temperature, on the thermal structure of the lithosphere.

Bodine et al. (1981) used the YSE and a cooling plate model to estimate the flexure as a function of time. They computed the bending moment that is associated with the YSE for different times and found, at each time, the equivalent flexure of a purely elastic plate.

The models used by Bodine et al. (1981) were only approximations for the flexure. The first numerical models of flexure using a YSE rheology were constructed by Beaumont (1979) and McAdoo et al. (1985) for the oceanic lithosphere, and Burov and Diament (1992) for the continental lithosphere.

Other workers have followed a simpler approach. Lago and Cazenave (1981), for example, assumed a power law rheology and a cooling half-space model and calculated how T_e and, hence, the flexure due to a load would change with time. These workers showed that T_e is initially large, decreases with time and then appears to stabilise after a certain time has elapsed. They estimated the time of stabilisation for oceanic lithosphere to be about 5 per cent of the age of the plate at the time of loading.

Courtney and Beaumont (1983), Willet et al. (1985), Zhong (1997) and Watts and Zhong (2000) developed a similar model to Lago and Cazenave. The main difference with the latter model was that they expressed the strain rate in the power law creep equation in terms of a viscosity. By using a thermal cooling model, these workers approximated the viscosity structure of oceanic lithosphere to several layers and solved for the flexure using a multilayered viscoelastic plate model.

Laboratory studies show that in the case of polycrystalline olivine, the general creep law (e.g., Karato and Wu, 1993) is given by

$$\dot{\varepsilon} = A(\tau_s/\mu)^n (b/d)^m e^{-(Q_p+PV)/R_g T}, \tag{6.27}$$

where A is a pre-exponential factor, τ_s is the shear stress, μ is the shear modulus, b is the length of the Burgers vector, d is the grain size, n is the stress exponent, m is the grain-size exponent, Q_p is the creep activation energy, P is ambient pressure, V is the activation volume, R_g is the universal gas constant and T is the temperature. The strain rate, $\dot{\varepsilon}$, is determined by the viscosity, η, and shear stress, τ_s, and so it follows from Eq. (6.27) that

$$\eta = \frac{\tau_s}{2\dot{\varepsilon}} = \eta_{ref} e^{\frac{(Q_p+PV)}{R_g T}}, \tag{6.28}$$

where

$$\eta_{ref} = \frac{\mu}{2A} (\tau_s/\mu)^{1-n} (b/d)^{-3}.$$

As Karato and Wu (1993) and others have pointed out, the rheological parameters for olivine vary widely for different creep mechanisms and volatile content. For example, in diffusion creep, $n = 1$, but it may be as high as 3–5 for dislocation creep. Similarly, the activation energy varies from 300–540 kJ mol^{-1} for 'dry' samples to 240–430 kJ mol^{-1} for 'wet' samples.

Because of these uncertainties, Courtney and Beaumont (1983) and Watts and Zhong (2000) assumed a linear rheology (i.e., $n = 1$). In this case, Eq. (6.28) shows that the viscosity can be expressed in terms of a reference viscosity, η_{ref}, Q_p, P, V, R_g and T. The parameters P, V, R_g and T can all be prescribed for the oceanic lithosphere: P is a pressure term that is given by $\rho g z$, where ρ is the average density of the lithosphere, g is the gravitational acceleration and z is depth; V is the activation volume; R_g is the universal gas constant; and T is the temperature in kelvin. Since pressure and temperature are known in the cooling lithosphere, the only unknowns are the parameters η_{ref} and Q_p.

Figure 6.36 shows the viscosity for different thermal ages of the oceanic lithosphere and for values of the parameter pair η_{ref} and Q_p of 10^{20} Pa s and 120 kJ mol^{-1}, respectively. The figure shows the viscosity to be very sensitive to temperature, decreasing by > 15 orders of magnitude with depth in the oceanic lithosphere. The maximum rate of decrease is greatest in the uppermost 25–50 km, with young seafloor showing the most rapid changes at shallow depths and old seafloor showing the least rapid.

Once the viscosity profile is specified, then the flexure can be computed either analytically or numerically using the multilayered viscoelastic plate model of Zhong (1997). Figure 6.37, for example, compares the predictions of the multilayered viscoelastic model with the same set of flexural rigidity estimates that were compared earlier to the predictions of the single-layer viscoelastic model in Fig. 6.6, the general linear model in Fig. 6.10 and the two-layer viscoelastic model in Fig. 6.17. The figure shows that a multilayered viscoelastic plate model with $\eta_{ref} = 10^{20}$ Pa s and $Q_p = 120$ kJ mol^{-1} explains the general trend of the flexural rigidity data with a number of individual estimates appearing to line up along the 70–110 Ma age of the lithosphere at the time of loading predicted curves. There are 'trade-offs', however. The data could be explained by either a higher η_{ref} and a lower Q_p, or a lower η_{ref} and a higher Q_p. If $\eta_{ref} = 10^{20}$ Pa s, as some post-glacial rebound studies suggest (e.g., Fig. 4.1), then $Q_p = 120$ kJ mol^{-1}. This is less than laboratory studies of olivine (which, based on Goetze and Evans's experiments on 'dry' olivine, is ~520 kJ mol^{-1}), probably because of the much lower strain rates involved in flexural loading. The main advantage of the multilayered model is that it predicts a rigidity that stabilises with load age, a behaviour that appears to be repeated in observations.

The final step, once η_{ref} and Q_p have been estimated, is to predict the flexure as a function of time. Figure 6.38 shows, for example, the flexure that would be expected for a seamount-type load that formed on 70-Ma-old oceanic lithosphere that responds to applied loads as a multilayered viscoelastic plate with a reference viscosity, η_{ref}, of 10^{20} Pa s and an activation energy, Q_p, of 120 kJ mol^{-1}. The figure shows that immediately following loading the

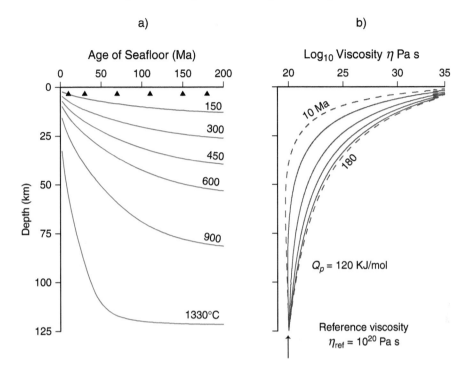

Figure 6.36 The viscosity structure of the oceanic lithosphere as a function of age. The calculations are based on a cooling plate model and a power law dependence of viscosity on temperature. (a) Depth to selected oceanic isotherms. (b) Viscosity structure of 10, 30, 60, 110, 150 and 180 Ma oceanic lithosphere. Reproduced from fig. 6 of Watts and Zhong (2000).

flexure is small and then it increases rapidly with time. By 10 kyr, the subsidence reaches nearly half of its final value. After 1 Myr, the flexure is small (< 20 per cent) but nevertheless significant.

We have thus far only considered the viscoelastic response to loading of oceanic lithosphere. Unlike oceanic lithosphere, there is no thermal model that describes the temperature structure of continental lithosphere as a function of age. We will therefore follow McKenzie et al. (2005) and assume a structure based on pressure and temperature data derived from the Jericho kimberlite in the North American craton that is consistent with surface heat flow data (Mareschal and Jaupart, 2004). Figure 6.39 shows a plot of the temperature and corresponding viscosity structure, assuming $n = 1$, a reference viscosity, η_{ref}, of 10^{20} Pa s and an activation energy, Q_p, of 200 kJ mol^{-1}. The figure shows that the viscosity decreases by several orders of magnitude with depth and compares well with the calculations for cratons of Dixon et al. (2004) who considered grain size, fugacity and water content. Dixon et al. (2004) also calculated the viscosity for tectonic western North America and, interestingly, it more closely resembles that of oceanic lithosphere than cratonic North America.

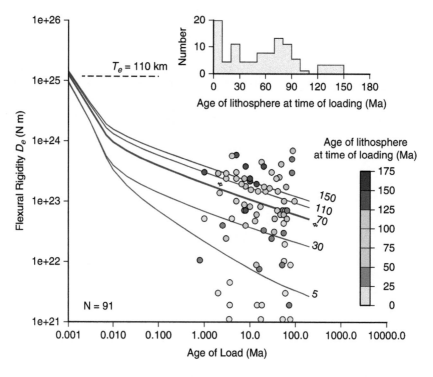

Figure 6.37 Comparison of observed and calculated flexural rigidity at seamounts and oceanic islands, excluding those in French Polynesia, that formed on >80 Ma oceanic lithosphere. The observed values are based on Table 6.1 and have been colour coded according to their age of the lithosphere at the time of loading. The calculated curves are based on a multilayered viscoelastic plate model with $Q_p = 120\,\text{kJ mol}^{-1}$ and $\eta_{ref} = 10^{20}$ Pa s and an age of the lithosphere at the time of loading, $t_{sf} - t$, of 5, 30, 70, 110 and 150 Ma. Note that the multilayer viscoelastic model better fits the general trends in the data than either the general linear model (Fig. 6.10) or the viscoelastic plate model (Fig. 6.19). Inset shows a histogram of ages of the lithosphere at the time of loading. Modified from fig. 10 of Watts and Zhong (2000).

The observed T_e is compared to the calculated T_e based on a multilayer viscoelastic model in Fig. 6.40. The observed T_e estimates are based on foreland basin and rim flank loads on young lithosphere (red filled circles) and old lithosphere (green filled circles), and the calculated T_e estimates are based on a cratonic temperature and viscosity structure and a range of values for η_{ref}, Q_p and water content (thick brown curves). Also shown for comparison are T_e estimates for oceanic lithosphere (blue shaded region), regions of glacial isostatic rebound (purple shaded region) and InSAR- and GPS-determined fault bottom depth and centroid depths of earthquakes (light gray shaded region, Wright et al. 2013) and GPS-determined thickness of the Pacific and Indo-Australian plates in New Zealand (dark gray shaded region, Cohen & Darby 2003). We see from the comparison that while the T_e associated with loads on

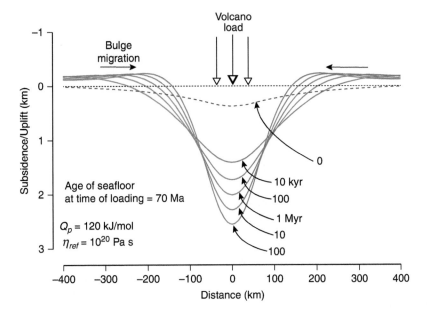

Figure 6.38 The flexure beneath the centre of a seamount load emplaced on a multilayered viscoelastic plate as a function of load age. Reproduced from fig. 12 of Watts and Zhong (2000).

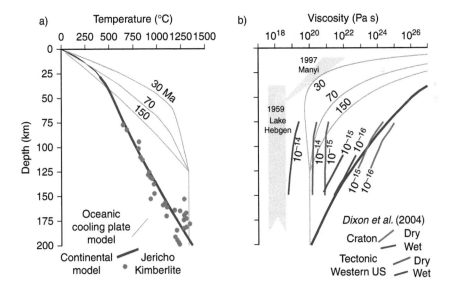

Figure 6.39 Plots of temperature and viscosity as a function of depth for oceanic (blue solid lines) and continental (brown solid lines) lithosphere. (a) Temperature versus depth. (b) Viscosity versus depth. Grey shaded region shows for comparison the viscosity structure determined from InSAR and geodetic measurements following the 1997 Manyi (Yamasaki and Houseman, 2012) and 1959 Lake Hebgen (Nishimura and Thatcher, 2003) earthquakes.

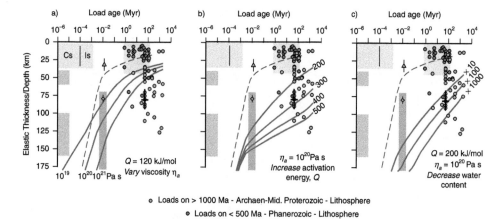

Figure 6.40 Plot of T_e estimates for foreland basins and rim flank uplifts against the load age. Green filled circles are for loads on old continental lithosphere, and red filled circles are for loads on young continental lithosphere. Calculated curves are based on multilayered visco-elastic models for oceanic lithosphere (blue dashed line) and continental lithosphere (thick brown lines). Reproduced from fig. 7 in Watts et al. (2013).

Phanerozoic continental lithosphere overlap those on oceanic lithosphere, loads on Archean to mid-Proterozoic lithosphere are associated with significantly higher T_e. Moreover, we see that the strength of continental lithosphere depends on the timescale of loading with the strength at co-seismic and interseismic and geodetic timescales following a significantly different relaxation 'path' than at glacial isostatic and geological timescales.

6.8 Relationship between Elastic Thickness, Curvature and Yielding

As pointed out earlier, flexure studies have mainly been based on elastic rather than viscoelastic models. The problem is that there is no limit to the amount of stress that an elastic plate can accumulate. The YSE, however, suggests that the strength of the lithosphere is limited by brittle and ductile deformation. If a load is applied that exceeds the strength of the plate, it will yield rather than flex (e.g., Fig. 6.41) and the plate will appear weaker than it actually is. By comparing the observations of flexure with calculations based on the YSE, it should be possible to quantify the amount of yielding and, hence, estimate the true T_e for a plate that has yielded.

Goetze and Evans (1979) were among the first to show that the amount of yielding in a flexed plate is related to its curvature. They pointed out that the stress distribution needed to explain the bending moment at deep-sea trenches was unknown. The elastic model predicts the magnitude of these stresses, but implicit in this model is the assumption that the bending moment depends linearly on the curvature of the flexure. Thus, as the moment increases, so does the curvature, which in turn increases the bending stresses. In contrast, the YSE,

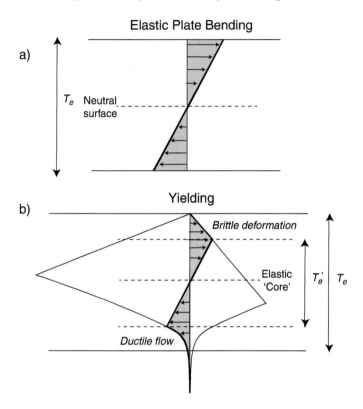

Figure 6.41 Schematic diagram showing the bending stresses that develop in (a) an elastic plate and (b) a plate whose strength is limited by brittle deformation in its upper part and ductile flow in its lower part.

because it includes failure concepts, implies a different stress distribution. Goetze and Evans showed that the bending moment generated in a YSE model depends non-linearly on the curvature. At first, the moments increase quite rapidly, but then later – as the curvature increases – they approach an asymptotic limit.

We can illustrate these results by comparing the behaviour of an elastic plate with one that deforms according to the YSE. We have from Section 3.4.2 that the bending moment in an elastic plate, $M_{elastic}$, is given by the vertical integral of the stress profile:

$$M_{elastic} = \int_{-\frac{T_e}{2}}^{-\frac{T_e}{2}} \sigma_x y_f dy \ ,$$

where T_e is the thickness of the plate and σ_x is a fibre stress a distance y_f from the neutral surface of the plate. It follows from Eqs. (3.9) and (3.10) that

$$M_{elastic} = \frac{-ET_e^3}{12(1 - v^2)} \frac{1}{r},$$

where r is the radius of curvature. Once the moment is known, then the curvature, K, can be computed using

$$\frac{1}{r} = K = \frac{-M_{elastic}12(1 - v^2)}{ET_e^3}. \tag{6.29}$$

In the case of the YSE, the moment is made up three main contributions: the upper part of the plate, M_{upper}, the lower part of the plate, M_{lower}, and a central elastic 'core', M_{core}. The total moment, M_{YSE}, is obtained from

$$M_{YSE} = M_{upper} + M_{core} + M_{lower}.$$

As several authors have shown (e.g., McAdoo et al., 1985 for oceans and Burov and Diament, 1992 for continents), the M_{YSE} can be calculated analytically by specifying the slope of the Byerlee curve, the Dorn law, the power law creep laws and the depth to the point of intersection of these curves on the YSE.

Figure 6.42a compares the moment generated by an elastic plate, $M_{elastic}$ with M_{YSE}, the moment associated with the YSE. Because of yielding in the upper and lower parts of the plate, $M_{YSE} < M_{elastic}$. This is seen well in the figure, which shows the relationship between the two cases for different curvatures of the plate. When the curvatures are small, the two moments are similar. However, as curvatures increase, the curves diverge. The elastic plate moments continue to increase with curvature,[8] while the YSE approaches a constant value which McNutt and Menard (1982) dubbed 'moment saturation'.

The strength of an elastic plate is determined by its thickness, T_e. It follows from Eq. (6.29), therefore, that the equivalent elastic thickness, T'_e, which generates the moment M_{YSE} and has the same curvature as an elastic plate, is

$$T'_e = \left[\frac{12(1 - v^2)M_{YSE}}{-KE}\right]^{1/3}. \tag{6.30}$$

Figure 6.41b shows the ratio of T'_e/T_e as a function of the curvature. We see that for low curvatures (i.e. $K > 10^{-7}$ m^{-1}), the ratio T'_e/T_e is 1, indicating little difference between the elastic thickness values. However, as curvature increases, the ratio decreases.

The importance of these results is that they quantify how yielding will change T_e for a particular curvature. For example, Fig. 6.42b shows that for curvatures greater than about 10^{-7} m^{-1}, the thickness of the plate that remains elastic decreases because of yielding in the upper and lower parts of the plate. Eventually, if the curvature is high enough, the plate may fail all the way through.

---

[8] This is best seen in the case that the *y*-axis in Fig. 6.42 is linear rather than logarithmic.

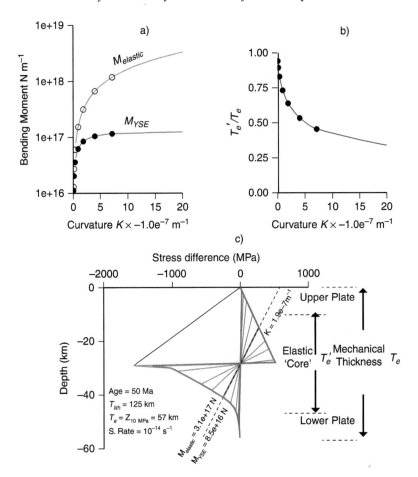

Figure 6.42 The relationship between (a) bending moment and curvature and (b) the ratio T'_e/T_e and curvature for an elastic plate and a plate whose strength is limited by brittle deformation in its upper part and ductile flow in its lower part. (c) the Yield Strength Envelope for each of the points plotted in (a) and (b) together with the thickness of the elastic core, T'_e, for a curvature, K, of -1.9×10^{-7} m^{-1}.

Although curvature is difficult to determine from observations (it depends on the slope and, hence, is sensitive to small irregularities in the surfaces of flexure), estimates have been made at deep-sea trench systems and the Hawaiian Ridge in the Pacific. Goetze and Evans (1979) estimated moments and, hence, curvatures by first fitting a smooth elastic curve to bathymetric profile data. They found curvatures at the Mariana, Bonin, Aleutian, Kuril and Tonga deep-sea trenches that are in the range of 1.0×10^{-6} to 3.0×10^{-7} m^{-1}. McNutt and Menard (1982) extended the analysis to the Japan, Middle America and Kermadec trenches, finding curvatures that were in the range of 1.8×10^{-7} to 8.2×10^{-7} m^{-1}. Finally, Wessel (1993) estimated the curvatures beneath the Hawaiian Ridge by smoothing the seismically constrained depth to the top of the oceanic crust (see Section 4.3). He found curvatures of the order of 5×10^{-7} m^{-1}, comparable in magnitude to some of the estimates from deep-sea trenches.

Figure 6.43 compares the observed curvatures at deep-sea trenches and Hawaii with the predictions of the YSE. We see that the observed curvatures imply T'_e/T_e ratios that are in the range of 0.8–0.7. Therefore, in both trench and oceanic island settings there has been sufficient yielding that the T_e determined from flexure studies may underestimate the actual elastic thickness of the lithosphere. The largest amounts of yielding appear to occur at deep-sea trenches. McNutt (1984), for example, found that when the values of Caldwell et al. (1976) at the Kuril Trench are corrected for curvature, the T_e increases from 20 to 40 km. Some yielding is in accord with seismic reflection and bathymetry data that suggest the presence of normal faults on the seaward wall of trenches (Fig. 4.30). Yielding may also occur at oceanic islands. Wessel (1993), for example, found that the value at Hawaii increases from 25 to 35 km, decreasing to 25 km (the preferred value of most previous studies) only immediately beneath the load. In the region flanking the Big Island, Hawaii, he found even higher values of 44 km.

6.9 Elastic Thickness and Earthquakes

According to the elastic plate model, the lithosphere is gently flexed into broad upwarps and downwarps in the region of large loads. The warping induces bending stresses that reach their maximum values in the top and bottom of a flexed plate. The YSE, however, implies that these stresses will be relieved by brittle faulting in the upper crust and by some form of ductile flow in its lower part. In between the brittle and ductile deformation fields there is an elastic 'core', which apparently can support the stresses induced by flexure on long geological timescales.

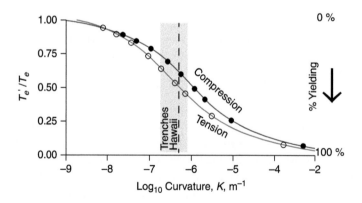

Figure 6.43 The relationship between T'_e/T_e and $\log_{10}(K)$. The circles are predicted values based on the Yield Strength Envelope. The grey shading shows the range of observed curvatures that have been estimated from the bathymetry of deep-sea trench–outer rise systems (McNutt and Menard, 1982) and seismic images of the top of the flexed oceanic crust beneath Hawaii (Wessel, 1993).

If the lithosphere behaves as a flexible plate, then the plate structure may be reflected in seismicity data. This is because seismicity is the consequence of frictional instabilities on sliding surfaces such as faults. The bending stresses induced by flexural loading might be one of several possible sources of the stress that could cause faulting and, hence, earthquakes. As Fig. 6.44 illustrates, a relationship might exist therefore between the depth interval over which earthquakes occur (i.e., the seismogenic layer, T_s) and the thickness of the 'core' that behaves elastically on long timescales, T_e. This is because as the curvature increases due, for example, to a load increase, T_e decreases. However, T_s and T_e have different meanings: T_s reflects the depth to which faulting and earthquakes can occur in the uppermost brittle part of the crust, while T_e reflects the integrated strength of the entire plate which includes the brittle layer, the ductile layer and an intervening layer or 'core' which supports loads essentially elastically on long timescales.

The main evidence for a link between the plate-like behaviour of the lithosphere and earthquakes is likely to come from the oceans rather than the continents. This is because the oceans have had a much simpler evolution than the continents and so it is more likely that the pattern of seismicity might reflect plate bending rather than other, more complex, geological processes. Unfortunately, it is difficult to use the pattern of seismicity around volcanic loads at seamounts and ocean islands because of the complications of magmatism, mass wasting and other local effects. However, one geological setting that is comparatively unaffected by such processes, and where the plate is gently flexed, is seaward of deep-sea trenches.

Figure 6.45 plots the depth of earthquakes seaward of deep-sea trenches in the Pacific Ocean against the thermal age of the approaching lithospheric plate. The

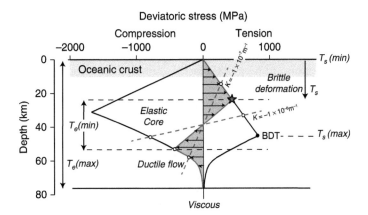

Figure 6.44 The relationship between T_e and the depth to the base of the seismogenic layer, T_s, in oceanic lithosphere. Note that T_e reflects the integrated strength of the lithosphere while T_s is the depth to which faulting and earthquakes occur. Reproduced from Watts and Burov (2003).

earthquake data are based on the compilation of Seno and Yamanaka (1996), which includes examples from the Tonga–Kermadec, Peru–Chile, Aleutian, Kuril, Japan and Bonin trenches. The figure shows an increase in the depth of earthquakes with thermal age. More significantly, there is a distinct pattern in the focal mechanisms of individual earthquakes: tensional events (open circles) characterise shallow regions of the crust and upper mantle, whereas compressional events (solid circles) dominate deeper regions. This pattern has been confirmed in more recent studies using body-wave inversion of teleseismic data from earthquakes on 36 Ma oceanic lithosphere offshore Chile (Clouard et al., 2007) and long recording Ocean Bottom Seismometers on 20–25 Ma oceanic lithosphere offshore Nicaragua (Lefeldt et al., 2009) and is exactly what would be expected from upwarping of the oceanic lithosphere seaward of a deep-sea trench, suggesting that the earthquakes compiled by Seno and Yamanaka are indeed the response to stresses that develop in a flexed oceanic plate.

If the earthquakes seaward of trenches are the result of stress release during faulting on the outer rise, then their maximum depth may be indicative of the transition between brittle and ductile flow. Rheological models suggest that this transition is controlled mainly by the temperature structure, but that it also depends on the strain rate and the magnitude of the stress that is applied. Since the temperature structure of the oceanic lithosphere can be predicted using cooling plate models, it is a relatively easy matter to use our knowledge of the brittle and ductile field to construct so-called 'failure' envelopes for the cooling oceanic lithosphere.

Figure 6.45 compares examples of failure envelopes for the cooling oceanic lithosphere to the earthquake data seaward of trenches. The envelopes have been computed assuming the strength of the oceanic lithosphere is limited by brittle (i.e., Byerlee's law) and ductile (i.e., olivine – power law) deformation, the strain-rates range from 10^{-14} to 10^{-20} s^{-1}, the applied stresses are 1, 10 and 100 MPa. The figure shows that most of the earthquake data compiled by Seno and Yamanaka (1996) fall within the elastic 'core' that separates the brittle and ductile deformation fields. Wiens and Stein (1983) have estimated the strain rate from the moment release of intra-plate earthquakes as in the range of 10^{-17}–10^{-19} s^{-1}. If we assume a strain rate of 10^{-18} s^{-1}, then Fig. 6.45b shows that a stress of about 10 MPa explains the thickness of the seismogenic layer seaward of deep-sea trenches. Lower strain rates require a lower stress, higher strain rates a higher stress.

The failure envelope also accounts for the focal mechanisms of earthquakes. The thick dashed blue line in Fig. 6.45b, for example, is the approximate 'half-way' between the upper surface of the failure envelope (as defined by a stress of 10 MPa and a strain rate of 10^{-18} s^{-1}) and a lower surface defining the approximate base of the seismogenic layer. The figure shows that a close agreement between the 'half-way' surface and the observed transition between tensile and compressional earthquakes seaward of trenches.

The bending stresses induced by flexure are not, of course, the only possible source of stress in the oceanic lithosphere. Other sources are generated by forces at the plate boundaries (e.g., slab pull, ridge push) that can apparently be transmitted for very large distances into the plate interiors (e.g., Bott and Dean, 1973). The stresses may be large enough to negate the flexural bending stresses in outer rises and cause, for

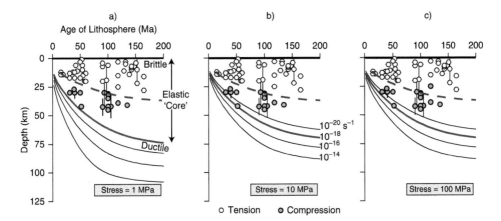

Figure 6.45 Comparison of the depth and focal mechanisms of earthquakes in deep-sea trench–outer rise systems with the predictions of the Yield Strength Envelope (YSE) model. The earthquake data are based on a compilation by Seno and Yamanaka (1996). The YSE model is based on a thermal structure determined by the cooling plate model, an olivine rheology and the range of strain rates and applied stresses shown.

example, a tensional stress in the uppermost part of the oceanic lithosphere to transition to a compressional stress (e.g., Clouard et al., 2007). Wiens and Stein (1984) have summarised the evidence for 27 earthquakes in the interior of the oceanic plates, the m_b values of which are > 5.1. Figure 6.46 shows that, like at trenches, these earthquakes fall within the failure envelopes. The main difference is that intra-plate earthquakes appear to require a higher stress to explain them (> ~ 100 MPa) for the same strain rate of 10^{-18} s^{-1}.

We have shown that the earthquakes seaward of trenches occur within the elastic 'core' that supports loads for long periods of geological time. This is, at first glance, a surprising result. The timescales associated with earthquakes and flexure are significantly different. Moreover, there may not be earthquakes below a certain depth simply because the confining pressure increases linearly with depth and becomes so high that fracture may not be able to occur.

The question then is what is the relationship, if any, between T_e and the thickness of the seismogenic layer? Figure 6.47 shows a good general agreement between the T_e at deep-sea trenches–outer rise systems and the depth of the brittle–ductile transition based on the failure envelope. Therefore, the failure envelope model appears to explain both the earthquake data and T_e, at least in this setting. There is a suggestion, however, from Fig. 6.45b, that while a stress of 10 MPa and a strain rate of 10^{-18} s^{-1} explains the thickness of the seismogenic layer (Fig. 6.45b), much of the T_e data in Fig. 6.47b appears to require a lower strain rate of 10^{-20} to 10^{-22} s^{-1} or less for the same stress. In this case, the thickness of the seismogenic layer is larger than the elastic thickness, T_e. However, when the entire oceanic T_e data set is compared with the earthquake data (e.g., Fig. 6.48), there is little disagreement

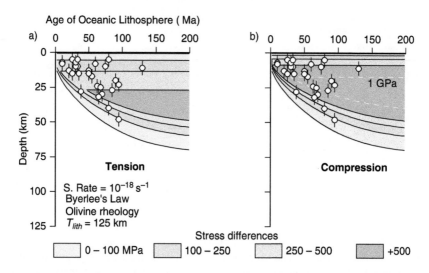

Figure 6.46 Comparison of the depth of large (i.e., $m_b > 5.1$) intra-oceanic plate earthquakes with the predictions of the Yield Strength Envelope (YSE) model. The earthquake data set is based on a compilation by Wiens and Stein (1984). The YSE models are based on a thermal structure determined by the cooling plate model, an olivine rheology, a strain rate of $10^{-18}\,\mathrm{s}^{-1}$ and show cases for both tensile and compressive bending stress.

between the two thickness estimates. In fact, the thickness of the seismogenic layer and oceanic T_e show a remarkably similar distribution in depth.

The relationship between T_e and the thickness of the seismogenic layer is not as clear in the continents. Figure 6.48 shows, for example, a clear 'peak' in the depth of continental earthquakes at relatively shallow depths of 5–10 km, with very few earthquakes at depths > 30 km. The T_e data, in contrast, shows a more subdued peak at 20–25 km and a more evenly spread distribution of values over the entire thickness of the lithosphere.

The source of the discrepancy between T_e and thickness of the seismogenic layer in the continents can be better understood if we consider different tectonic regions. Chen and Molnar (1983), for example, have already classified the earthquake data into 'rifts', 'active orogenic belts' and 'cratons'. In the case of the T_e estimates, we assume that 'rifts' include estimates from continental rift valleys and rift-type sedimentary basins, 'active orogenic belts' are localities where the culmination of thrust/fold loading is < 40 Ma and 'cratons' are the stable interiors of the continental lithosphere with a thermal age > 1,000 Ma.

Figure 6.49 shows some similarities exist in the distribution of T_e and earthquake data from rifts and young orogenic belts. Rifts are characterised by relatively shallow earthquakes and low T_e values that show two peaks, one at < 5 km and the other at 5–10 km. Most rift-related earthquakes occur at depths < 25 km. However, some deep events have been reported beneath the East Africa Rift (e.g., Shudofsky et al., 1987; Yang and Chen, 2010). Young orogenic belts are also characterised by several shallow earthquake events and low T_e

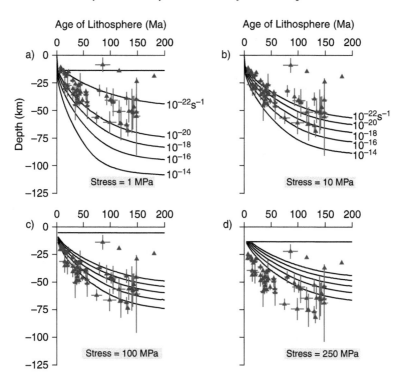

Figure 6.47 Comparison of T_e, corrected for the effects of curvature, at deep-sea trench–outer rise systems with the predictions of the Yield Strength Envelope (YSE) model. The T_e estimates are based on Table 6.1. The YSE model is based on a thermal structure determined by the cooling plate model, an olivine rheology and the range of strain rates and stresses shown.

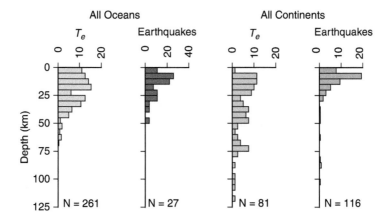

Figure 6.48 Comparison of T_e with the depth of earthquakes in oceans and continents. The T_e data are based on data in Tables 6.1 and 6.2. The earthquake data are based on compilations of intra-plate earthquakes compiled by Chen and Molnar (1983) and Wiens and Stein (1984).

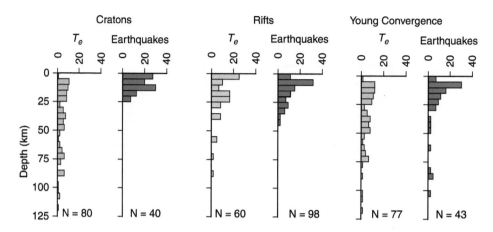

Figure 6.49 Comparison of T_e with the depth of earthquakes in continental cratons, continental and passive margin rifts, and regions of young (i.e., <40 Ma) continental plate convergence.

values. Like some rifts, these regions are also associated with some of the deepest earthquakes in the continents and high T_e values.

Probably the most striking result emerges from Fig. 6.49, however, is the difference between the elastic and seismogenic layer thickness in cratonic regions. Cratons show no obvious peak in T_e, being characterised by a wide range of values that span the entire thickness of the lithosphere. The depth to earthquakes, on the other hand, has a clear peak at depths of 10–15 km and there are no reported earthquakes at depths > 25 km in non-rift cratonic areas (e.g., Chen and Molnar, 1983; Gangopadhyay and Talwani, 2003). At present, we have no explanation for this discrepancy. It could be due to difficulties in the determination of T_e, in the topographically subdued cratonic interiors of the continents. Another possibility is that confining pressure linearly grows with depth and becomes so high at depths > 50 km that it inhibits faulting and, hence, earthquakes. Alternately, the sub-crustal cratonic mantle is simply too strong.

6.10 Summary

During the 1970s to 2000s, more than 180 studies of the elastic thickness of the lithosphere were published. The results of these studies have provided a wealth of new information on the long-term mechanical properties of the lithosphere and their relationship to plate and load age. Although the results of individual studies are subject to uncertainties, the analysis of large, global data sets tends to 'smooth' out local discrepancies and, hence, make it more likely they will reveal the main features that describe the long-term behaviour of the lithosphere.

Oceanic and continental flexure studies suggest that the long-term behaviour of the lithosphere can be modelled, to first order, as a thin elastic plate that overlies an inviscid fluid. The thickness of the elastic plate, T_e, varies both spatially and temporally, and this has provided information on the relationship between load and plate age.

An elastic model, however, is time-invariant and does not account for any changes that might occur in the rigidity, and the equivalent elastic thickness, with time. Perhaps the most well-documented change is the thinning that must occur during the initial stages of loading as the thickness of the lithosphere that supports a load decreases from its short-term seismic thickness to its long-term elastic thickness. The precise form of the decrease is not known, but in the oceans the elastic thickness is approximately one third of the seismic thickness determined from seismic tomography and there is some evidence that this 'rule of thumb' may also apply to the continents.

During the past two decades there has been much progress in modelling the temporal changes that occur within the lithosphere during loading and unloading. Multilayered viscoelastic models in which the viscosity structure of the lithosphere is determined from oceanic and continental geotherms are compatible, for example, with the observed relationship between T_e and the age of the lithosphere at the time of loading at seamounts and ocean islands. Such models, which are elastic initially and then become more viscous with time, predict that stresses induced by flexure migrate upwards from the relatively hot low-viscosity layers at the base of the lithosphere to the cold high-viscosity layers at the top of the lithosphere. This so-called 'stress relaxation' is initially fast and then slows and so may correspond to what the early isostatists called 'fast isostasy' and 'slow isostasy'. Observational evidence of the actual 'path' of relaxation is, however, limited although there is evidence from some seamounts and oceanic islands of a dependence of T_e, not only on the age of the lithosphere at the time of loading, but also on load age.

Data from experimental rock mechanics have been of much help in understanding the physical meaning of parameters such as T_e. The data show that, despite its long-term essentially elastic properties, the strength of plates is limited by brittle and ductile deformation in the upper and lower parts of the lithosphere. Therefore, loading of the plates may result in yielding rather than bending. Model studies suggest that yielding will be most significant in regions of high curvature, such as occurs in the seaward wall of all deep-sea trenches.

These and other studies have demonstrated the usefulness of T_e as a constraint on flow laws at lithospheric conditions. However, they still have not yet revealed which brittle or ductile flow law best accounts for the deformation occurring at the boundaries of or within an individual plate. The fact that one flow law works well in some tectonic settings and not in others is interesting and provides insight into the role of other factors such as variations in local temperature structure, stress regime and pre-existing structures.

There is evidence that oceanic T_e corresponds approximately to the thickness of the seismogenic layer as inferred from well-located intra-plate earthquakes. However, T_e and the seismogenic layer thickness are not the same parameter: T_e is the integrated thickness of the lithosphere that behaves essentially elastically on long geological timescales, while the seismogenic layer thickness defines the depth to which faulting and earthquakes occur.

In the continents, no simple relationship exists between T_e and the thickness of the seismogenic layer. The problem is that while continental T_e shows a wide range of values, most earthquakes occur in the uppermost 25 km of the continental lithosphere. Most low T_e estimates, however, occur in rifts and active orogenic belts where there is a generally good agreement between T_e and the thickness of the seismogenic layer. The main discrepancy is in cratonic regions, where earthquakes are limited to depths of 25 km and peak at 10–15 km, in contrast to the T_e data that show no obvious peak and a range of values of 0–120 km.

7

Isostasy and the Origin of Geological Features in the Continents and Oceans

Geology has to do with many cycles, but the greatest of these is the erosional flow of rock mass from the continents to the sea and the isostatic return flow of equivalent mass from beneath the sea to the land.

(Lawson, 1938, p. 416)

7.1 Introduction

In Chapter 4, it was demonstrated that lithospheric flexure is an important feature of the Earth's outermost layers. We developed a simple model for flexure and showed how the phenomenon contributes to our understanding of the evolution of relatively simple loading situations, such as ice, seamounts and oceanic islands, river deltas and deep-sea trench–outer rises. An advantage of looking at such systems first is to help our understanding of more complex ones which include rift valleys, mountain belts and strike-slip faults, transform faults and fracture zones. These features form in response to extensional and compressional and strike-slip forces that act on the lithosphere because of plate motions. Rift valleys, for example, are more often associated with plate boundaries than their interiors, and so their evolution is controlled not only by the flexural properties of the plates, but also by other processes such as those associated with the dynamics of global plate motions.

One part of the dynamics system is the accretion of oceanic crust at mid-ocean ridges and the coupled subduction of oceanic crust at deep-sea trenches. These processes give rise to gravity and bathymetry anomalies that may make it difficult, in some situations, to isolate the effects of flexure from surface observations.

Interactions of flexural and dynamic effects are not limited, however, to the plate boundaries. There is evidence that the two phenomena interact in both sedimentary and volcanic settings. One of the best examples, perhaps, are active volcanoes (e.g., Hawaii, Cape Verdes), both of which sit astride a broad 'swell' in seafloor bathymetry. These swells have been interpreted as the surface expressions of convective processes that are occurring in the mantle.

Because of the possible role played by dynamic effects, flexure should not be considered as the only contributor to the Earth's surface features. There are, however, certain aspects of the evolution of rift valleys, sedimentary basins, mountain belts and strike-slip faults, transform faults and fracture zones that can be attributed to flexure. In this chapter, we will attempt to isolate these features with the overall aim of determining the extent to which flexure is a global phenomenon.

7.2 Extensional Tectonics and Rifting

Rifts are one of the most striking features of the Earth's surface. A feature of both the continental and oceanic crust, rifts form continuous features that exceed some 10,000 km in length. While rifts were once thought of as the products of an expanding Earth, they are now more generally regarded as the surface manifestations of extensional processes that arise because of global plate motions.

7.2.1 Continental Rifts

Typically, continental rifts comprise 30–60-km-wide grabens with either two or more opposing border faults or half-grabens with a single main boundary fault. Although rifts show a large diversity in structural styles, there are two end-member types: those with thin sediments and abundant volcanics (e.g., East Africa) and those with thick sediments and sparse volcanics (e.g., Baikal). Both types of rifts are often associated with the development of broad rim flank uplifts.

7.2.1.1 Bullard's Hypothesis

One of the first studies of the state of isostasy at rifts was the one by Edward Crisp Bullard (1936). During November 1933 and April 1934, Bullard established 87 gravity stations between Mombasa, across the Eastern Rift near Lake Magadi to the Western Rift near Lake Albert (and back to Dar es Salaam), using a pendulum apparatus. By examining the relationship between gravity and topography over the rifts, he was able to develop a mechanical model to explain them.

Bullard (1936) showed that rift valleys in East Africa (e.g., Fig. 7.1) were charac-terised by free-air gravity anomaly 'lows' of up to −150 mGal and flanking highs of up to +50 mGal. After applying a Bouguer correction, Bullard found that a 'low' of up to −100 mGal still existed over some individual rifts. Furthermore, the low could not be accounted for when an Airy or Pratt isostatic correction was applied, suggesting that the rift might be overcompensated. Therefore, either the Moho was deeper (Airy) or the mean density of the crust and mantle was less (Pratt) than expected beneath the rift valley.

Because it was not possible to account for the isostatic anomaly 'low' using conventional models, Bullard used it to estimate the shape of the extra compensation. He showed that a strip of mass 28 km in length and 12 km in height beneath the central part of the rift with a uniform density contrast of −600 kg m^{-3} with the surrounding mantle, and two strips of mass of 28×1.2 km and 28×2.5 km with their centres 50 km to the east and west of the rift, could explain the observed low quite well.

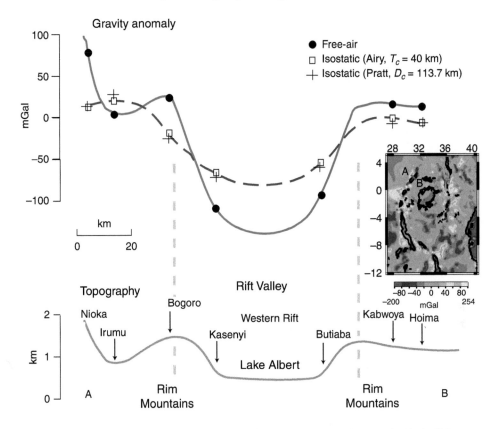

Figure 7.1 Topography, free-air anomaly (solid red line) and isostatic anomaly (dashed blue line) profile of Lake Albert in the East Africa Rift system. The isostatic anomaly has been computed assuming an Airy and Pratt model. Based on data in Table XXII of Bullard (1936). Inset shows a free-air gravity anomaly map of the East Africa region and the location of profile AB in the north-west part of the map.

Bullard (1936) argued that the relatively low-density strips of mass would 'cushion' the overlying crust and buoy it up unless some horizontal force was acting on the crust to prevent it. He suggested that this force was compressional in origin. A consequence of the compression would be to force the buoyant crust upwards and cause it to bend or break. Assuming it would break, then a crustal 'block' on one side of the fracture would 'ride up' on, and in essence load, the surface of the crustal block on the other side.

Aware of the earlier studies of Jeffreys (1926), Bullard used an elastic plate model (Fig. 7.2) to predict the depth and width of a rift valley that would result from the overthrusting of one crustal block on top of another. Specifically, he used a broken-plate model, where the break was located at the mid-point of one of the faults that bound the depression. He used an estimate of the magnitude of the force to calculate the flexure of the crust to the right of the break and the position where the plate curvature was greatest, and the bent plate was most likely to break again.

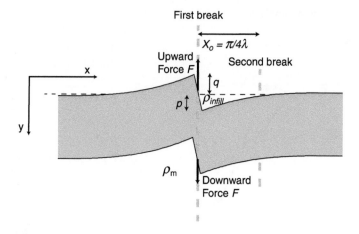

Figure 7.2 The broken-plate model that Bullard applied to continental rift valleys.

If the force/unit width that acts at the plate break is F and x is the distance from the plate break, then the flexure y is given by

$$y = \frac{2F\lambda}{\left(\rho_m - \rho_{infill}\right)} e^{-\lambda x} \cos \lambda x, \tag{7.1}$$

where

$$\lambda = \left[\frac{3\left(\rho_m - \rho_{infill}\right)g}{ET_e^3}\right]^{1/4}$$

and ρ_m and ρ_{infill} are the densities of the mantle and the material that infills the flexure, respectively. The second break, in Bullard's view, would occur where the bending stress is greatest; that is, at a distance x_o from the first break (see Section 3.5.2), given by

$$x_o = \frac{\pi}{4\lambda}. \tag{7.2}$$

Substituting Bullard's preferred values of $E = 7 \times 10^{10}$ N m^{-2}, $T_e = 40$ km, $\rho_m = 3{,}300$ kg m^{-3}, $g = 9.81$ m s^{-2} and $\rho_{infill} = 0$ (i.e., the material that infills the flexure is air) in Eq. (7.2) gives $x_o = 65$ km (Fig. 7.3). Low values of T_e would give smaller widths and higher values, wider rifts. The width of 65 km, as noted by Bullard, is remarkably similar to the observed width of the rift valleys of East Africa, a compilation of which is given in Table 7.1.[1]

---

[1] Table 7.1 is available on the Cambridge University Press server.

In the final stage of Bullard's model (i.e., after the second break), the centre wedge-shaped block is forced downwards at both ends of the crustal block by the force F. This force will be opposed by an equal and opposite force due to buoyancy in the subcrustal mantle. Assuming that the block is rectangular in shape, it follows that

$$2F = \rho_m x_0 pg, \tag{7.3}$$

where p is the vertical distance that the block has been forced down.

In addition to subsidence, there will be uplift of the flanks of the block. We can estimate the magnitude of this uplift using the same broken-plate model that was used to estimate the point of greatest curvature in the overthrust plate. If the uplift at the plate break (i.e., $x = 0$) is q, then it follows from Eqs. (7.1), (7.2) and (7.3) that

$$q = \frac{2F\lambda}{\rho_m g} = \frac{\pi}{4} p = 0.8p. \tag{7.4}$$

In other words, the total relief of the rift valley wall is 2.25 times the uplift. The East Africa rift walls rise some 500 m above the mean topography, suggesting a total relief that is of the order of 1,125 m. This is in reasonable agreement with the observed relief as inferred, for example, from the elevation of the rift walls in Lake Albert above and below the mean topography.

Finally, the width of the uplift is in reasonable agreement with observations. Using Bullard's preferred values for the elastic properties of the crust and noting that the width of the uplift is given by $\pi/4\lambda$ yields a width of 130 km.

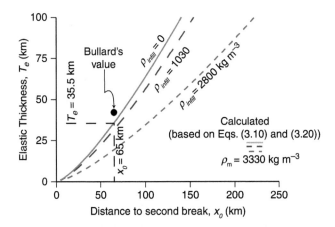

Figure 7.3 The relationship between elastic thickness T_e and x_o, the distance from the first break to the point of greatest curvature in a flexed plate, for different densities of the material that infills the flexural depression. It was Bullard's view (and later that of Vening Meinesz) that the position of the second break in the plate is determined by x_o.

Many geologists, however, questioned Bullard's suggestion that rift valleys were in a state of compression. They noted that the bounding faults of East African rift valleys generally sloped inwards rather than outwards. Bullard explained this by erosion, which removed material upslope from the fault and re-deposited it as talus downslope. However, the geologists preferred a model in which the inward dips were normal faults rather than thrust faults, and where the valley had formed in response to extensional, rather than compressional, forces. Bullard had, in fact, considered the possibility that rifts were in tension, as had been suggested earlier by Wegener (1966) but, understandably, he dismissed this on the basis that tension would thin the crust by necking and that this could not produce the mass deficiency beneath the rift required to explain the isostatic anomalies. To the contrary, it would produce a mass excess.

7.2.1.2 Vening Meinesz's Hypothesis

Vening Meinesz (1950) showed, using a similar mechanical model, that it was also possible to explain a rift valley in terms of tensional forces (Fig. 7.4). He argued that if a gradually increasing horizontal tensile force is applied to the crust, it must eventually lead to normal faulting along a tilted fault plane. If no friction exists across the fault surface, then faulting effectively leads to a removal of mass from one side of the fault and its addition to the other. We know from isostasy that the block that has had material removed from it will rise, while the block with additional material will sink.

Consider, as did Vening Meinesz, the mechanics of the blocks on either side of a normal fault. First, we need to consider the general case of removing material from the top of the crust. Since isostasy is disturbed, the removal of material leads to uplift and so the crust that remains rises. The magnitude of the rise can be estimated by balancing columns of the crust before and after material is removed. If h_o is the height above sea level of the column of crust that is removed by erosion and h is the height above sea level of the surface of the crust that remains, then it follows from Fig. 7.5 that

$$(h_0 + r + T_c)\rho_c g = (T_c + r)\rho_c g + \rho_m hg,$$

where ρ_c is the density of the crust, T_c is the zero-elevation thickness of the crust and ρ_m is the density of the mantle. The new height of the surface and, hence, the rock uplift that remains, h, is therefore given by

$$h = h_o \frac{\rho_c}{\rho_m}. \tag{7.5}$$

Using values of $\rho_m = 3,330$ kgm^{-3} and $\rho_c = 2,800$ kgm^{-3}, we see that an elevated region could have its height reduced to ~84 per cent of h_o because of erosion.

In the case of tensile forces that produce a normal fault within the crust, differences exist between the mechanical state of the two crustal blocks on either side of the fault. To the left, the crust has been unloaded, while to the right it has been loaded.

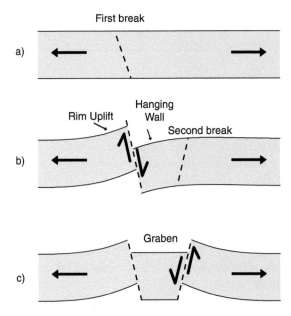

Figure 7.4 Vening Meinesz's model for the development of a graben. Reproduced from fig. 10D-1 of Heiskanen and Vening Meinesz (1958) with permission of McGraw-Hill Inc.

Consider first the crustal block to the left of A in Fig. 7.6. Here, h_o = constant and so there will be a certain surface within the block which is the height that the material that makes up the block would rise to if the block was removed, say by erosion. We call this the surface of unloading. If the height of this surface above the base of the block is h, then it follows that

$$h_o = h + h_u$$

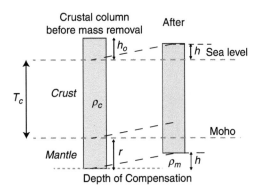

Figure 7.5 Schematic diagram showing the effects of removing a mass of height h from a column of the crust. (Left) A crustal column prior to material removal, (right) after uplift and removal.

$$\therefore \ h_u = \frac{h_o(\rho_m - \rho_c)}{\rho_m}, \tag{7.6}$$

where h_u is the distance below the surface to the surface of unloading.

We see from Eq. (7.6) that when $h_o = 0$, $h_u = 0$. This special case occurs at the point E. At D (a point vertically above E), h_u, will also be zero. The surface of unloading therefore extends from B (its height in the constant-thickness part of the crustal block to the left of the fault) to the point D, intersecting the fault plane at the point C.

Faulting disturbs the isostatic balance. The crustal block to the left of the fault, for example, has had material removed from it and so it will rise. We can estimate the magnitude of the load that has been removed by considering the load that would be needed to restore isostatic equilibrium. The latter is given by the mass of the material above the surface of unloading, which, in turn, is given by the product of the triangular area ACD (Fig. 7.6), the density of the crust, ρ_C, and average gravity.

The area ACD can be evaluated most easily by considering the two overlapping triangles, ABD and ABC. We have

$$\text{Area}_{\text{ABD}} = \frac{h_o \tan \theta h_u}{2},$$

$$\text{Area}_{\text{ABC}} \approx \frac{h_u^2 \tan \theta}{2},$$

$$\text{Area}_{\text{ACD}} = \text{Area}_{\text{ABD}} - \text{Area}_{\text{ABC}},$$

$$\therefore \ \text{Area}_{\text{ACD}} = \frac{h_o \tan \theta h_u}{2} - \frac{\tan \theta h_u^2}{2}. \tag{7.7}$$

Substituting Eq. (7.6) into Eq. (7.7) then gives

$$\text{Area}_{\text{ACD}} = \frac{h_o^2 \tan \theta (\rho_m - \rho_c)}{2\rho_m} - \frac{\tan \theta h_o^2 (\rho_m - \rho_c)^2}{2\rho_m^2},$$

$$\text{Area}_{\text{ACD}} = \frac{h_o^2 \tan \theta (\rho_m - \rho_c)}{2\rho_m} \left[1 - \frac{(\rho_m - \rho_c)}{\rho_m} \right].$$

The equivalent force/unit width, P, is given by

$$P = \text{Area}_{\text{ACD}} \rho_c g,$$

$$P = \frac{h_o^2 \ \tan \theta (\rho_m - \rho_c)}{2\rho_m} \left[1 - \frac{(\rho_m - \rho_c)}{\rho_m} \right] \rho_c g. \tag{7.8}$$

Substituting $\theta = 27°$, $g = 9.81 \ \text{ms}^{-2}$, $\rho_c = 2{,}840 \ \text{kg m}^{-3}$, $\rho_m = 3{,}270 \ \text{kg m}^{-3}$ and $h_o = 35$ km in Eq. (7.8), Vening Meinesz (Heiskanen and Vening Meinesz, 1958) obtained $P = 9.93 \times 10^{11} \ \text{N m}$. Vening Meinesz then applied this load to the end of a broken plate

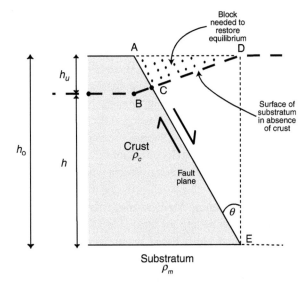

Figure 7.6 Schematic diagram showing the effective load that needs to be added to the crust to restore isostatic equilibrium following normal faulting. The thick black dashed line shows the surface of unloading. Modified from fig. 10D-2 of Heiskanen and Vening Meinesz (1958) with permission of McGraw-Hill Inc.

with $E = 10^{11}$ N m^{-2}, $v = 0.25$, $T_e = 35$ km (i.e. equal to the thickness of the crust) and $\rho_{infill} = 0$, and obtained an upward flexure of $y = 745$ m at $x = 0$.

The discussion thus far has been concerned with the crustal block to the left of the fault. We see from Fig. 7.6 that while the block to the left of the fault has been unloaded, the block to the right has been loaded. Assuming that the magnitude of this extra load is equivalent to the load removed, then there will be a subsidence of the block to the right of the fault that is equal in magnitude to the uplift due to unloading. The total relief at the edge of the fault could therefore be as large as $2 \times 745 = 1,490$ m, which is similar to the estimate of Bullard.

As in Bullard's model, Vening Meinesz then used the distance to the point of greatest curvature in the subsided crustal block to estimate the distance to the second break in the plate.

The final step in Vening Meinesz's model is the formation of a wedge-shaped crustal block. The block sinks because a narrowing downward wedge of crust would float in a liquid at a lower level than a rectangular block. Eventually, however, the wedge-shaped block will not be able to sink any further because of the inward slope of the bounding faults. However, if extensional forces continue to be applied, then space is created for the wedge-shaped block to subside. Either way, the displacement of the relatively low-density wedge-shaped block into the denser mantle would provide the underlying mass deficiency that

Bullard had needed earlier to explain the isostatic gravity anomaly lows that he had observed over rifts.

The dispute between the Bullard and Vening Meinesz models is now largely settled. In addition to the evidence for the style of faulting, there is the argument of Girdler (1963), who showed that for the same rift basin the two hypotheses would yield distinctly different-shaped gravity anomalies. In particular, Vening Meinesz's model would give a relatively narrow negative anomaly with an abrupt shallowing at the edges, compared with Bullard's model which would yield a relatively broad anomaly with a gentle shallowing at the edges. Girdler (1963) demonstrated that the gravity anomalies over Lake Albert, as well as other rift valleys such as the Rhine, were in better agreement with Vening Meinesz's model than with Bullard's model.

An important feature of the Vening Meinesz model is the rift flank uplift. The uplift arises from the mechanical unloading of one side of a crustal block by faulting. Figure 7.7 shows the calculated uplift based on the actual two-dimensional load that is removed, rather than a line load representation as Vening Meinesz had earlier assumed. The figure shows that the width of the rift flank uplift is determined by T_e, varying from about 80 km for $T_e = 15$ km to 200 km for $T_e = 50$ km. The amplitude of the uplift also depends on T_e as well as the angle between the fault plane and the vertical (i.e., the hade), θ_h, and the thickness of the crust, h_o.

As Table 7.1[2] and Fig. 7.7 show, many rift valleys are associated with rim flank uplifts. Typically, heights of the rim flank uplift range from 0.7 to 2.4 km and the widths from 75 to 160 km. These values suggest $15 < T_e < 30$ km, depending on whether the uplift or width is emphasised.

It is perhaps surprising that rift flank uplifts maintain their distinct flexural shape, even in the presence of erosion. Nevertheless, they do, as is clearly seen in East Africa.

The Vening Meinesz model has been widely applied to models of rift basin evolution. Bott (1976), for example, applied it to active rift systems in East Africa as well as to ancient rift basins. These included the 'syn-rift' basins that form in response to extension during the early stages of continental break-up and seafloor spreading. However, as more deep seismic data has accumulated over these and other types of rift basin, it has become increasingly clear that the Vening Meinesz model cannot explain all the details of the gravity anomaly, topography, sediment distribution and crustal structure of continental rifts.

To address this problem, the Vening Meinesz model has been modified, most notably by Weissel and Karner (1989) and Kusznir and Egan (1990). One modification has been to consider how extension might be distributed throughout the entire lithosphere, in contrast to Vening Meinesz who considered only the crust.

Other modifications concern the mechanical model, the geometry of the border faults and the inclusion of sediment loading and erosional unloading. Weissel and Karner (1989), for example, consider the footwall (i.e., the rift flank uplift on the unloaded side of the border fault) and the hanging wall (i.e., the rift valley subsidence on the loaded side of the border fault) as integral parts of a continuous elastic plate, rather than as two independent broken

---

[2] Table 7.1 is available from the Cambridge University Press server.

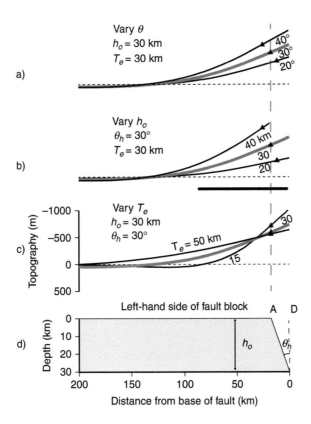

Figure 7.7 The footwall uplift that results from normal faulting and its dependency on (a) the angle between the fault plane and the vertical, θ_h, (b) the thickness of the elastic layer, h_o, and (c) the elastic thickness, T_e. The solid red line shows the case $\theta_h = 30°$, $h_o = 30$ km and $T_e = 30$ km. Solid black lines show the ranges of heights and widths of observed rim flank uplifts.

plates (Fig. 7.8a), as Vening Meinesz did. Conceptually, such a model implies that after a border fault slips, and before isostatic restoring forces act, the two crustal blocks on either side of the fault are in effect locked and a coupling exists across them. In addition, Weissel and Karner (1989) modified the style of faulting by including both planar and curved faults, the thermal and mechanical effects of crustal thinning, the loading effects of the sedimentary material that might infill the rift basin as the hanging wall subsides and erosion of the rim flank uplifts.

Ebinger et al. (1991) showed that the Weissel and Karner (1989) model accounts for many of the details of the observed topography and gravity anomalies over the Western Rift in East Africa (Fig. 7.8b). These include variations in the shape of the hanging wall and the footwall uplift. They found the best fits between calculated and observed gravity anomalies were for

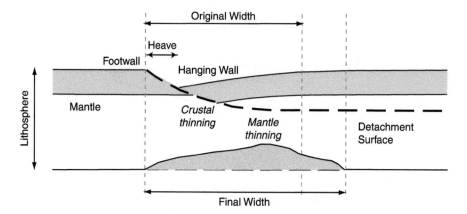

Figure 7.8a Weissel and Karner's kinematic model of instantaneous extension of the lithosphere involving simple slip with heave along a detachment surface, combined with pure shear at depth in the region of the hanging wall and footwall. Reproduced from fig. 6 of Weissel and Karner (1989), copyright by the American Geophysical Union.

models in which the border faults were curved, rather than planar as Vening Meinesz had earlier assumed. The faults themselves were associated with relatively small amounts of extension in the range of 3–10 km. Moreover, they demonstrated that the density of the material that infills

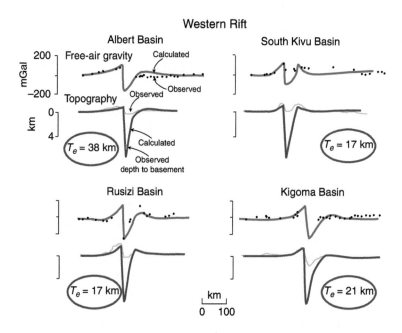

Figure 7.8b Comparison of the predictions of the Weissel and Karner (1989) model with free-air gravity anomaly, depth to basement and topography profile data over the Western Rifts of East Africa. Numbers in circles are the best-fit estimates of T_e. Reproduced from fig. 11a–f of Ebinger et al. (1991), copyright by the American Geophysical Union.

the rift valleys is in the range of 0–2,500 kg m^{-3}, and that T_e varied in the range of 17–38 km along the length of the rift valley.

One type of a curved fault is the listric fault. These faults sole out at depth on a surface known as a detachment surface. According to a model proposed by Wernicke (1985), detachment surfaces act as relays that partition extension in the upper and lower lithosphere. Ebinger et al. (1991), in their models, assumed a detachment surface at the base of the crust. They showed, however, that neither variations in the depth of the detachment surface nor variations in the distribution in extension below it significantly affected their conclusions concerning T_e.

Kusznir and Egan (1990) presented a model that combines a Wernicke-type simple shear in the upper crust with a McKenzie-type pure shear in the lower crust and sub-crustal mantle (Fig. 7.9a). Dubbed the flexural cantilever model, it incorporates the thermal effects of both crustal and sub-crustal thinning and heating – as well as the possibility of erosion of the rift flank uplifts and sedimentation in the rift valleys.

Magnavita et al. (1994) showed that the Kusznir and Egan (1990) model can account for several features of the rifts that formed in eastern Brazil during drifting apart of South America and Africa. These include the main geometrical features of the Recôncavo–Tucano–Jaotobá rift system (Fig. 7.9b). Simple shear in the brittle part of the crust produces one or more basins with a hanging wall and flank uplift. Initially, the basins remain unfilled. However, as time passes, the rim uplift is eroded, and this leads to more uplift and material infilling the basin. Finally, there is a thermal subsidence as the ductile part of the crust and sub-crustal mantle cools. The net result is a fault-controlled basin with a thickness of syn-rift sediments that exceeds that of the post-rift.

Both the modifications of the Vening Meinesz wedge model discussed earlier predict a gravity anomaly low over the central rift. The Weissel and Karner (1989) model is dominated by hanging wall subsidence and footwall uplift. The Moho is therefore depressed more beneath the rift than the flanking regions, and as a result the rift is associated with a gravity low. The Kusznir and Egan (1990) model, however, involves lower crustal thinning that takes the form of an elevated Moho. It is more difficult, therefore, for this model to predict gravity lows, unless the lower crustal thinning is laterally displaced from the upper crustal thinning, as has been proposed by Ussami et al. (1986) and Bell et al. (1988) for the Tucano and Newark rift basins, respectively. Hence, both models can account for the gravity lows over rifts.

The Vening Meinesz wedge model and its modifications has played an important role in the development of mechanical models for continental rifts. Despite this success, however, there are several problems that have been cited in various papers. These are as follows:

- The rift is not always associated with a gravity low. Sometimes there is an axial high that seems to widen along the rift (Baker and Wohlenberg, 1971). This has been interpreted as the result of magmatic intrusions along the rift axis.
- There is often no single border fault: throws are often distributed along several parallel faults (Bosworth, 1985).

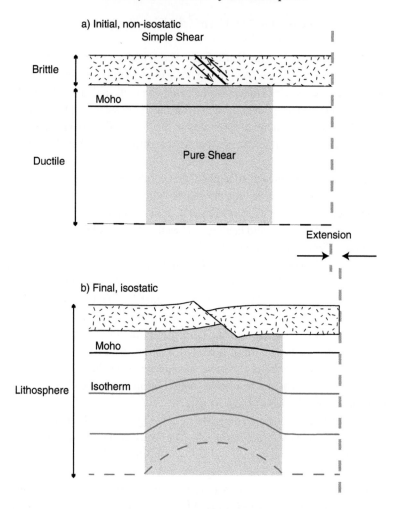

Figure 7.9a Kusznir and Egan's flexural cantilever model of extension, which occurs by a combination of simple-shear faulting in the upper crust and pure shear plastic deformation in the lower crust and mantle. Reproduced from fig. 2 of Kusznir and Egan (1990).

- There is little evidence from seismic refraction data for a simple downdropped wedge in the region of a rift valley. The crustal structure of the Kenya Rift, for example, shows significant lateral variations in the structure along and across the rift (KRISP Working Party, 1991). And the crust, on some profiles, thins beneath the rift rather than thickens.
- The uplifts that flank rifts are not always interpreted in terms of flexure. Steckler (1985) has proposed, for example, that in the Gulf of Suez flank uplifts may be a consequence of a small-scale convection in the underlying mantle that augments the thermal and mechanical effects of rifting.
- The second break in the Vening Meinesz model may not always occur in the hanging wall. Rather, it may occur in the upper crustal layer beneath the footwall uplift (Bott, 1997).

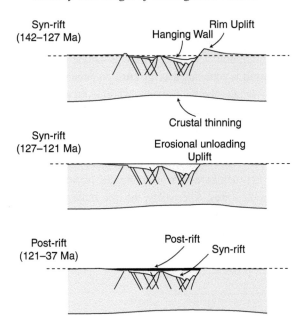

Figure 7.9b Stratigraphic development of the Recôncavo Basin in north-east Brazil, according to the predictions of the Kusznir and Egan (1990) model. Reproduced from fig. 9 of Magnavita et al. (1994), copyright by the American Geophysical Union.

Another important observation is that continental rifts do not always form discrete narrow zones. Some rifts, like those in the Basin and Range province, western United States and the Aegean Sea, are characterised by extension that is distributed over regions as wide as 500–800 km.

Buck (1991) reviewed the observations for narrow and wide rifts. Simple flexure considerations (e.g., Fig. 7.3) suggest that narrow rifts are regions of weak continental lithosphere, while wide rifts are strong regions. Whether the lithosphere is weak or strong depends, however, on strain rate, initial crustal thickness, and thermal setting (Fig. 6.29). Buck pointed out that there are no clear differences in strain rate between narrow and wide rifts. There are, however, differences in the initial crustal thickness (wide rifts > 50 km, narrow < 50 km) and heat flow (wide rifts > 90 mW m^{-2}, narrow < 60–70 mW m^{-2}). Buck showed that any crustal rheology can explain narrow rifts in those situations where the lithospheric strength is dominated by the mantle. If the lower crustal rheology is weak, however (i.e., there is a 'notch' in the YSE), then we cannot get wide rifts. Wide rifts are therefore compatible with a coupled and strong upper and lower crust.

The need to develop a self-consistent model has led to a new generation of models for the development of wide continental rifts. Like the models for discrete narrow rifts, flexure plays an important role.

Consider, for example, the flexural rotation model of Wernicke and Axen (1988), Buck (1988) and Lavier et al. (1999). This model, like the Vening Meinesz model, considers the offset of a high-angle normal fault. The main difference is that (e.g., Fig. 7.10) the deformation of the fault surface itself is considered as the crust on either side of the fault responds to the new loads. At shallow depths the fault rotates, while at greater depths the fault remains straight. Once the active fault has rotated by some amount, a new fault is cut that dips at the original angle of the fault, intersecting it at some depth of nucleation. As extension continues, one or more low-angle normal faults form and the new faults are abandoned and flattened. Buck found that a low T_e was required to explain the geometry of abandoned faults that have been observed in diffuse rifts such as the Basin and Range. He speculated that the low T_e was caused by high crustal temperatures. Alternately, the low values could be explained by high bending stresses that lead to yielding.

The rifts that develop in response to the break-up of continents and the formation of new ocean basins also appear to be of narrow or wide type. For example, crustal restorations

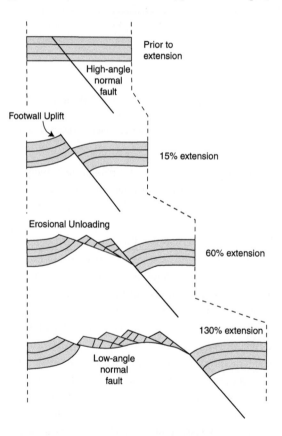

Figure 7.10 Buck's 'rolling-hinge' flexural rotation model for the origin of narrow and wide rift valleys, detachment surfaces and low-angle normal faults. Reproduced from figs. 7 and 8 of Buck (1988) and fig. 4 of Wernicke and Axen (1988).

based on seismic and gravity modelling show that the Vøring and Goban Spur margins are narrow rifts, while the East Coast, United States, and the Brazil/Gabon margins are of wide type. Bassi et al. (1993) suggested that whether the rift that develops during the early stages of break-up is narrow or wide depends not only on the initial conditions of the lithosphere, but also on buoyancy forces, mechanical instability and temperature changes. They found, for example, using a finite element approach rather than an analytical approach, that an initially cold, dry lithosphere can lead to a localisation of strain and narrow rifts, while an initially warm, dry lithosphere favours cooling during rifting and the migration of the maximum region of thinning from the ocean towards the continent.

Since the work of Bassi et al. (1993) there has been a rapid increase in the application of finite element models to continental break-up and rifting, many of which have emphasised the role of crustal rheology. Behn et al. (2002), for example, used the models to show that when crustal thickness is large, both mantle rheology and the vertical temperature gradient play a significant role in the deformation of the upper crust, but when the thickness is small it is the vertical temperature gradient that plays the major role. Their results, which were irrespective of lower crustal rheology, predict a relationship between the width of a rift and vertical temperature gradient, a result compatible with the T_e calculations shown in Fig. (7.11). Pérez-Gussinyé and Reston (2001) constructed Yield Strength Envelope curves during the evolution of rifts and used finite element models to show that the entire crust may become brittle when the amount of extension is sufficiently large. The faulting leads to hydration and serpentinisation, which further weakens the lithosphere and helps localise rifting. Finally, Brune et al. (2017) and Pérez-Gussinyé et al. (2020) have emphasised how crustal rheology controls the style and architecture of rifted margins, especially when it is accompanied by a seaward migration of the rift.

7.2.2 Oceanic Rifts

One of the major results of lead line bathymetric surveys at the turn of the twentieth century was the discovery of a submarine ridge in the Atlantic Ocean, segments of which appeared to continue into the Arctic and Indian Oceans. Because of its location approximately midway between Europe and North America and Africa and South America, the feature was dubbed the mid-oceanic ridge system. Rising some 3–4 km above the deepest parts of the ocean floor, the mid-ocean ridge forms a region of relatively shallow seafloor a few thousand kilometres wide and several thousands of kilometres long.

In the late 1950s, Bruce Charles Heezen, Marie Tharp and colleagues at what was then the Lamont–Doherty Geological Observatory showed that mid-ocean ridge crests are often associated with a continuous rift valley and flanking highs[3] (Heezen et al., 1959). These oceanic rift systems share many of the features of continental rifts, with their deep central grabens, steep inward-facing border faults and rim flank uplifts. Furthermore, there are

---

[3] This may not have been the first observation. A rift valley and a rim flank uplift appear to have been discovered on cruises of the German vessel R/V *Meteor* to the mid-ocean ridge which were carried out shortly after the development of the first echo sounder in the late 1920s (Stocks and Wüst, 1935).

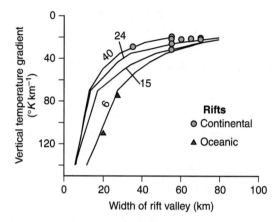

Figure 7.11 Comparisons of observations of relationship between rift valley width and the vertical temperature gradient to calculations based on finite element model. Rift valley width increases with decrease in temperature gradient, which is compatible with the flexure model in Fig. 7.3 showing an increase in rift valley width with increase in T_e, a proxy for the temperature structure, with low T_e indicative of a weak, hot lithosphere and high T_e indicative of a strong, cold lithosphere. Numbers indicate the cases for different crustal thickness in kilometres. Modified from table 3 and fig. 10 in Behn et al. (2002).

localities (e.g., the Afar triple junction) where individual 'arms' of a mid-oceanic ridge system extend onshore into a continental rift. This suggested to Heezen et al. (1959) that oceanic rifts, like continental rifts, were also sites of extension. To accommodate extension on such a large scale, these workers suggested that the Earth must be expanding.

While Heezen's views on an expanding Earth were not widely accepted, his observations helped focus future marine geophysical surveys on the mid-ocean ridge system. These surveys included the mapping of marine magnetic anomalies, which yielded the first evidence that mid-oceanic ridges were the sites of seafloor spreading and the formation of new oceanic crust. They also included the mapping of the marine gravity field, which showed that the mass excess of the mid-oceanic ridge was compensated at depth by a broad region of low-density material. The combined results of these potential field studies suggested that mid-ocean ridges were the sites of hot, upwelling mantle material that had subsequently cooled to form new oceanic crust.

Selected bathymetry and free-air gravity anomaly profiles of the mid-ocean ridge system are shown in Fig. 7.12. The profiles show an axial rift valley that is 1–2 km deep, 15–30 km wide and has flanking rim uplifts. The rim uplifts are characterised by a rough and faulted topography. There is a close correlation between free-air gravity anomaly and bathymetry such that the rift valley and rim flank uplifts are associated with a large-amplitude gravity 'low' and flanking 'high', respectively.

As we demonstrated for one of the East Africa rifts, the free-air gravity anomalies over oceanic rifts are not removed when a Bouguer and isostatic correction are applied. The rift valley on the crest of the mid-Atlantic ridge, for example, correlates with an isostatic anomaly 'low' of up to –25 mGal (Fig. 7.12), suggesting that it is overcompensated. In other words, a greater mass deficiency is required beneath the rift than is predicted by an Airy model, which

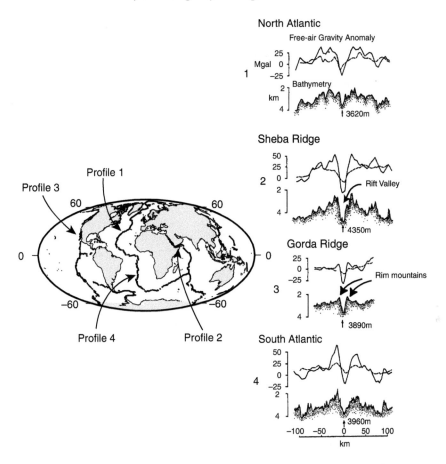

Figure 7.12 Free-air (solid lines) and Airy isostatic (dashed lines) gravity anomalies and bathymetry profiles of selected active oceanic rifts in the North and South Atlantic, Indian (Sheba Ridge) and Pacific (Gorda Ridge) oceans. Reproduced from fig. 1 of Watts (1982a).

predicts thin crust. This indicates that (a) the crust is thicker beneath oceanic rifts than the Airy model would predict, (b) the crust is as thick as the Airy model would predict and there is a mass deficiency in the underlying mantle or (c) some combination of these possibilities.

Because of the similarity of their topography to continental rifts, early ideas on the origin of oceanic rifts focused on the role played by extensional forces. Tapponier and Francheteau (1978), for example, suggested that oceanic rifts were the consequence of a steady-state necking of an elastic lithosphere that produces an uncompensated water-filled depression. The mass deficiency of the water-filled depression is associated with an upward-acting force and equilibrium is restored by flexural uplift of the surrounding crust and mantle (Fig. 7.13). The gravity anomaly associated with the model is the superposition of two effects: one a low due to the uncompensated water-filled depression, and the other a high due to the uplift of the top and bottom of the crust. As a comparison of Figs 7.12 and 7.13 shows, the calculated

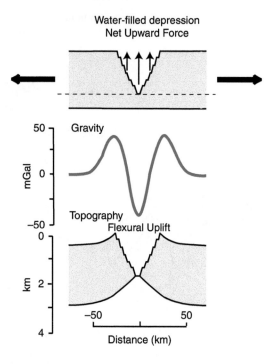

Figure 7.13 Schematic diagram illustrating Tapponier and Francheteau's model for the origin of oceanic rift valleys and their flanking rim uplifts by necking and flexural unloading of a relatively strong crust.

gravity anomalies closely resemble the amplitude and wavelength of observed gravity anomalies.

Spectral studies confirm the hypothesis of Tapponier and Francheteau that mid-ocean ridge crest bathymetry is flexurally supported. Cochran (1979), for example, showed that the admittance (see Section 5.6) based on stacking several 400-km-long free-air gravity and bathymetry profiles of the mid-Atlantic ridge could not be explained by an Airy model, since such a model implied a mean crustal thickness at the ridge crest in the range of 20–40 km. This is significantly larger than the 6–8 km estimated from seismic refraction studies. He therefore preferred a flexure model. The best-fitting values of T_e, assuming a mean crustal thickness of 6 km, were in the range of 7–13 km. These values of T_e were significantly lower than estimates at intra-plate volcanoes (e.g., Hawaii), and so Cochran attributed them to the formation of ridge crest bathymetry in the high-temperature axial region of mid-ocean ridges.

Although a flexure model accounted well for gravity anomalies over the ridge flank, it failed to account fully for the free-air gravity anomaly 'low' over the rift valley. Subsequent two-dimensional studies (Bowin and Milligan, 1985) that examined data along carefully selected flow lines of the Mid-Atlantic Ridge and three-dimensional studies (McNutt, 1979) that used a 1,500 × 800 km grid of data centred on the Juan de Fuca Ridge were able, however, to explain the bathymetry of the rift valley as well as the flank topography in terms of a flexure model.

The notion that, despite the high temperatures, oceanic rifts have a certain flexural strength is supported by gravity anomaly and bathymetry data at extinct ridges. Remarkably, free-air gravity anomalies in the Labrador Sea (Kristoffersen and Talwani, 1977) and Coral Sea (Weissel and Watts, 1979), which ceased spreading 42 and 60 Ma, respectively, resemble those of active rifts (Fig. 7.14). This is true even though the seafloor in the Labrador and Coral seas continued to subside following the cessation of rifting and so is typically deeper than the actively spreading mid-Atlantic ridge. Watts (1982a) argued that

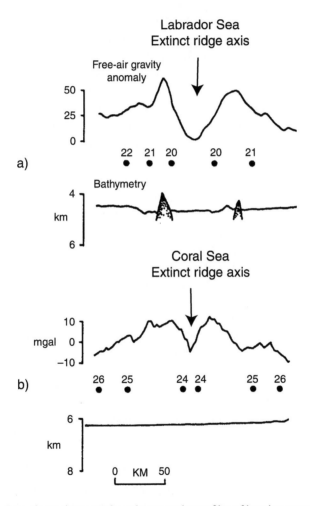

Figure 7.14 Free-air gravity anomaly and topography profiles of inactive oceanic rifts in the (a) Labrador and (b) Coral seas. Numbers correspond to marine magnetic anomalies associated with Paleocene and Eocene oceanic crust. Reproduced from fig. 4 of Watts (1982a).

the similarity in gravity anomalies between inactive and active rifts is a strong argument that stresses that are induced in the lithosphere by flexure are maintained elastically for long periods of time.

Further support for a finite strength at the ridge crest has come from microseismicity studies (Huang and Solomon, 1988). These data indicate that the thickness of the seismogenic layer at the mid-Atlantic ridge crest is 10–12 km, which is in close agreement with the T_e results. Although there is no reason to suspect an agreement because of the different timescales that are involved in seismic and flexure studies, it is interesting to note that a similar correlation between the seismic and elastic thickness as occurs at the ridge crest is observed in deep-sea trench–outer rise regions and in the oceanic plate interiors (see Section 6.9).

One problem with the flexure model, however, is that it predicts the crust thins beneath the rift valley and thickens beneath the flanking uplifts. Although still sparse, seismic refraction data suggest that the thickness of the oceanic crust is remarkably uniform in the vicinity of both slow- (Purdy and Detrick, 1986) and fast- (Vera et al., 1990) spreading ridges. Parmentier and Forsyth (1987) therefore preferred a model in which the rift valley and its flanking uplifts are uncompensated.

A model of an uncompensated rift was supported by Dalloubeix et al. (1988). These workers analysed several 356-km-long profiles of the mid-Atlantic ridge crest and showed that the admittance could be explained by T_e in the range of 10–20 km, similar to the results of Cochran (1979). They then stacked the profiles, together with their mirror images. By mirroring, Dalloubeix et al. (1988) introduced additional smoothing that enhanced common features such as the rift valleys and flanking highs at the expense of local irregularities such as the ridge flank topography. The mirrored profiles yielded admittances that approached the predictions of an uncompensated, rather than a flexure model. They found two values at the longest wavelengths, however, that could not be explained. When the topography and free-air gravity anomaly along each profile were corrected for the thermal effects of the cooling ridge, the admittance followed closely the predictions of an uncompensated model. However, the admittance was unusually low at $0.05 < k < 0.15$. Their preferred model, therefore, was a combined one in which there is a high admittance due to the uncompensated rift, superimposed on which is a low admittance at long wavelength because of cooling and a low admittance at short wavelength due to some combination of surface or sub-surface loading of a very weak plate.

Based on these and other results, Kuo and Forsyth (1988) developed an isostatic model that they used as a first step in interpreting the pattern of free-air gravity anomalies observed at a mid-ocean ridge. The model they chose considered the results from both gravity modelling, which suggested some degree of uncompensation, and seismic refraction data, which suggested oceanic crust was of uniform thickness and density. They showed that when the free-air anomaly data were corrected for the gravity effects of the bathymetry and a Moho that was calculated from the topography assuming a constant-thickness crust, significant anomalies of up to –80 mGal remained. Because they thought that the gravity effect of the top and bottom of the crust had been correctly accounted for, they dubbed them Mantle Bouguer Anomalies (MBA).

A surprising result of early studies was that the MBA did not vary linearly along-strike of the ridge crest. Rather, the anomalies formed a pattern of circular lows, or 'bulls'-eyes' (Kuo and Forsyth, 1988), which gave the ridge crest a distinct along-strike segmentation. The close correlation between the MBA and regions of unusually shallow seafloor suggested to Kuo and Forsyth (1988) and others that it represents a region of low-density, high-temperature upwelling in the underlying mantle. The existence of zones of elevated temperature beneath the crest of mid-ocean ridges has now been confirmed from seismic studies. *P*-wave inversion studies (Purdy et al., 1992) show, for example, that the depth of the low-velocity and hence elevated-temperature region is as shallow as 3–4 km beneath the crest of the mid-Atlantic Juan de Fuca and Lau basin ridges.

These considerations suggest that oceanic rifts are not simply the isostatic response of the crust to extensional forces. Phipps Morgan et al. (1987), for example, generally agreed with the statements of Parmentier and Forsyth (1987) on the degree of isostatic compensation, Kuo and Forsyth (1988) on dynamics, and Watts (1982a) on the permanence of ridge crest gravity anomalies, concluding that oceanic rifts were dynamically maintained, stress-supported features. They considered three models (Fig. 7.15): (a) mantle flow stresses due to plate spreading, (b) topography supported by moments due to horizontal stresses in a thickening lithosphere and (c) asthenospheric upwelling in a narrow steady-state conduit. They dismissed (a) because it required a mantle viscosity that was two or three orders of magnitude higher than generally accepted values based on post-glacial studies, and (c) because it predicts that axial topography should disappear after spreading stops, contrary to the observations at extinct rifts.

The preferred model of Phipps Morgan and co-workers was model (b), in which the topography is supported by the finite strength of the lithosphere. However, rather than use a surface loading model, which earlier workers had found hard to accept, they developed a model that considered the loads that are applied to the crust because of the increase in thickness of the plate with age. These loads take the form of a bending moment that is balanced by topographic forces resulting from flexural deformation. Using a continuum idealisation of deformation due to faulting, rather than a simple Vening Meinesz–type wedge model, Phipps Morgan and co-workers showed that the axial valley width depends on the plate thickness such that a 10-km-thick plate produces a 40-km-wide rift valley.

Not all mid-oceanic ridge crests are associated with a rift valley and flanking uplifts. Some (e.g., East Pacific Rise) are characterised by an axial horst rather than a rift valley. The horst is typically 2–10 km wide, 200–400 m high and is flanked by smooth sea-floor topography. The axial horst has led to models of mid-ocean ridges that have focused on the role of the underlying mantle. One class (Sleep and Rosendahl, 1979) considers the viscous interaction between the ascending hot mantle material and the adjoining lithosphere. According to these models, rift valleys result from a loss of hydraulic head, whereas axial highs indicate regions where magmatic material can freely flow to the surface. The other class (Anderson and Noltimier, 1973) considers the imbalance between the width over which hot mantle material is

Upward force due to water-filled depression

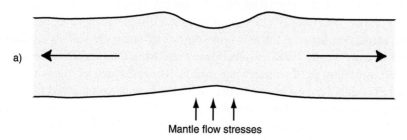

a)

Mantle flow stresses

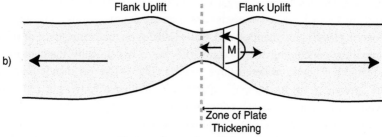

Central Depression

Flank Uplift   Flank Uplift

b)

Zone of Plate
Thickening

Moment due to plate thickening with age

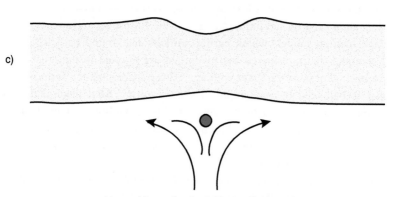

c)

Upward force due to melting and separation

Figure 7.15 The origin of ridge crest topography by (a) mantle flow stresses due to plate spreading, (b) bending moments due to a thickening of the lithosphere with age and (c) mantle upwelling in a narrow, steady-state conduit. Reproduced from figs. 3, 5 and 7 of Phipps Morgan et al. (1987), copyright by the American Geophysical Union.

intruded and the distance that is required for newly erupted material to accelerate from zero horizontal velocity to the spreading velocity.

The axial horsts on fast-spreading ridges correlate with free-air gravity anomaly highs (about 20–30 km wide and 15–25 mGal in amplitude), which are flanked by low-amplitude negative anomalies. Cochran (1979), in his spectral study, showed that free-air gravity anomaly and topography data over the flank of the East Pacific Rise could be best explained by a flexure model with a T_e that was in the range of 2–6 km. Significant discrepancies between observed and calculated anomalies, however, were found in the region of the axial horst. Observed anomalies were generally 10 mGal higher in amplitude than those calculated based on an elastic model, suggesting that the axial horsts, unlike the rift valleys, were undercompensated. In other words, Moho is shallower than predicted by the flexure model, which treats the horst as a constructional feature and would predict thicker-than-normal crust.

Magde et al. (1995) has shown that the East Pacific Rise crest is associated with a MBA 'low' of about 10 mGal. This suggests that the axial horst is overcompensated with Moho being deeper than would be predicted by a constant-thickness oceanic crustal model. Alternatively, the low is caused by lateral density changes in the crust and mantle. They argue that about 70 per cent of the MBA at the East Pacific Rise is caused by a region of partial melt and elevated temperatures in the mid-to-lower crust beneath the rise axis, the top of which occurs at depths of about 1.5 km below the sea floor. This is confirmed by seismic data that show a *P*-wave velocity inversion (Purdy et al., 1992) and a large-amplitude, flat-lying, axial magma chamber reflector at shallow depths beneath the ridge crest. The remaining 30 per cent of the MBA they attribute to lateral changes in the density of the crust and mantle due to thermal cooling of the lithosphere with age.

The elevated temperature zone beneath the crest of the East Pacific Rise axis may act as a low-density buoyant zone that supports the axial high. Madsen et al. (1984), for example, interpreted the axial high as primarily an isostatic feature that resulted from upward flexure of a thin elastic plate by buoyancy forces at the ridge crest. They ruled out surface loading because that would produce significant crustal thickening and, besides, such a model could not explain the steady-state nature of the high.

Magde et al. (1995), however, suggested that the buoyancy associated with lower crustal densities and thermal expansion of a hot upwelling mantle may not be sufficient to fully explain the axial high. They therefore propose that the high is supported, in addition, by a narrow partial melt conduit that extends to depths of 50–70 km beneath the ridge axis. This melt signature, they argue, would have a negligible gravity contribution, but could explain up to half the height of the axial high. An independent constraint is the diverse and depleted chemistry of the lavas from young axial seamounts. Wilson (1992) suggests that these lavas reflect originally enriched magmas generated at depths that bypass the main zone of melting and interact with strongly depleted material at intermediate depths.

Another possibility is that the axial high is, at least in part, a constructional feature. This is suggested by the increase in abundance of axial seamounts in regions of axial highs, together with the evidence at fast-spreading ridges for large supplies of melt at shallow depths beneath these regions, which is presumably more able to reach the surface of the crust. Figure 7.16 shows, for example, the residual bathymetry (i.e., the observed bathymetry minus the calculated one based on a cooling plate model) of an axial high on the slow-spreading, hot-spot-influenced crest of the Reykjanes Ridge. The dashed line is the bathymetry calculated by Owens (1996), assuming the axial high represents a load on the surface of an elastic plate with T_e = 9 km. The predicted bathymetry explains well the 'moat' that is observed around the axial high.

There are examples where the axial topography of a particular mid-ocean ridge system changes along its strike: from a rift valley to an axial horst and vice versa. For example, the crest of the Reykjanes Ridge changes from a rift valley in the south to an axial horst at about 59°N (Searle et al., 1998). A similar transition from a rift to a horst occurs along the east–west-trending Galpágos Spreading Centre (Canales et al., 1997).

It was originally thought that differences in axial topography reflected differences in the age of the ridge. However, we now know that the presence or absence of a rift valley depends on several factors, most notably the local rate of seafloor spreading. Rift valleys, for example, are usually found on ridges that are spreading at full rates of < 50 mm a^{-1}, while axial highs are a special feature of fast-spreading ridges (> 80 mm a^{-1}). There are exceptions, of course. For example, the Reykjanes Ridge is associated with an axial high despite a spreading rate of only 20 mm a^{-1} while the Australian–Antarctic discordance has an axial valley even though this section of the South-East Indian Ridge is spreading at 70 mm a^{-1}. Since the Reykjanes Ridge and Australian–Antarctic discordance are unusually shallow and deep, respectively, axial morphology may therefore ultimately be controlled by the thermal and mechanical structure of the ridge axis, as well as other factors such as crustal composition, thickness and magma supply.

Support for this suggestion comes from consideration of the relationship between T_e and spreading rate, since both parameters should reflect the thermal and mechanical structure at

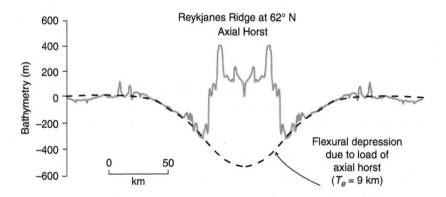

Figure 7.16 The origin of the axial 'horst' on the crest of the Reykjanes Ridge crest by flexural loading of a relatively strong lithosphere. The solid line is the observed bathymetry at 62° N, corrected for the effects of lithospheric cooling. Note the bathymetry is from one side of the horst and has been reflected about the horst axis. Reproduced from fig. 5.4 of Owens (1996) with permission.

the axis of a mid-ocean ridge. Cochran (1979), for example, suggested from an admittance analysis of gravity anomaly and bathymetry profiles from the mid-Atlantic ridge and East Pacific Rise that T_e decreased with increase in spreading rate. More recently, Luttrell and Sandwell (2002) confirmed this relationship from an analysis of the bathymetry, critical stress limit and gravity anomalies from more than 25 sub-regions of the global mid-ocean ridge system. The results of the Cochran and Luttrell and Sandwell studies show (Fig. 7.17) that the decrease in T_e with increase in spreading rate, S_{rate}, can be described approximately by

$$T_e = 11.5 - (S_{rate})^{0.5} \qquad (7.9)$$

The square root dependency of T_e on S_{rate} demonstrates that a link might exist between the bathymetry and the thermal and mechanical structure of a mid-ocean ridge and, importantly, that both estimates of the elastic thickness and spreading rate directly reflect this structure. Interestingly, there is a decrease in the depth to the top of the axial low-velocity zone with increase in S_{rate} (e.g., Purdy et al. 1992), suggesting a link may also exist between the long-term thermal and mechanical structure of a mid-ocean ridge and its short-term seismic structure.

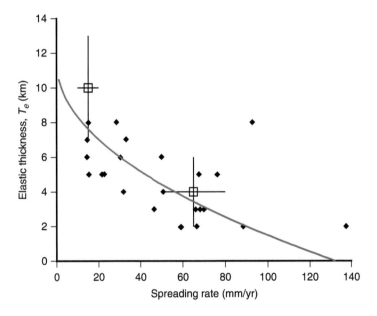

Figure 7.17 The relationship between elastic thickness, T_e, and spreading rate at a mid-ocean ridge. The open squares (with error bars) are based on Cochran (1979) and the filled diamonds are based on Luttrell and Sandwell (2002). The solid red line is based on Eq. (7.9).

That flexure has a significant role to play at mid-ocean ridges despite the high temperature gradients has been suggested in subsequent studies. For example, Buck (2001) proposed the axial high at fast spreading ridges was the consequence of flexural loading due to accretion of magmatic material at the axis and flexural unbending as the newly formed oceanic lithosphere moved away from the axis, thereby flattening the plate. A flexural unbending model was subsequently used at the fast-spreading East Pacific Rise at 8° S and 9° N by Shah and Buck (2003) to explain abyssal hills and faulting and by Sohn and Sims (2005) to explain off-axis volcanism. These studies assumed a lithosphere thickness of 5–6 km (as defined by the depth to the 700°C isotherm) and an elastic thickness, T_e, of ~2.5 km, respectively.

7.2.3 Rift-Type Basins

Seismic reflection and refraction profiling suggest that many sedimentary basins are underlain by rift-type structures such as half-graben, tilted fault blocks and thinned continental crust. Therefore, rifts may be an important precursor to sedimentary basin formation.

Perhaps the most obvious type of depression produced by rifting is the ocean basins themselves. Ocean basins form by thermal contraction of the oceanic crust as it cools, increases in density, and subsides away from a mid-ocean ridge crest. Plate cooling models (e.g., Parsons and Sclater, 1977) provide a good description of the form of this subsidence. The sediment that accumulates in these deep-water basins is of modest thickness, however, when compared with the vast accumulations that are now found beneath the continental margins.

Seismic data at Atlantic-type or 'passive' continental margins show that many are underlain by rift-type basins. Reflection data at the Iberia margin, for example, show, despite its large present-day water depth, spectacular images of tilted fault blocks (Montadert et al., 1979), listric normal faults and mid-crustal undulating reflectors (Reston et al., 1996), which represent detachment surfaces.

The rift structures in many margins, however, are obscured by large thicknesses of sediments. The East Coast, Canada and US margin, for example, is associated with > 10 km of sediment (LASE Study Group, 1986; Chian et al., 1995). Seismic refraction data, together with magnetic anomaly reconstructions, show that a large portion of these sediments were deposited on continental rather than oceanic crust.

We have already demonstrated that flexure is an important contributor to the 'architecture' of continental and oceanic rifts. It is also likely to be an important controlling factor on the post-rift stratigraphic development of continental margins. The difficulty in isolating the role of flexure is that many of the sediments that accumulate at margins were formed in relatively shallow water. This means that sediment loading of water-filled depressions cannot be the only cause of basin subsidence, and other factors must be involved.

Isostatic arguments (e.g., Sleep, 1971, 1973; Foucher and Le Pichon, 1972; Kinsman, 1975), however, suggest that irrespective of the cause of their subsidence, sediment and water loading will be an ongoing process that contributes to rift basin development. The role played by loading is likely to be smaller, of course, for the sediment-starved margins than for the well-sedimented ones. Nevertheless, if techniques could be developed to 'correct' the stratigraphic record for the effects of sediment and water loading, then we would have gone some way towards better understanding the form of the subsidence and, hence, the origin of rift-type basins.

7.2.3.1 'Backstripping'

One of the first attempts to quantitatively analyse the subsidence history of rift basins was by Sleep (1971). He plotted the subsidence in several wells in New Jersey and normalised them with respect to the depth to a single stratigraphic horizon (the Late Cretaceous Woodbine Group). Although we now know that many of the wells analysed by Sleep were in the flexural depression that flanks the main sediment load, he demonstrated that the subsidence of rift-type basins was exponential in form and resembled that of a midocean ridge. This similarity led him to suggest that passive continental margins originate by heating of the lithosphere at the time of rifting.

By normalising the subsidence of a rift basin with respect to a particular stratigraphic horizon, Sleep was able to 'track' the cumulative sediment accumulation in the basin through geological time. The sediment accumulation, however, is the consequence of not only the tectonic processes that are responsible for basin formation, but also the effects of water and sediment loading.

Watts and Ryan (1976) therefore developed a technique to 'correct' the stratigraphic record for the disturbing effects of water and sediment loading and to isolate the form of the unknown tectonic driving forces responsible for rift basin subsidence. By restoring the sediment thickness at the time of deposition, considering compaction and water depth changes, and then isostatically unloading it from the crust, these workers were able to determine the depth at which that basement would be in the absence of water and sediment loading. Because the technique involved the removal of loads, they referred to it as 'backstripping'.

The backstripping technique presupposes that we know how water and sediments load the crust, which of course we do not. Despite this limitation, the technique has proved a useful method to analyse stratigraphic data, whether it be in the form of well logs, a seismic reflection profile or restored geological cross-sections.

There are two main types of backstripping which differ in the way that sediment loading is treated. These are Airy and flexural.

The concept behind Airy backstripping is illustrated in Fig. 7.18. The figure shows two columns of the crust and upper mantle that before and after loading are in local isostatic equilibrium. Balancing the pressure at the base of the two columns gives

$$\rho_w g W_{di} + \bar{\rho}_{si} g S_i^* + \rho_c g T = Y_i \rho_w g + \rho_c g T + x \rho_m g,$$

where W_{di}, S_i^* and Y_i are the water depth, de-compacted sediment thickness and tectonic subsidence of the i th stratigraphic layer, respectively, T and ρ_c are the mean thickness and density of the crust respectively (both of which are assumed to be constant during loading

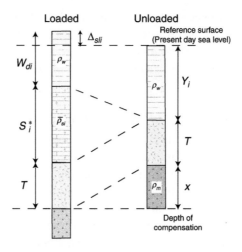

Figure 7.18 The principles of Airy-type backstripping. The loaded and unloaded sediment columns are shown together with the depth of compensation on which the pressure at the base of each column is equated.

and unloading) and ρ_w, ρ_m and $\bar{\rho}_{si}$ are the densities of the water, mantle and de-compacted sediment, respectively. It follows from Fig. 7.18 that

$$x = W_{di} + S_i^* + T - (Y_i + \Delta_{sli} + T),$$

where Δ_{sli} is the height of mean sea level above (or below) the reference surface,

$$\therefore Y_i = W_{di} + S_i^* \left[\frac{(\rho_m - \bar{\rho}_{si})}{(\rho_m - \rho_w)} \right] - \Delta_{sli} \frac{\rho_m}{(\rho_m - \rho_w)} \qquad (7.10)$$

Equation (7.10) is known as the 'backstripping equation'. This equation allows Y_i to be determined directly from stratigraphic data. The first term is a water-depth term, the second a sediment loading term, and the third a sea level loading term.

The water-depth term refers to the depth of deposition of a particular stratigraphic unit. For a unit deposited at the present day, it is simply the depth of water. Estimating the water depth in the past is a more difficult problem, however. Fortunately, there are certain 'depth indicators'. These include benthic micro-fossils, sedimentary facies and distinctive geo-chemical signatures. Perhaps the most widely used are benthic micro-fossils. Some suites of benthic organisms are known to inhabit certain water depths as an adaption to local hydrostatic pressures. Cita and Zocchi (1978), for example, demonstrated that benthic foraminifera in the Mediterranean Sea are distributed in a systematic way with depth. Since present-day species are also represented in the ancient record, it is possible to estimate a water depth of deposition with some precision, at least for Tertiary shelf and upper slope environments where there has been little 'reworking' of sediment. However, water depths are still difficult to resolve, especially in lower slope and rise environments and in Mesozoic and older sediments.

The sediment loading term in Eq. 7.10 describes the way that the sediment load deforms the crust and mantle. It includes a thickness and a density term. The thickness and density, however, are not as we would measure them today. This is because sediments are modified after their deposition by processes such as compaction, diagenesis and so on.

To account for the effect of compaction on sediment thickness and density, we will assume a simple model in which compaction is a mechanical, non-reversible process that involves the expulsion of pore water from voids in the sediment (e.g., Perrier and Quiblier, 1974). In other words, the effects of any chemical alteration of the sediment or diagenesis are not considered. We then can write that

$$S_i^* = \frac{S_i(1 - \phi_{si})}{(1 - \phi_{si}^*)}$$
(7.11)

$$\bar{\rho}_{si} = \rho_w \phi_{si}^* + \rho_{gi}(1 - \phi_{si}^*),$$
(7.12)

where S_i is thickness of the compacted layer (i.e. the present-day thickness), S_i^* is the thickness of the de-compacted layer, ϕ_{si} is the porosity of the compacted layer (i.e. the present-day porosity), ϕ_{si}^* is the porosity of the de-compacted layer and ρ_{gi} is the sediment grain density.

Equations (7.10) and (7.11) show that the calculation of the thickness and density at the time of deposition requires an assumption about the porosity of the layer, prior to compaction. One approach is to construct a porosity-depth curve from 'empirical' curves that are based on sonic or density logs from a single well. The porosity of any de-compacted layer can then be estimated by 'sliding' the present-day thickness of the unit up the porosity-depth curve. Alternatively, some porosity-depth function is assumed that is based on basin-wide compilations of data for individual lithologies.

In practice, backstripping is carried out at times during the evolution of a basin and so may involve the removal of more than one stratigraphic unit. To do this, we need to restore all the units that formed during a particular time-step – de-compacting the younger units and compacting the older ones. The important point here is that it is the total thickness, S^*, and average density, $\bar{\rho}_s$, of the restored sequence that is required at any one time. Since the total mass of the sequence must be equal to the sum of the mass of each restored unit, it follows that

$$\bar{\rho}_s S^* = \sum_{i=1}^{n} \left[\rho_w \phi_{si}^* + \rho_{gi}(1 - \phi_{si}^*) \right] S_i^*,$$
(7.13)

where n is the total number of stratigraphic units at a particular time. The average density of the column of sedimentary rock at the time of deposition is then given by

$$\bar{\rho}_s = \frac{\sum_{i=1}^{n} \left[\rho_w \phi_{si}^* + \rho_{gi}(1 - \phi_{si}^*) \right] S_i^*}{S^*}.$$

(7.14)

The corresponding total tectonic subsidence or uplift, Y, is

$$Y = W_d + S^* = \left[\frac{(\rho_m - \bar{\rho}_s)}{\rho_m - \rho_w} \right] - \Delta_{sl} \frac{\rho_m}{(\rho_m - \rho_w)}$$

(7.15)

where W_d is the water depth and Δ_{sl} is the height of sea level (with respect to the present day) at a particular time.

The final term in the backstripping equation is water loading, which is due to global sea level changes. Some localities (e.g., the Mediterranean Sea during the Neogene) were isolated from the global ocean, and so a sea level correction may not be necessary. However, many of the rift-type basins that develop as a result of continental break-up are eventually linked to the global ocean.

The problem is that there is no general agreement on the magnitude of sea level changes in the past. Fig. 7.19 shows five of the most used sea level curves in backstripping studies. The figure shows two types of curves: oscillatory (e.g., Vail et al., 1977; Haq et al, 1987) and smooth (e.g., Pitman, 1978; Watts and Steckler, 1979; Watts and Thorne, 1984). The Vail and Haq et al. curves were derived from the stratigraphic record using seismic sequence analysis techniques. These techniques use reflector terminations on seismic reflection profiles to estimate former positions of the shoreline and the rise and fall of sea level in the past.

The Pitman curve, on the other hand, is based on consideration of the changes in volume of the ocean basins that are caused by changes in the spreading rate and length of the world's mid-ocean ridge system. These changes are significant and would be enough to displace large volumes of water onto the flanking continents.

Consider that, at some time in the past, an increase in the volume of the ocean basins causes a certain sea level rise, h. The rise in sea level results in an additional load on the crust, which subsides according to isostasy. What is more important geologically, however, is the height, f, which is the new height of the sea after isostasy relative to the original reference surface. The height f (see Section 1.13) is called the freeboard and is the equilibrium position of the sea surface that follows an increase or decrease in volume of the ocean basin.

We can find the value of f from Fig. 1.37. Isostatic balancing, for example of columns 3 and 5, gives

$$\rho_w W_d g + \rho_c T_c g + \rho_m y g = \rho_w h g + \rho_w W_d g + \rho_c T_c g$$

$$\therefore f = \frac{(\rho_m - \rho_w)h}{\rho_m},$$

(7.16)

which assuming $\rho_m = 3330$ kg m^{-3}, $\rho_w = 1030$ kg m^{-3}, gives $f = 0.69h$.

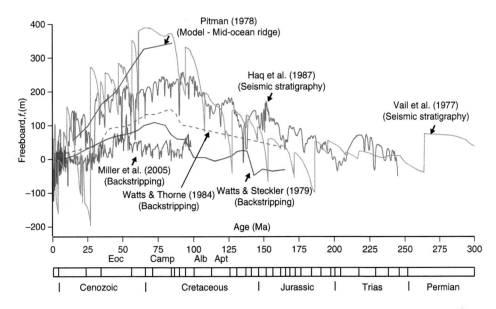

Figure 7.19 Changes in freeboard during the past 300 Myr. The curves are based on modelling the change in volume of mid-ocean ridges, seismic stratigraphy and backstripping.

We can estimate f by considering how the water that is displaced by a change in the volume of the ocean basin – due, for example, to changes in the spreading rate and length of a mid-ocean ridge – is distributed across a continental margin. The change in water volume is the product of the new surface area of the ocean and the sea level rise. The new surface area of the ocean, however, is the surface area of the ocean prior to the volume change in the ocean basin plus the surface area of water that is spilled onto the margin of the continent.

According to global hypsometric curves, approximately one-sixth of the Earth's surface area of 5.1×10^8 km^2 is located at depths of 0–0.5 km below mean sea level. If we assume that the water associated with a change in ocean basin volume increases linearly over this water-depth range, then the change in surface area/km of water depth is given by

$$\frac{1}{6}\left(5.1 \times 10^8\right)\frac{1}{0.5} = 1.7 \times 10^8 \text{ km}^2 \text{ km}^{-1}$$

The new surface area, A, considering the freeboard is then

$$A = A_o + 1.7 \times 10^8 f,$$

where A_o is the present surface area of the oceans. The corresponding volume, V, of water is given by

$$V = A\,h$$
$$\therefore \quad V = A_o h + 1.7 \times 10^8 0.69 h^2$$
$$\therefore \quad V = 3.6 \times 10^8 h + 1.2 \times 10^8 h^2$$
$$\therefore \quad V = 1.2 \times 10^8 h (3 + h). \tag{7.17}$$

Given that V can be estimated from changes in ridge volume, Eq. (7.17) can be used to calculate h, and hence f. Pitman (1978) found that f reached a maximum value of about 330 m during the late Cretaceous and then steadily decreased during the Tertiary.

We see from Fig. 7.19 that Vail et al. superimposed their oscillatory curve on the Pitman curve. The reason for this is that seismic sequence analysis tends to emphasise the short-term rather than the long-term shifts in the stratigraphic response to sea level changes. While this is acceptable, there has been much controversy over the Vail et al. curve and whether the short-term variations in sea level are real (e.g., Miall, 1990). The main problem has been the lack of a causative mechanism. While it has been demonstrated that parts of the Vail et al. curve correlate with isotopic evidence of major changes in the temperature of the oceans and, hence, with times of glacials and interglacials, other parts do not. The most problematic has been the origin of the short-term variations in the Vail et al. curve during a 'greenhouse' world (e.g., during the Mesozoic) when there is little evidence for the presence of ice.

The three other sea level curves shown in Fig. 7.19 were also derived from the stratigraphic record. However, these curves (Watts and Steckler, 1979; Watts and Thorne, 1984; Miller et al., 2005), unlike the Vail et al. or the Haq et al. curves, were derived by backstripping and so they explicitly consider the exponential form of rift basin subsidence, changes in water depth through time and de-compaction and compaction effects.

The backstripping technique has now been widely applied to rift-type basins, both modern and ancient. The well-sedimented margins of the western Mediterranean Sea (Watts and Ryan, 1976; Steckler and Watts, 1980), the East Coast, United States (Watts and Steckler, 1979) and eastern Canada (Royden and Keen, 1980), and Australia (Hegarty et al., 1988) were among the first to be studied. Since then, a notable application of the technique has been to restored cross sections of ancient passive margin sequences in the Rocky Mountains (Armin and Mayer, 1983; Bond and Kominz, 1984), the Betics (Peper and Cloetingh, 1992), the Alps (Wooler et al., 1992) and the Zanskar Himalayas (Corfield et al., 2005).

Figure 7.20 illustrates the application of the backstripping technique to bio-stratigraphic data at the COST B-2 well, which was drilled by a consortium of oil companies in the outer continental shelf of the East Coast US continental margin in 1978. The well penetrated a total of 4.8 km of Mesozoic and Tertiary sediments. The sediment backstrip is based on downhole porosity-depth data, paleobathymetry determined from bio-stratigraphy and the Watts and Steckler sea level curve. The results show that the sediment accumulation at the well can be divided into two parts: one represents the contribution of sediment and water loading to the sediment accumulation, the other the unknown tectonic subsidence and uplift.

One of the most important results to have emerged from backstripping studies (e.g. Fig. 7.20) is that rift-type basins are associated with a tectonic subsidence that is typically fast early on and then decreases. The curve is exponential in form and resembles the subsidence curve of a mid-oceanic ridge. As we will see later, it is the distinct exponential form of their backstripped curves that allow extensional basins (Fig. 7.21) to be distinguished from other types of basin (e.g., those of compressional and strike-slip types; Xie and Heller, 2009).

The backstripping technique described so far is based on the Airy model. When applied to geophysical data from downhole logs, this form of backstripping is probably the most accurate. However, it is limited because it provides no information on the tectonic subsidence and uplift on either side of a well. One solution to this problem is flexural backstripping. In this form of backstripping, sediment thickness data – say, as determined on a depth-converted seismic reflection profile – are used to determine the tectonic subsidence along a profile or from an isopach map of a basin. The main limitation with this technique is that it requires information on the spatial variations that may occur in the strength of the underlying lithosphere.

Flexural backstripping follows a similar procedure as for Airy except that the sediments are now flexurally unloaded from the lithosphere. Consider the case of sediment loading, but this time for the case of $k \to 0$ (i.e., a wide sediment load). We then have

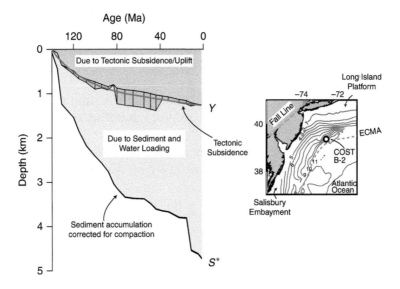

Figure 7.20 The results of Airy backstripping at the COST B-2 stratigraphic test well in the outer continental shelf, offshore New Jersey. Reproduced from fig. 14 of Watts (1981). The inset shows contours of the sediment thickness (at 1-km intervals) based on compilations by the United States Geological Survey (USGS) at Woods Hole.

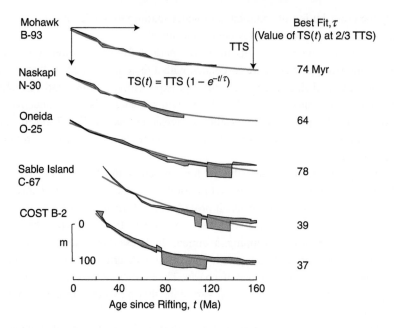

Figure 7.21 The backstrip at the COST B-2 well, offshore New Jersey, US, compared to backstrips at other deep wells offshore Nova Scotia, Canada. The numbers show the best-fit exponential depth constant that describes the tectonic subsidence. Note the best-fit exponential varies along-strike the East Coast US and Canada passive continental margin, as does the water depth of deposition.

$$y \rightarrow \frac{(\rho_l - \rho_w)h}{(\rho_m - \rho_{infill})}, \qquad (7.18)$$

where h is the height of the load, y is flexure, ρ_l is the density of the load, ρ_w is the density of the material displaced by the load, ρ_{infill} is the density of material infilling the flexure and ρ_m is the density of the mantle.

Let us now replace h by S^*, which is the total sediment thickness at a particular time, and ρ_l by ρ_s, the density of the sequence. Also, we will assume that $\rho_{infill} = \rho_w$ rather than ρ_s. Equation (7.17) then becomes

$$y \rightarrow \frac{(\rho_s - \rho_w)S^*}{(\rho_m - \rho_w)}. \qquad (7.19)$$

Since $S^* = y + h$ (Fig. 7.22), it follows that

$$h \rightarrow \frac{(\rho_m - \rho_s)S^*}{(\rho_m - \rho_w)}, \qquad (7.20)$$

which is Airy backstripping, for example Eq. (7.15), with zero water depth and no sea level changes.

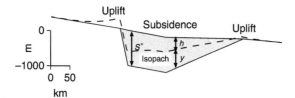

Figure 7.22 The principles of flexural backstripping. The yellow shaded region shows a wedge of sediments that overlies 'basement'. The dashed line is the position basement would be in the presence of flexure and the absence of sediment loading. Note that basement would need to form a region of subsidence that is flanked by uplift to explain the shape of the sediment wedge.

Flexural backstripping therefore amounts to first computing the flexure due to the total sediment load with $\rho_{infill} = \rho_w$, and then subtracting the resulting flexure, y, from the observed thickness, S^*, to obtain the tectonic subsidence and uplift.

It follows from Eq. (7.19) that

$$Y(k) = \frac{(\rho_s - \rho_w)S^*(k)}{(\rho_m - \rho_w)}\phi''_e, \qquad (7.21)$$

where $Y(k)$ is the frequency domain equivalent of $y(x)$, $S^*(k)$ is the frequency domain equivalent of the sediment thickness $S^*(x)$, and ϕ''_e is the wavenumber parameter (e.g., Eq. 5.25) that, when multiplied by the Airy response, gives the flexure. The backstrip $h(x)$ is then calculated from[4]

$$h(x) = S^*(x) - y(x).$$

In practice, flexural backstripping is carried out layer by layer with the option of assigning each layer a different density and T_e. The flexural backstrip for a particular time is obtained by summing the effects of each layer, beginning with the first layer. Any water that remains must also be added since it represents the unfilled portion of the basin. The effects of compaction do not need to be explicitly considered, other than through the density that is assumed for each layer.

Figure 7.23 shows the application of the flexural backstripping technique to stratigraphic data in the Valencia Trough, western Mediterranean (Watts and Torné, 1992b). The Early Miocene to Recent strata that have accumulated in the trough have been divided into three units, each of which has been flexurally backstripped with a different density and either a constant T_e or a temporally varying T_e. The cumulative tectonic subsidence reaches 3.5 km in the centre of the basin. In regions flanking the basin, there is flexural bulge. Although there is scatter, especially in the centre of the basin, the overall form of the backstrip resembles that of an East Africa rift valley with a central rift and flank uplifts. The Valencia

---

[4] A MATLAB® script, chap7_backstrip2d.m, to flexurally backstrip an arbitrary-shaped two-dimensional sedimentary load for different T_e is available from the Cambridge University Press server.

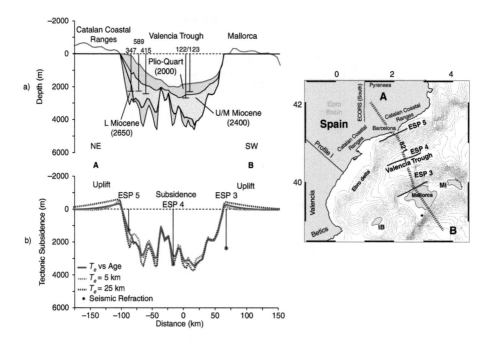

Figure 7.23 The application of the flexural backstripping technique to the Valencia Trough, a young rift basin in the western Mediterranean that underwent extension during the Lower Miocene. (a) Stratigraphic cross-section. (b) Backstripped tectonic subsidence and uplift. Reproduced from fig. 5 of Watts and Torné (1992b).

Trough was therefore most probably a continental rift that began extending in the Early Miocene but failed to generate oceanic crust.

The flexural backstripping method can easily be extended to three dimensions by replacing k in Eq. (7.21) by $\sqrt{k_x^2 + k_y^2}$, where k_x is the wavenumber in the x direction and k_y is the wavenumber in the y direction. The application of unloading (which includes backstripping) and loading in three dimensions to the sediments of the Valencia Trough and to the desiccation and flooding of the entire Mediterranean is described by Watts and Torné (1992a) and Govers et al. (2009), respectively.

7.2.3.2 McKenzie's Model of Crustal and Mantle Extension

The observation from backstripping that rift-type basins are characterised by an exponentially decreasing tectonic subsidence is a strong argument that they form by some sort of thermal contraction. Although Bott (1976) had earlier recognised through his application of the Vening Meinesz wedge model that extension results in a faulting stage followed by a 'stretching' and 'thinning' stage, it was McKenzie (1978) who first used observations of the deep structure of rift-type basins together with the exponential nature of their subsidence to construct a thermal model for rift-type basin evolution (Fig. 7.21). The model predicts subsidence and uplift, crustal thinning and heat flow for different times since rifting. He showed that because of extension, there is an initial subsidence due to crustal thinning followed by a thermal subsidence as the underlying stretched lithosphere cools.

Consider some initial thickness of the crust, T_c, and lithosphere, a, a steady-state geothermal gradient in the lithosphere (which includes the crust) and an isothermal sub-lithospheric mantle (Fig. 7.24). If the temperature at the surface of the lithosphere is 0°C and the temperature at its base is T_1, then the average temperature in the crust is given by

$$\frac{T_1 T_c}{2a} \tag{7.22}$$

Similarly, the average temperature in the sub-crustal mantle is

$$\frac{\left((T_1\, T_c)/a\right) + T_1}{2} = \frac{T_1 (T_c + a)}{2a}. \tag{7.23}$$

The pressure at the base of the unstretched column is given by

$$\rho_c g T_c + \rho l g (a - T_c) \tag{7.24}$$

where

$$\rho_c = \rho_{co}\left[1 - \frac{\alpha_v T_1 T_c}{2a}\right], \quad \rho_l = \rho_{lo}\left[1 - \frac{\alpha_v T_1 (T_c + a)}{2a}\right] \tag{7.25}$$

and α_v is the coefficient of volume expansion, ρ_{lo} is the temperature of the lithosphere at 0°C and ρ_{co} is the temperature of the crust at 0°C. Substituting Eq. (7.24) into Eq. (7.23) gives the following for the pressure:

$$\rho_{co}\left[1 - \frac{\alpha_v T_1 T_c}{2a}\right] g T_c + \rho_{lo}\left[1 - \frac{\alpha_v T_1 (T_c + a)}{2a}\right] g (a - T_c). \tag{7.26}$$

The effect of rifting is to thin the crust and lithosphere and to raise the geotherm. We can write the thickness of the thinned crust as T_c/β_s and the thickness of the sub-crustal mantle as

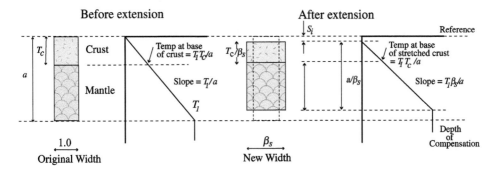

Figure 7.24 Schematic model showing the elements of the McKenzie (1978) thermal model for rift basin evolution.

$(a/\beta_s - T_c/\beta_s)$, where β_s, is the stretching factor. The pressure at the base of the stretched column then becomes

$$
\rho_w S_i g + \rho_{co}\left[1 - \frac{\alpha_v T_1 T_c}{2a}\right]g\frac{T_c}{\beta_s} + \rho_{lo}\left[1 - \frac{\alpha_v T_1(T_c + a)}{2a}\right]g\left[\frac{a}{\beta_s} - \frac{T_c}{\beta_s}\right]
$$
$$
+ \rho_{lo}(1 - \alpha_v T_1)g\left[a - S_i - \frac{a}{\beta_s}\right],
$$
(7.27)

where S_i is the initial subsidence.

Combining Eqs. (7.25) and (7.26), we get

$$
S_i = \frac{a\left\{(\rho_{lo} - \rho_{co})\frac{T_c}{a}\left(1 - \frac{\alpha_v T_1 T_c}{2a}\right) - \frac{\rho_{lo}\alpha_v T_1}{2}\right\}\left[1 - \frac{1}{\beta_s}\right]}{\left(\rho_{lo}(1 - \alpha_v T_1) - \rho_w\right)}.
$$
(7.28)

Equation (7.28) gives the initial subsidence for different values of the initial thickness, temperature and density of the crust and sub-crustal mantle and the stretching factor.

To get an idea of the magnitude of the initial subsidence, it is useful to substitute some typical values:

Lithospheric thickness, $a = 125\,\text{km}$
Crustal thickness, $T_c = 31.2\,\text{km}$
Density of lithosphere at 0°C, $\rho_{lo} = 3{,}330\,\text{kg m}^{-3}$
Density of crust at 0°C, $\rho_{co} = 2{,}800\,\text{kg m}^{-3}$
Density of water, $\rho_w = 1{,}030\,\text{kg m}^{-3}$
Temperature at base of lithosphere, $T_1 = 1{,}330°\text{C}$
Coefficient of volume expansion, $\alpha_v = 3.28 \times 10^{-5}°\text{C}^{-1}$
Stretching factor, $\beta_s = 3.0$

Substituting these parameters into Eqs. (7.25) and (7.28) gives $\rho_c = 2{,}785\,\text{kg m}^{-3}$, $\rho_l = 3{,}239\,\text{kg m}^{-3}$ and $S_i = 2.273\,\text{km}$. So, when the crust is stretched, the densities of the crust and mantle decrease, and the lithosphere subsides.

The final subsidence can be obtained by balancing the pressure at the base of the unstretched column with the final column. The pressure at the base of the unstretched column is given by Eq. (7.26). The pressure at the base of the final column is given by

$$
\rho_w g S_f + \rho'_c \frac{T_c}{\beta_s}g + \rho'_l, \left(a - S_f - \frac{T_c}{\beta_s}\right)g,
$$
(7.29)

where ρ'_c is the density of the cooled stretched crust and ρ'_l is the density of the cooled sub-crustal mantle. Equating Eqs. (7.26) and (7.29) and re-arranging gives

$$S_f = \frac{a\left(\rho_l' - \rho_i\right) + T_c\left(\rho_l + \frac{\rho_c'}{\beta_s} - \frac{\rho_l'}{\beta_s} - \rho_c\right)}{\left(\rho_l' - \rho_w\right)}, \tag{7.30}$$

where ρ_c' is the density of crust before stretching and ρ_i' is the density of the sub-crustal mantle before stretching; these have been defined earlier. The values of ρ_c' and ρ_l' can be expressed in terms of the coefficient of volume expansion and the new, cooled temperature structure; that is,

$$\rho_c' = \rho_{co}\left(1 - \frac{\alpha_v T_1 T_c}{2\beta_s a}\right),$$

$$\rho_l' = \rho_{lo}\left(1 - \frac{\alpha T_1 T_c}{2a\beta_s} - \frac{\alpha_v T_1}{2}\right).$$

Using the same parameters as before, we get $\rho_c' = 2{,}795$ kg m^{-3}, $\rho_l' = 3{,}251$ kg m^{-3} and $S_f = 4.927$ km. Therefore, the density of the crust and lithosphere increases from the initial conditions and the lithosphere subsides.

The thermal subsidence, S_t, is given by the difference between the final and initial subsidence:

$$S_t = S_f - S_i. \tag{7.31}$$

Substituting the values for S_i and S_f into Eq. (7.31) gives $S_t = 2.654$ km. The initial and thermal subsidence are therefore similar.

We have so far only computed the subsidence at the two 'end-member' points: one corresponding to the onset of rifting, and the other after a long amount of time has elapsed. Intermediate values of the subsidence can be computed by using the one-dimensional heat flow equation that relates the temperature, T, at a time since rifting, t, to y, the depth measured downwards, and κ, the thermal diffusivity of the lithosphere:

$$\frac{\partial T}{\partial t} = \kappa \frac{\partial^2 T}{\partial y^2} \tag{7.32}$$

We can solve Eq. (7.32) for the temperature distribution in the cooling plate by Fourier expansion. And since the temperature distribution determines the density structure, Eq. (7.32) can be used to calculate the subsidence as a function of time. The elevation, $e(t)$, above the final depth to which the surface of the lithosphere (which includes the crust) sinks is given, for example by McKenzie (1978), as

$$e(t) = E_0 r e^{-t/\tau},$$

where

$$E_o = \frac{4a\rho_{lo}\alpha_v T_1}{\pi^2\left(\rho_{lo} - \rho_w\right)}$$

and

$$r = \left(\frac{\beta_s}{\pi}\right)\sin\left(\frac{\pi}{\beta_s}\right).$$

The thermal subsidence, $S_t(t)$, since rifting can then be obtained from

$$S_t(t) = e(0) - e(t),$$

where $e(0)$ is the elevation at $t = 0$. $e(0)$ is simply $E_o r$; therefore,

$$S_t(t) = E_o r\left(1 - e^{-t/\tau}\right). \tag{7.33}$$

A plot of $S_t(t)$ versus $\left(1 - e^{-t/\tau}\right)$ will have a slope $E_o r$. This function depends on thermal parameters of the lithosphere as well as β_s, the amount of extension. Since S_t can be estimated from either Airy or flexural backstripping, the slope can be used to estimate β_s directly. This procedure has now been used to quantitatively estimate the amount of extension at a number of rift-type basins in various tectonic settings.

7.2.3.3 The Crustal Structure Produced by Rifting

The value of β_s gives the amount of thinning and, hence, the amount of heating that has occurred in a rift-type basin. It also allows an estimate of the crustal thickness. At a well, for example, the β_s that best fits the backstrip data as a function of time can be used to estimate the amount of crustal thinning and, hence, the crustal structure.

The advantage of flexural over Airy backstripping is that the backstrip can be determined along a profile of a basin. The disadvantage is that we only obtain the total backstrip – not the backstrip as a function of time. The crustal structure can be estimated from this back-strip, however, provided some assumption is made about the state of isostasy during rifting. Two cases will be considered here: Airy and flexure.

According to Airy, the backstrip will be compensated locally by changes in the thickness of the crust. Therefore, regions of tectonic subsidence will be associated with thin crust, while regions of tectonic uplift will have thick crust. We can estimate the amount of thinning and thickening by considering the pressure at the base of a crustal column before and after thinning (e.g., Fig. 7.25a). We then have

$$T_c \rho_c g = Y \rho_w g + T_s \rho_c g + x \rho_m g,$$

where

$$x = T_c - T_s - Y.$$

Rearranging gives

$$T_s = T_c - Y\frac{(\rho_m - \rho_w)}{(\rho_m - \rho_c)}.$$

Since

$$\beta_s = \frac{T_c}{T_s}$$

$$\therefore \beta_s = \frac{(\rho_m - \rho_c)T_c}{(\rho_m - \rho_c)T_c - Y(\rho_m - \rho_w)}. \qquad (7.34)$$

The relationship between β_s and the backstrip Y is shown in Fig. 7.25b. The dashed line shows the predicted relationship based on an Airy model with $\rho_w = 1,030\,\text{kg m}^{-3}$, $\rho_w = 2,800\,\text{kgm}^{-3}$ and, $\rho_m = 3,330\,\text{kg m}^{-3}$. The solid lines show the predictions of McKenzie's thermal model assuming an age of rifting of 30, 60, 120 and 180 Ma. In both models, the thickness of the unstretched crust is assumed to be 31.2 km. The figure shows that the Airy model is a good approximation to McKenzie's thermal model, provided that the basin is 120 Ma or older. This is because thermal effects are important early on and so consideration should be given when backstripping to changing density of the cooling crust and upper mantle since a rifting event.

We have assumed so far that when the crust is extended it produces a water-filled depression that is compensated for by a thinning of the crust, as predicted by an Airy model. The problem with this assumption is that it ignores the finite strength of the lithosphere. As several workers (e.g., Braun and Beaumont, 1987) have pointed out, strong zones in the lithosphere might significantly modify the crustal structure of extended lithosphere. This is because the response of the lithosphere to extension is not just a thermal problem but a mechanical one.

We know from the results of experimental rock mechanics (see Section 6.6) that the strength of the materials that make up the crust and mantle increases and then decreases with depth. There will therefore be a strength maximum at some depth in the lithosphere. Sensitivity studies (e.g., Fig. 6.33) show that the depth of the strength maxima depends on the geothermal gradient, the initial crustal thickness, and the strain rate. When the lithosphere is subject to extension, we may expect that the strength maxima will act as a resistant zone that vertically partitions the strain into shallower and deeper levels within the crust and mantle.

Weissel and Karner (1989) argued that the depth of the basin depression, S_t, which would result from extension, is related to the depth of the strength maxima, Z_{neck}, by

$$S_t = \left(1 - \frac{1}{\beta_s}\right) Z_{neck}. \qquad (7.35)$$

This relationship suggests that for a particular β_s, the depth of the basin increases with the depth to the strength maxima (Fig. 7.26). For a shallow strength maximum, there is only a shallow basin. Because the crust is extended by some amount, the Moho relief will, however, be large to balance it. For a deep strength maximum, there is a large basin and small Moho relief.

Because the depression and Moho relief depend on both the amount of the extension and the depth of the strength maxima, the stretched crust will not necessarily initially be in

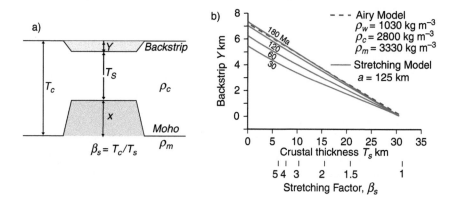

Figure 7.25 The relationship between (a) the amounts of stretching and crustal thinning and (b) the magnitude of the sediment-derived backstrip and crustal thickness.

isostatic equilibrium. The force, q_{net}, that attempts to return the depression and Moho relief to a state of isostatic balance is given by the sum of all the density contrasts within the lithosphere (which includes the crust), compared with that of undeformed lithosphere. These include an upward force, q_{up}, due to replacement of crust by water, and a downward force, q_{down}, due to replacement of crust by mantle. Ignoring the thermal effects on the density of the sub-crustal mantle, we get

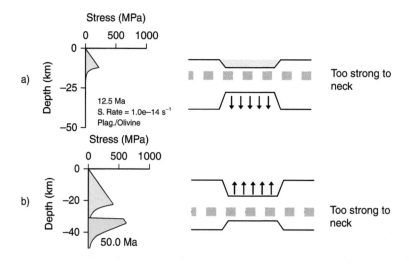

Figure 7.26 Schematic diagram showing the role that a strong zone in the crust might play in determining the geometry of rifting. (a) The strength maxima are at relatively shallow depths and (b) the strength maxima are at relatively deep depths. In both cases, the strength maxima act to re-distribute the extension with depth, producing an undersized basin in case (a) and an oversized basin in case (b). Based on the works of Braun and Beaumont (1987), Weissel and Karner (1989) and Kooi et al. (1992).

$$q_{net} = q_{up} + q_{down}$$
$$q_{net} = (S_t \Delta \rho_1 + S_m \Delta \rho_2)g \tag{7.36}$$

where $\Delta \rho_1 = (\rho_w - \rho_c)$, $\Delta \rho_2 = (\rho_m - \rho_c)$ and g is average gravity. S_m is the Moho upwarp, which is given (e.g., Fig. 7.27b) by

$$S_m = T_c \left(1 - \frac{1}{\beta_s} \right) S_t. \tag{7.37}$$

The deformation associated with q_{net} will be distributed according to the strength of the lithosphere during rifting. If the lithosphere has no strength, then the restoring forces will return the depression to a state of Airy-type isostatic equilibrium. The final depth of the basin, Y, is then obtained by re-arranging Eq. (7.33). This gives

$$Y = T_c \left(1 - \frac{1}{\beta_s} \right) \frac{(\rho_m - \rho_c)}{(\rho_m - \rho_w)}. \tag{7.38}$$

If, on the other hand, the crust is strong during rifting, then the basin shape will be modified by flexure as it returns to a state of equilibrium. As shown in Fig. 7.27, the initial basin depression, S_t, depends critically on Z_{neck}. For a shallow Z_{neck} (Fig. 7.27c), the depression is small in comparison with the magnitude of the Moho upwarp and a downward state of flexure results, predicting subsidence of the basin margins. For deep depths (Fig. 7.27a), an upward state of flexure prevails, predicting an upwarp in the basin centre and uplift on the basin flanks. For one particular Z_{neck} (Fig. 7.27b), the upward and downward forces exactly balance, and the final basin shape is the same as would be predicted by an Airy model.

The final basin shape after considering both Z_{neck} and flexure is the position at which the basement would be in the absence of sediment and water loads. This is also the depth that would be obtained from backstripping, Y. We therefore require an expression that uses the observable Y to determine the amount of crustal thinning that considers both Z_{neck} and the strength of the lithosphere during rifting.

We have, by combining Eqs. (7.36) and (7.37), that

$$q_{net} = \left(S_t (\Delta \rho_1 - \Delta \rho_2) + \left(1 - \frac{1}{\beta_s} \right) T_c \Delta \rho_2 \right) g \tag{7.39}$$

$$= \left(\left(1 - \frac{1}{\beta_s} \right) Z_{neck} \Delta \rho_3 + \left(1 - \frac{1}{\beta_s} \right) T_c \Delta \rho_2 \right) g \tag{7.40}$$

$$= \left(Z_{neck} \Delta \rho_3 + T_c \Delta \rho_2 - \frac{Z_{neck} \Delta \rho_3 + T_c \Delta \rho_2}{\beta_s} \right) g \tag{7.41}$$

$$= q_0 - q_0 \beta_s^{-1},$$

where $\Delta \rho_3 = (\rho_w - \rho_m)$ and $q_0 = (Z_{neck} \Delta \rho_3 + T_c \Delta \rho_2)g$.

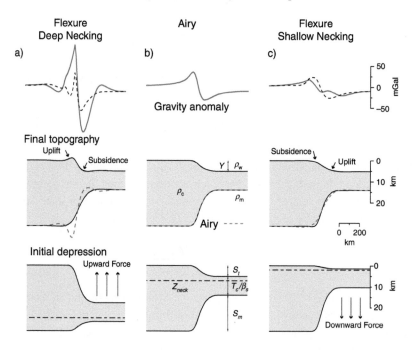

Figure 7.27 The forces associated with an (a) oversized and (c) undersized basin and their effects on the subsidence and uplift patterns and gravity anomalies. The calculations are based on a T_e during rifting of 20 km. (b) The case of Airy isostasy (i.e., no strength during rifting) for comparison. Reproduced from fig. 3 of Watts and Stewart (1998).

For the densities listed in Fig. 7.25 and $T_c = 31.2$ km we see that $q_{net} = 0$ if $Z_{neck} = 7.2$ km. Necking at this depth will therefore give a final basin geometry identical to that predicted by the Airy model of isostasy, despite the possibility that the lithosphere may possess some strength during rifting.

The flexure of the lithosphere, y_{net}, that results from the load q_{net} is obtained from the solution of the elastic plate equation; that is,

$$D\frac{d^4 y_{net}}{dx^4} + (\rho_m - \rho_w)y_{net}g = q_0 - q_0\beta_s^{-1}, \tag{7.42}$$

where D is the flexural rigidity. The solution to Eq. (7.42) is obtained by taking the Fourier transform of both sides:

$$\boldsymbol{Y}_{net}(k)(Dk^4 + (\rho_m - \rho_w)g) = \boldsymbol{Q}_0(k) - q_0\ \beta_s^{-1}(k),$$

$$\boldsymbol{Y}_{net}(k)\boldsymbol{\psi}(k) = \boldsymbol{Q}_0(k) - q_0\beta_s^{-1}(k), \tag{7.43}$$

where the bold uppercase variables represent the Fourier transform of the lowercase variables and $\psi(k)$ is given by $(Dk^4 + (\rho_m - \rho_w)g)$.

Noting that $y_{net}(x)$ is the response of the lithosphere to the downward- or upward-acting loads and so is the difference between the final basin shape, Y, and the initial basin depression, S_t, Eq. (7.43) simplifies (e.g., Watts and Stewart, 1998) to

$$\left(\mathbf{Y}^*(k) + Z_{neck}\beta_s^{-1}\right)\psi(k) = \mathbf{Q}_0 - q_0\beta_s^{-1},$$

here $\mathbf{Y}^* = \mathbf{Y} - Z_{neck}$.

$$\therefore \beta_s^{-1} = \frac{\left(\mathbf{Q}_0(k) - \mathbf{Y}^*(k)\psi(k)\right)}{\left(Z_{neck}\psi(k) + q_0\right)}. \tag{7.44}$$

The usefulness of Eq. (7.44) is that it allows the β_s distribution and, hence, the crustal structure to be determined directly from the backstripped profile Y, considering the possibility of strength during rifting.[5]

7.2.3.4 The Strength of Extended Continental Lithosphere and Process-Oriented Gravity and Flexure Modelling

It should be obvious that once the backstrip has been determined, it is possible to estimate the thickness of the extended crust. This can be carried out with or without considering strength during rifting. The main problem is that we do not really know what T_e value should be used to backstrip the sediments and restore the configuration of the crust at the time of rifting.

One parameter that may be used to constrain T_e is the gravity anomaly. The gravity anomaly in a rift basin is sensitive to both the initial crustal structure and the processes, such as sedimentation, magmatic underplating and erosion, that may have modified it since rifting. By calculating the gravity anomaly caused by the initial crustal structure and these modifying processes and comparing it with the observed anomalies, it should be possible to constrain T_e.

The question is, what method should be used to calculate the gravity anomalies? We could use an object-oriented approach based on a line-integral technique, but, while this would determine the density structure of the crust that is required to explain the gravity anomalies, it would ignore the geological processes that had shaped the margin through time. Nor would it correctly consider the effects of temporal or spatial changes in T_e through time.

Watts (1988) therefore suggested a process-oriented gravity modelling (POGM)[6] approach instead. This is a three-step technique. First, the gravity anomaly due to the backstripped, water-filled basin and its compensation are computed. This anomaly, which can be computed considering strength during rifting, is known as the rifting anomaly (Fig. 7.28 a,b). Second, the gravity anomaly due to the sediment load and its compensation

---

[5] A MATLAB® script, chap7_strength_during_rifting.m, to restore the crust from the backstrip, taking into account strength during rifting, is available from the Cambridge University Press server.

[6] A MATLAB® script, chap7_POGM_2d.m 3, to carry out two-dimensional process-oriented gravity modelling is available from the Cambridge University Press server.

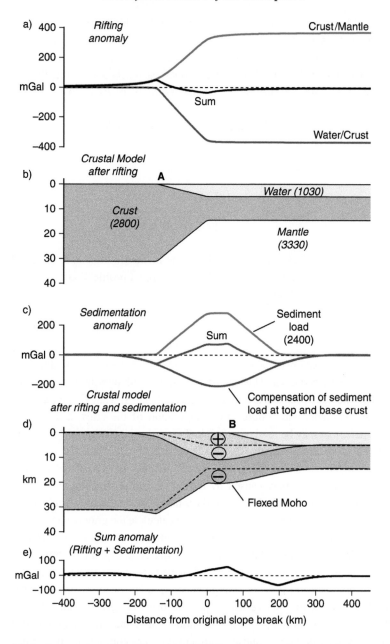

Figure 7.28 The process-oriented gravity modelling (POGM). When applied to a rifted continental margin, the technique considers the observed gravity anomaly as the sum of two main processes: rifting and thinning of the crust (the rifting anomaly) and sedimentation and flexure of basement and the Moho (the sedimentation anomaly). The circled '+' highlights the positive contribution of the sediment load to the gravity anomaly, and the circled '−' indicates the negative contribution of the load-induced flexure. Reproduced from fig. 2 in Watts (2022).

are computed. This anomaly is known as the sedimentation anomaly (Fig. 7.28c). The final step is to compute the sum anomaly due to rifting and sedimentation (Fig. 7.28d,e) and compare it with the observed anomaly over a rift basin.

Figure 7.29 illustrates an application of POGM to the East Coast, US margin in the region of the Baltimore Canyon Trough. The thick red dashed lines in Fig. 7.29b show the water-filled backstrip (upper curve) and the backstrip Moho (lower curve), considering Airy isostasy (i.e., no strength during rifting). The rifting anomaly due to the backstrip and its compensation is then computed. The difference between the water-filled backstrip and the present-day topography yields the total sediment load at the margin. The sedimentation anomaly is then computed from the sediment load and its compensation assuming a particular value of T_e. Figure 7.29a shows that the sum anomaly, obtained by addition of the rifting and sedimentation anomalies, is characterised by a high and a flanking low that resembles that of the observed anomaly. The sum anomalies based on a model where the sediment load is incrementally added and T_e is given by the depth of the 450°C isotherm explain the anomaly seaward of the shelf break, whilst a model based on the 150°C isotherm better accounts for the anomaly landward of the shelf break. These results suggest that the present day shelf break marks the transition between oceanic and continental crust and that the seismically defined region of thin crust beneath the shelf is also a zone of weakness.

The question that may be raised is whether the requirement of a weak crust is justified, given that Airy isostasy was assumed to calculate the backstrip Moho. Recent studies have concluded, however, that rifted crust is weak (i.e., $T_e < 10$–11 km) even when the backstrip Moho is computed considering the possibility of strength during rifting (Cloetingh et al., 1995; Watts and Stewart, 1998). This is not to suggest that all rifted margins are characterised by low T_e values. The Western Platform, New Zealand (Holt and Stern, 1991) and the Labrador and West Greenland margins (Keen and Dehler, 1997) are associated with T_e values of > 20 km. Moreover, the Atlantic margins offshore Nova Scotia (Keen and Dehler, 1997) and Maryland (Pazzalglia and Gardner, 1994) are associated with values of 20–60 km and 40 km, respectively. The latter results are of interest because the Nova Scotia and Maryland margins are located to the north and south of the Baltimore Canyon Trough, which gravity modelling suggests (e.g., Fig. 7.29) is a low T_e margin. This suggests that continental margins may be segmented with regard to their flexural strength, which may, in turn, explain the along-strike variations in the gravity anomaly that have been observed at some margins (e.g., Gabon/Angola; Watts and Stewart, 1998).

7.2.4 Seaward Dipping Reflectors and Magmatic Underplating

Many rift-type basins are associated with magmatism. Probably the best examples are the basins that form the so-called 'volcanic margins' (Eldholm et al., 1995) of the plume-influenced North Atlantic. The magmatism is expressed on seismic reflection profiles (Hinz,

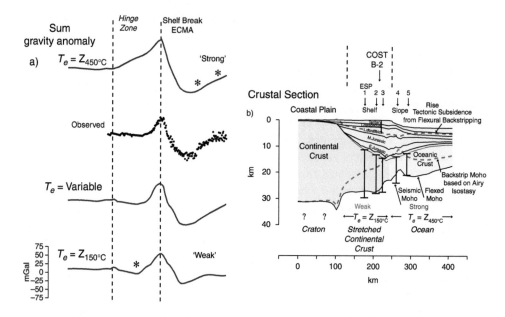

Figure 7.29 The application of the 'process oriented' approach to seismic and gravity anomaly data at the East Coast, US, rifted margin. In this approach, sediments in a rift basin are progressively backstripped through time and the gravity anomalies due to the combined effects of rifting and sedimentation are computed for different assumed values of T_e. By comparing observed and calculated gravity anomalies it is possible to constrain T_e and, in some cases, its spatial and temporal variations. (a) Observed (black dots) and calculated gravity anomalies. (b) Crustal structure deduced from backstripping and gravity modelling. Reproduced from figs. 6 and 7 of Watts and Marr (1995).

1981) as a series of seaward dipping reflectors (SDRs) and on seismic refraction data (Fowler et al., 1989; White, 1992; Kelemen and Holbrook, 1995) as unusually high P-wave-velocity lower crustal bodies.

Typically, SDRs show a reflector pattern dominated by downdip thickening. The pattern closely resembles the clinoform geometry of large prograding river systems (see Section 4.4, Fig. 7.30). Ocean Drilling Project (ODP) and seismic reflection profile data together with analogous exposures on land (e.g., Lebombo – Cox, 1980; East Greenland – Wager and Deer, 1938; Snake River – Gibson, 1966) suggest SDRs comprise thick sequences of extrusive basalts. According to Mutter et al. (1982), the extrusive basalts represent sub-aerial lavas which are fed by vertical dykes at depth. If this is correct, then the sigmoidal geometry is probably a consequence of the combined effects of thermal contraction and flexure of the underlying lithosphere by volcanic basaltic loads, as the margin separates further from the main magma source at the future site of the mid-oceanic ridge.

To test this possibility, Buck (2017) and Morgan and Watts (2018) independently constructed simple models of flexure to explain the reflector pattern observed in seaward dipping

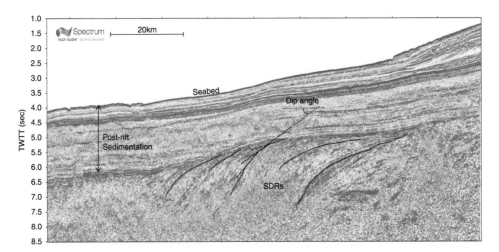

Figure 7.30 Seismic image of the Seaward Dipping Reflectors (SDRs) offshore Namibia in the Orange basin. Key reflectors within the SDRs are annotated in black. The dashed line tangential to the SDR indicates the method of dip calculation; multiple tangents are taken along the length of the SDR and a moving average calculated. Reproduced with permission from Spectrum from Morgan and Watts (2018).

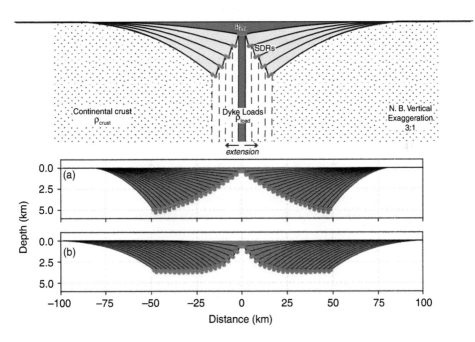

Figure 7.31 Simple model of flexure for the origin of seaward-dipping reflectors. Subsurface dykes intrude the crust, load, and cause a surface flexure that is subsequently infilled by volcanoclastic and basaltic lava flows. The red circles show the tilting of the dyke tops that occurs because of progressive dyke loading. Based on figs. 1 and 7 in Morgan and Watts (2018).

reflectors (Fig. 7.31). The 'driving' loads in each case were assumed to be sub-surface vertical dykes which intrude the stretched continental crust and cause a surface flexure that is infilled with basaltic flows and, possibly, intercalated volcanoclastic and pelagic sediments. Buck showed, using analytical techniques and line loads, that progressive loading models predict a symmetric pattern of onlap and that ridge jumps could explain the asymmetric patterns observed in conjugate margins. Morgan and Watts showed, using finite element techniques and distributed loads, that the patterns depend on the T_e structure and whether the SDRs were formed during the weakening syn-rift phase of rifting or the post-rift strengthening phase. Comparisons of model predictions with seismic and gravity data suggested SDR dips of up to ~12 degrees and T_e in the range of 1–3 km.

The lower crustal body, which is spatially related to the SDRs, has not yet been drilled, but its P-wave velocity structure (e.g., White, 1992; White et al., 2008) suggests that it comprises magmatic material that has either underplated or intruded the crust. The origin of the magmatic material is not known, but most workers suggest that it was generated at great depth in the mantle, has risen due to buoyancy, and has either become trapped below Moho or intruded into the crust as sills. Where it has been imaged on seismic profiles, the crust between the SDRs and the lower crustal body is extremely thin, and, in at least one locality, it has been attributed to stretched continental crust.

Backstripping in volcanic margins shows that the tectonic subsidence has a similar overall form to non-volcanic margins. However, it is generally smaller in amplitude (Clift et al., 1995). This is to be expected because the emplacement of both SDRs and magmatic material below or within crust effectively thickens it so causes uplift. Indeed, the thickness of oceanic layer 2 increases fourfold in the region of the SDRs off Norway (Mutter et al., 1984). This uplift will compete with the subsidence due to thermal cooling of a rifted margin.

The addition of magmatic material within or below the crust disturbs the state of isostasy of a region, so we can estimate the amount of uplift that might result by balancing a column of crust that has been thickened with one that has not. We have for the case of magmatic underplating (Fig. 7.32) that

$$u\rho_w + T_c\rho_c + T_m\rho_m + r\rho_a = T_c\rho_c + v\rho_x + T_m\rho_m,$$

where u is the amount of uplift, v is the thickness of the underplate, $r = v - u$, T_c is the crustal thickness and T_m is the thickness of the sub-crustal mantle lithosphere,

$$\therefore\ u\rho_w + r\rho_a = v\rho_x,$$

$$u\rho_w + \rho_a(v - u) = v\rho_x,$$

$$\therefore\ u = v\frac{(\rho_a - \rho_x)}{(\rho_a - \rho_w)}. \tag{7.45}$$

We see from Eq. (7.45) that, if $v = 15$ km (the approximate thickness of the underplated body at Hatton Bank according to Fowler et al., 1989), $\rho_a = 3,200$ kg m^{-3}, $\rho_x = 2,900$ kg m^{-3} and $\rho_w = 1,030$ kg m^{-3}, then $u = 2.07$ km.

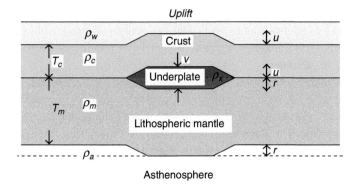

Figure 7.32 Simple model of the addition of magmatic material to the base of the crust by underplating. The pre-existing crust is assumed here to be of uniform thickness and density.

Equation (7.45) assumes, of course, an Airy-type response of the crust to underplating. In fact, the strength of the crust may limit the amount of uplift. We can compute the flexural effects by defining the appropriate wavenumber parameter that modifies the Airy response to produce the flexure.

$$\therefore U(k) = V(k) \frac{(\rho_a - \rho_x)}{(\rho_a - \rho_w)} \phi_e(k), \tag{7.45}$$

where $U(k)$ and $V(k)$ are the Fourier transforms of the uplift and the thickness of the underplate, respectively, and $\phi_e(k)$, the wavenumber parameter, is given by

$$\phi_e(k) = \left[1 + \frac{Dk^4}{(\rho_a - \rho_w)g} \right]^{-1}.$$

The addition of magmatic material to the crust, whether by underplating or intrusion, implies a re-distribution of mass that should be associated with gravity anomalies. Beneath a continental slope with no flexural strength, for example, underplating would widen the edge-effect low and reduce its amplitude (Watts and Fairhead, 1997). If the crust has flexural strength, however, then the low narrows and increases in amplitude. The spatial extent of underplating should be revealed in Airy isostatic anomalies, although they will be clearest in crust with a high flexural strength.

We have assumed so far that the addition of magmatic material causing uplift is less dense than the surrounding mantle. This would be the case if, for example, hot basaltic melt had risen and ponded beneath or intruded the crust as sills. However, as Maclennan and Lovell (2002) have pointed out, as a melt crystallises, cools and solidifies, its density would increase, causing subsidence. Indeed, examples of light crustal rocks which replace dense mafic rocks that cause excess subsidence have already been proposed by Royden et al. (1980) for Cape Hatteras, East Coast, US and by Shi et al. (2005) for the Baiyun Sag, South China Sea margins. Therefore, the addition of magmatic material to the crust involves a coupled uplift and subsidence, which might, as Maclennan and Lovell suggest, trigger a stratigraphic response and help explain transgressions and regressions that have previously been interpreted in terms of sea level change.

The question that remains is the extent of seaward dipping reflectors and the underplated or intruded magmatic bodies at rifted margins. Certainly, some margins, such as in the North Atlantic that were influenced by the Iceland plume in the Paleogene (e.g., Clift et al. 1995), are strongly magmatic. The magmatism is reflected in the subsidence and uplift history and the relatively thin post-rift sediment cover in these margins. Others, however, show little evidence of magmatism.

7.2.5 The Post-Rift Stratigraphy of Rift Basins

Backstripping, gravity and seismic modelling studies show that at some rift-type basins (e.g., East Coast, United States), the seismic Moho is at the depth that would be expected based on the margin's sediment loading and thermal subsidence and uplift history. If we know how sediment loading and thermal subsidence and uplift contribute to the seismically constrained crustal structure, it should be possible to go the next step and forward model the stratigraphy of a rift basin.

One problem in stratigraphic modelling is in specifying the distribution of sediment loads as a function of time. Ideally, we need palaeobathymetry information, since this, together with the sediment thickness at the time of deposition, gives the geometry of the upper and lower surfaces of an individual sedimentary layer. In practice, this information is rarely available and so we need to set up some simple models for the mechanics of sediment fill.

We have already discussed one such model, the clinoform model (see Section 4.4). In this model, sediments build outwards into a pre-existing water-filled basin by progradation. The model produces offlap on the landward side and onlap on the oceanward side, in the direction of progradation. An alternative model, in which sediments build up vertically by aggradation, is the steer's head model (Watts et al., 1982).

The steer's head model assumes that accommodation space is provided by a tectonic subsidence that is exponential in form, increasing from high rates early in basin evolution to low rates at later times. The tectonic subsidence is spatially restricted in the model such that the hinge line that separates normal thickness crust from stretched crust does not migrate with time (Fig. 7.33). Following an increment in tectonic subsidence, sediments are assumed to infill up to a horizontal surface. The amount of sediment added defines the load, which is then used, together with T_e, to calculate the flexure of the lithosphere.

Figure 7.34 shows the predicted stratigraphy of a 100-Ma-old so-called 'steer's head' basin[7]. The model is based on an exponentially decreasing subsidence and a T_e that is given

---

[7] The name 'steer's head' apparently originated in the oil and gas industry because of the resemblance of the model to Texas Longhorns.

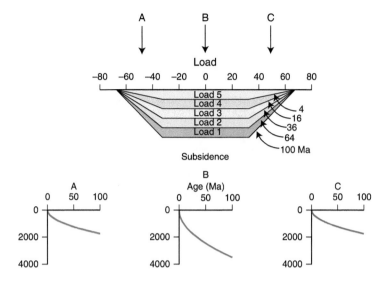

Figure 7.33 Simple model for the tectonic subsidence of a sedimentary basin as a function of age. The subsidence, which occurs about a fixed hinge line, is assumed to vary linearly with the square root of age. Sediments infill the water-filled depression up to a horizontal surface. Reproduced from fig. 3 of Watts et al. (1982) with permission of The Royal Society.

by the depth to the 300, 450 and, 600°C oceanic isotherms (as suggested, for example, by the oceanic T_e data plotted in Fig. 6.11). The thickness of a particular sedimentary unit is calculated by summing the load and the flexure at each subsidence increment. Because each increment corresponds to a particular time, the top and bottom surfaces of the unit can be regarded as a timeline or chron. The stratigraphy is then computed by summing the thicknesses of subsequent units, ensuring that their flexural effects on underlying units are considered. Sedimentary units that find themselves in the flexural depression of a new sediment load, for example, will be depressed to greater depths, while units that are in the bulge will be uplifted and may be eroded. Since the steer's head model assumes T_e increases with time, the net result is to produce an onlap of younger strata onto older strata at the edge of the basin.

While onlap is a feature of all steer's head models, the overall 'architecture' of the basin that is produced varies. The weak lithosphere case, for example (i.e., $T_e = Z_{300°C}$), produces a basin that is relatively narrow and deep compared with a strong case (i.e., $T_e = Z_{600°C}$) which produces a relatively wide and shallow basin.

Although erosion may contribute, it is the assumption that T_e increases with age, following basin initiation that causes onlap, since this is the only mechanism in the model that is able to produce a depocentre that is, in effect, widening with time.

The steer's head model makes several predictions concerning the stratigraphy of sedimentary basins. The most important are that the basin depocentre widens with time, the youngest strata occur at the edges of the basin and the edges of a basin are characterised by onlap.

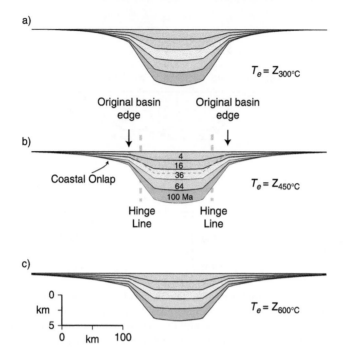

Figure 7.34 The predicted stratigraphy of a basin, based on the tectonic subsidence model in Fig. 7.27. The stratigraphy has been computed by assuming that the sediments represent a load on the underlying lithosphere, which subsides under their weight. The calculations assume a T_e that increases with the age of loading, as defined by the depth to the (a) 300, (b) 450 and (c) 600°C oceanic isotherms. Each model is associated with 'onlap' of strata at their edges that is due to the increase in T_e with age. Modified from fig. 4 of Watts et al. (1982) with permission of The Royal Society.

Some rift-type basins, for example Gippsland Basin (Fig. 7.35; James and Evans, 1971), Malay Basin (Madon, 2007) and Porcupine Bank (Tate et al., 1993), show some of the characteristics of a steer's head basin well. Others, for example Michigan Basin (Fig. 7.35; Sleep and Snell, 1976), Middle Amazon (Nunn and Aires, 1988) and Parnaíba Basin (Tozer et al., 2017; Watts et al., 2018), however, do not. These basins are characterised by offlap at the basin edges and, significantly, the youngest sediments are confined to the centre of these basins.

The observation that the youngest sediments in some rift-type sedimentary basins are confined to the basin centre led Sleep and Snell (1976) and Beaumont (1978) to suggest that the lithosphere should be modelled with a viscoelastic rather than an elastic plate. Such models predict that a basin depocentre would continue to deepen and narrow as viscous processes dominate over elastic ones and weaken the plate. Sleep and Snell (1976), for example, showed that the decrease in the width of the Michigan Basin during the Mid-Ordovician to Pennsylvanian could be explained by a viscoelastic model with a Maxwell relaxation time of 1 Myr and a viscosity that was

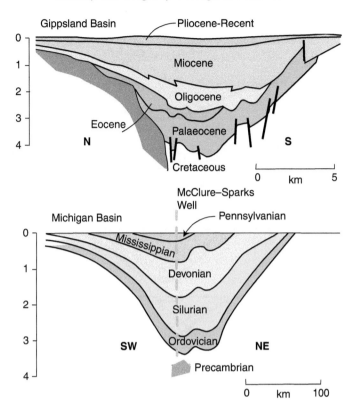

Figure 7.35 Stratigraphic patterns at the edge of the (a) Gippsland Basin, south-east Australia and (b) Michigan Basin, north-central United States.

in the range of 4×10^{23} to 4×10^{24} Pa s. Beaumont (1978) obtained a similar result using the decrease in width of the North Sea Basin during the Late Cretaceous to Recent.

In a related study, Sweeney (1977) argued that stratigraphic patterns in the Sverdrup Basin, which underwent distinct phases of tectonic subsidence in the late Palaeozoic and Mesozoic, were also consistent with the predictions of the viscoelastic model. There was evidence during the Late Palaeozoic and Early Mesozoic that peripheral bulges had migrated towards the basin centre and that the magnitude of the subsidence had increased through time.

Sloss and Scherer (1975), however, argued that despite showing offlap at their edges, many basins do, in fact, widen with time, as the steer's head model would predict (Fig. 7.36). They based their arguments on a detailed analysis of facies, lithology and the geometry of individual sedimentary units at the basin edges. The Michigan Basin in North America and the Moscow Basin in the Russian Federation were two basins singled out by Sloss and Scherer as showing a widening of the depocentre with time. The fact that the width of some basins appears to decrease with

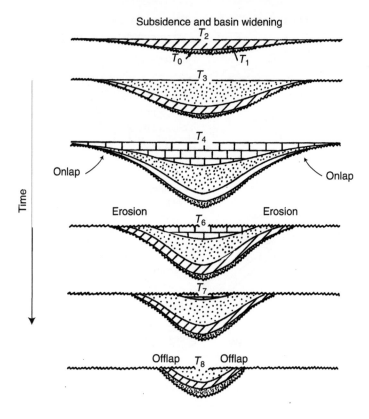

Figure 7.36 The cycle of deposition and erosion in a sedimentary basin in the cratonic interior. Reproduced from fig. 1 of Sloss (1963).

time was attributed by them to the effects of a basin-wide erosional event that had 'cut' to a deep stratigraphic level within the basin, thereby removing the evidence of onlap.

Haxby et al. (1976) showed the subsidence history of the Michigan Basin could be explained by an elastic model and did not require a viscoelastic model. They examined the geometry (i.e., depths and radial distance from the basin centre) of the Trenton Limestone (Mid-Ordovician) and Dundee formation (Devonian) and argued that the best-fitting T_e was 45 km, with a smaller value of 30 km earlier on and a higher value of 60 km later. Haxby et al. (1976) attributed the change in T_e during basin evolution to a high heat flow associated with the initiation of the basin.

The steer's head model assumes that sediments infill a basin up to a horizontal surface. However, at continental margins, sediments may infill only up to a particular palaeobathymetry. For example, in river deltas sediments prograde into a basin, and because they are unable to fill it completely, they build up a shelf, slope and rise.

Figure 7.37 shows the predicted stratigraphy for a wedge of sediments that load a preexisting continental margin. Like the steer's head model, accommodation space is created

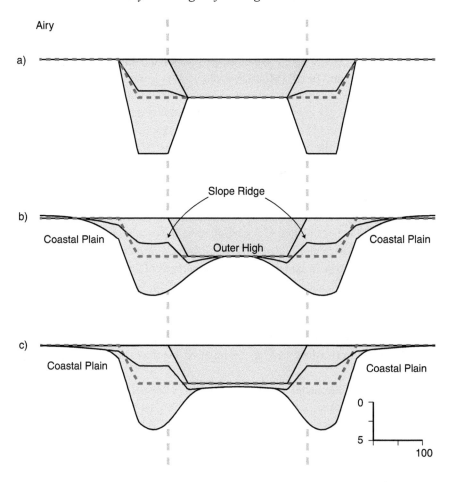

Figure 7.37 Simple models illustrating the stratigraphic response to a two-dimensional wedge-shaped load that is added to the edge of a continental margin. The initial configuration of the margin is shown as a red dashed line. (a) Airy type (i.e., $T_e = 0$), (b) flexure with a constant T_e and (c) flexure with a T_e that increases with time.

by a tectonic subsidence that produces a water-filled basin. However, in order to illustrate the role of flexure, three types of mechanical models have been assumed: an Airy-type (i.e., weak) crust (Fig. 7.37a), a strong lithosphere with constant T_e (Fig. 7.37b) and a T_e that increases with time (Fig. 7.37c). The figure shows that flexure is responsible for the bending of the crust beneath the sediment wedge, the formation of a bulge in flanking areas and, in the case of a T_e that increases with time, stratigraphic onlap at the basin edges.

Continental margins form by cooling of the lithosphere following heating at the time of break-up, and so their stratigraphic 'architecture' should reveal some of the features predicted by the steer's head model. We might expect them to show, for example, a gentle coastal plain, evidence of an outer stratigraphic high and, most importantly, stratigraphic onlap at their edges. Evidence for onlap is seen particularly well in the Jurassic post-rift sequences of the Atlantic margins of the East Coast, United States (Klitgord et al., 1988) and Brazil (Mohriak et al., 1990). As was the case for the interior rift-type basins, however, there are departures from these observations. For example, the Late Cretaceous to Recent sequences offshore south-west Africa (Gerrard and Smith, 1982) and the East Coast, United States show offlap rather than onlap. Again, Sleep and Snell (1976) argued that the offlap offshore the East Coast, United States was because of viscoelastic stress relaxation.

There is still a question, then, concerning the rheology of the lithosphere in rift basin settings. Is the lithosphere best modelled as an elastic or viscoelastic plate? We saw that in the case of the Michigan Basin, there is evidence from stratigraphic data that the basin has widened with time. Such a situation also appears to be the case for the Hudson Bay Basin, which like Michigan is circular in planform and involves Late Palaeozoic sediments. The main problem appears to be in explaining the offlap patterns offshore the East Coast, United States, south-west Africa, and in the North Sea. The offlap pattern usually involves Late Cretaceous to Recent sequences, and this, interestingly, was a time of global sea level fall. As Pitman first demonstrated, a fall in sea level of ~300 m since the Late Cretaceous could cause offlap on a slowly subsiding margin, which may explain why some sequences are dominated by offlap even when they have formed on margins that increase their flexural strength with age.

Since the steer's head model was first published, several increasingly sophisticated forward models have been constructed to explain the post-rift stratigraphic development of rift basins. Most of these models (e.g., Fig. 7.38; Lawrence et al., 1990; Reynolds et al., 1991; Steckler et al., 1999) incorporate some form of sediment loading and tectonic subsidence. And most include flexural isostasy, but some (e.g., White and McKenzie, 1988) assume an Airy-type crust together with a depth-dependent extension. They all consider the role of sea level changes. The main differences between the models are in the way that sediment transport and other factors such as carbonate productivity and slope stability are treated and their impact on the paleobathymetry and the 'architecture' of stratigraphic sequences.

7.2.6 Rift Flank Uplifts and Erosional Unloading

Many rift-type basins in passive margins, irrespective of whether they are of volcanic or non-volcanic origin, are flanked by uplifts in the adjacent continent. Figure 7.39 from Weissel and Karner (1989), shows that rift flank uplifts occur on the edges of each of the Gondwana continents, irrespective of their age. Furthermore, there is evidence that the rift flank uplifts of some young margins (e.g., Red Sea) are characterised by a sharp edge and a flank that rises gradually towards the coast. In contrast, old margins (e.g., south-east Brazil) are associated with a gentle edge and a flank that increases and then decreases towards the coast.

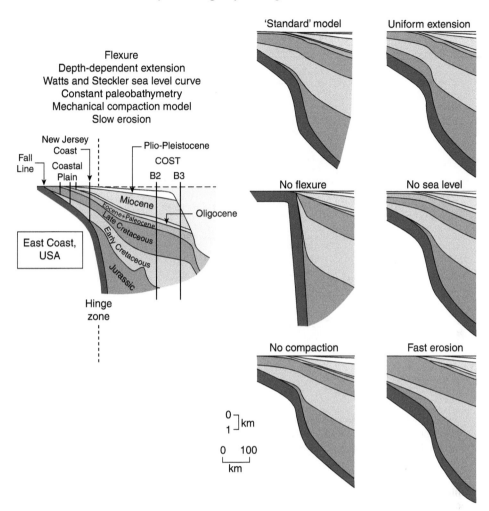

Figure 7.38 Comparison of the modelled stratigraphy of the coastal plain, shelf and slope offshore New Jersey to observations. The modelled stratigraphy shows a 'standard' model and a set of other models with different input parameters removed. Note that flexure is the main control on the basement configuration in the region of the hinge zone and the development of a coastal plain 'wedge'. Depth-dependent extension, however, is the main control on the onlap of Jurassic strata onto basement. Reproduced with modification from figs. 7 and 9 in Watts and Thorne (1984).

The presence of rift flank uplifts at margins with different break-up ages has led workers to consider models such as magmatic underplating (Cox, 1993) and flexure (Weissel and Karner, 1989; Gilchrist and Summerfield, 1990; ten Brink and Stern, 1992; van der Beek and Cloetingh, 1992). The advantage of these models is that they permanently raise the crust, which helps explain the persistence of rift flank uplifts for long periods of time. Thermal models, for example, can be ruled out in many cases because the uplifts are

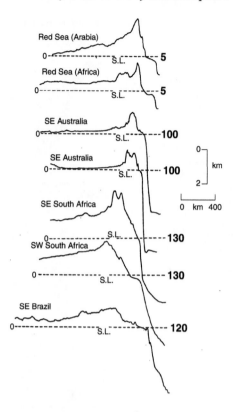

Figure 7.39 The uplifts that flank the margins of the Gondwana continents. Numbers in bold are approximate ages of rifting. Reproduced from fig. 1 of Weissel and Karner (1989), copyright by the American Geophysical Union.

maintained for much longer times than the thermal time-constant of the lithosphere. Although we know that some mid-plate swells (e.g., Hawaii) associated with deep mantle plumes are long-lived, it seems unlikely that dynamic effects in the mantle could explain the uplifts.

The characteristic shape of rift flank uplifts is a strong argument for a flexural origin. There is a difference of opinion, however, concerning the source of the loads required to explain the flexure. Weissel and Karner (1989), for example, suggest that the uplift is caused by flexural rebound following the mechanical unloading of the lithosphere during normal faulting, while van der Beek and Cloetingh (1992) attribute the uplift to variations in the depth of necking. Finally, Gilchrist and Summerfield (1990) and ten Brink and Stern (1992) have considered the flexural effects of erosional unloading due to contrasts in denudation rate once a permanent rift flank has formed.

It is difficult to envisage a situation whereby rift flank uplifts, once they are formed, are not subject to erosion. The question is, what is the precise form of the law that controls the erosion?

Studies of the sediment yield from drainage basins (e.g., Schumm, 1963) suggest that the rate of erosion is approximately proportional to the average elevation. Hence,

$$\frac{\partial h_o}{\partial t} = k_d h_{av},$$ (7.47)

where h_o is the initial elevation, h_{av} is the average surface elevation and k_d is the denudational coefficient (Hirano, 1976). The isostatic uplift, h, that results say from an unloading event of an initial elevation, h_o, is given (e.g., Eq. 7.5) by

$$h = h_o \frac{\rho_c}{\rho_m},$$

from which it follows that the rate of isostatic uplift is

$$\frac{\partial h}{\partial t} = \frac{\partial h_o}{\partial t} \frac{\rho_c}{\rho_m}.$$ (7.48)

The rate of change in the average surface elevation is given by the difference between the rate of isostatic uplift, which tends to raise the crust,[8] and the rate of erosion, which lowers it. The isostatic uplift rate depends on how quickly the crust and mantle respond to the removal of loads on its surface, while the rate of erosion depends on many factors, including climate, vegetation and lithology. Assuming that uplift and erosion are independent processes, then

$$\frac{\partial h_{av}}{\partial t} = \frac{\partial h}{\partial t} - \frac{\partial h_o}{\partial t}.$$ (7.49)

Substituting Eq. (7.48) in Eq. (7.49) gives

$$\frac{\partial h_{av}}{\partial t} = \frac{\partial h_o}{\partial t} \left[\frac{\rho_c}{\rho_m} - 1 \right]$$

or

$$\frac{\partial h_{av}}{\partial t} = -k_d \left[1 - \frac{\rho_c}{\rho_m} \right] h_o.$$

Hence,

$$h_{av_{(t)}} = h_{av_{(t=0)}} e^{-[1-(\rho_c/\rho_m)]k_d t},$$ (7.50)

where $h_{av_{(t=0)}}$ is the initial average elevation and $h_{av_{(t)}}$ is the average elevation at subsequent times. It follows from Eq. (7.50) that an initial isostatic uplift will, when subject to erosion, decay exponentially with time.

Equation (7.50) gives, of course, the Airy response to erosional unloading. The flexural response can be derived by defining a wavenumber parameter, which, when modified by the

---

[8] For this reason, the isostatic uplift is sometimes referred to (e.g., Molnar and England, 1990) as the rock uplift.

Airy response, gives the flexural response. If we assume an elastic model, then it follows that

$$h_{av_{(t)}} = h_{av_{(t=0)}} e^{-[1-(\rho_c/\rho_m)]k_d t \phi_e'''}, \qquad (7.51)$$

where ϕ_e''' is given by Eq. (5.41). We see that when $\phi_e''' \to 0$, $D \to \infty$ and the lithospheric plate is so rigid that there is no uplift. When $\phi_e''' \to 1$, $D \to 0$ and we get the weak plate or Airy case.

The observed rift flank uplifts that flank continental margins do not appear, however, to represent upwarps of a successively eroded flexural bulge. There are two observations that need to be accounted for when applying Eq. 7.51 to these margins (e.g., Fig. 7.39). Firstly, several flank uplifts (e.g., south-east and south-west South Africa) are superimposed on a broad topographic high. Secondly, the topography between the coast and uplift often slopes towards the sea.

A model that accounts for these observations is one where the rate of erosion is proportional to the slope of the elevation (Hirano, 1976). If the initial slope is steepest on the oceanward side (which would be the case for a flexural bulge), then erosion will result in the retreat of the initial slope into the continental interior.

Figure 7.40 shows an example of how the retreat of a pre-existing steep slope could account for rift-flank uplift at the margin of continents. The model assumes an initial topography that comprises a bulge that is superimposed on a topographic high. The bulge is bounded on its seaward side by a scarp that will retreat inland because the highest rates of erosion are in the region of the scarp. As the scarp retreats and material is removed from the bulge, the crust that remains experiences a flexural rebound. The flexural rebound results in uplift that is centred over the region where the maximum material has been removed.

Gilchrist and Summerfield (1990) have shown that a model of scarp retreat combined with the unloading of a plate with $T_e = 15$ km can explain topographic profiles of the Great Escarpment and flanking regions in south-west Africa. A similar model was used by ten Brink and Stern (1992) to explain the Great Escarpment of the Transantarctic Mountains and of southern Africa, and by Campanile et al. (2008) to explain the Western Ghats in Western India. The main difference with the Gilchrist and Summerfield model and the model illustrated in Fig. 7.40 was the introduction by ten Brink and Stern (1992) of spatial variations in T_e and by Campanile et al. (2008) of the effects of offshore sediment loading.

The presence of uplifts flanking rift-type basins has important geomorphological and sedimentological implications (Gilchrist and Summerfield, 1990). The presence of rift flank uplifts may explain the deflection of drainage systems away from rift basins and into the continental interiors. Moreover, sediment-charged rivers may only be able to breach a flexural uplift in certain, specific localities along-strike a continental margin. This may explain why the sediment supply to many rift basins, particularly in the margins of the South Atlantic, is dominated by a few widely spaced outlets of large river systems.

The Transantarctic Mountains, the largest rift flank uplift on Earth, formed in an extensional setting at the boundary between West and East Antarctica. This 3,000-km-long, 4,500-m-high uplift is unusual in that it has been subject to repeated glacial erosion events which have incised the surface of uplift to depths of up to 1.5 km below sea level and widths of up to 35 km. Stern et al. (2005) have used the surface of uplift to estimate the amount of material that has been removed from the incised valleys, and calculated the flexural rebound that has resulted. They found that climate-induced erosional unloading in this polar region accounts for ~50 per cent of the topography of the rift flank.

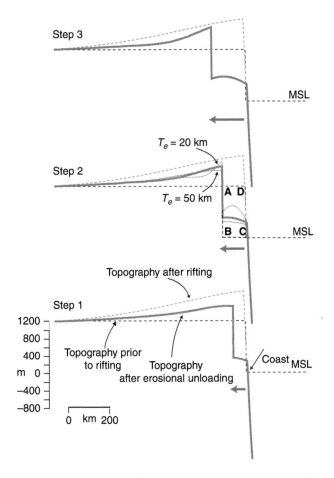

Figure 7.40 Simple elastic plate model for fault scarp retreat at a continental margin. Scarp retreat due to, say excavation, unloads the crust, causing a flexural rebound over a broad region. Thin dashed green lines show the topography after rifting, which was calculated for a stretching model with $\beta_s = 1.5$, $Z_{neck} = 25$ km and $T_e = 30$ km. Thick brown lines show the subsequent flexure assuming progressive unloading and an erosion time and constant of 10 and 25 Myr, respectively. The fine dotted brown lines in Step 2 show the flexure for $T_e = 20$ and 50 km, and ABCD indicates the total amount of the topography that has been removed at this step.

An outstanding question is why some rift flank uplifts in low-latitude settings show evidence of recent uplift long after rifting has ended. For example, the coast of Morocco shows raised marine terraces which are a few thousand years old and form part of a 'staircase' of terraces. One possibility is sea level fall (Pedoja et al, 2011), but it is difficult to explain the development of a staircase of terraces unless sea level has fallen steadily through the Pleistocene.

Another possibility is erosional unloading due to large-scale slope failures offshore. McGinnis et al. (1993) and Dykstra (2005) have shown the effects of flexural unloading on the slope on the shelf which causes uplift over a broad area and a flanking depression. When T_e is high and the amount of material large, then the uplift could extend onshore. Many margins are, in fact, characterised by steep, deep canyons, so it is instructive to see the effects of repeated incision events on the flexure. Figure 4.14, for example, shows the uplift onshore Morocco that resulted from removal of material from the canyons in the margin offshore.

7.3 Compressional Tectonics and Orogeny

As in the case for extension, compression is a major contributor to the Earth's surface features. In the Andes, Himalayas and Rockies, for example, compressional forces have built mountain belts up to 5 km in relief, 100–500 km wide and 10,000 km in length. Other, equally impressive expressions of compression are seen in the accretionary wedges of island arc–deep-sea trench systems in the Western Pacific, along segments of the Eastern Pacific and in the Caribbean. While orogenic belts were once thought of as the products of mainly vertical movements, they are now more generally regarded as the surface manifestations of compressional forces that arise because of plate motions.

7.3.1 Crustal Thickening

According to isostasy, mountain belts are underlain by deep, low-density crustal 'roots' that project downwards into the dense underlying mantle. Evidence for these roots has come from gravity anomaly and seismic observations. Mountain belts are associated with regional Bouguer gravity anomaly 'lows' that are indicative of local crustal thickening (e.g., Jiménez et al., 2012). Furthermore, seismic refraction data (e.g., Mooney et al., 1998) suggest that, in some cases, the Moho is deeper by as much as 20–30 km beneath mountain belts than under surrounding regions.

An important question is, what is the degree of isostatic compensation at orogenic belts? Undoubtedly, there is large variation in the state of isostasy between orogenic belts. Some are characterised by Bouguer anomaly 'lows' that correlate inversely with the topography, indicating that they are more or less completely compensated. Examples include the Andes, Zagros, Himalayas and western Alps. Others, however, are characterised by Bouguer anomaly 'highs' superimposed on the 'lows'. Examples include the Appalachians, Caledonides and the Eastern Alps. The presence of these highs suggests departures from complete compensation. More quantitative statements on the state of isostasy at orogenic belts, however, need to be based on seismic refraction data, and these are still surprisingly sparse in these regions.

One orogenic belt where the crust and mantle structure are reasonably well known is the 'bend' region of the central Andes (James, 1971; Whitman, 1994). Seismic and Bouguer anomaly data (e.g., Fig. 7.41b) are in general accord with the predictions of the Airy model. In particular, the crustal thickness and the amplitude of the Bouguer anomaly 'low' both increase as the topography increases. There is a hint, however, of departures. Topography in the range of 1–3 km tends to be associated with thicker-than-expected crust, indicating a degree of overcompensation. Similarly, topography > 3 km tends to be associated with smaller Bouguer anomaly 'lows' than expected, indicating a degree of undercompensation.

Irrespective of the degree of isostatic compensation, it is clear from the Central Andes that the continental crust is significantly thickened at orogenic belts. A key question, therefore, is, what is the origin of the thickening? According to early workers, the similarity between the seismically constrained crustal thickness and the predictions of the Airy model suggested continental crust is a fundamentally weak material that responds to external forces by flowing and thickening locally. If this were true, then we would expect some agreement between the amount of shortening deduced from field measurements in the highly deformed 'hinterland' of orogenic belts, and that implied by the Airy model of isostasy.

7.3.1.1 Shortening

The amount of shortening implied by the Airy model can be evaluated by considering crustal columns. We see, for example (Fig. 7.42), that if isostasy prevails before and after shortening, then

$$T_c a = T_t b. \tag{7.52}$$

Also,

$$r = \frac{h\rho_c}{\rho_m - \rho_c}, \qquad T_t = T_c + h + r,$$

$$\therefore T_t = T_c + h\left(\frac{\rho_m}{(\rho_m - \rho_c)}\right). \tag{7.53}$$

Substituting Eq. (7.53) into Eq. (7.52) and re-arranging gives

$$a = b\left(1 + \frac{h\rho_m}{(\rho_m - \rho_c)T_c}\right),$$

from which it follows that

$$a = b\left(\frac{T_c(\rho_m - \rho_c) + h\rho_m}{T_c(\rho_m - \rho_c)}\right).$$

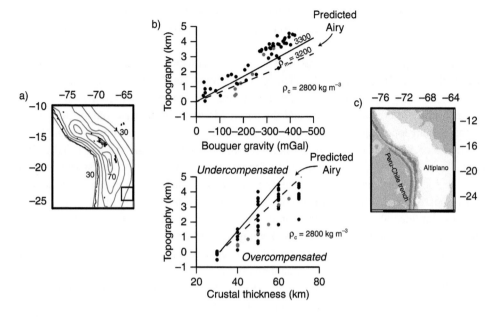

Figure 7.41 Topography, gravity anomalies and crustal structure of the central Andes. (a) Contours of crustal thickness (red solid lines) at 10-km intervals. Reproduced from James (1971), copyright by the American Geophysical Union. (b) Comparison of observed Bouguer gravity anomalies and crustal thickness with the predictions of an Airy model with $\rho_c = 2{,}800$ kg m^{-3} and a mantle density of $3{,}300$ kg m^{-3} (solid black line) and $3{,}200$ kg m^{-3} (dashed black line). (c) Topography.

The amount of shortening is $(a - b)$ and is related to the cross-sectional area, bh, by

$$a - b = \frac{bh\rho_m}{T_c(\rho_m - \rho_c)}. \tag{7.54}$$

The strain, ε, in the horizontal direction is given by the change in the width of the compressed region divided by the original width. In terms of a percentage strain, we then get

$$\varepsilon = \left(\frac{a - b}{a}\right)100.$$

Hence,

$$\varepsilon = \left(\frac{h\rho_m}{\left(T_c(\rho_m - \rho_c) + h\rho_m\right)}\right)100. \tag{7.55}$$

Equation (7.55) shows that a mountain belt of height $h = 5$ km, $T_c = 35$ km, $\rho_c = 2{,}800$ kg m^{-3} and $\rho_m = 3{,}330$ kg m^{-3} would be associated with a total amount of

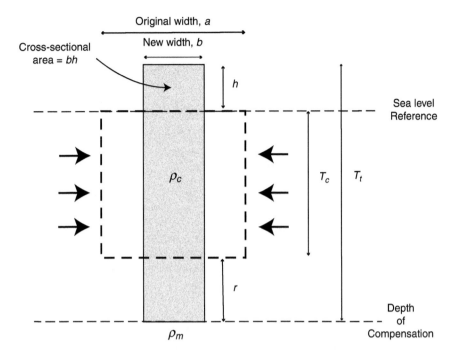

Figure 7.42 Schematic diagram showing how shortening of a column of crust by horizontal compression thickens the crust, which leads to an upward movement of the surface and a downward movement of Moho with respect to mean sea level.

shortening of $\varepsilon = 47.3\%$. Shortening of this magnitude has been reported in some orogenic belts, for example, New Guinea (Hill, 1991).

The central Andes, however, suggest complexities that cannot be explained by a shortening model based on Airy's model. Lamb and Hoke (1997) argue that, despite their high elevation, shortening in the central Andes since the late Cretaceous could be as small as 17 per cent. As Kley and Monaldi (1998) demonstrated, the shortening is less than expected, based on the Airy model. The differences are greatest for the central high area, where they are up to 30 per cent less (Fig. 7.43).

The question is, what is the cause of the discrepancy between observed and predicted shortening? Assuming the observations to be reliable, then the most likely explanation is that the Airy model, which invokes weakness, does not provide an adequate description of how the crust thickens in orogenic belts. Possible reasons for this are that other processes, such as magmatic underplating, crustal doubling due to overthrusting, and movements along deeply penetrating faults, also contribute.

Lamb and Hoke (1997) pointed out that the present-day topography of the central Andes in the 'bend' region is not the product of a single shortening event. Rather, the surface elevation was generated during two distinct time periods: one in the late Cretaceous to late Oligocene, and the other since the late Miocene. Geological reconstructions suggest that the

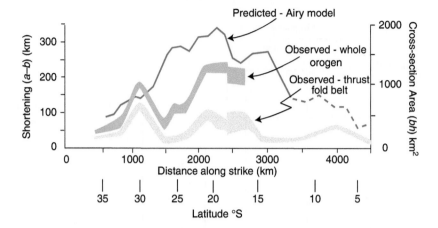

Figure 7.43 Comparison of observed and calculated shortening along-strike of the central Andes. The observed shortening (orange and yellow bands) is based on field data compiled in Kley and Monaldi (1998). The calculated shortening (solid red line) is based on an Airy model.

increase in surface elevation, by the end of the first period, was about 2 km. This suggests a shortening of 21–24 per cent, which is in reasonable agreement with the observed shortening of > 15 per cent. Since the late Miocene, however, there has been a further increase of 2 km in the surface elevation, yet the observed shortening has been < 2 per cent. This is more than 10 times less than the expected amount, suggesting that some process other than compressional thickening has contributed to the uplift of the central Andes region since the late Miocene.

One possible explanation for the surface uplift, suggested by Lamb and Hoke (1997), is magmatic underplating. Underplating since the late Miocene would, as demonstrated in Fig. 7.32, provides the means of elevating the pre-existing surface of the crust without shortening it.

While the central Andes region is an example of a high orogenic belt with a small amount of shortening, cases exist (e.g., DeCelles et al., 2002) of a high orogenic belt with a large amount of shortening. For example, the amount of shortening in the approximately 280-km-wide fold and thrust belt of the Himalaya in northern India and western Nepal has been estimated as ~600 and ~675 km respectively, which implies a shortening of ~68–71 per cent, well in excess of that expected for a mountain belt with an average height of 6.0 km (52 per cent) to 8.8 km (61 per cent).

7.3.1.2 Rheology

One possible explanation for this discrepancy between the observed and predicted shortening is that the Airy model of isostasy does not provide a full description of how Earth's crust deforms under horizontal compression. While an Airy model accounts for the first-order variation in the Bouguer gravity anomalies and the seismically constrained crustal thickness at orogenic belts, it ignores the influence on the deformation of strength inhomogeneities in the crust and upper mantle.

One of the first to link the topography of orogenic belts to the rheology of the crust was Chapple (1978). He developed a critical wedge model that includes the effects of body forces as well as externally applied compressional forces. He demonstrated that a sequence of thrust sheets will deform internally, building a topographic slope until a critical slope is reached, at which time the entire deforming mass would be at failure.

A difficulty with the critical wedge model, however, is its assumption of uniform plasticity. The continental lithosphere is rheologically stratified such that its strength varies with depth (e.g., Section 6.6). Therefore, a critical wedge model is unlikely to provide an adequate description of how an entire orogenic belt deforms.

To address this problem, England and McKenzie (1982) proposed a model in which the lithosphere responds to a horizontal compressional stress in a similar manner as would a thin viscous sheet that overlies an inviscid fluid. In their model, the rate of flow of the materials that make up the thin sheet is determined by the viscosity, which, in turn, reflects the integrated strength of the lithosphere. They found that a rate of flow that is proportional to the applied stress (which is the case for a Newtonian fluid with $n = 1$ in Eq. 6.22) could produce a 55-km-thick crust after ~32 Myr. A power law rheology (i.e., $n > 1$), on the other hand, involved a more restricted flow that locally increased the maximum crustal thickness to 90 km. England and McKenzie (1982) estimate that to explain a crustal thickness of this order, the integrated strength of the lithosphere in an orogenic belt must be high enough that it could support stresses of at least 30 MPa at strain rates of 10^{-15} s^{-1}.

Several workers have pointed out, however, the limitations of assuming a uniform rheology for the lithosphere. Seismic reflection profile data show, for example, that thrust and fold belts at the edge of orogenic belts are underlain by a basal décollement that dips in towards the highly deformed interior of the orogen. This vertical stratification in structure could explain why a shallow, thin-skin type of deformation characterises the thrust and fold belt, and deep-seated deformation characterises the interior of an orogenic belt.

Harry et al. (1995) constructed a model based explicitly on a vertically stratified rheology. Using a quartz diorite for the composition of the crust and dunite for the mantle, they showed that major décollements would be expected, on rheological grounds, to develop in orogenic belts at middle crustal and perhaps lower crustal levels.

Royden (1996) constructed a model that combined elements of both the England and McKenzie (1982) and Harry et al. (1995) models. She showed that a uniform-viscosity plate produces a flow in which deformation in the crust is coupled with that in the mantle (Fig. 7.44a). Partitioning of strain occurs as lower crust and mantle is entrained by motions at the Moho. The model produces a topography that is triangular in its shape, and there is a significant thickening of the crust. When a low-viscosity layer is introduced in the lower crust, however, shortening is focused at the edge of an orogenic belt, which causes a flattening of the topography in the hinterland (Fig. 7.44b). Royden argued that these two end-member cases could account for the present-day topography of the Alps and Himalayas.

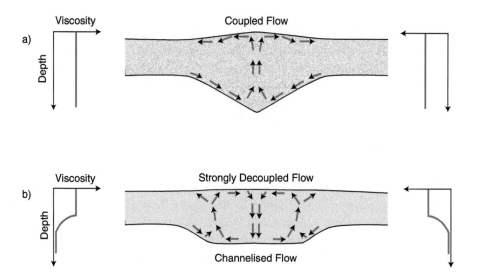

Figure 7.44 The formation of the topography and root of a mountain belt by viscous flow in the crust. Reproduced from fig. 7 of Royden (1996), copyright by the American Geophysical Union.

More recent studies have emphasised the contribution to the topography of processes such as those associated with a low viscosity (i.e., weak) mid-crustal channel flow that extrudes partially molten material from the underthrusting plate crust between two major ductile shear zones in the converging plate (Searle et al., 2006), a high-viscosity (i.e., rigid) indenter into a low-viscosity crust with a high-viscosity uppermost layer (Bendick and Flesch, 2007), and underplating of the crust by underthrusting of a converging plate (Nabelek et al., 2009). While these processes are probably not mutually independent, they provide insight into the mechanics of crustal thickening and clues as to how isostatic equilibrium might eventually be achieved in orogenic belts.

7.3.2 *Orogenic Belts and Erosion*

Once a mountain belt is formed, it is vulnerable to erosion. Erosion involves the removal of material from one part of the orogen and its deposition in flanking regions. Therefore, erosion disturbs the isostatic balance so that those portions of the crust that had material removed would experience uplift whilst portions that were loaded would subside.

Consider the situation, illustrated in Fig. 7.45, of an orogenic belt that is initially in isostatic equilibrium such that the mass excess of its surface elevation, h_0, is compensated for by the mass deficiency of a deep crustal root, r. Erosion reduces the surface elevation, disturbs the isostatic balance and results in uplift (the rock uplift in

Fig. 7.45) of the crustal material that remains. The uplift is given, for example, by Eq. (7.5):

$$h = \frac{h_o \rho_c}{\rho_m},$$

where ρ_c and ρ_m are the densities of the crust and mantle, respectively.

If erosion is a prolonged event, then the reduced surface elevation will, in turn, also be subject to erosion (Fig. 7.45). Eventually, enough material will have been removed that there will be no further uplift. The condition for this is when the total amount of material that is removed, S_e, reaches an amount given by $h_o + r$.

The Airy root r is related to the height h_0 by

$$r = \frac{h_o \rho_c}{(\rho_m - \rho_c)}.$$

Hence, it follows that

$$S_e = h_o \left[1 + \frac{\rho_c}{(\rho_m - \rho_c)} \right]. \tag{7.56}$$

We see from Eq. (7.56) that if $h_o = 5$ km, an orogenic belt that is initially in isostatic equilibrium could have upwards of 27.5 km removed from it by erosion! This amount of erosion would be sufficient to expose deeply buried, possibly lower crustal rocks (Fig. 7.45). In the case considered here, rocks that had been subject to overburden pressures of about 100 MPa would now be exposed at the surface.

There is evidence from geobarometry that large amounts of material have indeed been removed from the highly deformed interiors of orogenic belts. In the Alps, for example, exposures of amphibolite facies and high-pressure greenschist rocks suggest that erosion

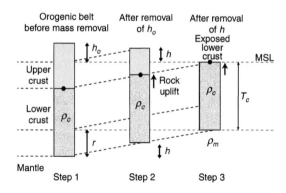

Figure 7.45 The isostatic adjustments that follow erosion of a orogenic belt. Erosion reduces the thickness of the crust, and so the surface of the Earth moves downward with respect to mean sea level (i.e., the geoid). Individual rocks remaining on or below the surface, however, move upward with respect to mean sea level.

has removed upwards of 16 km of overburden from the deformed crust (England, 1981). And Champagnac et al. (2007, 2008) have estimated that a significant fraction (~50 per cent) of the present-day geodetically measured vertical movement in the western Alps and of the present-day abandonment surfaces of foreland basins in the western Alps result from the isostatic response to enhanced erosion during Pliocene–Quaternary time.

The erosion of the Alps cannot, however, have been a simple response to crustal thickening. Rather, it appears to have been a complex time-dependent process. Erosion in the Alps began with the culmination of the orogeny about 30–40 Ma. At the present day, they rise approximately 1–2 km above flanking regions, suggesting, if we assume an Airy model, that a deep crustal root of some 5–11 km remains in the region. Therefore, there is the potential here for a further 6–13 km of erosion before the crustal thickness returns to the values that underlie its flanks.

The outstanding question is what are the factors that determine how erosion changes with time? When discussing rim flank uplifts in Section 7.2.6, we discussed a model in which the erosion rate was proportional to the average elevation. In this model, changes in average elevation are driven by differences in the rates of rock uplift and the rates of erosion. Average elevation increases, for example, when the rate of rock uplift exceeds the rate of erosion. This appears to be the case in the Swiss Alps (Gubler et al., 1981) and the central Himalaya (Jackson and Bilham, 1994).

Molnar and England (1990), however, have questioned whether changes in the average elevation are simply the result of differences between the rates of rock uplift and erosion. They point out that rock uplift and erosion rates may not be independent. Rather, increases in erosion rate (due, for example, to climate change) could themselves induce increases in the rate of rock uplift. Furthermore, they argue that in terrains where material has been removed by river valley incision, rock uplift could lead to absolute increases in surface elevation of flanking ridges, while the average elevation of the region may actually decrease.

According to Gilchrist et al. (1994), isostatic uplift due to river valley incision will be limited. They argue that the highest ridges in orogenic belts are therefore more likely to be the result of the tectonic mass flux that is added to an orogenic belt because of convergence, rather than isostatic uplift of a heavily dissected plateau. In their model, dissection accompanies an increase in average elevation due to tectonics. Initially, incision rates are high enough that there is little erosion of the ridges and isostatic uplift works in concert with tectonics to raise them. Eventually, however, the ridges are eroded, and in the absence of tectonics, the elevation of these features declines. If tectonic uplift is maintained, then a steady state may exist such that the tectonic mass flux that is added to the orogenic belt by plate convergence is balanced by the flux that is removed by erosion. This appears to be the case for the southern Alps, where, despite high rates of denudation (up to 5–8 mm a^{-1}), topography appears to have been stable for the past 2 Ma (Kamp and Tippett, 1993).

Several workers (e.g., Moretti and Turcotte, 1985) have suggested that, when considering erosion, it is important to consider the conservation of mass of a topographic surface. One approach is to assume that erosion is a linear diffusion process such that there is a relationship between the topographic surface and the transport velocity of material from an eroded region to neighbouring basins.

If the initial topography is h_o and the topography after diffusion is h, then we can write for the relationship between erosion rate and the gradient of the slope (e.g., Flemings and Jordan, 1990) that

$$\frac{\partial h_o}{\partial t} = \kappa_s \frac{\partial^2 h}{\partial x^2}, \tag{7.57}$$

where κ_s is the subduing coefficient (Hirano, 1976).

Consider a particular wavenumber of the topography such that $h = h_o \cos kx$. We then have

$$\frac{\partial h}{\partial x} = -h_o k \sin kx$$

$$\frac{\partial^2 h}{\partial x^2} = -h_o k^2 \cos kx$$

$$\therefore \frac{\partial h_o}{\partial t} = -h_o \kappa_s k^2 \cos kx.$$

This is the change in the initial topography due to diffusion. The change in average elevation is given by the difference between the isostatic uplift and erosion rates. Hence,

$$\frac{\partial h_{av}}{\partial t} = \frac{\partial h}{\partial t} - \frac{\partial h_o}{\partial t}.$$

.It follows from Eq. (7.5) that

$$\frac{\partial h}{\partial t} = \frac{\partial h_o \rho_c}{\partial t \rho_m}$$

$$\therefore h_{av_{(t)}} = h_{av_{(t=0)}} e^{-[1-(\rho_c/\rho_m)]\kappa_s k^2 t}. \tag{7.58}$$

This is the Airy response. The flexural response is given by

$$h_{av}(t) = h_{av}(t=0) e^{-[1-(\rho_c/\rho_m)]\kappa_s k^2 \phi_e'''' t}, \tag{7.59}$$

where ϕ_e'''' is defined in Eq. 5.46.

The effects of diffusion are illustrated in Fig. 7.46. The figure shows the effect of varying time, subduing coefficient and T_e on the evolution of an orogenic belt, 3 km high and 200 km wide, assuming that erosion is a diffusive process.[9] In general, the degree of modification increases with time, diffusion constant and T_e. Since the diffusion process involves mass transport, there is a region where material is removed and an adjacent one where it is deposited. The net result is to smooth out the edge of newly formed orogenic belt such that its profile is now more exponential in form.

It is the case that many orogenic belts have a distinct exponential form. The eastern edge of the Rocky Mountains, for example, is characterised by gentle topographic dip to the east that resembles the form of the model calculations (Fig. 7.47). The culmination of orogeny in

---

[9] A MATLAB® script, chap7_diffusion_of_topography.m, to calculate the flexural response to diffusion for different values of T_e, κ_s and t is available from the Cambridge University Press server.

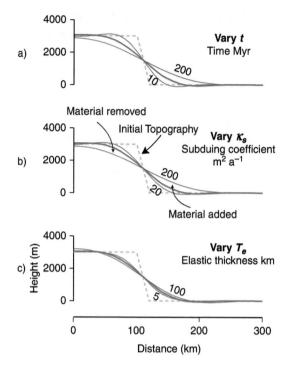

Figure 7.46 The modification of topography by slope diffusion. Dashed line shows the assumed initial topography. Solid lines show the modified topography and its dependency on variations in (a) time, t, (b) subduing coefficient, κ_s, and (c) elastic thickness, T_e. The thick red line shows a 'standard' model in which $t = 50$ Myr, $\kappa_s = 50$ m^2 a^{-1} and $T_e = 50$ km.

this mountain belt was in the Late Cretaceous, and so it is possible that slope diffusion has been a dominant factor in its subsequent evolution. Leeder (1991), however, has criticised the diffusion model because most applications are two-, not three-dimensional, and the parameterisation ignores sediment dynamics, including the proportion of suspended and washload sediment and dissolved load. Nevertheless, the main feature of a diffusion model of a region of erosional unloading flanked by a region of sediment loading is quite well replicated in the minimum and maximum topography profiles of the edges of orogenic belts, for example, the Longmen Shan (Densmore et al. 2005) and the Colorado and Wyoming Rockies (Fig. 7.47).

7.3.3 Foreland Basins

One of the most important manifestations of flexure in sedimentary basin evolution are the foreland basins that develop in front of advancing thrust and fold loads (e.g., Price, 1973). They are found in different tectonic settings, including regions of ocean/continent (e.g., Andes) and continent/continent collision (e.g., Himalayas). Because foreland basins are underlain by a variety of type of basement, their characteristic asymmetric wedge shape

Topography

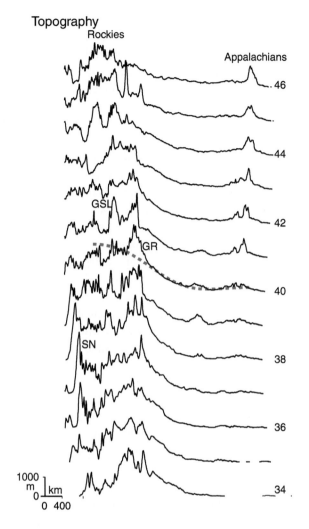

Figure 7.47 Topographic profiles of North America between latitudes 34° N and 46° N. The red dotted line is a diffusion-type exponential curve. GSL = Great Salt Lake, GR = Gore Ranges, SN = Sierra Nevada.

provides some of the best evidence that we have on the thermal and mechanical properties of the continental lithosphere.

Unlike submarine volcanoes, where the regional depth of the seafloor provides a reference, the loads that drive foreland basin subsidence are not always visible. We know, however, that the highly deformed edge of the orogenic belt that flanks a foreland comprises a complex stack of individual thrust sheets. The rocks that make up individual thrusts are transported across the Earth's surface by a complex process that involves the 'telescoping' of large blocks over 'ramps' in the underlying topographic surface (e.g., Elliot, 1976; Johnson, 1981). The thrusts are usually denser than the medium they displace and, therefore, will act as a load that flexes the underlying crust. The problem is that, during their transport, an individual thrust sheet may lose some of its capacity to load because of erosion, the consequences of which would be to cause crust that was once depressed, and maybe infilled by sediment, to rebound.

Ideally, the loads that make up a thrust and fold belt need to be restored to their original position before any attempt is made to use flexural models to predict the shape of the resulting foreland basin.

One of the first attempts to reconstruct the loading history of an orogenic belt was that of Jordan (1981). By matching hanging walls and footwall 'cut-offs' she was able to restore the pre-thrust geometry of the rocks that now make up the Late Jurassic to Early Eocene Idaho-Wyoming thrust and fold belts of the eastern Rocky Mountains. The pre-thrust geometry was then used to estimate the magnitude of the loads which, in turn, were used to calculate the flexure of a continuous elastic plate. By comparing calculations of the cumulative flexure of all the thrust loads with observations of the depth and width of the Cretaceous foreland basin that flanks the thrust belts, she estimated a T_e for the continental lithosphere of ~22 km.

Unfortunately, it is not always possible to restore accurately the pre-thrust geometry of orogenic belts. Although computer-based techniques of cross-section balancing using 'pin points' exist, they need to be constrained where possible by field-based as well as sub-surface information.

A simpler approach is to initially assume that the load is a surface load and is given by the present-day topography. The reasoning behind this is that the topography of the highly deformed outer edges of orogenic belts is, in large part, made up of successive thrust sheets. The main problem is that a single topographic load will yield a single depression, which – while suitable to compare with stratigraphic and Bouguer gravity anomaly data – will yield a T_e that is an average of the entire loading history, rather than one that reflects its variations, if any, through time.

It has been known for some time that while the interiors of orogenic belts are generally in a state of isostatic equilibrium, their edges show large departures. Airy isostatic anomalies, for example, are > +60 mGal high over the Lesser Himalaya and < −100 mGal over the flanking Ganga foreland basin (Qureshy, 1969). Similar belts of positive and negative isostatic anomalies have been mapped over the edges of the Zagros (Snyder and Barazangi, 1986), Andes (Watts et al., 1995), Appalachians (Taylor, 1989), Rockies (Simpson et al., 1986) and Oman (Ali and Watts, 2009) mountains. Positive isostatic anomalies indicate that the Moho is shallower than expected, and therefore that the crust beneath the edge of an orogenic belt is undercompensated. Negative anomalies, on the other hand, indicate that the Moho is deeper than expected, and therefore that the crust beneath foreland basins is overcompensated.

These departures from Airy-type isostatic equilibrium can be explained if we consider the topography is a load that has been emplaced on the surface of a thin elastic plate (Fig. 7.48). This is because elastic plate flexure, irrespective of whether the plate that is loaded is continuous or broken, can spread the compensation over a broader area, such that

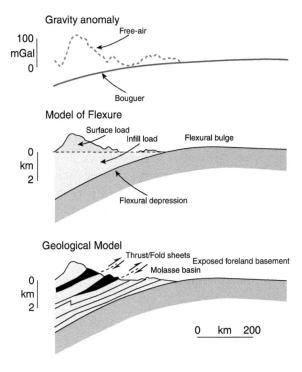

Figure 7.48 Gravity anomalies and flexure of the lithosphere caused by surface loading at the edge of an orogenic belt. Reproduced from fig. 8 of Karner and Watts (1983).

less compensation is provided beneath the orogenic belt edge and more beneath the flanking foreland basin.

A surface-loading approach has been adopted in several studies. Karner and Watts (1983), for example, used the topography of the Lesser and Greater Himalaya to calculate the flexure of the Indo-Australian plate. They used a broken-plate model, similar to the one used earlier by Hanks (1971) and Watts and Talwani (1974) at island arc–deep-sea trench systems (see Section 4.5), showing that such a model could explain the dip of the underthrust plate, the width and depth of the Ganga foreland basin, and the position of an 'outer topographic rise' in central India. The best fit to the Bouguer gravity anomaly data was for T_e of the underthrust plate in the range of 80–100 km and a plate break that is located about 250–300 km to the north of the Main Boundary Fault, in the region of the Indus-Tsangpo suture.

Similar conclusions were drawn by Lyon-Caen and Molnar (1983). These workers questioned, however, whether the underthrust plate extended beneath the entire Himalaya. They found that the load of both the Lesser Himalaya and the Greater Himalaya (which includes Mt. Everest) was too large to explain the observed gravity anomalies and Moho depth. They argued that an additional force was required to limit the deformation of the underthrust plate caused by the surface topography. Lyon-Caen and Molnar (1983) attributed this additional load to a bending moment that is applied to the end of the plate by cold, dense material beneath southern Tibet and that, in the process, elevates the underthrust plate.

Other orogenic belts where it has been demonstrated that surface loading plays a major role in contributing to foreland basin flexure are Papua New Guinea (Abers and Lyon-Caen, 1990; Haddad and Watts, 1999), the western Tien Shan (Burov et al., 1990), Zagros (Snyder and Barazangi, 1986) and the central Andes (Watts et al., 1995). We illustrate one of these cases in Fig. 7.49 which shows that surface loading with a plate break about 200 km west of the outermost thrust in the Bolivian Andes and a T_e of the underthrust Brazilian Shield of about 75 km explains the observed Bouguer anomaly over the adjacent foreland basin.

If surface loading is the main cause of foreland basin flexure, then Fig. 7.49 suggests that the Bouguer gravity anomaly should be generally negative over an orogenic belt. Indeed, many orogenic belts (e.g., Rockies, Andes, Himalayas) are dominated by Bouguer anomaly

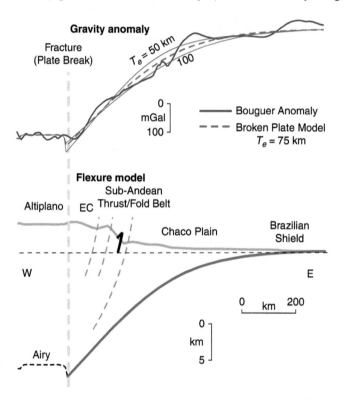

Figure 7.49 Simple model for the flexure of the lithosphere due to the load of the eastern Cordillera, central Andes. The effect of flexure is to broaden the Bouguer gravity anomaly 'low' associated with the orogenic belt and to depress the Brazilian Shield for distances up to 350 km inland of the edge of the thrust and fold belt. The calculations assume a broken-plate model with a break at the eastern boundary of the Altiplano. Reproduced from fig. 6 of Watts et al. (1995).

'lows' that, because of flexure, often extend over a broader area than the orogenic belt and into the flanking foreland.

Other orogenic belts, however, are characterised by Bouguer anomaly 'highs' (e.g. western Alps, Appalachians, Caledonides) that are generally superimposed on a longer wavelength low (e.g., Brooks, 1970). Perhaps the best-known example is the Appalachian gravity high in the eastern United States and Canada. This anomaly extends from Tennessee in the south-west, through New England, to the Gaspé Peninsula in the north-east. The high then continues offshore, across the Gulf of St. Lawrence and into Newfoundland. In Gaspé, the high correlates with the Shickshock Mountains and is interpreted as being due to both uplifted metamorphic rocks and high-level intrusions of ultra-basic rocks (Tanner and Uffen, 1960).

Gibb and Thomas (1976) noted similar Bouguer anomaly highs over the Precambrian Shield areas of northern Canada, as did Wellman (1978) over the shield areas of central Australia. These workers pointed out that many of the highs were flanked by a low and that the resulting positive–negative anomaly 'couple' generally coincided with a major tectonic boundary that separated different-age structural provinces. They interpreted the gravity anomaly couple as a result of the juxtaposition of high-density, normal-thickness crust against overthickened crust.

Karner and Watts (1983) suggested, however, that the Bouguer positive–negative anomaly couple was more likely to be a result of flexure. In particular, the positive anomaly represented a buried load, while the negative anomaly indicated the downward flexure of the crust by the load (Fig. 7.50). Using Gauss's theorem, Karner and Watts (1983) estimated the

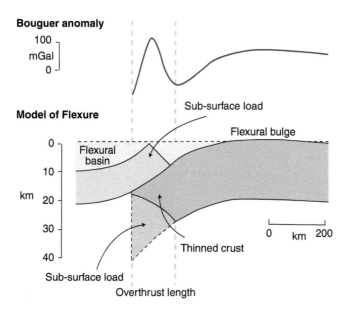

Figure 7.50 Gravity anomalies and flexure of the lithosphere caused by sub-surface loads within an orogenic belt. The diagram shows two buried loads: one due to the overthrusting of one crustal block over another, and the other due to crust and mantle that had been thinned prior to orogenesis. Reproduced from fig. 10 of Karner and Watts (1983).

mass of the buried load from the gravity high and then computed the corresponding flexure of a semi-infinite elastic plate. By comparing the combined gravity effect of the load and flexure with observations, they showed T_e varied from a low value of 25 km at the western Alps, through to intermediate values of 50 km at the eastern Alps, to values more than 100 km at the Appalachians.

Subsequent studies showed that buried loads were required to explain the pattern of Bouguer anomalies at several other orogenic belts. These included the Apennines (Royden and Karner, 1984), Carpathians (Royden and Karner, 1984), Transverse Ranges (Sheffels and McNutt, 1986), Urals (Kruse and McNutt, 1988), eastern Tien Shan (Burov et al., 1990), New Zealand Alps (Holt and Stern, 1994) and the Karakorum (Caporali, 1995).

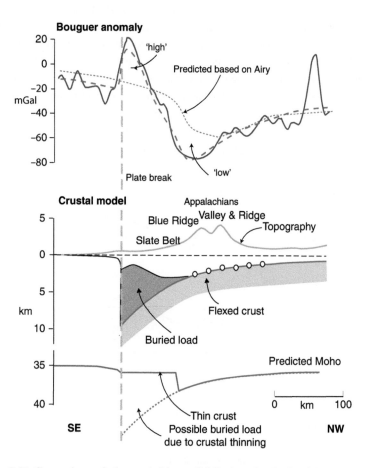

Figure 7.51 Comparison of observed (blue solid line) and calculated Bouguer gravity anomalies (red dashed line) along a profile that intersects the Appalachians in the region of Harrisonburg, Virginia. The calculated anomalies are based on a broken-plate model that is subject to both surface (i.e., topographic) and sub-surface loads. The steep gradient to the Bouguer 'high' requires that at least part, if not all of the buried load resides in the uppermost part of the crust. Reproduced from fig. 5 of Stewart and Watts (1997).

Although the concept of buried loads has been useful in modelling the gravity anomaly associated with orogenic belts (e.g., Stewart, 1998; Fig. 7.51), the geological significance of these loads is not always clear. The loads have been variously attributed, for example, to a dense downgoing slab (Royden and Karner, 1984; Sheffels and McNutt, 1986), to slivers of continental crust that were obducted onto the underthrust plate (Kruse and McNutt, 1988), ophiolite obduction (Ali and Watts, 2009) and intra-crustal thrust loads (Holt and Stern, 1994; Caporali, 1995; Lin and Watts, 2002). Finally, Burov et al. (1990) attributed the buried loads to either a statically supported thick lithospheric root or to dynamically induced viscous stresses. In those cases where the source was unknown, the buried load was simply replaced by a terminal vertical force or bending moment applied at the end of the under-thrust plate (Nunn et al., 1987; Royden, 1988a; Ruppel and McNutt, 1990).

In some orogenic belts (e.g., Appalachians, Oman), the Bouguer anomaly high is associated with outcrops of basic and ultra-basic rocks, confirming that the buried loads are located within the crust. Others, however, suggest a sub-crustal dense downgoing slab origin. Royden (1993) found, for example, that the Appenines, Carpathians and Hellenides lack the narrow, steep-gradient Bouguer anomaly high that would be expected if the required buried load was located within the crust. She showed that in the case of the Apennines the best fit to the observed gravity anomaly data was for a model in which the source of the sub-surface load was a dense downgoing slab associated with active

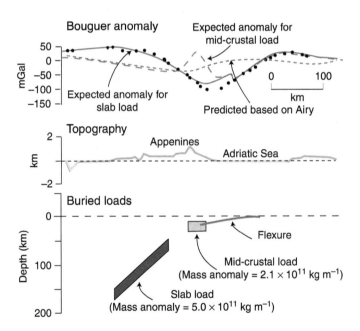

Figure 7.52 Comparison of observed and calculated Bouguer gravity anomalies along a profile that intersects the Apennines in the region of Bologna, Italy. The shallow gradient to the Bouguer 'high' cannot be explained by a buried load within the mid-crust and requires a much deeper source. Reproduced from figs. 5 and 12 of Royden (1993), copyright by the American Geophysical Union.

subduction of an Adriatic plate beneath Italy (Fig. 7.52). In addition, subduction of dense downgoing slabs in the past (e.g., the Farallon plate beneath a westward migrating North American plate during the Late Cretaceous) may have directly impacted the subsidence of foreland basins (e.g., Liu et al., 2008). Related loads that might influence surface topography in mountain belt regions and, potentially, the gravity anomaly and the subsidence associated with foreland basins, include mantle delamination (Bird, 1979) and convective removal of mantle lithosphere 'drips' (Gögüs and Pysklywec, 2008).

The discussion so far has focused on modelling the Bouguer gravity anomaly and, where observed, the base of the foreland sequence. Another observation that contains information on flexure is the stratigraphic architecture of the foreland basin fill. Foreland basins have a certain internal stratal geometry (e.g., onlap/offlap patterns) and basin-wide unconformities that also need to be explained.

One of the first to consider the stratigraphic response to foreland basin flexure was Beaumont (1981). He showed that flexure due to a series of migrating thrust and fold loads in the Canadian Rockies could explain the Upper Jurassic to Eocene evolution of the neighbouring Alberta foreland basin. His preferred model, however, was based on a viscoelastic plate rather than an elastic plate, as had been used by Jordan. He preferred this model because he found that an elastic model did not predict enough subsidence, predicted too large a topographic relief and, lastly, could not explain the dominance of offlap over onlap in the basin.

Beaumont argued that the best fit to stratigraphic data in the Alberta foreland was for a viscoelastic model with an initial flexural rigidity of 10^{25} N m (which corresponds to a T_e of ~67 km) and a Maxwell relaxation time of 27.5 Myr. The main feature of this model is that once the underlying lithosphere is loaded by sediment, it undergoes a stress relaxation. The result is to produce a foreland basin depocentre that increases in subsidence and decreases in width through time. This explained the large amount of subsidence, the absence of significant topographic relief in the orogenic belt and the dominance of offlap over onlap (Beaumont, 1981).

The viscoelastic model was applied by Beaumont to other foreland basins. In particular, Quinlan and Beaumont (1984) showed that the model could explain the stratigraphic architecture of the Appalachian foreland basin as well as details of its internal structure such as unconformity-bounded stratigraphic sequences. The sequences can be correlated over large areas of the eastern United States and had been variously interpreted as the result of global sea level change (Vail et al., 1977) or widespread tectonic events (Sloss, 1963). Quinlan and Beaumont (1984) argued that they were the products of stress relaxation in the underlying lithosphere. According to their model (Fig. 7.53), the unconformities were cut in the time interval, between the end of one stratigraphic sequence, the underlying lithosphere of which had already undergone stress relaxation, and that of the next one, which formed because of a renewed episode of thrust and fold loading.

An interesting feature of the work of Quinlan and Beaumont was their suggestion that flexure in one basin could influence the stratigraphic development of neighbouring basins. They showed, for example, that flexural interactions between the Appalachian foreland basin and the Michigan and Illinois intra-cratonic basins could explain the existence of the Kanakee, Findly-Algonquin and Cincinnati arches as well as the Jessamine and Nashville domes in the

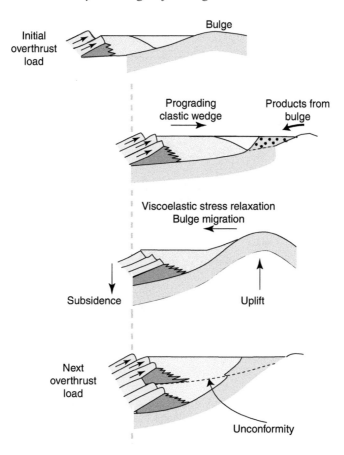

Figure 7.53 The model of Quinlan and Beaumont for the origin of basin-wide unconformities in foreland basins by the superposition of successive thrust and fold loads on the surface of a viscoelastic plate. Reproduced from fig. 18 of Quinlan and Beaumont (1984) with permission from the National Research Council, Canada Press.

eastern United States. These arches and domes, according to their model, existed in fluctuating conditions, alternately yoking together, and decoupling the foreland basin from one or both intracratonic basins in the interior of the North American plate (Fig. 7.54).

While Beaumont's work raised many issues concerning the role of flexure in the foreland basin evolution, the problem remains that the loads in their best-fit models are largely unconstrained. Quinlan and Beaumont (1984) estimated the load configuration in the Canadian Rockies and Appalachians, for example, by 'trial and error to explain the stratigraphic record'. Therefore, the loads have not been independently verified, say by gravity modelling. Quinlan and Beaumont (1984) recognised that '[r]esolution of the details of the necessary loads is poor' and that it should be possible to verify the loads 'through examination of their gravity signature'. Although they showed that, in the case of the Appalachians, the edge of the loads was coincident with the western boundary of the Appalachian gravity high,

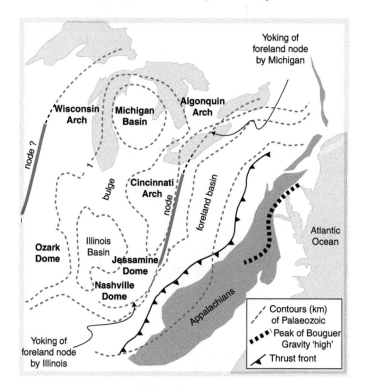

Figure 7.54 Flexural features of North America illustrating the yoking of the Appalachian foreland basin bulge by the Michigan and Illinois basins. Modified from fig. 22 of Karner and Watts (1983).

they did not calculate the gravity effect of these loads, which in places are up to 18 km high. There will always be some uncertainty about their conclusions, therefore, especially with regard to the applicability of the viscoelastic plate model.

Interestingly, most workers who have used an elastic plate to model foreland basins have used the gravity anomaly as a constraint on the magnitude and distribution of the driving loads. An elastic model can explain many of the details of the gravity anomaly of foreland basins. McNutt et al. (1988), in their review of the gravity-derived T_e estimates published up to 1988, do not bring up the issue of elastic versus viscoelastic plates at forelands, nor is there any reference to the stratigraphic studies by Beaumont and colleagues that are based on the use of a viscoelastic rather than an elastic model.

An outstanding question then is, how well does an elastic model explain the stratigraphic patterns that develop in foreland basins? A series of papers (Flemings and Jordan, 1990; Coakley and Watts, 1991; Sinclair et al., 1991) showed that when more-realistic models of basin fill are used, an elastic plate can still explain the observed stratigraphy of several foreland basins.[10] The

---

[10] A Matlab file, chap7_foreland_strat.m, to calculate the stratigraphy of a foreland using an elastic model and successive rectangular thrust/fold loads with one edge on the end of a semi-infinite (i.e. broken) plate is available from the Cambridge University Press server.

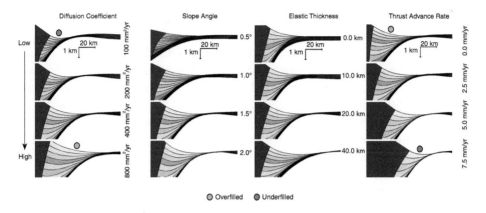

Figure 7.55 Stratigraphic models for the origin of a foreland basin by elastic plate flexure in front of an advancing thrust wedge. Basins tend to be underfilled if the diffusion constant is low and/or the thrust advance rate is high and overfilled if the diffusion constant is high and/ or the thrust advance rate is low. Reproduced from figs. 8–12 in Sinclair et al. (1991).

Sinclair et al. model is noteworthy because it demonstrated how parameters such as the diffusion coefficient, thrust wedge slope angle and advance rate, and, importantly, T_e combine to produce the stratigraphy of a foreland basin (Fig. 7.55). The model also demonstrates how temporal variations in these parameters, for example, the thrust wedge slope and advance rate, can produce unconformities within the basin (Fig. 7.56).

The diffusion model is of particular interest to foreland basin modelling since it allows the transport of sediment from source to sink to be specified, even in basins where the sediment supply is poorly known. For example, the difference between the slope after diffusion and the original slope gives the amount of material removed by erosion and the amount that is added by sedimentation for a particular time (Fig. 7.46b). Diffusion causes the area of the material that is removed and added to broaden with time. This effect would have important consequences for the development of stratigraphic patterns at the basin edge – causing offlap in the region of erosion and onlap in the more distal region.

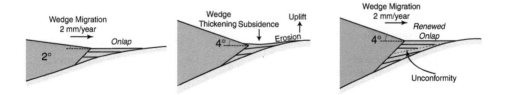

Figure 7.56 Simple model for the origin of basin-wide unconformities in foreland basins by thrust and fold load migration and wedge thickening on an elastic plate. Reproduced from fig. 17 in Sinclair et al. (1991).

Flemings and Jordan (1990), Coakley and Watts (1991), and Sinclair et al. (1991) have shown that an elastic plate flexure based on a diffusion model can explain the stratigraphic architecture of many foreland basins. These include basins that developed in front of fold–thrusts in the Brooks Ranges and western Alps, both of which are associated with gravity anomaly highs. In the case of the Colville and Swiss Molasse basins, the same T_e as had been deduced in earlier gravity modelling studies by Nunn et al. (1987) and Karner and Watts (1983) was used in the stratigraphic modelling. Thus, at least in these basins, an elastic model provides a self-consistent explanation of both the gravity anomaly and stratigraphic data.

The elastic model has now been widely accepted, and so foreland basin research has begun to focus on other issues. One of these concerns the influence that foreland basin flexure might exert on the structural styles that develop in a thrust and fold belt. For example, Watts et al. (1995) suggest that T_e of the underthrust foreland lithosphere may determine whether thin-skin or thick-skin tectonics dominate the style of deformation of

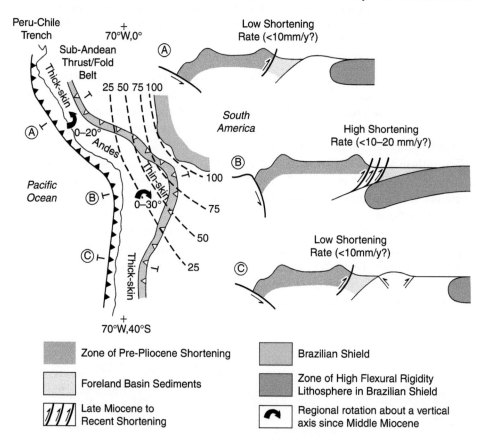

Figure 7.57 Relationship between tectonic style in the sub-Andean thrust and fold belt and the T_e structure of the underlying Brazilian shield. Thin-skin tectonics characterise regions of high T_e, while thick-skin tectonics characterise regions of low T_e. Sections A, B and C schematically illustrate how strain is partitioned along different segments of the Pacific/South America plate boundary. Reproduced from fig. 7 of Watts et al. (1995).

the thrust and fold belt. Observational support for this exists in the Andes of southern Bolivia and northern Argentina. Here, regions of high T_e appear to correlate with thin-skin tectonics, and those of low T_e with more basement involvement and, hence, thick-skin tectonics (Fig. 7.57).

Another issue is the influence that flexure might have on the drainage networks that develop in foreland basins. Garcia-Castellanos (2002) showed, for example, that although the Himalaya (Ganges), Alpine (Swiss Molasse, Po) and Betic (Guadalquivir) foreland basins show a diversity of drainage networks, most rivers are axial (i.e., parallel to the orogenic belt) and flow along the distal margin of the basin. This is also seen in the 'bend' region of the sub-Andean foreland basin (Fig. 7.58). North of the bend, the rivers that exit the central Andes have been directed into an axial river which flows sub-parallel to the steep western edge of the flexural bulge. South of the bend, however, exiting rivers flow across the basin and over the flexural bulge, possibly as the result of the construction of large megafans which have prograded out onto the flexural bulge overfilling the flexural depression. In both cases, rivers flow east off the crest of the bulge, several of which are directed into the Paraguay River, an axial river which flows south sub-parallel to the more gradual eastward slope of the bulge.

And another is the issue first raised by Desegaulx et al. (1991) concerning the inheritance of foreland basin flexure from a previous rifting event. A cornerstone of

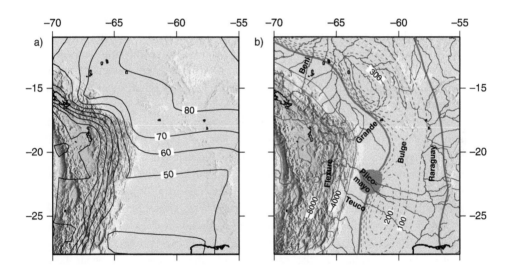

Figure 7.58 Relationship between the flexural bulge and drainage patterns in the bend region of the central Andes. The flexure has been calculated using a finite-difference method and assumes that the topography of the Andes represents a load on the surface of a spatially varying T_e plate. (a) T_e structure based on Stewart and Watts (1997). Contour interval, 10 km. (b) Flexure. Contour interval, 2 km. Blue lines show permanent rivers. Red lines show the flexure. Solid lines show the flexural depression. Dashed lines highlight the flexural bulge. Contour interval, 50 m. Filled brown area shows the extent of the Pilcomayo megafan, as mapped by Leier et al. (2005).

plate tectonics is that mid-ocean ridges, passive continental margins, and orogenic belts are genetically linked (Dewey and Bird, 1970). The Wilson Cycle, for example, predicts that as oceans open and close and continents collide, passive continental margins transition to active Andean-type subduction zones. The exact mechanisms responsible for the initiation of subduction are not clear, but sedimentary loading of weak, young (i.e., < 20 Ma) oceanic lithosphere (Cloetingh et al., 1982), fault reactivation (Erickson, 1993) and the proximity of some deep-sea trenches to passive margins (e.g., Ryukyu Trench and East China Sea margin) probably all play a role. Several lines of evidence support the transition. First, the highly deformed nappe structures of orogenic belts, when restored, reveal a subsidence history that bears a close similarity to modern passive margins (e.g., Wooler et al., 1992). Second, seismic reflection profiling of thrust and fold belts reveals thin crust and rift structures beneath some foreland basin sequences (Lillie, 1985; Flinch et al., 1996; Lin and Watts, 2002). Finally, backstripping studies of deep wells in foreland basins (e.g., Haddad and Watts, 1999; Ali and Watts, 2009; Watts, 2012) often show passive margin tectonic subsidence signatures in the stratigraphic sequences that immediately underlie foreland basins.

If passive margins and orogenic belts are genetically linked, as these observations suggest, and the lithosphere can store stresses for long periods of time, then foreland basins may, in some cases, inherit the thermal and mechanical properties of an underlying passive margin. As was pointed out in Section 7.2.3.4, the T_e structure of modern rifted margins is variable, with some margins having low values and others having high values. The cratonic interiors generally have very high values. If thrust and fold loads sequentially encroach on an underlying passive margin, as the numerical models of Stockmal et al. (1986) suggest, then it is not surprising that a foreland might inherit the flexural properties of the margin.

The inheritance of the T_e structure of passive margins may explain the wide range of widths of foreland basins that are observed. For example, low T_e at the Appenine foreland may reflect stretched lithosphere that was weakened during Tethyan rifting, while high values at the Appalachian foreland (> 100 km) may reflect encroachment of fold and thrust belts onto the rigid continental interiors. There is evidence that forelands reflect some of the details of rifted margin T_e structure. Stewart and Watts (1997) found evidence for lateral changes in T_e along profiles of the sub-Andean thrust belt and Chaco Plain foreland basin (Fig. 7.59). These changes, from low values in the west to high values in the east, resemble the changes that occur between stretched continental crust and oceanic crust at the East Coast, US rift-type margin and so the Tertiary-Quaternary foreland basin may have inherited the T_e structure of an underlying Cambrian–Lower Ordovician Pre-cordillera passive margin. Harry and Mickus (1998), on the other hand, have argued that the T_e structure of the Ouachita thrust belt and Arkoma foreland basin better reflects the transition between continental and oceanic crust at a transform-type margin.

Despite the inheritance of a rifted margin T_e structure that sometimes involves strong cratonic lithosphere, it is perhaps surprising that so few foreland basins are

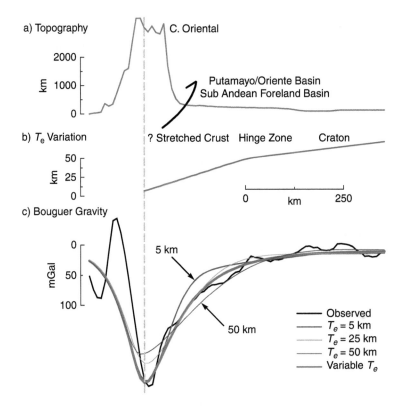

Figure 7.59 Comparison of observed and calculated Bouguer gravity anomalies along a profile of the sub-Andean thrust and fold belt in Ecuador. The observed anomalies are based on a compilation by GETECH. The calculated anomalies are based on a broken-plate model which is subject to a surface (topographic) load and an additional shear force of $0.50 \times 10^{12}\,\mathrm{N\,m^{-1}}$ that has been applied at the plate break. (a) Topography, (b) variation in T_e and (c) Bouguer gravity anomalies. Reproduced from fig. 7 of Stewart and Watts (1997).

preserved in the geological record. Some orogenic belts (e.g., the Caledonides of Scotland) show no evidence for a foreland basin at all. As pointed out by Burbank (1992) and Cederbom et al. (2004), however, orogenic belts are susceptible to weathering, and when it is a surface topographic load that causes the flexure that is removed, a foreland basin may rebound and be eroded. Cederbom et al. estimate, for example, that upwards of 2 km of material has been removed from the Swiss Molasse basin due to post-orogenic intensification of the moisture-bearing Atlantic Gulf Stream. The fact that some foreland basins are preserved in the Paleozoic and earlier record may therefore be more a testament to sub-surface than surface loads which remain within the crust and, hence, they and their associated forelands are protected from climate-induced erosion.

7.4 Strike-Slip Tectonics

We have so far been concerned with the role of flexure at extensional- and compressional-type plate boundaries. The other major type of plate boundary that we still need to consider is the strike-slip boundary. We see evidence of this type of boundary where the plates are sliding past each other in the oceans in the form of transform faults, and in the continents in the form of strike-slip faults.

7.4.1 Transform Faults

Transform faults are offsets that separate two segments of a spreading mid-ocean ridge system. They mark the site where two oceanic plates are actively slipping past each other. The slip varies from a few millimetres per year for some Atlantic transforms to > 15 mm a^{-1} for some fast-slipping transforms in the east-central Pacific. The main evidence for slip is restricted to within the transform fault region, not the region beyond it.

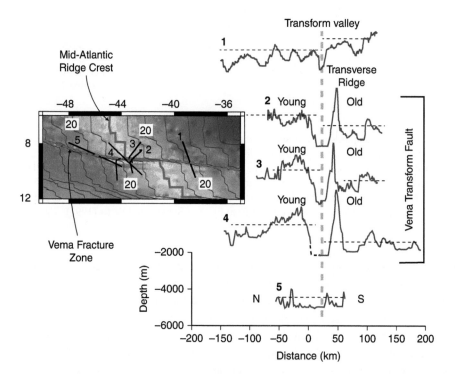

Figure 7.60 Bathymetry profiles of the Vema Transform Fault and Fracture Zone, Equatorial south Atlantic mid-oceanic ridge crest. The data were acquired during cruises V1603, V2502 and V2202 of R/V *Vema* and Leg 39 of M/V *Glomar Challenger*. The inset shows the bathymetry and magnetic isochrons (thin red lines) at 10-Ma intervals. Thick solid red line shows the ridge crest and transform fault, while the dashed red line shows the fracture zone.

Bathymetry maps show that transform faults are characterised by a highly irregular seafloor relief (Fig. 7.60). Typically, they comprise a transform valley that is flanked by one or more transverse ridges. Since they generally separate oceanic lithosphere of different age, transform faults are characterised by a 'step' in the regional depth of the seafloor. The transform valley is often infilled with sediments, and segments are generally deeper than expected for the age of the lithosphere, suggesting subsidence. Dredges from the flanks of the transverse ridges yield rocks of lower crust and upper mantle composition (i.e., gabbros and peridotites), suggesting uplift. Some transverse ridges are high enough to form islands (e.g., St. Peter–Paul islets in the central Atlantic), while others were once islands that have since subsided below sea level. The Romanche transverse ridge (Equatorial Atlantic), for example, rises some 4 km above the normal depth of the seafloor and is capped by a few hundred metres of thick shallow-water limestone (Bonatti et al., 1994).

Transverse ridges have been interpreted (e.g., Pockalny et al. 1996) as flexural features that form in response to normal faulting, as in the Vening Meinesz model illustrated in Fig. 4.7b. Indeed, flexure, as we will see in the next section, progressively accumulates as the offset plates cool and thicken with age along the length of a transform fault.

Transform faults form an integral part of the mid-ocean ridge system. They were attributed by Turcotte (1974) to thermal contraction cracks. He calculated the along-ridge tensile stress that accumulates as the lithosphere ages, cools and contracts. His calculations show that where thermal stresses exceed the strength of the lithosphere, cracks develop as transform faults.

Sandwell (1986) supported the view that transform faults are contraction cracks. Moreover, he showed that thermal contraction could explain the correlation between ridge crest segment length and spreading rate. The model predicts that crustal fractures should develop rapidly (i.e., within the first million years), because the thermal stress exceeds lithospheric strength, consistent with earthquake activity.

While crack models explain the existence of transform faults, they fail to account for the variations in bathymetry seen along them. The prevailing view, for example Collette (1974), is that the valley and transverse ridge are the products of stress changes across the transform fault. Changes in the direction of seafloor spreading – due, for example, to plate re-organisations – might give rise to changes in the extensional and compressional stress across a transform fault. This, in turn, could give rise to constructional volcanic features such as seamounts. 'Leaky' transforms are faults where such features are particularly well developed.

The notion, however, that transform faults are conservative plate boundaries in which magmatic material is carried along a single fault has recently been challenged by Wolfson-Schwehr et al. (2017) and Grevemeyer et al. (2021). For example, Grevemeyer et al. (2021) pointed out that transform faults are generally deeper than their fracture zone counterparts (which is unexpected from plate cooling and flexure models) and that the tectonics of the intersection of mid-ocean ridges and transform faults are strongly asymmetric, with an 'outside corner' associated with shallow relief and extensive magmatism and an 'inside corner' associated with deep nodal basins

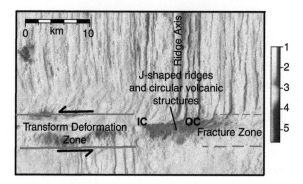

Figure 7.61 The Eastern Clipperton mid-ocean ridge – transform fault intersection at the fast-spreading (103 mm yr^{-1}) northern East Pacific Rise as revealed by swath bathymetry data. **IC** = inside corner, **OC** = outside corner. Color bar indicates the seafloor depth at 1-km intervals. Reproduced from fig. 1b in Grevemeyer et al. (2021).

and a lack of magmatism. The primary evidence for these observations, which are indicative of extension and crustal thinning within the transform, is swath bathymetry data, an example of which is shown in Fig. 7.61.

One feature that has formed by extension perpendicular to a transform fault is Atlantis Bank on the south-west Indian Ocean mid-oceanic ridge. The bank is described as an oceanic core complex, and comparisons with continental core complexes suggest it formed in an extensional setting by detachment faulting which resulted in the removal of upper crustal rocks and the exposure of lower crustal and mantle rocks in lineated domes. The dome region is punctuated by normal faults, and Baines et al. (2009) have proposed that some 66 per cent of the uplift of the dome is caused by flexure that followed mechanical unloading on one or more of these faults.

7.4.2 Fracture Zones

It has been known for some time that bathymetric features within a transform fault could be traced beyond it, along the projection of the offset. These features are called fracture zones. Fracture zones extend for up to several hundreds of kilometres from a mid-ocean ridge and, in some cases, can be traced into the continental margin, where they are obscured by large thicknesses of sediments (e.g., Sibuet and Veyrat-Peinet, 1980) and even into the continents, where they appear to align with zones of weakness which were reactivated by continental rifting (Sykes, 1978).

Geoid data derived from NASA's SEASAT mission show that fracture zones are characterised by a step of up to a few metres. Modelling studies show the step to be the consequence of contrasts in the density structure, due to differences in thermal age, on either side of a fracture zone. Detrick (1981) showed that the magnitude of the geoid step increased linearly with the average age of the offset. He also showed that the geoid step at the Mendocino Fracture Zone decreased systematically along the

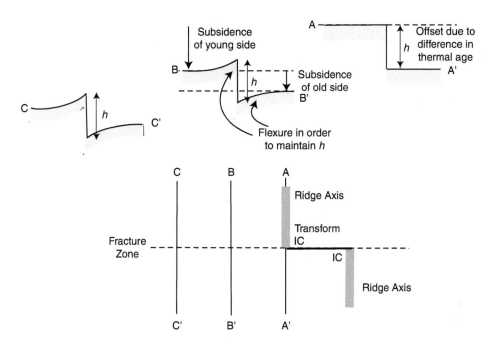

Figure 7.62 Sandwell and Schubert's model for the evolution of a fracture zone by the differential subsidence of young and old oceanic lithosphere and flexure. IC = Inside corner. Reproduced from fig. 1 of Sandwell and Schubert (1982), copyright by the American Geophysical Union.

fracture zone offset, from > 4 m at the eastern end, to 2–3 m over the older portion of the fracture zone, west of 150° W where the offset was only 1–2 m. Detrick (1981) cited this as evidence for a plate cooling model rather than a half-space model, the latter of which would predict a constant offset.

Sandwell and Schubert (1982) pointed out that because there is no slip at fracture zones, the height of the scarps that define the offsets should not vary with age. The oceanic lithosphere either side of a fracture zone, however, subsides at different rates. To maintain the same scarp height, therefore, they proposed that the lithosphere on either side of a fracture zone experiences flexure (Fig. 7.62). They found that while the flexure increases with age, its rate of increase decreases such that its contributions are most significant while the fracture zone is young. Also, they found that the flexure was asymmetric – the flexural wavelength being shorter on the younger side of the fracture zone than on the older side – and that the best fit to observed bathymetry profiles of the fracture zone was for a model in which T_e is given by the depth to the 450°C oceanic isotherm, although the data could also be explained by 300°C or 600°C.

Parmentier and Haxby (1986) and Haxby and Parmentier (1988) argued that, in addition to differential subsidence, thermal stresses could also contribute to the flexure

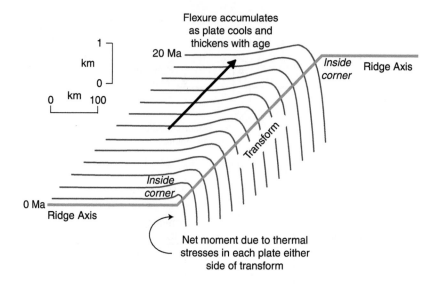

Figure 7.63 Parmentier and Haxby's model for the accumulation of flexure along a transform fault due to bending moments generated as the plates on either side cool and thicken with age. Reproduced from fig. 4 of Parmentier and Haxby (1986), copyright by the American Geophysical Union.

at fracture zones. They separated thermal stresses into two parts: the first due to changes in the temperature variation with depth, and the second, due to changes in the vertically averaged temperature. The first gives rise to thermal bending stresses and the second to thermal contraction stresses. The most important are the thermal bending stresses that arise due to the differential cooling of the plate (the top of the plate has a lower temperature than the base of the plate and therefore has cooled more) because these exert a moment on the plate that will flex in response. They argued that thermal bending stresses could account for the ridge–transform intersection deep and the associated topographic high that occurs in the ridge–transform inside corner. Also, they pointed out that flexure will progressively accumulate as the plate cools and thickens with age along a transform fault (Fig. 7.63).

Wessel and Haxby (1990) constructed a model for fracture zones that included flexure due to both differential subsidence and thermal stresses (Fig. 7.64). They showed that a combined model could explain bathymetry and geoid anomaly data at both slow- and fast-slipping fracture zones. By including the effects of lateral heat flow, they showed that the geoid anomaly with the isostatic edge effect removed can be explained by a flexure model with a T_e structure given by the depth to the 600°C oceanic isotherm (Fig. 7.65). Interestingly, this is the same oceanic isotherm that appears to control the depth of earthquakes along the Romanche and Chain transform faults in the Equatorial Atlantic, where Abercrombie et al. (2007) interpreted it as the maximum depth of brittle failure in the transform lithosphere.

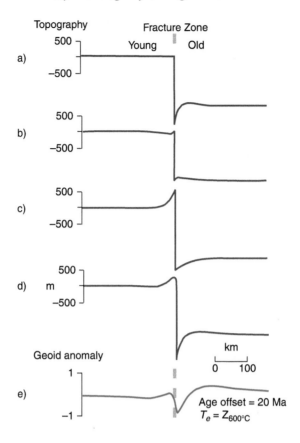

Figure 7.64 The contributions of (a) thermal stress, (b) lateral heat flow and (c) differential subsidence to the topography of a fracture zone with an age offset of 20 Ma. Also shown is (d) the total contribution and (e) the geoid anomaly, with the isostatic edge effect removed. T_e is given by the depth to the 600°C oceanic isotherm. Reproduced from fig. 10 of Wessel and Haxby (1990), copyright by the American Geophysical Union.

These studies therefore show that an elastic plate model in which differential subsidence and thermal stresses flex the lithosphere provides a simple explanation for the main bathymetric features of a transform fault. The use of an elastic model also provides an explanation for the preservation of bathymetry in fracture zones, which in the cases of some large offset fracture zones in the Pacific has lasted for long periods (> 100 Myr) of geological time.

7.4.3 Transform Margins

The break-up of continents does not always lead to the formation of linear trending rift-type margins that are separated by a mid-ocean ridge. At some margins, the break-up involves an active ridge crest that has one or more rectilinear offsets. Thus, rather than always

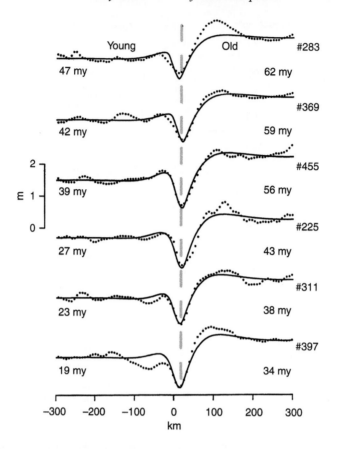

Figure 7.65 Comparisons of observed and calculated geoid anomalies across the Udintsev Fracture Zone in the south-western Pacific Ocean. The calculated anomalies are based on a model that combines the flexural effects of differential subsidence and thermal bending moments. T_e is given by the depth to the 600°C oceanic isotherm. Reproduced from fig. 12e of Wessel and Haxby (1990), copyright by the American Geophysical Union.

separating from the continent, a mid-ocean ridge may first pass by a lateral offset in the continental crust before separating. This process produces a particular type of continental margin – the so-called shear or transform-type margin.

Transform margins are usually found in conjugate pairs on opposite sides of an ocean basin. They are characterised (e.g., Scrutton, 1982; Lorenzo and Wessel, 1997; Clift and Lorenzo, 1999; Greenroyd et al., 2008; Parsiegla et al., 2009) by a steep continental slope and rise, thin sedimentary cover, marginal highs on both the oceanic and continental side of the initial offset and an abrupt change in the depth of Moho.

Several workers (Todd and Keen, 1989; Lorenzo and Vera, 1992; Gadd and Scrutton, 1997; Lorenzo and Wessel, 1997; Clift and Lorenzo, 1999) have considered the role of plate flexure at transform margins. Gadd and Scrutton (1997), for example, constructed a model

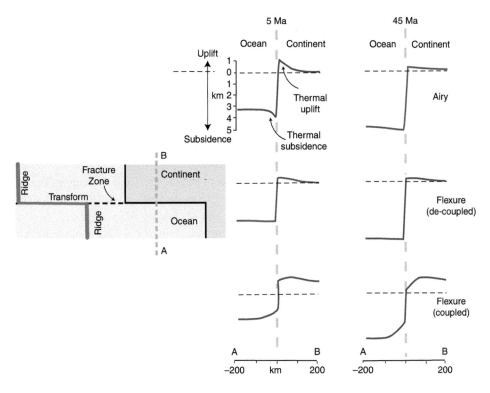

Figure 7.66 Flexural effects due to the contrast in thermal structure at a transform continental margin. Reproduced with permission from fig. 8 of Gadd and Scrutton (1997).

that considered both the thermal structure and flexure (Fig. 7.66). They showed that the temperature contrasts across a transform margin were large enough that on the cold, continent side of the margin, thermal uplifts of up to 1 km could form due to thermal expansion, while on the hotter, ocean side, thermal subsidence of up to a few hundred metres could occur due to thermal contraction. They also showed that the amplitude and wavelength of the thermal uplift and subsidence depended on T_e as well as the degree of mechanical coupling between the continental and oceanic crust and lithosphere.

An important result of modelling is that, although the thermal uplifts and subsidence at a transform margin are transient, elastic plate flexure ensures they persist for some time after an active ridge crest passes by a lateral offset of the continent. These tectonic movements would be expected to influence any sediments that make their way to the transform margin, for example, from the continents. We would expect, for example, that the stratigraphic response would be dominated by onlap as bulges decay and sediment that is initially deposited in relatively narrow depocentres is overstepped by younger deposits. Such interactions between sedimentation and tectonics might help to explain the stratigraphic data on the continent side of transform margins such as the one offshore Cote D'Ivoire–Ghana, where onlap is observed onto a marginal ridge in seismic reflection profiles (Basile et al., 1993; Peirce et al.,1996; Clift and Lorenzo, 1999)

7.4.4 Strike-Slip Faults

Strike-slip faults are an important structural feature of the continental crust. They are generally linear in outline but contain bends. Where two strike-slip faults overlap, there is either extension or compression, which results in 'pull-apart' and 'pop-up' structures, respectively.

Some pull-aparts contain a substantial thickness of sediments. The Los Angeles Basin, for example, is a pull-apart that is located within the San Andreas strike-slip fault system and is infilled by > 10 km of sediments. Turcotte and McAdoo (1979) suggested that the basin, based on its subsidence history, formed by thermal contraction. This is supported by the lack of a large-amplitude Bouguer anomaly low centred over the basin. The observed Bouguer anomaly is consistent with a degree of crustal thinning beneath the basin, and therefore we would expect to see some flexural bulges due to thermal stresses and/or sediment loading at its edge.

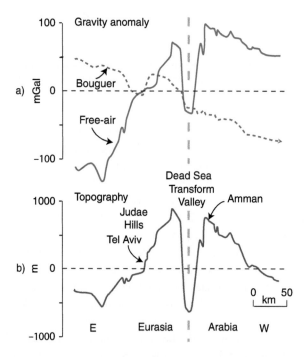

Figure 7.67 Bouguer and free-air gravity anomalies (a) and topography (b) associated with the Dead Sea transform valley along a transect between Tel Aviv and Amman. Reproduced from ten Brink et al. (1993), copyright by the American Geophysical Union.

Several workers have pointed out, however, that most pull-aparts are characterised by large-amplitude gravity lows, suggesting that they may not be compensated by crustal thinning. ten Brink et al. (1993), for example, have shown that the Dead Sea Basin, which is part of the Dead Sea transform that separates the Levantine and Arabian plates, is uncompensated. They point out that a strong correlation exists between the free-air anomaly and topography ($\sim$100 mGal km^{-1}) across this basin (Fig. 7.67), suggesting that the Moho is not involved. Indeed, the Bouguer anomaly low of >-100 mGal over the centre of the basin can be almost entirely explained by the sedimentary fill. The large sediment thickness ($\sim$10 km) suggests, however, that while the Moho may not be involved, there must have been some thinning of the lower crust. ten Brink et al. (1993) therefore regard the Dead Sea Basin as a mechanical basin rather than a thermal basin, and they therefore consider the rift flank–like topography, which forms the Judae and Hebron Hills in the west and the East Bank Plateau in the east, as reflecting stresses that act across the plate boundary.

A similar pattern of Bouguer anomaly lows as occur over the Dead Sea Basin have been reported from pull-apart basins in ancient strike-slip fault systems. For example, the Magdalen Basin (Watts, 1972) in the Gulf of St. Lawrence and the Orpheus Basin (Loncarevic and Ewing, 1963) on the inner continental shelf offshore Nova Scotia, which formed in response to Variscan–Alleghenian strike-slip movements, are both characterised by Bouguer anomaly lows of >-50 mGal. Seismic and gravity modelling show that these lows can be explained by the thickness of sediments and do not require that the crust has been thinned (or thickened) beneath the basins. These basins are therefore unlikely to be extensional basins with an elevated Moho, as was proposed by Dewey and Pitman (1982). Rather, extension beneath these basins appears to have been limited to the middle and lower crust, which has resulted in a flat Moho.

This view has generally been confirmed by Stern and McBride (1998), who reviewed seismic exploration of active and Paleozoic continental strike-slip zones and concluded that rather than being elevated beneath a pull-apart, Moho is either vertically offset or depressed downwards. Numerical models (e.g., Petrunin and Sobolev, 2006) suggest that pull-apart basins develop by growing along with strike-slip displacement on the main border faults: the main constraint on the thickness of sediment is not extension or compression but the depth to the detachment surface that separates the brittle upper crust from the ductile lower crust, or the brittle-ductile transition.

7.5 Intra-plate Deformation

We have so far been concerned with the role of flexure at the three main types of plate boundary. However, some of the best examples that we have of flexure are from the plate interiors, in regions of ice sheet, sediment and volcanic loads. The problem is that even in the tectonically quiet interiors of the plates, flexure may not always be acting alone. There is evidence, for example, that at some volcanic loads, other dynamic effects that have their sources deep in the mantle may contribute to the same observations (e.g., topography or bathymetry and gravity anomaly) that are used to constrain flexure.

7.5.1 *Intra-cratonic and Intra-continental Basins*

While some examples of intra-cratonic sedimentary basins, such as the North Sea, are clearly associated with rifting, others are not. The Michigan, Illinois, Williston, Hudson Bay, Congo and Paris basins, for example, differ from rift-type basins in several important respects. These include the long-time interval between rifting and the initiation of basin subsidence, the absence of uplift immediately prior to basin subsidence, the concentric nature of their outcrop, the lack of seismic and gravity anomaly evidence for crustal thinning, and a subsidence history that appears linear, rather than exponential, in form.

The Michigan Basin is considered by many as the arch-typical intra-cratonic basin. The basin is circular in outline, has a generally linear subsidence and has had a long geological evolution. McGinnis (1970) argued, based on the gravity anomaly, that Michigan is a flexural sag basin that formed in response to downward-acting loads in the crust (Fig. 7.68). Haxby et al. (1976) examined stratigraphic profiles of the basin edge and concluded that the average T_e of the basin was 40–50 km. These values are relatively low when compared with the shield areas of North America, and so they therefore attributed them to high heat flow. However, rather than attributing the high heat flow to heating and crustal thinning at the time of initial rifting, Haxby et al. (1976) suggested that it was caused

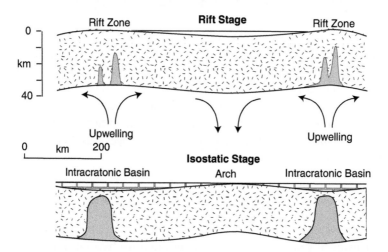

Figure 7.68 McGinnis's model for the origin of intra-cratonic basins: mantle upwelling and the intrusion of dense magmatic rocks into the lower crust. Reproduced from fig. 9 of McGinnis (1970), copyright by the American Geophysical Union.

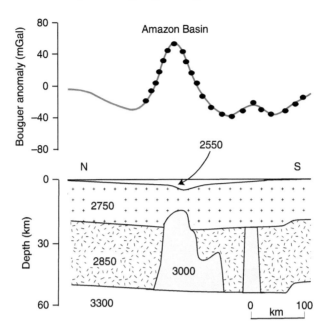

Figure 7.69 Bouguer gravity anomaly profile of the Middle Amazon Basin in Brazil. The basin correlates with a large-amplitude gravity 'high' that is interpreted as being due to a dense body in the lower crust. Reproduced from fig. 6 of Nunn and Aires (1988), copyright by the American Geophysical Union.

by a diapiric intrusion of hot asthenospheric material that metamorphosed the gabbroic lower crust to eclogite. The subsidence, in their view, was caused by the resulting increase in the density of the lower crust.

A similar model was proposed by Fowler and Nisbet (1985) to explain the subsidence of the Williston Basin. Here, the model is supported by seismic refraction data (Hajnal et al., 1984), which show a high-velocity upper mantle beneath the basin, and seismic reflection profile data (Baird et al., 1995) that show a poorly defined Moho. The seismic data also show dipping reflectors in the lower crust that, in many cases, continue uninterrupted across the depth-converted reflection Moho and into the upper mantle.

Both the Michigan and Williston basins are associated with a central Bouguer anomaly 'high' that is flanked by 'lows'. The Williston Basin has the broadest high (> 45 mGal), while the Michigan Basin has the narrowest. These highs probably reflect a dense lower crust (e.g., Hinze et al., 1992).

The Illinois Basin is also associated with a central gravity high, suggesting a similar origin to the Michigan and Williston basins. Likewise, the Amazon Basin has a wide central high of up to +50 mGal that is flanked by lows of up to –45 mGal (Fig. 7.69). Nunn and Aires (1988) have modelled the high over this basin as dense material that has intruded or replaced lower continental crustal material.

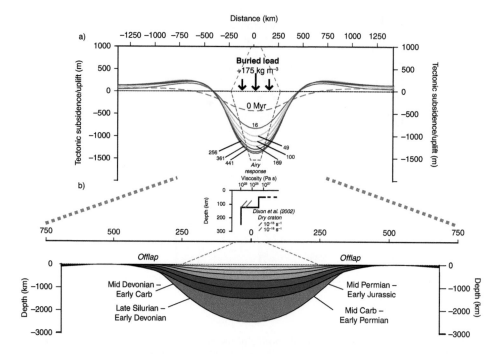

Figure 7.70 Simple model for the origin of the Parnaíba cratonic basin in north-east Brazil by sub-surface (buried) loading of a multi-layered viscoelastic plate. (a) Tectonic subsidence and uplift as a function of age since basin initiation. The water-filled basin is initially wide and then narrows as viscous effects become important. However, the basin does not approach the Airy case even after 440 Myr have elapsed. (b) Assumed viscosity structure (solid black line) compared to the theoretical results of Dixon et al. (2004) for a dry craton and modelled stratigraphy. Reproduced from fig. 14 in Watts et al. (2018).

One of the best surveyed intra-cratonic basins which has a central gravity 'high' flanked by 'lows' is the Parnaíba basin in north-east Brazil. The basin is of Silurian–Carboniferous age and is flanked on its western, eastern and southern sides by Neoproterozoic and Archaean cratons. Seismic reflection data reveal no obvious extensional structures such as a rift or its associated hanging wall and footwall uplifts beneath the basin centre, and seismic refraction data show a crust that is as thick or thicker beneath the basin centre than either of its flanks (Tozer et al., 2017). The sediment thickness is modest (~3 km) and at a depth of 17–25 km beneath the basin centre is a prominent, laterally continuous, mid-crustal reflector which has been interpreted as the top of a high P-wave velocity and dense magmatic body formed following intrusion of melt in the lower crust, cooling and densification. Backstripping of well data reveals what appears to be an exponentially decreasing tectonic subsidence, which Tozer et al. (2017) interpreted as the result of flexural loading by the dense magmatic body of a multi-layered viscoelastic plate. Watts et al. (2018) showed such a model to be consistent with the amplitude and wavelength of the sediment-corrected Bouguer gravity anomaly and with stratigraphic data which show offlap at the edge of the basin (Fig. 7.70).

Other intra-cratonic basins lack a central high and correlate with Bouguer gravity anomaly 'lows'. Some (e.g., Central Australia, Paraná) have basin-wide lows. Others are mainly 'lows' with either a narrow (e.g., Hudson Bay, Paris) or a wide (e.g, Congo) central high. The 'lows' appear to rule out models in which the lower crust has been intruded by dense rocks, since this model would produce a gravity high that, while competing with the low associated with the water, would work with and enhance the high due to the sediments. We may also rule out crustal thinning, since such a model would have a similar effect on the gravity as would a dense lower crust. Therefore, either the base of the crust beneath these basins is flat lying or, in the case of the largest lows, it is depressed downwards into the dense underlying mantle.

The Hudson Bay Basin (Fig. 7.71) has a particularly well-developed gravity 'low'. The basin, which is mainly Ordovician to Devonian in age, is elongate in shape and its subsidence appears to have been linear through time (Quinlan, 1987). Given its location on the North American plate, the gravity low might be a result, at least in part, of incomplete isostatic rebound that followed the Laurentide de-glaciation event. According to James (1992), however, > 70 per cent of the gravity anomaly over Hudson Bay is due to processes unrelated to glacial rebound.

One possible explanation for the gravity 'lows' is that some intra-cratonic basins form as a result of cold, downwelling, convective currents in the underlying mantle (e.g., McGinnis, 1970; Fig. 7.68). Pari and Peltier (1996) have proposed, for example, that the lithosphere beneath Hudson Bay is dynamically depressed by convective downwelling in the under-lying mantle.

A similar interpretation has been advanced by Hartley and Allen (1994) to explain the prominent 'low' observed over the Congo Basin. Unlike Hudson Bay, which, except for a central horst, shows no evidence of rift structures, the pre-Mesozoic history of the Congo Basin has been interpreted to be the result of Late Proterozoic rifting followed by a Palaeozoic thermal subsidence (Daly et al., 1992). Downey et al. (2009) estimated the long wavelength admittance over the basin to be ~50 mGal km^{-1} and so interpreted the 'low' (e.g., Fig. 5.34) in terms of the downward flow of a dense body in the lithosphere, consistent with the anomalously high shear-wave velocity data beneath the basin.

The Paris Basin shows similarities to both the Hudson Bay and Congo basins with its strongly concentric outline and its broad gravity low with a narrow or absent central high. Yet, as Brunet and Le Pichon (1982) have shown from backstripping, the tectonic subsidence is exponential in form and can be explained by a thermal model in which the continental crust was thinned during the Permo-Triassic by a β_s factor of 1.3. Moreover, there was significant rift flank topography, similar to what is observed in continental rifts, during the Permo-Triassic when sediment deposition was limited to a relatively narrow region of the basin. Subsidence data suggest a crustal thickness of 24 km beneath the centre of the basin, assuming a normal (i.e., unstretched) thickness of the crust of 31.2 km. Yet, seismic refraction data show that the Moho is at a depth of about

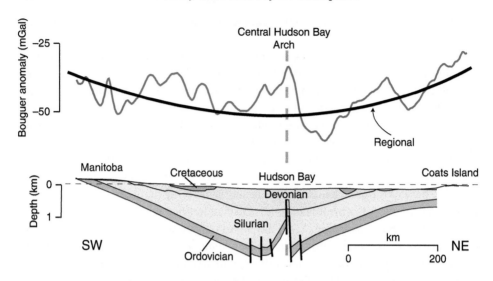

Figure 7.71 Bouguer gravity anomaly profile of the Hudson Bay cratonic basin. The basin correlates with a central gravity 'high' that is superimposed on a broader gravity 'low'. The summary geological cross section is based on Sanford (1987).

35 km in the region of the basin, which is significantly deeper than the depth of Moho predicted from backstripping.

A similar situation of a seismic Moho deeper than the backstrip Moho exists for several other basins in the south-west approaches to the United Kingdom. For example, the North and South Celtic Sea basins are both associated with large-amplitude gravity anomaly lows. As Dyment (1990) has shown, both the seismic-reflection Moho and the gravity-constrained Moho are deeper than that predicted from backstripping wells in the basin, and thermal and sediment loading considerations.

These considerations indicate complexities in intra-cratonic basin evolution. While some basins may well be due to sub-surface loading in the lower crust or some form of convective downwelling in the upper mantle, others show evidence of a rifting origin. Normally, rift basins – because of their association with thin crust – correlate with near-zero or positive gravity anomalies. Basins such as the Paris Basin, which has both a rift component and a gravity low, therefore still need to be explained. Either the initial crustal thickness was thicker than normal beneath the basin or the seismically constrained Moho is a relatively young feature. The latter possibility could arise if, following rifting, the thinned crust had been underplated by magmatic material, or the weak lower crust had spread horizontally, at the expense of mantle material.

Another possible explanation for the gravity lows and, hence, the basins is some form of large-scale folding. Weissel et al. (1980), for example, have mapped a series of gravity 'highs' and 'lows' in the Central Indian Ocean that correlate with 'highs' and 'lows' in the bathymetry and the underlying oceanic basement (Fig. 7.72). The relationship between gravity anomaly and bathymetry is consistent with a warping of the crust, and Weissel et al. (1980) interpreted the alternating 'highs' and 'lows' as the consequence of folding of the oceanic lithosphere.

There is evidence that the deformation mapped in the central Indian ocean continues to the east, across a region of high seismicity south of the Java–Sumatra Trench and towards the Australian continent (Sykes, 1970).

In central Australia, there is a striking belt of sub-parallel east–west-trending gravity anomaly lows and highs. The lows correlate with deep sedimentary basins of Meso- to Neo-Proterozoic age, whilst the highs correlate with intervening basement blocks.

Lambeck (1983) suggested that the intra-cratonic basins of Central Australia were, like the deformation in the central Indian Ocean, the products of far-field stresses driving lithosphere-scale folding (Fig. 7.73). Rather than using an elastic plate model, which would predict large bending stresses, he used a viscoelastic plate model like the one used earlier by Sleep and Beaumont to model rift-type sedimentary basins. The advantage of this model is that the compressive stresses required to produce the folding are much less than in the elastic model since the elastic deflection, while initially small, would grow in any event as the viscous contribution to the deflection becomes more important with time.

One difficulty with Lambeck's model, however, is a trade-off between the viscosity and the compressional stresses. If, for example, the subsidence is too large, then either the viscosity needs to be increased or the stress reduced. Another problem concerns the origin of the unconformities within the basins. Lambeck (1984) suggested that they are caused by fluctuations in the stresses and the subsequent isostatic re-adjustment of the basin depo-centres and basement highs. The problem is that upwards of five unconformity-bounded stratigraphic sequences have now been identified (Lindsay and Leven, 1996) in seismic reflection profiles of the central Australia basins, and it is difficult to see how externally applied, fluctuating, compressional stresses could be the only factor controlling all these sequences.

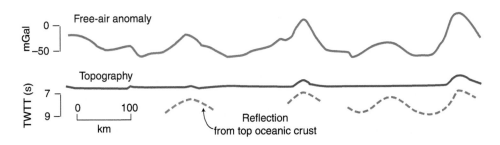

Figure 7.72 Free-air gravity anomaly, topography and line drawing showing the top of the oceanic crust along a profile of the Bengal Fan, south of Sri Lanka. Reproduced from fig. 3 of Weissel et al. (1980).

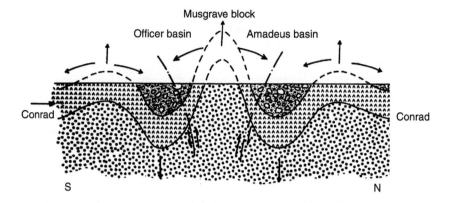

Figure 7.73 The formation of the Officer and Amadeus basins in central Australia by crustal compression. Reproduced from fig. 6 of Lambeck (1984) with permission.

Despite these problems, the existence of gravity lows that have persisted for very long periods of geological time is a strong argument that the central Australia basins are the product, at least in part, of compressional deformation. Gravity lows are a characteristic feature of basins in compressional settings. These include not only the foreland basins that flank the Himalayas and Alps, where one continental plate has underthrust another, but also the basins that flank the western edge of the Rockies and Andes, where deformation has occurred within the same plate. In both types of setting, the gravity lows are attributed to a downwarping of a low-density crust into the denser mantle by flexure.

A foreland setting that involved a downwarping of the crust has been proposed by Haddad et al. (2001) for the Officer basin in central Australia, on the basis of its asymmetric sediment distribution, subsidence and uplift history, and intense negative gravity anomalies. The Mesoproterozoic basement block (Musgrave, Fig. 7.73) that separates Officer from the Amadeus basin to the north (also interpreted as a foreland-type basin) may therefore represent the eroded remnant of an ancient mountain belt associated with these forelands.

Another wider class of basin are the intra-continental basins, many of which are associated with gravity lows. Burov and Molnar (1998), for example, showed that in the case of the Ferghana Basin in the western Tien Shan (which received most of its sediment during the Neogene), sediments contribute only about 30 mGal to the observed low of 120 mGal over the basin. This suggests that the basin is underlain by a thicker crust than flanking regions. Burov and Molnar interpreted the gravity low as the product of two events: an early (Jurassic?) rifting event that thinned the crust locally and a later Neogene folding that produced a downward buckling of the crust beneath the basin and an upward deflection of its flanks. Burov and Molnar (1998) had considered the role of convective downwelling beneath the Ferghana Basin but ruled it out on the basis that it could not explain the short-wavelength gravity anomalies observed at the edge of the basin. Cloetingh and Burov (2011) have reviewed the role of lithosphere-scale folding in the Ferghana basin as well several other compressional basins in intra-continental settings in Iberia and on the Russian Federation and north-west European platforms, pointing out the significant role that a vertically stratified rheology may play in controlling the stratigraphic architecture of these basins.

7.5.2 Intra-plate Volcanism, Hotspots and Mid-Plate Swells

Probably the best example that we have of the interactions of surface-loading and plate flexure effects with dynamic effects in the mantle is found in the interior of the oceanic plates.

Studies of the long-wavelength gravity field, derived from satellite tracking, terrestrial and altimeter-derived measurements, show that the ocean basins are characterised by several positive and negative anomalies with wavelengths that exceed 1,000 km. Some of these anomalies correlate with departures from the expected depth of the seafloor based on thermal cooling models. A correlation between gravity and bathymetry is suggestive of flexural loading models. Isostatic considerations suggest, however, that gravity and bathymetry anomalies that have wavelengths > 1,000 km are unlikely to be supported by flexure. This is because at these wavelengths the oceanic lithosphere, irrespective of its T_e, appears as a weak rather than a strong structure in its response to loads, and so there is little contribution to the gravity field.

One of the most prominent long-wavelength gravity and bathymetry anomalies in the oceans is located at the south-eastern end of the Hawaiian Ridge in the central Pacific Ocean. The existence of large areas of seafloor which are unusually shallow or deep for their age has been known since the pioneering work of Menard (1973), who dubbed them 'depth anomalies'. He recognised depth anomalies in maps of the north-east Pacific and East Pacific Rise and their expression in free-air gravity anomaly profiles of a number hotspot chains in the Pacific, including the Hawaiian Ridge. The first systematic examination of long-wavelength gravity and bathymetry around Hawaii was by Watts (1976), who showed a correlation between residual depth and gravity anomalies of ~22 mGal km^{-1} at wavelengths of ~1,100 km and higher.

Figure 7.74 shows a 1,700-km-long shipboard gravity anomaly and bathymetry profile that intersects the Hawaiian Ridge between Molokai and Oahu. The ridge is associated with a short-wavelength high and flanking lows that are superimposed on a longer-wavelength high. The short-wavelength anomalies correlate with the Hawaiian Islands, and the flanking Hawaiian moat and bulge have been attributed to flexure (see Section 4.3). The long-wavelength anomaly correlates with a ~1,500-km-wide bathymetric swell that rises some 1.5 km above the surrounding depths of the seafloor.

The close correlation between long-wavelength gravity and bathymetry anomalies is a strong argument that the Hawaiian swell is supported by some form of convection in the mantle. This interpretation is supported by numerical experiments of a constant-viscosity fluid that is heated from below (McKenzie et al., 1974). Because of thermal expansion, the fluid at the bottom is lighter than the fluid at the top. This 'top-heavy' arrangement is unstable, and the fluid will tend (depending on its viscosity) to re-distribute itself. Cold material will detach and sink from the surface to be replaced by hot material from below. If the Earth's mantle has similar regions of hot, upwelling and cold, downwelling material, then we would expect them to be reflected in surface observations. Hot, upwelling material, for example, would cause both a long-wavelength gravity and a bathymetry anomaly (Fig. 7.75). This is because this type of flow will be associated with low-density regions in the mantle that uplift the overlying crust, causing both a gravity anomaly high and a shallower-than-expected topography.

We can estimate the depth of a hot, upwelling region by considering its loading effect on the overlying lithosphere. Consider, for example, Eq. (5.27), which gives the free-air gravity anomaly admittance that is associated with a buried load. Since the swell is of long wavelength, the contributions to the gravity anomaly and topography of the crustal layer (i.e., its thickness and density) and the strength of the lithosphere can be ignored. Equation (5.27) then simplifies to

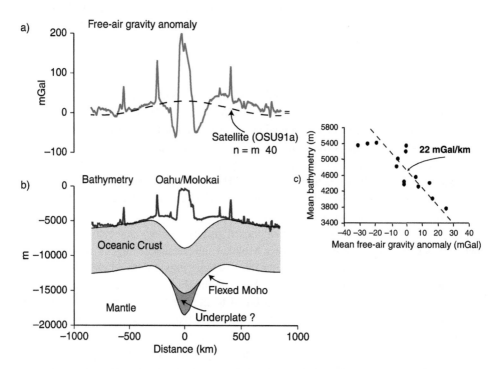

Figure 7.74 Gravity anomaly, bathymetry and crustal structure of the Hawaiian swell along a profile that intersects the Hawaiian Ridge between Oahu and Molokai. (a) Free-air gravity anomaly. Red solid line shows the shipboard gravity anomaly based on data acquired during cruises C1209 of *R/V Robert D. Conrad* and V2105 of *R/V Vema*. Black dashed line shows the satellite-derived gravity anomaly field of Rapp and Pavlis (1990) complete to degree and order 40. (b) Bathymetry (dark blue solid line) and simplified crustal structure. (c) Plot of the mean free-air gravity anomaly and topography along 14 profiles of the Hawaiian–Emperor chain shown in the inset in Fig. 5.20.

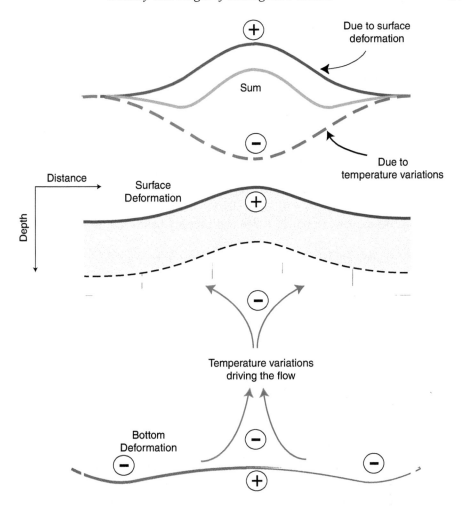

Figure 7.75 Schematic diagram showing the relationship between gravity anomalies, bathymetry and deep mantle convection. The '+' and '−' indicate either a positive or a negative contribution to the gravity anomaly respectively.

$$Z(k)_{buried} = 2\pi G e^{-kd}(\rho_m - \rho_w)(1 - e^{-kZ_L}), \qquad (7.60)$$

where d is the mean depth of the swell, Z_L is the mean depth of the hot, upwelling region and k is the wavenumber where $k = 2\pi/\lambda_w$. We see from Fig. 7.74a that the mean depth and wavelength of the Hawaiian swell are about 4.5 km and 2500 km (i.e., 2 × width), respectively. Hence, $kd \ll 1$ and so e^{-kd} in Eq. (7.60) → 1. The e^{-kZ_L} term contains the parameter that we are interested in, and so taking its series expansion then gives

$$Z(k)_{buried} = 2\pi G(\rho_m - \rho_w)\left(1 - (1 - kZ_L) + (kZ_L)^2 \ldots\right).$$

Retaining terms in $k\,Z_L$, but ignoring higher terms such as $(k\,Z_L)^2$, which will be small at long wavelengths, we then get

$$Z(k)_{buried} = 2\pi G(\rho_m - \rho_w)kZ_L. \tag{7.61}$$

We see from Fig. 7.74c that $Z_{observed}$ for the Hawaiian swell is ~22 mGal km^{-1}, close to the ~30 mGal km^{-1} estimated by Crosby et al. (2006) using more recent data sets. Substituting $Z(k)_{buried}$ = 22 mGal km^{-1} into Eq. (7.61), together with λ_w = 2,500 km, ρ_m = 3,330 kg m^{-3} and ρ_w = 1,030 kg m^{-3}, gives Z_L = 91 km, which is ~76 km below the base of the thickened oceanic crust beneath the Hawaiian Ridge and possibly below the base of the seismic lithosphere.

While a hot, upwelling region may therefore contribute to the gravity anomaly and bathymetry signature of some oceanic volcanic loads, we see that it is relatively easy to separate surface-loading and dynamic effects in the observations. Flexure, for example, is associated with shorter-wavelength and larger-amplitude deflections of the crust than mantle convection. Moreover, the gravity anomaly associated with flexural loading is of shorter wavelength and larger amplitude. The significance of this is that the gravity and topography signature associated with seamounts and oceanic islands can be used, with some confidence, to deduce the mechanical properties of the lithosphere.

An important result to emerge from oceanic flexure studies is that T_e mainly reflects the age of the lithosphere at the time of loading, not the load age. This is true even though the lithosphere continues to thicken with age as it cools. Therefore, T_e is effectively 'frozen in' at the time of loading. This implies that T_e will retain a 'memory' of the tectonic setting of a particular feature. and so we should be able to use it to deduce information on features of unknown origin on the seafloor.

Consider, for example, two seamounts that are of the same size (i.e., they have the same loading capacity), but one forms in an on-ridge setting while the other forms off-ridge. Oceanic flexure studies suggest that because the two seamounts formed on different-age lithosphere, they would have a different T_e: the feature that formed on-ridge would be associated with a T_e of about 5 km, while the feature that formed off-ridge would be associated with a value of about 25 km.

Figure 7.76 shows the gravity anomalies that would be expected for a seamount with a base radius of 45 km that loads a lithosphere with $T_e = 5$ and 25 km. There are significant differences in the gravity anomalies between the two cases: the $T_e = 5$ km case has much lower-amplitude anomalies and shorter wavelengths than the $T_e = 25$ km case.

The gravity anomalies over seamounts depend not only on T_e but also on the height and width of the loads. Figure 7.76c shows, for example, the dependence of the gravity

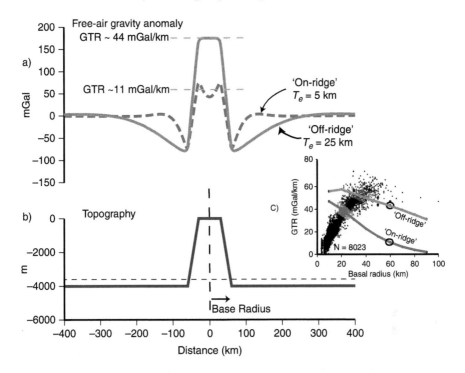

Figure 7.76 (a, b) Free-air gravity anomaly, bathymetry and gravity-to-topography ratio (GTR) for a seamount that forms on-ridge and off-ridge. (c) Compares the observed and calculated GTR for seamounts in the Pacific Ocean. The observed data (black filled circles) are based on the compilation of Wessel and Lyons (1997). The calculated profiles are based on simple two-dimensional elastic plate models. The intersection of the observed and calculated curves can be used to locate those bathymetric features that formed on-ridge and off-ridge. Open circles show the cases modelled in (a) and (b).

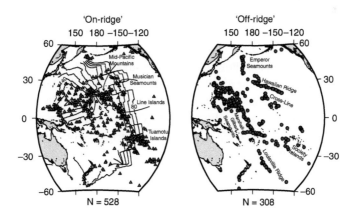

Figure 7.77 The distribution of on-ridge and off-ridge seamounts in the Pacific Ocean. The magnetic isochrons on the on-ridge map are shown at 10 Ma intervals.

anomaly of the seamounts that form on-ridge and off-ridge, normalised for their height, on their basal radius. The normalised gravity anomalies, which are derived from the ratio of the gravity anomaly to the bathymetry, or gravity to topography ratio (GTR), are largest for the smallest loads and least for the largest loads. This is because as a feature increases in size, it will appear more locally compensated, irrespective of whether it formed on-ridge or off-ridge.

The usefulness of the GTR is that it can also be derived from observations. Wessel and Lyons (1997), for example, have compiled the peak gravity anomaly, peak topographic height, basal radius and location of more than 8,000 seamounts and oceanic islands in the Pacific Ocean. The solid dots in Fig. 7.76c show the GTR based on their data. The observed GTR generally increases and then decreases with increase in basal radius. This trend probably reflects the systematic change that occurs in the basal radius of seamounts as a function of plate age. Narrow seamounts tend to characterise young seafloor, where the lithosphere is weak and the GTR is small, while wide seamounts tend to occur on old seafloor, where the lithosphere is strong and the GTR is high.

Figure 7.76c shows that the calculated and observed GTR intersect in two well-defined regions. The calculated GTR based on seamounts that form on or near a ridge intersects the data at a basal radius of about 20 km, while the off-ridge curve intersects in the data at a basal radius of about 35 km. This suggests that seamounts with a basal radius of about 20 km have a GTR that is similar to what would be expected if they formed on or near a mid-oceanic ridge crest. In contrast, seamounts with a basal radius of 35 km have a GTR that is similar to what would be expected if they formed off-ridge.

Figure 7.77 shows the spatial distribution of those seamounts that appear, based on their GTR, to have formed on-ridge and off-ridge. We see that the distribution of the off-ridge seamounts follows the expected pattern. The off-ridge map clearly shows the Hawaiian–Emperor, Louisville Ridge and Marshall-Gilbert seamounts that available age data suggest formed in an off-ridge setting. Several other clusters of off-ridge seamounts occur in the Society Islands and Cross-Line region and in what is now the oldest part of the western Pacific. The distribution of the on-ridge seamounts is not so expected, however. We see that several features in the central Pacific Ocean formed at or near a ridge crest. These include the Musician seamounts as well as a cluster of seamounts that connect the Mid-Pacific Mountains, Line Islands and the Tuamuto Islands. While most of these seamounts are aligned along chrons, there is evidence that others appear to align themselves along the traces of large offset fracture zones.

7.6 Dynamic Topography

We have alluded at several places in this book to the important role that mantle dynamics may play in controlling Earth's surface features. Morgan (1965) first pointed out how a rising or sinking body in a viscous mantle could perturb Earth's surface layers and gravity and geoid field. He applied his ideas to downwelling at trenches and upwelling at mid-ocean ridges, but they could equally have been applied to hotspots and cold spots in the plate interiors.

Identification of motions in the mantle in observations made at Earth's surface is, of course, complicated by surface processes, for example mountain building, sedimentary basin formation and volcanism and their tendency to approach a state of isostatic equilibrium. Nevertheless, early oceanic studies recognised that the contributions of mantle dynamics could be separated from long-wavelength gravity and bathymetry data. For example, Sclater et al. (1975), Watts (1976) and Cochran and Talwani (1977) subtracted the calculated bathymetry based on cooling half-space and cooling plate models from observed shipboard bathymetry data and then constructed $5 \times 5°$ and higher averages of the difference which they defined as 'residual depth anomalies'. The ratio of free-air gravity anomaly to the residual depth anomaly was then used to derive an observed admittance which McKenzie et al. (1977) were able to compare to the calculated admittance associated with numerical models of mantle convection.

By correcting the observed bathymetry using thermal models, these early studies were, in effect, considering the effects of local isostasy and its contribution to seafloor depth. The admittance associated with the Airy and Pratt models of isostasy, for example, approaches zero at long wavelengths, and while the uniform-density Airy model is a poor predictor of crustal thickness at a mid-ocean ridge, the Pratt model describes well the lateral changes in mantle density that are required to isostatically compensate the ridge. Indeed, the Pratt model was used as a basis to calculate the change in depth away from a mid-ocean ridge crest, derive the cooling half-space and plate models (Sclater and Francheteau, 1970; Parsons and Sclater, 1977) and compute their gravity effects (Lambeck, 1972).

Hager et al. (1985) built on this work by considering geoid anomalies, since they are known to have a better signal-to-noise ratio at long wavelengths than gravity anomalies. They constructed different models for mantle viscosity structure, finding that convection could contribute geoid anomalies of up to ±100 m and deformations of Earth's surface and the core–mantle boundary of up to ±1 and ±3 km respectively. They dubbed the surface deformation that could be attributed to convection 'dynamic topography'. They recognised, however, the difficulty of calculating dynamic topography, since account would first need to be taken of the crustal structure and upper mantle heterogeneities, both of which could also contribute to Earth's surface topography.

The work of Hager and colleagues led to further research on how dynamic topography contributed to Earth's gravity and geoid fields as well as to surface processes such as those associated with sedimentary basin formation, drainage networks and eustatic sea level change. Since the early studies had already emphasised the importance of correcting topography and bathymetry for the effects of crustal structure and isostasy, Panasyuk and Hager (2000) and Pari and Peltier (2000) redefined dynamic topography as the difference between the observed topography[11] and the expected topography based on the crustal structure and Airy-type isostasy.

---

[11] Here (and throughout this section) we are referring to 'topography' as the topography on land and the bathymetry in the oceans.

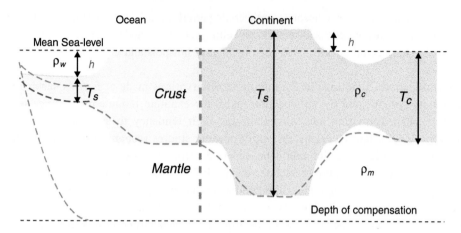

Figure 7.78 Schematic showing the relationship between topography on land and bathymetry in the oceans and the crustal thickness according to the Airy model of isostasy. The orange dashed and blue dashed lines illustrate that Moho may not be as predicted by an Airy model because of the effects of plate flexure and upwelling at a mid-ocean ridge respectively.

Figure 7.78 schematically illustrates the relationship between surface topography, h, and the expected crustal thickness, T_s, as determined, for example, from active source seismology. We can write from the balancing of a mountain continental column with a zero-elevation continental column that

$$T_s \rho_c g = T_c \rho_c g + r \rho_m g,$$

where r is the crustal 'root' beneath a mountain and is given by

$$r = T_s - h - T_c.$$

Substitution and rearranging then gives

$$h = \frac{(T_s - T_c)(\rho_m - \rho_c)}{\rho_m} \text{ for continents} \tag{7.62}$$

$$h = \frac{-(T_c - T_s)(\rho_m - \rho_c)}{(\rho_m - \rho_w)} \text{ for oceans} \tag{7.63}$$

We see from Eqs. 7.62 and 7.63 that the expected topography at any point on Earth's surface can be computed, provided the crustal thickness and the crustal and mantle density are known. The crustal thickness has now been compiled from active-source seismic experiments and combined into $5 \times 5°$, $2 \times 2°$ and, most recently, $1 \times 1°$ grids together with density information (e.g., Bassin et al., 2000).

Figure 7.79 shows a map of dynamic topography computed from the difference between the observed topography and the expected topography based on the crustal thickness. In the continents, an Airy model has been assumed. In the oceans, the Airy model, as we have

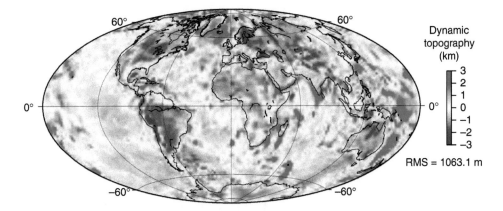

Figure 7.79 Dynamic topography as derived from a 2 × 2° grid of crustal thickness data (CRUST 2.0, Bassin et al. 2000) and an Airy model of isostasy. Blue regions have lower topography than expected for their crustal thickness. Red regions have higher topography than expected for their crustal thickness. The RMS has been computed for the region – 60° S to 60° N.

already pointed out, is a poor indicator of crustal structure at the mid-ocean ridge, and so we have used instead the cooling plate model, which is based on the Pratt model. Finally, in both oceans and continents a correction has been applied for the height that basement would have had it not been for sediment loading.

The figure shows dynamic topography of up to ±3 km. Continents are generally associated with negative dynamic topography, which means they have lower elevation than expected for their crustal thickness, suggesting mantle downwelling. There are exceptions, notably in the western US, Afar and Iceland where dynamic topography is positive, suggesting mantle upwelling. The oceans, however, are associated with a more even distribution of dynamic topography with negative dynamic topography, suggesting mantle downwelling between Australia and Antarctica, the Argentine basin and the West Philippine basin, and positive dynamic topography, suggesting mantle upwelling in the North Atlantic and western Pacific and locally in the region of ocean islands, including Hawaii, Reunion, Kerguelen, Cape Verde and the Canaries.

We caution, however, that while the general patterns of dynamic topography in Fig. 7.79 are consistent with those of previous studies, absolute values remain uncertain. This is because we have assumed local models of isostasy and uniform mantle density. However, plate flexure and heterogeneities in the mantle may significantly alter the height of topography.

Nevertheless, several workers have used the dynamic topography derived by assuming an Airy model to constrain numerical models of convection. Panasyuk and Hager (2000) and Pari and Peltier (2000), for example, derived an observed dynamic topography from a 5 × 5° grid of crustal thickness data (CRUST 5.0; Mooney et al., 1998) and compared it to the calculated dynamic topography based on numerical models of mantle convection. They showed that

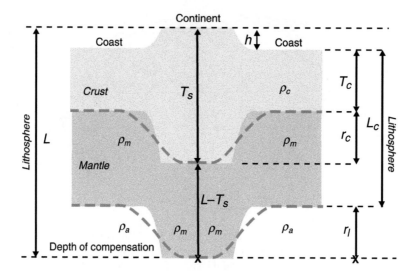

Figure 7.80 The Whole Lithosphere Isostasy model of Lamb et al. (2020) in which the depth of compensation is assumed to occur at the base of the conductive lithosphere rather than the crust. Topography is supported by the buoyancy of both a crustal and a lithospheric root.

dynamic topography was in the range of ±1 km to ±2 km for wavelengths > 5,000 km and that such an amplitude is generally consistent with the predictions of mantle convection models.

Lithgow-Bertelloni and Silver (1998) and Moucha et al. (2008a), on the other hand, focused on smaller regions such as southern Africa and the Colorado Plateau, concluding that that dynamic topography was in the range of ±500 to ±750 m. While these estimates are significantly smaller than those of Panasyuk and Hager (2000) and Pari and Peltier (2000), Molnar et al. (2015) criticised them on the basis that the Airy isostatic anomaly in these regions was small (0–10 mGal), which suggests, assuming an admittance for convection of 50 mGal km^{-1} (e.g., Fig. 5.34), dynamic topography of no more than ~±200 m. They attributed the anomalously high topography of southern Africa and the Colorado Plateau not to dynamic topography but to isostatically compensated density contrasts within both the crust and mantle lithosphere.

In a recent study, Lamb et al. (2020) modified the Airy model to specifically include the effects of isostatically compensated density contrasts in the sub-crustal mantle. Their model, dubbed the 'Whole Lithosphere Isostasy' model, is illustrated in Fig. 7.80. Balancing the left-side Continent column with the right-side Coast column gives

$$T_s\rho_c + (L - T_s)\rho_m = T_c\rho_c + r_c\rho_m + \left(L_c - (T_c + r_c)\right)\rho_m + r_l\rho_a.$$

Substituting $r_l = L - h - L_c$ then gives

$$T_s\rho_c + L\rho_m - T_s\rho_m = T_c\rho_c + (L_c - T_c)\rho_m + L\rho_a - h\rho_a - L_c\rho_a$$

or

$$\Delta C\rho_c - \Delta C\rho_m + \Delta L\rho_m = \Delta L\rho_a - h\rho_a,$$

where

$$\Delta C = \text{change in crustal thickness} = T_c - T_s$$
$$\Delta L = \text{change in lithosphere thickness} = L - L_c$$

or

$$h = \frac{\Delta C}{\alpha} - \frac{\Delta L}{\beta}, \tag{7.64}$$

where

$$\alpha = \frac{\rho_a}{(\rho_m - \rho_c)} \quad \text{and} \quad \beta = \frac{\rho_a}{(\rho_a - \rho_m)}.$$

This is the form of the equation obtained by Lamb et al. (2020) in their eq. 2. They calculated the expected height of topography, h, for the Whole Lithosphere Isostasy model, assuming $T_c = 32$ km and $L_c = 100$ km, T_s and L from CRUST1.0 (Laske et al., 2013) and seismic tomography (Priestley and McKenzie, 2013) respectively, and the best-fitting density terms α and β from observations.

Figure 7.81 shows a map of dynamic topography computed from the difference between the observed topography and the calculated topography based on the Whole Lithosphere Isostasy model. The main difference from the Airy model in Fig. 7.79 is the significantly smaller amplitude of the dynamic topography. The RMS of the dynamic topography deduced from the Whole Lithosphere Isostasy model is, for example, nearly a factor of three smaller than that for the Airy model. This is well seen in the case of the continents, for example in Australia, South and North America, which are associated with a uniformly negative dynamic topography in Fig. 7.79, but smaller amplitude and a more evenly distributed positive and negative dynamic topography in Fig. 7.81. The reason for this is that in the Whole Lithosphere Isostasy model $\rho_m > \rho_a$ the lithospheric root acts as a type of buried load which lowers the topography compared to what would be expected for its crustal thickness and an Airy model with $\rho_c < \rho_m$. Hence, isostatically compensated lateral densities in the lithosphere are capable of explaining subdued topography in the regions of thick or thin crust, and so there is no need to appeal to mantle downwelling or upwelling, for example, to explain them. The negative dynamic topography over much of the Russian Federation and north-west China or the positive dynamic topography along the mid-ocean ridge crest is still not explained but could be due, at least in part, to uncertainties in the crustal thickness and the inability of seismic tomography to resolve shallow (<50 km) lithospheric thicknesses respectively, as has been pointed out by Lamb et al. (2020).

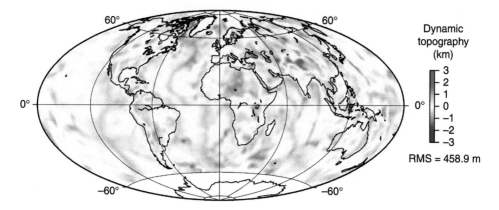

Figure 7.81 Dynamic topography as derived from a $1 \times 1°$ grid of crustal thickness (CRUST 1.0; Laske et al., 2013) and seismic tomography (Priestley and McKenzie, 2013) and the Whole Lithosphere Isostasy model (Lamb et al. 2020). Continents still show large tracts of negative dynamic topography, but the amplitude is much than smaller than that derived from the Airy model. Oceans are dominated more by positive dynamic topography, especially at mid-ocean ridge crests. Note that the colour bar range is the same as in Fig. 7.79 and so demonstrates how subdued dynamic topography is in the Whole Lithosphere Isostasy model compared to that of the Airy model. The RMS has been computed for the region – 60° S to 60° N. Modified from fig. 8 of Lamb et al. (2020).

It is certainly true that studies such as those illustrated in Figs. 7.79 and 7.81 have first considered the state of isostasy and then deduced dynamic topography. However, it has been the local rather than the global isostatic studies that arguably have been the most persuasive. Coakley and Gurnis (1995), for example, back-stripped sediments of the Michigan cratonic basin, showing that during the Mid-Ordovician there had been a departure from a 'bull's-eye' pattern of the tectonic subsidence to an 'up-to-the-west', 'down-to-the-east' tilt. They attributed the tilt to dynamic effects associated with slab subduction beneath the western United States. Similarly, Pang and Nummedal (1995) and Liu and Nummedal (2004) backstripped the Cretaceous stratigraphy of the western interior foreland basins, noting there was a 'background' subsidence that extended from the foothills of the Rockies, across the basins to >1,500 km eastward. They attributed the subsidence to subduction of the Farallon plate beneath a westward-migrating North American plate. Finally, Burgess et al. (1997) examined the Sloss stratigraphic sequences that can be correlated over large tracts of the North America continental interior and suggested that they might be caused by a model which combined Airy-type sediment loading and eustatic sea level changes with alternating dynamic topography 'highs' and 'lows' that were generated by thermal insulation and a downgoing slab respectively.

In a global backstripping approach, Heine et al. (2008) used a $1 \times 1°$ sediment thickness grid and compared the observed tectonic subsidence to the calculated tectonic subsidence based on the best-fit stretching factor, β_s, derived from a $2 \times 2°$ crustal thickness grid. They

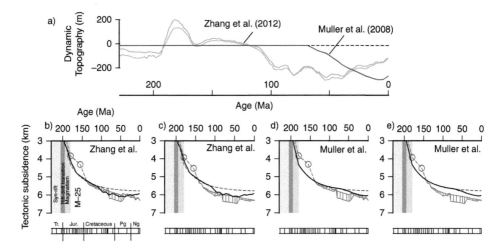

Figure 7.82 Comparison of the observed tectonic subsidence derived from backstripping at the COST B-2 well, based on different sea level curves with the calculated subsidence predicted by finite rifting models and different models of dynamic topography. (a) The Zhang et al. (2012) and Müller et al. (2008) models of dynamic topography in the vicinity of the COST B-2 well. (b) The case of Haq et al. (1987) sea level and Zhang et al. dynamic topography. Red solid lines show the tectonic subsidence derived from backstripping (red dashed line – uncertain). Blue dashed line shows the subsidence predicted by a finite rifting model, assuming a thickness of crust beneath the well of 9.3 km, an initial crustal thickness of 40 km and a $\beta = 4.21$. Black solid line shows the subsidence predicted by the finite rifting and dynamic topography models. (c) The case of Watts and Steckler (1979) sea level and Zhang et al. dynamic topography. (d) The case of Haq et al. (1987) sea level and Müller et al. dynamic topography. (e) The case of Watts and Steckler (1979) sea level and Müller et al. dynamic topography.

attributed the difference between the observed and calculated subsidence in the region of intracratonic basins to dynamic topography, which they found to be in the range +250 m to – 1,700 m. The study highlights, however, some of the challenges in using global grid files which include the isolation of individual basin types, that all basins might not be of rift origin, and the tendency to ignore the loading and unloading effects of sediment now missing through erosion.

Nevertheless, Fig. 7.82 shows an example, from the East Coast, United States continental margin, of how consideration of the state of isostasy is useful and can, in some instances, be used to constrain dynamic topography. The red lines in Fig. 7.82b–e show the observed tectonic subsidence curves derived from backstripping biostratigraphic data at the COST B-2 well (see Figs. 7.20 and 7.29 for well location and crustal thickness) which consider the thickness of Triassic and Jurassic sediment imaged seismically beneath the well by the LASE study group (1986) and the ages of rift initiation and duration determined by Withjack et al. (1998). Corrections have also been included for both the Haq et al. (1987) and Watts and

Steckler (1979) sea level curves. The dashed blue line shows the expected tectonic subsidence based on a finite rifting model with a stretching factor, β_s, determined from the seismically constrained crustal thickness. While the observed tectonic subsidence fits the expected tectonic subsidence reasonably well during the early post-rift, it starts to exceed it after ~150 Ma. The black lines in in Fig. 7.82b-e show the calculated tectonic subsidence derived by summing the predictions of the finite rifting model and the dynamic topography models of Müller et al. (2008) and Zhang et al. (2012). There is now a good agreement between the observed and expected tectonic subsidence for the case of the Haq et al. sea level curve and the Zhang et al. dynamic topography.[12] The main discrepancy is since ~50 Ma, when the backstrip curves reveal a tectonic subsidence that is increasing to the present day whereas the finite rifting and dynamic topography predict a decrease. The Zhang et al. dynamic topography is based only on the long-wavelength subsidence since the break-up of the Pangea supercontinent and so does not include the effect of the any subducting slabs beneath a North American plate that is migrating west. The Müller et al. dynamic topography, however, only includes the effect of the Farallon subducting plate beneath a westward-migrating North American plate and predicts a decrease in tectonic subsidence since ~50 Ma, in good agreement with the observed backstrip for the case of the Haq et al. sea level curve (Fig. 7.82d).

These results suggest that since ~100 Ma there has been an ongoing tectonic subsidence at the COST B-2 well, which is more than that predicted by finite rifting models. The tectonic subsidence can be accounted for if there is a contribution from dynamic topography that is negative in sign and does not exceed ~250 m as predicted by the Zhang et al. dynamic topography model. The main discrepancy is since ~30–50 Ma, when the tectonic subsidence continues to increase, yet the Zhang et al. model predicts a small uplift. One possibility, therefore, is some combination of the Zhang et al. and the Müller et al. dynamic topography models. Although the Zhang et al. model predicts uplift since 30–50 Ma, it is more than negated by the subsidence in the Müller et al. model. Certainly, we see little evidence in the COST B-2 well backstrip data of an uplift of up to 300 m since 30 Ma as reported by Moucha et al. (2008b) and attributed by them to a steepening and deepening of the Farallon plate.

7.7 Plate Flexure and Landscape Evolution

We have noted elsewhere in this book (e.g., Figs. 4.53, 7.58) how plate flexure through the topography that it creates can influence drainage patterns, river networks and 'staircase' patterns of marine terraces. For example, sediment loading offshore north-east Brazil has clearly influenced drainage systems onshore through its creation of flexural bulges and depressions. While the Amazon and Pará rivers have been

---

[12] The author is grateful to Shijie Zhong for providing the dynamic topography curve of Zhang et al. at the location of the COST B-2 well.

able to traverse these features and exit into the Atlantic Ocean, the bulges and depressions that have been created by offshore sediment loading (Fig. 4.53) have resulted in unique onshore ecosystems with tropical rain forests on the bulges and savannah and 'moist forests' which flood daily in the depressions. Plate flexure is therefore capable of influencing both the landscape and its ecological habitats (e.g., Bishop, 2007), independently of any anthropogenic or climate-induced changes.

Several studies have focussed on the evolution of flexural uplifts that form in the interiors of the plates following sediment unloading. One such study is by Lane et al. (2007) on flexural uplift in the Cotswold Hills of south-central England, which they believe formed because of the excavation of sediment by meltwater-charged rivers following the last major ice advance (the Anglian) ~425 kyr. The example is useful in illustrating the 'family' of rivers that might develop on a flexure profile with its steep scarp slope, asymmetric crest and the gentle slopes of its long tail. Critical is the development of a drainage divide which follows the trend of the asymmetric crest and separates dip streams that develop on the tail from scarp streams, which develop on the steep slope (Fig. 7.83b). Other landscape features that formed following sediment unloading include the 'staircases' of river terraces that formed following the excavation of sediments in the Severn and Thames valleys that flank the Cotswold Hills as well as the staircases of marine terraces that formed following the excavation of sediments in submarine canyons in the Atlantic passive continental margin (e.g., Fig. 4.14).

Other studies have focussed on the flexural uplifts created in regions of extensional and compressional tectonics. For example, Garcia-Castellanos et al. (2003) showed how drainage has evolved in the Catalan Coastal Ranges of north-east Spain which formed following the initiation of crustal extension in the Early Miocene in the Valencia Gulf[13] to the south-east. They constructed a simple model of how a lake that forms in the flexural depression inland of a rift flank uplift evolves with time. As the lake evaporates and erosion and uplift modify the rift flank, a drainage system develops that leads to sediment deposition in the distal part of the lake and the proximal part of the Gulf. Eventually, as erosion proceeds, the crest in the flexure profile is subdued, the lake is captured and a complex river network develops which includes a scarp stream. They found that the length of time that it takes to capture the lake depends on T_e: lower values increase the vertical isostatic movements and predict a slower capture, higher values decrease the movements and predict a faster capture.

It is clear, therefore, from Fig. 7.84 that erosion plays a significant role in the development of drainage networks on a fault-bounded rim flank uplift through its ability to remove the flexural peak on the footwall, and in allowing rivers to flow directly into the water-filled hanging wall, where they form a delta. Olive et al. (2014, 2020) have reassessed the capacity of erosion to modify a fault (and hence the flexural profile) before being abandoned in favour of a new fault. They have found that

---

[13] The same rift and footwall uplift was isolated from stratigraphic data in the Gulf by flexural backstripping in Fig. 7.23.

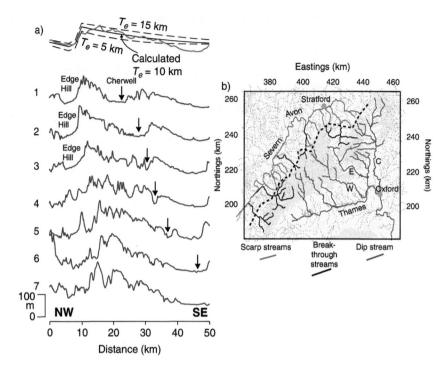

Figure 7.83 Topography profiles and 'families' of rivers in the Cotswold Hills of south-central England. The Cotswolds are interpreted as a heavily incised rim flank uplift that formed following excavation of sediments by meltwater-charged rivers in the English Midlands to the north. (a) Comparison of the ensemble average[14] of the topography (red line) to the calculated profile based on different models of flexure (black dashed line). (b) Map showing the 'families' of Cotswold rivers. The dotted black line outlines the drainage divide which separates north-flowing rivers on the scarp from south-flowing rivers on the dip slope. Modified from Lane et al. (2007).

erosion can enhance the amount of horizontal extension on a fault, especially when erosion acts at rates comparable to tectonic slip on a fault.

The role of surface processes compared to those of tectonics in forming topography is probably no better manifest than in orogenic belts. That tectonics in the form of plate flexure can impact the landscape is shown by consideration of the simple model of a Gaussian-shaped mountain in Fig. 7.85. The figure shows that, if the rock erodibility and time in a landscape evolution model are fixed, then it becomes apparent that T_e is a major control on the drainage networks that develop. This is because low T_e favours a more localised isostatic uplift, hence more eroded material and enhanced topographic relief and average elevation, while high T_e favours more

---

[14] Ensemble averaging is a spectral technique that subdues features not common to individual profiles and enhances features that are common.

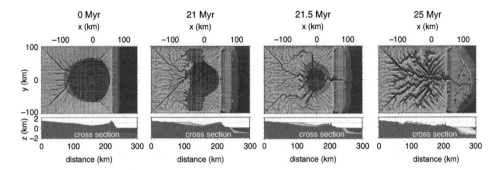

Figure 7.84 Evolution of a synthetic numerical model of lake capture and drainage opening on a rift flank uplift, the steep slopes of which are subject to erosion. Topography, drainage network and cross section of the model along $y = 0$ are shown at four stages. Rivers are drawn with line width proportional to their predicted water discharge. $T_e = 10$ km. Lake capture and drainage opening occurs between 21 and 21.5 Myr. Reproduced from fig. 8 in Garcia-Castellanos et al. (2003) with permission.

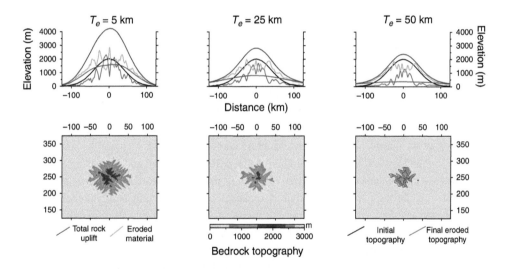

Figure 7.85 Simple model showing the sensitivity of a landscape to elastic thickness, T_e. The calculations are based on an initial 2-km-high, 200-km-wide gaussian-shaped topography subject to precipitation and the landscape evolution model of Braun and Willet (2013). The key parameters are K, which determines rock erodibility, and t, the scaled length of time a model is run. The models assume $K = 10^{-5}$ m$^{1/3}$ y^{-1}, $t = 2$ Myr and $T_e = 5$, 25 and 50 km. Modified from Paxman (2015).

regional isostatic uplift, hence less eroded material and subdued topographic relief and average elevation.

The role of crustal strength in an eroding mountain belt has been considered, for example, by Avouac and Burov (1996). They pointed out that while mountains are supported by the strength of the crust, laboratory experiments predict their compensating roots should flow with time such that the topography of mountains will be reduced, perhaps by 50 per cent in a few tens of millions of years. They suggest that the maintenance of mountains for long periods of time is a consequence of the interactions of surface processes which can work to prevent such a collapse. The removal of material by erosion and its deposition and flexural loading in adjacent foreland basins, for example, may prevent lateral spreading of a mountain's compensating roots and result in a focusing of uplift in the orogen, thereby maintaining topographic relief for long periods of time. They showed, however, the final topography to be sensitive to lower crustal rheology, with weaker rheologies (e.g., quartz-dominated) favouring lower average-elevation mountains than stronger rheologies (e.g., Quartz-Diorite).

The topography of orogenic belts is rarely smooth, as the models of Avouac and Burov predict, except perhaps in central Tibet and the Altiplano, where the topography is remarkably flat (e.g., Fielding et al., 1994; Hartley, 2003). These regions are currently arid and have been for the past few million years, so erosion rates associated with river incision are low. Precipitation is now focused at the edges of Tibet in the Karakorum, Himalaya and eastern Tibet and the Altiplano in the western and eastern Cordillera, where deep river incision of the topography correlates with the highest rainfall.

It is in the rivers that exit high mountains where the effects of flexure are, perhaps, most manifest. Simpson (2004), for example, pointed out the close association between transverse river systems and the crests of double plunging anticlines.[15] The association suggests the anticlines may be the consequence of river incision, erosion and isostatic uplift. Numerical modelling of an elastic plate with a yield stress confirms this and suggests that increasing regional compression increases the number of non-cylindrical folds, and increasing the yield stress increases the wavelength of folding. Therefore, large, incised transverse rivers can unload the crust and, through the flexure that is created, facilitate folding. And in their study, Montgomery and Stolar (2006) discuss the incised river systems that flow parallel to and down the axis of double plunging anticlines. They also suggest these anticlines are the consequence of river incision, erosion and isostatic uplift. Analytical modelling of an elastic plate with $2.5 < T_e < 5$ km, for example, showed eroded river valley widths of 10–20 km to be associated with rock uplift with wavelengths of 50–90 km, which is comparable with the anticline wavelengths observed along major Himalaya rivers. Erosion rates must, as they point out, outpace the isostatic uplift rate, otherwise there would only be rock uplift and no incision. Hence, they conclude that river anticlines are the result of the interaction of highly erosive rivers and a relatively weak crust and that rivers are not just the source of their own incised valleys, but the phenomena of plate flexure dictate

---

[15] Otherwise referred to as domes or periclines.

that it is also responsible for some of the structural geological features of the surrounding mountains.

7.8 Plate Flexure, Inheritance and the Wilson Cycle

We have so far focussed on the role of flexure and its ability to create features in the topography and bathymetry at individual plate boundaries, the interior of the plates and in those locations where other competing processes may be operative, such as those involved in landscape evolution and mantle dynamics. But we have so far made little mention of one of Earth's greatest cycles – the Wilson Cycle – in which continents disperse and reassemble and oceans open and close. The cycle, which has dominated Earth evolution since at least ~1 Ga with the assembly of Rodinia, is a key part of plate tectonics and is responsible above all else for the rich variety of structural geology, igneous activity and metamorphism that we see in the Proterozoic and Phanerozoic rock record. What, may we ask, is the role played by flexure in this cycle?

We know from flexure studies that Earth's outer layers respond to superimposed long-term loads essentially as an elastic plate. While viscoelastic stress relaxation may be significant initially, the response to loads eventually stabilises, which means that large-scale structures associated with flexure with a past tectonic event may be preserved in the rock record and, in some cases, inherited by a subsequent rifting or orogenic event. This is well seen in the tendency of sedimentary basins to 'morph', for example, from a rift-type basin to a foreland-type basin, and for a foreland basin to 'inherit' the thermal and mechanical properties of an underlying rifted continental margin, as seems to have occurred in the west Taiwan basin, in the United Arab Emirates and Oman foreland basins and in the Colville Trough (Alaska). In each case a passive margin has morphed into an active margin – an important orogenic stage of the Wilson Cycle when a continent is brought into contact with another continent, an island arc or an aseismic ridge as a consequence of subduction and the closure of an ocean.

One manifestation of orogeny is ophiolite obduction in which high-density oceanic crust and mantle is superimposed on and loads relatively low-density continental crust. Because the plates are essentially elastic in their response to long-term loading, the obduction of mantle and then crust ophiolite onto a pre-existing rifted continental margin (e.g., Ali et al. 2020) can be modelled using a process-oriented gravity modelling approach (i.e., POGM) by summing the gravity anomaly associated with the rifted margin and each ophiolite loading event (e.g., Fig. 7.86), calculating the associated change in crust and mantle structure (e.g., Fig. 7.86) and comparing the sum gravity gravity anomaly and associated crust and mantle structure to observations at the present day. The flexure models make several predictions, for example, which can be compared to observations including an unconformity in flanking regions of the load that separates the basal detachment surface from the pre-existing crust, a lateral landward offset of the load and any topography that might be associated with it from the deepest Moho, and a wide flat-topped gravity

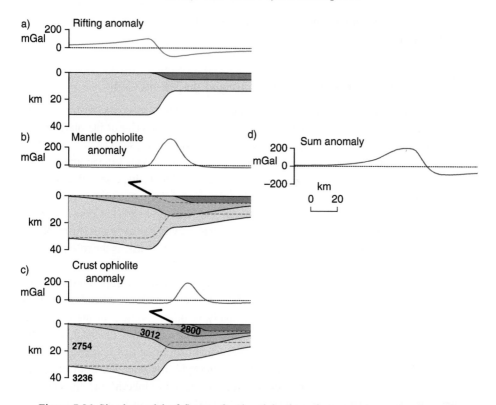

Figure 7.86 Simple model of flexure for the obduction of an oceanic crust and mantle ophiolite onto a pre-existing rifted continental margin. The half-arrow symbolises the detachment surface that separates the 'driving' load (above) from its 'infill' load (below).

anomaly high with a flanking low on its oceanward side. Other examples of ophiolite obduction such as in forearcs, between a trench and the magmatic arc (e.g., Stern et al. 2012), and at spreading centres and transform faults (e.g., Morris and Maffione, 2016) may, however, be difficult to model using simple models of flexure, and the complexity of their tectonic settings most likely will require numerical rather than analytical models in order to explain them.

7.9 Summary

We have demonstrated in this chapter that flexure plays a major role in the evolution of a number of the Earth's main surface features. Although not as easy to document as the simple cases presented in Chapter 4, flexural isostasy is an important process that helps the lithosphere to re-adjust where it is extensively strained at extensional, compressional and strike-slip plate boundaries.

In extensional regions, flexure manifests itself in the width and amplitude of continental and oceanic rifts and the characteristic 'tail' of the uplifts that flank them. Sedimentary basins that form in extensional regions are strongly influenced by flexure, both during the rifting phase and following it. The flexural rigidity of extended continental crust is generally lower than that of surrounding regions, presumably because extension involves thinning and heating of the lithosphere at the time of rifting. There is evidence that some rift-type basins are weak during rifting and remain weak thereafter. Others, however, appear to increase their strength with time.

In compressional regions, the stresses and strains that are involved in oceanic and continental collision tend to obscure the effects of flexure. However, there is evidence that foreland basins that flank the edges of orogenic belts develop because of flexure in front of advancing thrust and fold loads. The 'architecture' of foreland basins is controlled, however, by both surface and buried loads. Flexure controls the width and amplitude of foreland basins, the stratigraphic patterns within them and, maybe, the structural styles that develop in orogenic belts. There is evidence that the flexural properties of foreland-basin lithosphere have been inherited from ancient rifted continental margins that underlie them.

In strike-slip regions, flexure manifests itself in the topography of oceanic transform faults and fracture zones. The large contrasts that exist in the age and, hence, thermal structure of the lithosphere across these features give rise to thermal bending moments and differential vertical loads that cause topography in the immediate vicinity of the offset to flex. Flexural features include unusually shallow transverse ridges and deep transform valleys, both of which display the characteristic 'tail' shape of flexure. The role of these loads in the vicinity of strike-slip faults in the continents is not, however, as clear. Many 'pull-apart' basins appear to be uncompensated and, although they are often flanked by rim uplifts, they appear to have formed more by mechanical processes in the lower crust than by flexure.

There is evidence in both the continents and the oceans that flexural features at the plate boundaries are maintained for long periods of geological time. Some of the best examples are seen at extinct ridges and fracture zones where the topography bears a striking resemblance to active ridges and their transform offsets. Other locations where flexural features are maintained are at compressional plate boundaries, where outer rises persist despite the cessation of subduction, and foreland basins are found even though the adjacent orogenic belt has long since disappeared.

There is evidence that flexure in intra-plate settings is modulated by other processes, such as plate boundary forces and dynamic topography. These processes can deform the lithosphere into broad folds with amplitudes of up to a few kilometres and wavelengths of a few hundred to a few thousand kilometres. The relative role that plate boundary forces and dynamic topography have played in contributing to the Earth's topography and gravity field is difficult to determine, although the long wavelength of these deformations in some regions (for example, in the region mid-plate swells) separates them from the effects of flexure. In other regions it is necessary to first evaluate the role of isostasy fully in contributing to topography before estimating dynamic topography because it has been shown that models of isostasy which include both a crustal and a lithospheric root are more than capable of explaining the present-day topography, for example in the subdued continental interiors.

Irrespective of dynamic topography, there is evidence that flexure influences landscape evolution in a wide range of tectonic settings. The characteristic shape of a flexural rim flank uplift formed by loading or unloading of the crust, for example, is sufficient to induce a 'family' of river systems in the landscape, including dip streams, scarp streams and breakthrough streams. Arguably, the strongest evidence of the influence of flexure, though, has come from studies of large river systems in fold and thrust belts which incise to sufficient depths and widths to unload the crust and form anticlines, and in foreland basins where rivers that exit a fold and thrust belt may be redirected by flexure and flow along structure rather than across it.

Perhaps the most important implication of flexure is for structural inheritance and the Wilson Cycle. The presence on Earth of an essentially elastic outermost layer (the lithosphere) that has the capability to retain a 'memory' of a past geological event, to appear both strong and weak on long geological timescales, and to support stresses of up to a few hundreds of MPa before yielding ensures that flexure will play a role in explaining structural inheritance, a widely reported phenomenon in the rock record. Inheritance plays a major role in the evolution of sedimentary basins as they morph from one basin type to another, in orogenic belts where 'thin-skin' and 'thick-skin' tectonic styles may vary along-strike a fold and thrust belt, and in the Wilson Cycle where continents disperse and reassemble and oceans open and close.

8

Isostasy and the Terrestrial Planets

I certainly am a strong supporter of the universality of the tendency towards isostasy; in fact, I do not think any part of the crust of any extension could resist it.

(Vening Meinesz, 1932, p. 27)

8.1 Introduction

We have demonstrated in this book that isostasy is an important concept that has helped our understanding of the evolution of the Earth's main surface features. The Earth is part of the solar system, and therefore it is reasonable to suppose that some degree of isostasy may also exist on other planetary bodies.

Of the nine planets in the solar system, the inner four have been orbited by spacecraft and all have been studied using Earth-based observations. We now know that the planets divide into two types based on their size, density and distance from the Sun. The major planets, Jupiter, Saturn, Uranus and Neptune, contain most of the mass in the solar system, are of large diameter (49,244–142,984 km), low average density (700–1,600 kg m^{-3}), and range from 7.78×10^8 to 4.494×10^9 km in their distance from the Sun.[1] These large outer planets are also rich in volatiles, especially hydrogen and helium, and have low surface temperatures and well-developed satellite systems. The inner planets, Mercury, Venus, Earth and Mars, are, in contrast, small in diameter (4,878–12,756 km), high average density (3,900–5,500 kg m^{-3}), and range from 5.8×10^7 to 2.28×10^8 km in their distance from the Sun. These so-called terrestrial planets (Fig. 8.1), which are composed chiefly of rock, are comparatively poor in volatiles, and have few satellites.

During the last three decades of the twentieth century, a vast amount of satellite-derived data has been acquired by the United States National Aeronautics and Space Administration (NASA) during 'fly-by' and 'orbiter' missions to the terrestrial planets. Of particular relevance to this book are the gravity anomaly and topography data sets. As we know from studies of Earth, these data are key, in the absence of independent seismic information on the crustal structure, to our understanding of planetary isostasy. There is now sufficient

---

[1] Distance in this case is the semi-major axis of the planet's orbit around the Sun.

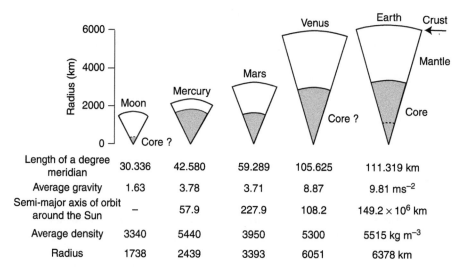

Length of a degree meridian	30.336	42.580	59.289	105.625	111.319 km
Average gravity	1.63	3.78	3.71	8.87	9.81 ms^{-2}
Semi-major axis of orbit around the Sun	–	57.9	227.9	108.2	149.2×10^6 km
Average density	3340	5440	3950	5300	5515 kg m^{-3}
Radius	1738	2439	3393	6051	6378 km

Figure 8.1 Principal facts for the terrestrial planets, including the Moon. Data compiled from various sources.

high-resolution gravity anomaly data of the Moon, Mars, Mercury and Venus to be able to assess, in a preliminary way at least, the state of isostasy on these planets.

In this chapter, the current state of knowledge concerning the state of isostasy on the terrestrial planets is reviewed. By focusing on these planets, we do not mean to imply that some degree of isostasy does not prevail on the larger, outer planets of the solar system. The satellites of Jupiter (e.g., Io, Europa, Ganymede and Callisto) and Saturn (e.g., Titan, Enceladus and Tethys), for example, show a variety of surface features, including liquid oceans, active volcanoes, mountains and basins, and extensional and compressional features. The problem is that we lack both gravity anomaly and topography data over some of these features, so it may be some time before we fully understand the state of isostasy on these planets.

8.2 Moon

The Moon is nearest to Earth and is the planet about which we know the most. There is evidence that the Moon has a low-density, feldspar-rich crust that overlies a denser, silicate-rich mantle. The attenuation of S waves generated by natural and man-made sources, together with more recent seismic and electromagnetic studies (e.g., Hood et al., 1999; Weber et al., 2011), suggest that the Moon, like the Earth, has a partially molten core. Unlike, the Earth, however, the Moon lacks an atmosphere, and there is little evidence on its surface of erosion, sedimentation and mass transport.

The most striking feature of the Moon's near side, clearly visible to telescopic observers on Earth, is the circular basins (Fig. 8.2). These flat-floored topographic depressions have diameters in the range of 100–300 km, with some exceeding 800 km. In reflected sunlight,

Figure 8.2 Camera image of the near-side of the Moon. The grey shades indicate brightness. The dark maria are basaltic lava flows. The maria are flanked by elevated, rugged highlands. Credit: Goddard Space Flight Centre/University of Arizona.

the basins appear dark. It was originally thought that these dark regions were pockets of water and, hence, they were referred to by the early astronomers as maria.[2]

The maria are now considered to be basaltic lava flows that have infilled giant impact craters (Fig. 8.3). One example, Mare Humorum, is illustrated in Fig. 8.4. The mare is circular in outline, about 300 km across and is characterised by a smooth surface that appears to have been punctuated by a number of smaller, younger, impact craters. Other prominent maria on the Moon's near-side are Serenitatis, Imbrium, Crisium and Nectaris. Of these, Mare Imbrium has the 'freshest' appearance and for this reason is believed to be among the youngest.

It was G. K. Gilbert, the early isostasist, who first suggested that Imbrium was a giant impact crater that formed by the collision of a large meteorite with the Moon (Gilbert, 1893). Subsequent studies have shown that the crater has been infilled by at least three sets of lava flows. The lavas appear to have erupted near the outermost edge of the crater and to have flowed inwards towards its centre. The surfaces show typical basaltic flow features such as steep flow fronts, well-developed lava channels bounded by levées, and sinuous depressions that have been interpreted as collapsed lava tubes.

The maria are flanked by topographic highs that rise a few metres to a few kilometres above the floor of surrounding basins. These features are more rugged than the maria

---

[2] Mare (plural, maria) is the Latin word for sea.

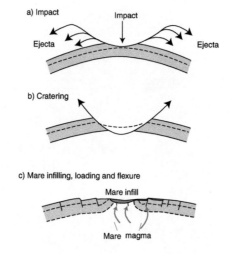

Figure 8.3 The formation of a lunar basin by (a) impact, (b) cratering and (c) mare infilling, loading and flexure. Shaded orange region schematically illustrates the crust and the dashed line a layer within the crust. The red arrows schematically illustrate magma pathways. Reproduced with permission from fig. 4.6 of Greely (1994).

(Fig. 8.4) and comprise mainly ejecta, large blocks and remnants of crater rims. The material is highly brecciated. Samples obtained from the south-west rim of Imbrium during the manned Apollo 15 mission suggest an age of 3.9 Ga. Since the Moon is believed to be 4.7 Ga old, these dates suggest that the impact that formed Imbrium's crater occurred relatively early in the Moon's history.

Muller and Sjogren (1968) using tracking data of the Lunar Orbiter, were among the first to construct a gravity field of the Moon. One of their first results was to show that Imbrium, Serenitatis, Crisium, Humorum and Nectaris were associated with large-amplitude, positive free-air gravity anomalies of up to +300 mGal. Muller and Sjogren (1968) dubbed these anomalies 'mascons'. They suggested that mascons were caused by dense planetesimal material of an iron-nickel composition that had fallen to the Moon's surface during a period of heavy bombardment early in its history.

The discovery of mascons triggered a vigorous debate about the state of isostasy on the Moon. Initially, discussions focused on whether the mascons were isostatically compensated. Two opposing views were advanced: one is that they were uncompensated (Conal and Holstrom, 1968; Gilvarry, 1969), and the other that they were compensated (O'Keefe, 1968; Hulme, 1972).

According to the uncompensated view, the mascons are supported by the strength of the lunar lithosphere. Givarry (1969) for example, proposed that the load of the material that infilled[3] the pre-mare basin was too small to depress the crust and mantle beneath the basin. Hence, the basin would appear uncompensated. Conel and Holstrom (1968), however,

---

[3] Gilvarry (1969) suggested that the maria were infilled by sedimentary rocks rather than lavas.

Figure 8.4 Camera image of Mare Humorum (Fig. 8.2) showing cratering of its smooth surface. Reproduced from fig. 8 of Golombek and McGill (1983), copyright by the American Geophysical Union.

suggested the mascons were due to high-density material that had become 'embedded' into the Moon's surface at the time of impact. Both models imply that the Moon's outermost layers did yield when loaded and, therefore, that it is a relatively rigid and, hence, cold planet.

An alternative view is that mascons are compensated at depth. O'Keefe (1968), for example, found that, when the free-air anomalies over the mascons were corrected for topography, large-amplitude positive and negative Bouguer anomalies resulted. He therefore argued that the basins that underlie the mascons and the highlands that flank them are isostatically compensated. The form of the compensation could not be determined precisely, but O'Keefe (1968) suggested that the Bouguer anomalies were in accord with the predictions of either an Airy or a Pratt model of isostasy. He recognised, however, that local isostatic models would not be able to explain all the anomalies and that 'small' deviations might occur, particularly in the maria where the material that infilled the basins had arrived after their formation.

Hulme (1972) agreed with O'Keefe that the maria were close to being in isostatic equilibrium. He constructed a model in which the maria are made of layers of lava that have been 'set' into the lunar crust. Because the lava is denser than its surroundings, Hulme argued that it would act as a load that depresses the low-density lunar crust into the denser mantle. Isostatic equilibrium, in his view, is achieved by a balancing between the

Isostasy and Flexure of the Lithosphere

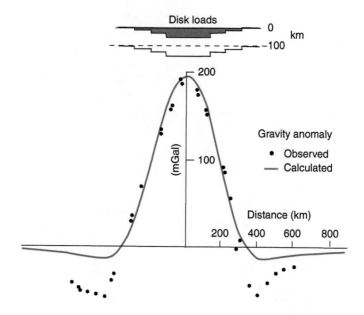

Figure 8.5 Comparison of observed and calculated gravity anomalies over Mare Serenitatis. The observed anomaly is based on the gravity data of Muller and Sjogren (1968) along a North–South transect of the mare. The calculated anomalies are based on a model in which the mass excess of the mare load is underlain locally by a mass deficiency. Reproduced from fig. 2 of Hulme (1972) with permission of *Nature*, London.

downward-acting load of the lava and the upward buoyancy of the depressed crust. Hulme went on to show that the gravity anomaly over Mare Serenitatis was generally in accord with such a model (Fig. 8.5).

The work of O'Keefe and Hulme was significant because it was the first to make the case that surface features on the Moon might be isostatically compensated. The existence of isostasy has important consequences for the origin of the Moon. It implies that rather than being a cold and rigid body, the Moon's interior was more fluid, and perhaps hotter, than had been previously thought.

If the mascons are compensated, as O'Keefe and Hulme suggest, then they should be flanked by gravity anomaly lows. This is the case at Mare Serenitatis, as is indicated in the observed free-air gravity anomaly profile in Fig. 8.5 and the map in Fig. 8.6a.

Kunze (1976) noted that the area under gravity anomaly high over mascons is approximately equal to that under the flanking lows. This implies, according to Gauss's theorem, that the mass excess represented by the high is similar to that of the mass deficiency represented by the lows. He therefore speculated that the mascon gravity anomaly pattern was the result of the superposition of a positive anomaly associated with the mare load and a negative anomaly caused by a downwarping of the low-density crust by the load into a dense underlying mantle. However, he did not go the next step and propose a regional model of isostasy. He pointed out that such a model would imply stress differences that the

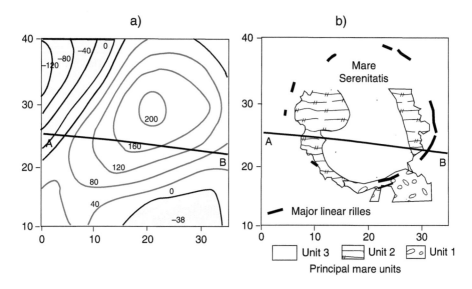

Figure 8.6 Mare Serenitatis. Profile AB locates the crustal model of Janle (1981) shown in Fig. 8.7. (a) Radial accelerations (in mGal) generated by Moon's gravitational field at 100 km altitude. Reproduced from fig. 4 of Ferrari (1977), copyright by the American Geophysical Union. (b) Geology and tectonic structure. Unit 1 is the oldest basaltic flow and Unit 3 the youngest. Concentric rilles are interpreted graben-like features that form in the region of maximum flexural bending stress due to mare basalt loading. Reproduced from fig. 4a of Solomon and Head (1980), copyright by the American Geophysical Union.

lunar crust might not be able to support on long timescales. Rather surprisingly, therefore, Kunze opted for Hulme's local model of isostasy instead.

It is clear from Hulme's results, however, that it is difficult to explain the amplitude and wavelength of the lows that flank the mascons using only a model in which the maria loads are locally compensated. As Fig. 8.5 shows, a model in which the mass deficit is located immediately beneath the pre-mare basin explains the high, but it predicts a low that has a longer wavelength and smaller amplitude than is observed.

One explanation for the gravity anomaly pattern is that the maria are underlain by crust that is denser and thinner than beneath the surrounding highland regions. Janle (1981) showed, for example, that gravity anomalies over Mare Serenitatis could be explained by a model in which the maria are underlain by relatively high-density, thin crust that abuts relatively low-density, thick crust beneath the Caucasus and Taurus mountains (Fig. 8.7).

But, like Hulme, Janle was unable to account for both the high and flanking lows. While the thicker crust beneath the surrounding highlands produces a gravity low, their effect is partly negated by a gravity high due to the dense, thin crust beneath the pre-mare basin. This suggests use of the model that was considered and then rejected by Kunze: namely, that the maria are compensated by a regional rather than a local model of isostasy.

The first to apply the ideas of flexural isostasy to the lunar maria was Kuckes (1977). He showed that the pattern of gravity 'highs' and 'lows' over the maria could be explained by

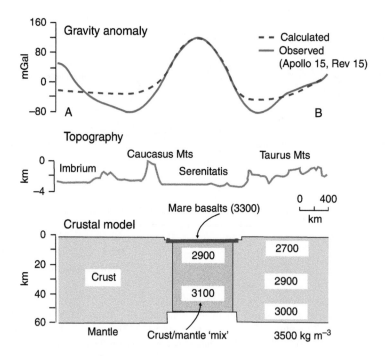

Figure 8.7 Crustal model of Mare Serenitatis based on gravity modelling. Profile AB is located in Fig. 8.6. Reproduced from fig. 4a–c of Janle (1981).

a model in which the basaltic lavas that infilled the impact basins act as a load that flexed the underlying lithosphere downwards due to their weight. He estimated the magnitude of the load from the gravity high and used it to calculate the flexure and, in turn, the gravity anomaly. Kuckes argued that the best fit to the observed gravity anomalies was for an elastic plate model with a T_e of 50–100 km. He showed that the bending stresses induced by flexure were of the order of 100 MPa and argued that, since stresses of this order could apparently be supported by the strength of the Earth's lithosphere over timescales of tens of millions of years, they might also be supported on the Moon.

The work of Kuckes triggered a search for other manifestations of flexure. Solomon and Head (1979), for example, proposed that the mare 'ridges' and sinuous 'rilles', described earlier by Strom (1971) and others, might be of flexural origin. The ridges form complex, radiating patterns that extend for hundreds of kilometres within the maria (e.g., Fig. 8.8a). The rilles, on the other hand, are narrow concentric graben-like depressions that are located outside the mare (e.g., Fig. 8.8a). Solomon and Head (1979) pointed out that both types of features are consistent with a model in which maria loads flex the underlying lithosphere. Such a model predicts compressional features in the upper part of the plate within and beneath the mare load and extensional features in flanking regions (Fig. 8.8b).

Solomon and Head (1979) used an elastic plate model to estimate the bending stresses associated with the basaltic loads of Mare Serenitatis. They argued that the rilles had formed

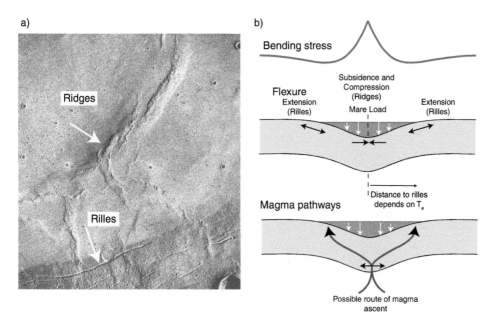

Figure 8.8 Simple model of flexure for the origin of ridges and rilles associated with mare. (a) Camera image of the south-east side of Mare Serenitatis showing ridges and rilles. Reproduced with permission from fig. 4.19 of Greely (1994). (b) Schematic diagram showing the bending stresses, flexure and magma pathways associated with the loading of an impact crater by basaltic lavas. The 'magma pathways' link regions of extensional stress in the flexed plate. Reproduced from fig. 14 of Solomon and Head (1979). Copyright of (a) and (b) by the American Geophysical Union.

in response to flexure of the lithosphere by the mare loads. By comparing the predicted position of the stress maxima with the distance of the rilles from the load centres (e.g., Fig. 8.9), they estimated a T_e of 25–50 km. The ridges, however, were attributed to horizontal compression of the flexed crust after the completion of mare loading. By comparing the calculated position of the stress maxima with the locations of the mare ridges, they estimated a T_e of 50–100 km.

Solomon and Head (1979) argued that the differences in T_e at the rilles and ridges in Mare Serenitatis was significant and indicated an increase in T_e between the time shortly after basin formation and the time when mare volcanism was ending. In their view, such an increase was reasonable since it implied that temperatures were higher during the early history of the Moon than they are at the present day.

Solomon and Head (1979) used the distance to the rilles from the load centres to constrain T_e at other maria: notably Grimaldi, Humorum, Imbrium, Nectaris, Smythii and Crisium (Fig. 8.2). Their results are summarised in Table 8.1.[4]

Pullen and Lambeck (1981), however, were somewhat critical of the results of Solomon and Head. In their view, several factors would need to be considered before they could be accepted.

---

[4] Table 8.1 is available from the Cambridge University Press server.

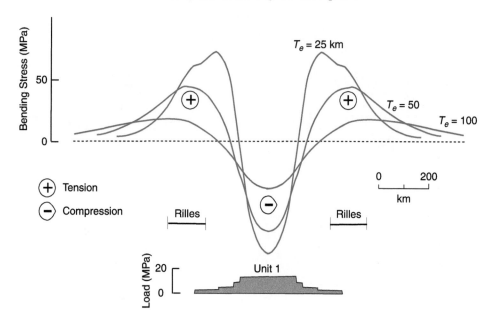

Figure 8.9 The tensional and compressional bending stresses that develop in the upper part of the flexed lunar lithosphere because of mare loading. Reproduced from fig. 9 of Solomon and Head (1979), copyright by the American Geophysical Union.

Firstly, the T_e estimates were only effective values: they did not represent the actual depth in the Moon to which materials were behaving elastically on long timescales. Secondly, the flexure models ignored the effect that the basin-forming impact would have had on the stress state prior to mare loading. For example, they queried whether the rilles were the result of the flexure of the lunar lithosphere. Pre-existing zones of weakness that were associated with the impact structure might be sufficient, in their view, to explain the rilles. Finally, and perhaps most importantly, is the fact that their flexure analysis was based on an elastic rather than a viscoelastic model. The models were therefore time-invariant and ignored the effects of viscous relaxation, which might be important on the long timescales that it has taken for the Moon to cool.

There is little question that the elastic plate model used by Solomon and Head is a simple one. The first two points made by Pullen and Lambeck are well taken, but they could equally apply to many flexure studies on Earth. The point is that while the elastic model is a highly idealised one, it has proved a useful 'reference' with which to compare the observations of flexure at different types of geological feature.

The final point made by Pullen and Lambeck concerning the role of viscous relaxation is, however, an important one. The role of viscous relaxation on the Moon has been examined in several earlier studies. Baldwin (1971), for example, applied Haskell's formula (see Section 1.11) to the Moon. He argued that if the Moon had a similar viscosity structure as Earth's mantle, a lunar crater, 200 km in diameter, would be compensated within about 1 Myr. Large craters would relax more quickly than small ones. He concluded that since many of the

Moon's largest craters are believed to date from the Moon's early history, its outermost layers must be more rigid than the Earth's. Otherwise, he argued, craters on the Moon would have disappeared long ago.

Shaber et al. (1977) agreed with Baldwin. They pointed out that although the Moon was bombarded by meteorites during a relatively short period of time, soon after its formation some craters appeared 'fresh' with a deep basin floor and a high rim while others were 'degraded' with a shallow basin floor and a low rim. They noted that it was the larger, rather than the smaller, craters that showed the signs of the greatest degradation. They therefore found the evidence 'undeniable' that the large lunar craters had been modified by isostatic processes that were very efficient, at least during the early evolution of the Moon.

Solomon et al. (1982) quantitatively examined the role that viscoelastic relaxation might play in accounting for the topography of a lunar mare and its flanking rim. They selected Orientale, on the far-side of the Moon, for the study because it is one of the best preserved, and probably youngest, of all lunar basins. First, they showed that if the Moon behaved as a viscous half-space with $\eta = 10^{24}$ Pa s, then Orientale should have fully relaxed after about 300 Myr. They then considered a model of a uniform viscosity layer ($\eta = 10^{24}$ Pa s) that overlies an inviscid half-space. They found that the relaxation depended on layer thickness but could be complete in times as short as 3 Myr. Finally, they examined a model of a uniform viscosity layer ($\eta = 10^{24}$ Pa s) that was separated from the underlying inviscid fluid by a density interface. They found that this model, where the surface topography and density interface are coupled, relaxed so slowly that significant topography remained, even after 3 Gyr (Fig. 8.10).

Based on these results, Solomon et al. (1982) used the topography of Orientale as a model for the initial topography of other, older, lunar basins. They compared, for example, topographic profiles of Mare Tranquillitatis (Fig. 8.2) and the South Pole–Aitken with a relaxed profile of Orientale's topography. They showed that while Tranquillitatis had undergone significant relaxation, Mare South Pole–Aitken had apparently undergone none. Solomon et al. (1982) speculated that these differences in relaxation reflected regional variations in the lunar viscosity, and hence, temperature structure.

The role of viscous relaxation in modifying lunar impact basins is therefore unresolved. There is evidence that some form of viscous relaxation may be required to explain the morphologic differences between large lunar basins, with young basins appearing less degraded than older ones. However, many smaller lunar craters do not appear to have undergone any significant viscous relaxation.

Williams and Zuber (1998) re-examined the problem by using a much higher resolution topographic data set than had been available to earlier workers. They used the new data to measure the depth and diameter of 29 large craters, including Tranquillitatis, and found a systematic relationship between the depth and diameter of lunar craters. In particular, log–log plots reveal a linear relationship with a change in slope at diameters of 100–200 km. Williams and Zuber suggested that the slope change reflects the transition as a crater evolves from a complex crater into a multi-ring basin. Since the slope change appears to be a characteristic feature of all the craters, they argued that significant viscous relaxation was unlikely. In fact, they considered the relationship between crater depth and diameter as

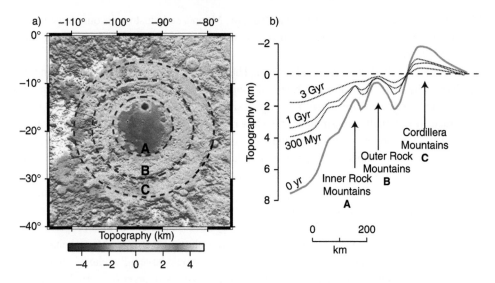

Figure 8.10 Topography of Mare Orientale on the Moon's far-side. (a) Topography derived from a Japanese LAser ALtimeter (LALT) mission (Araki et al., 2009). Grid resolution is approximately 0.0625 degree or about 1.9 km. The black dashed circles outline the Inner Rock Mountains (A), Outer Rock Mountains (B) and the Cordillera Mountains (C). (b) Representative topography profile. Dotted lines show how the topography would appear at different ages if the lunar lithosphere underwent relaxation assuming a model of a viscous layer overlying an inviscid fluid half-space with layer viscosity 10^{25} Pa s, thickness 50 km, and fluid density of 500 kg m^{-3} higher than the layer. Compensation of 50 per cent of the topography was also assumed as an initial isostatic condition, prior to mare loading. Reproduced from fig. 10 of Strom (1971) and fig. 9 of Solomon et al. (1982), copyright by the American Geophysical Union.

so diagnostic that they used it to estimate the thickness of basalts in the lunar basins, finding a thickness that varied from 0.6 km for Orientale to 5.2 km for Imbrium.

What is clear from this discussion is that the elastic plate (flexure) model has been applied with some success to the lunar mare. The results in Table 8.1, for example, reveal that $12.5 < T_e < 100$ km for the lunar lithosphere with a mean value of 59.3 km. NASA's Lunar Orbiter Laser Altimeter (LOLA; Smith et al., 2010) and Gravity Recovery and Interior Laboratory (GRAIL; Zuber et al., 2013) missions have significantly improved the resolution of Moon's topography and gravity fields such that topography has been resolved to spherical harmonic degree and order, n, of 720 and 420, which corresponds to wavelengths of 15 and 26 km respectively. As a result, new information has been obtained on both the density distribution of lunar crust (Wieczorek et al., 2013) and lunar plate flexure (e.g. Huang et al., 2014; Zhong et al., 2018).

Figure 8.11 shows a Bouguer gravity anomaly map of the of the Moon based on GRAIL data. The gravity field, which has been calculated from measurements of the change in distance between two orbiting spacecraft, has been reduced to free-air anomalies using the LOLA topographic data and to Bouguer anomalies assuming a uniform density of

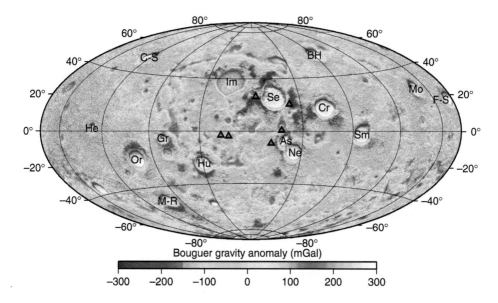

Figure 8.11 Bouguer gravity anomaly map of the Moon based on data obtained during the GRAIL mission (Zuber et al., 2013). The map, which shows the near side of the Moon in the middle and its far side to the left and right, is dominated by the gravity 'highs' and flanking 'lows' associated with the maria. Labels are as follows: Im = Imbrium, Se = Serenitatis, Cr = Crisium, Sm = Smythil, Mo = Moscoviense-N, F-S = Freundlioh-Sharonov, BH = Belkovitch Humboldtianum, As = Asperitatis, Ne = Nectaris, Hu = Humorum, Gr = Grimaldi, Or = Orientale, M-R = Mendel-Rydberg, He = Hertzsprung, C-S = Cruger-Sirsajis.

topography of 2,560 kg m^{-3}. The map is dominated by several large-amplitude gravity 'highs' over the maria and 'lows' over their flanks. Some of the 'lows' are, in turn, flanked by 'highs'. By analogy with mountain belts on Earth this pattern of Bouguer anomalies suggests the maria are associated with sub-surface rather than surface (topographic) loads. The pattern supports the earlier suggestion of Solomon and Head (1979) based on stress considerations: maria are impact craters that have subsequently been infilled by extrusive basalts which have loaded and flexed downwards the underlying lunar lithosphere.

We may test the applicability of the flexure model by focussing on the new topography and gravity field of Serenitatis Mare. Fig. 8.12a shows the mare (filled black circles indicate its approximate edge), together with some of the ridges and rilles seen earlier (e.g., in Fig. 8.8). The free-air gravity anomaly 'high' (Fig. 8.12b) reaches a maximum amplitude of about +450 mGal and is confined to a region within the edge of the mare. The 'high' is therefore significantly larger in amplitude (by about +200 mGal) and more confined to the mare than the 'high' shown in Figs. 8.5–8.7 which were generated from earlier radial acceleration data. The flanking 'lows' are also more subdued.

Figure 8.13 shows the observed free-air gravity anomaly (solid red line) and topography (solid blue line) of Serenitatis Mare along a profile that intersects the Apollo 15

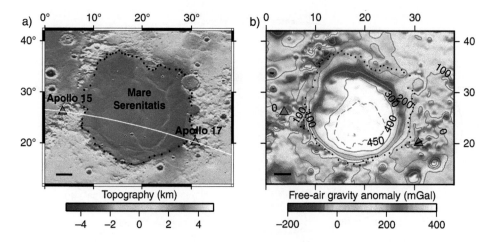

Figure 8.12 Topography and free-air gravity anomaly of Serenitatis Mare. Black filled circles outline the edge of the mare used in the flexure calculations. (a) Topography based on LALT data (Araki et al., 2009). Red filled triangles show Apollo 15 and 17 landing sites. Solid white line locates the profile shown in Fig. 8.13. (b) Free-air gravity anomaly derived from the Bouguer anomaly of Zuber et al. (2013) assuming a uniform density of topography of 2,650 kg m^{-3}. The free-air anomaly has been adjusted to agree with measurements acquired during the Lunar Gravity Traverse Experiment at the Apollo 17 landing site (Talwani, 1972; Urbancic et al., 2012). The contour interval is 100 mGal, except for values > 400 mGal where it is 50 mGal.

and 17 landing sites. The figure shows the mare is associated with a ~600-km-wide generally flat topographic depression, ~2.2 km below the reference elevation, and a >450 mGal gravity anomaly 'high'. The 'high' cannot be explained by the topography of the mare, and its steep gradients suggest it is caused by a mass excess in the uppermost crust. To test this possibility, we consider a simple 'process-oriented' model (See Section 7.2.3.4) in which the gravity anomaly measured today is considered as the sum of the gravity effect of the crater-forming process and the basalt infilling and loading process. The crater is considered to have formed by a single oblique impact (Wieczorek and Zuber, 2001) of a ~45-km-wide body with velocity ~20 km s^{-1} (Sugano and Heki, 2004). Such an impact would likely have locally thinned the crust and lithosphere, resulting in nearly complete isostatic compensation of the crater (Sugano and Heki, 2004). The negative gravity effect of the basin-forming crater would therefore be balanced by the positive gravity effect of the thinning, resulting in near-zero gravity anomalies. We may therefore assume that the main contribution to the gravity anomaly we see today arises not from the crater-forming process but from the mass excess associated with the emplacement of basaltic material that subsequently infilled the depression created by the crater. The crater, which was assumed to be 'bowl-shaped' and extend to different depths below the mare surface, was used to calculate the amount of the load and the gravity effect of the load and its flexural compensation at the top and base of the lunar crust calculated assuming different values of T_e. Observed and calculated gravity anomalies are compared in

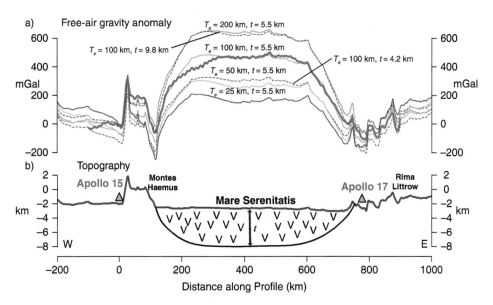

Figure 8.13 Free-air gravity anomaly and topography profile (Fig. 8.12) of Mare Serenitatis. (a) Solid colour curves show the calculated gravity based on a simple three-dimensional model of flexure in which an approximately 5.5 km thickness of basalt (density = 3,300 kg m^{-3}) infills an impact crater and loads the underlying lithosphere with T_e of 25 (purple), 50 (orange), 100 (green) and 200 (light blue) km. The dashed grey lines show the case of an approximately 4.2 and 9.8 km infill on a lithosphere with T_e of 100 km. (b) Topography showing Apollo landing sites and crater infill.

Fig. 8.13a, which shows that a calculated anomaly based on a crater depth of ~5.5 km and a T_e of ~100 km can generally account for both the amplitude and the wavelength of the observed anomaly. Higher values of T_e for the same crater depth predict too large an amplitude, while lower values for the same crater depth predict too small an amplitude. The largest influence on the gravity anomalies, however, is the crater depth with large crater depths (i.e., 9.8 km) predicting too large an amplitude, and small crater depths (i.e., 4.2 km) predicting too small an amplitude for the same T_e.

One implication of the comparisons in Fig. 8.13 are the limits they provide on the thickness of basalts in the maria. Previous morphological, composition and spectral studies show a wide range of values. Gong et al. (2016) for example, used a density spectrum analysis to suggest thicknesses in the range of 0.9–3.9 km with an average thickness of 2.9 km. Figure 8.13 suggests, however, a 'trade-off' between elastic thickness T_e and the crater depth. Previous work has suggested a minimum and maximum T_e for Serenitatis Mare of 25 km (Huang et al. 2014) and 100 km Solomon and Head (1979) respectively. As Fig. 8.13 and our calculations here show, if $T_e = 100$ km, the thickness of basalt infilling the impact crater is 5.5 km, but if $T_e = 25$ km, then the thickness would increase to > 9.8 km.

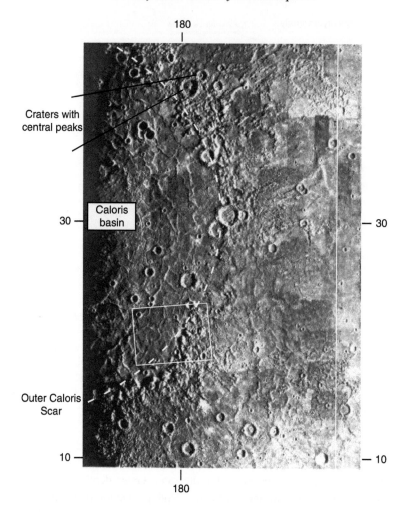

Figure 8.14 Mariner 10 camera image of eastern edge of Caloris basin in the northern hemisphere of Mercury. The area of the white box is shown in more detail in Fig. 8.16. Reproduced from fig. 11b of Soha et al. (1975), copyright by the American Geophysical Union.

8.3 Mercury

Mercury is the planet in the solar system that most closely resembles the Moon. Like the Moon, the surface of Mercury is heavily cratered (Fig. 8.14). It also has no atmosphere. Unlike the Moon, however, Mercury has a large metallic core.

The most common topographic feature on the surface of Mercury is the crater (Trask and Guest, 1975). Camera images from the Mariner 10 'fly-by' mission show that the craters range in diameter from a few tens of kilometres to several hundreds of kilometres, the largest being the Caloris basin, which is some 1,300 km across. The craters range in type from old, degraded craters with scalloped edges to young craters with well-developed central peaks, flat floors and terraced walls. Like the Moon, many of the craters on Mercury are surrounded by a blanket of ejecta material.

Between the craters are large areas of 'plains'. According to Trask and Guest (1975), the plains can be sub-divided into two main morphological types: intercrater and smooth. Intercrater plains are distinguished by their high density of superimposed small craters and are believed to be among the oldest terrains on Mercury. Smooth plains, on the other hand, are relatively sparsely cratered and are considered one of the youngest terrains on the planet.

A special feature of the smooth plains is ridges and scarps, which give them a 'rolling' appearance. The ridges resemble those found in the lunar maria. According to Strom et al. (1975), the scarps may be either curved faults or flow fronts. One of the largest, Discovery Rupes, is a lobate scarp (Fig. 8.15) that has been interpreted by Strom et al. (1975) as a thrust fault that formed in response to compressional stresses.

Shaber et al. (1977) noted that the morphology of the floor of two large craters in the equatorial region of Mercury resembled that of the intercrater plains. This observation, together with their subdued rims, suggested to Shaber et al. (1977) that the craters might be in isostatic equilibrium. They assumed that the craters had reached this state over the past 4 Ga and used Haskell's equation (Eq. 1.2) with a density of the sub-surface of 2,900 kg m^{-3} to estimate that the viscosity of the uppermost part of the Mercurian lithosphere was approximately $10^{25.5}$ Pa s.

The largest morphological feature on Mercury is Caloris Basin. This feature resembles the Orientale (compare Figs. 8.10 and 8.14) Basin on the Moon and, like Orientale, is thought to have formed by the impact of a large meteorite during the early history of the solar system. The impact site is surrounded by ring-like features within relatively smooth plains that are probably made up of volcanic material that flooded into the region after

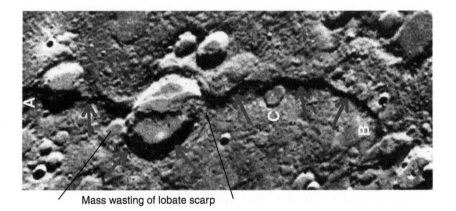

Mass wasting of lobate scarp

Figure 8.15 Camera image of Discovery Rupes in the southern hemisphere of Mercury. The lobate scarp has been interpreted as a thrust fault that formed in response to compressional stresses. Reproduced from fig. 11 of Strom et al. (1975), copyright by the American Geophysical Union.

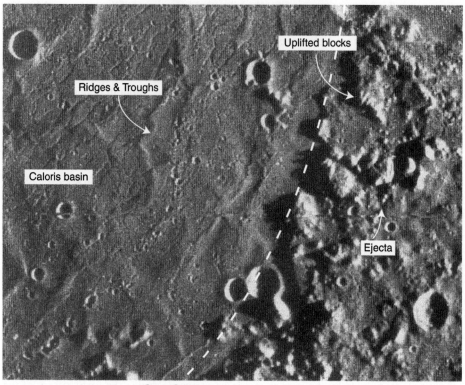

Figure 8.16 Camera image of the south-east side of Caloris Basin (see location in Fig. 8.14) showing the ridges and troughs in the otherwise smooth surface of the basin and the rugged topography of its flanks. Both the basin and flanks are heavily cratered. Reproduced with permission from fig. 5.27 of Greely (1994).

impact (Fig. 8.14). The Caloris Basin is flanked by topographic highs, 30–50 km across and several kilometres high, which comprise thin deposits of ejecta material (Fig. 8.16).

Antipodal to Caloris Basin is a region where there is a chaotic mix of hills and fractures. Schultz and Gault (1975) suggested that this terrain may have been produced in response to shock waves that travelled to this point on the surface following the impact that formed the Caloris basin.

Although there is evidence of filled craters, there have been only a few features on Mercury that have been attributed to isostasy. The problem is that the Mariner 10 mission only acquired radar altimeter and tracking data and, hence, gravity anomaly and topography data over less than half of the surface of the planet. Nevertheless, Watters et al. (2002) were able to estimate the depth of thrust faulting at Discovery Rupes (Fig. 8.15) from measurements of the width[5] of the fault 'back scarp' and 'dip'. Nimmo and Watters (2004) used these estimates, together

---

[5] Estimated from a topographic depression on the trailing edge of the scarp.

with estimates from other lobate scarps, to construct a Yield Strength Envelope (YSE) for the brittle-ductile transition assuming a dry anorthosite and olivine rheology for the crust and mantle respectively and a strain rate of 10^{-17} s^{-1}. By superimposing the curvature of the back scarp on the YSE they found $25 < T_e < 30$ km for the Mercurian lithosphere at the time of lobate scarp formation ~4 Ga. Subsequent studies (e.g., Williams et al. 2011) have argued that the depth to the brittle-ductile transition would increase as the planet cooled such that T_e would increase to ~76-77 km at the present day.

NASA's MErcury Surface, Space ENvironment, GEochemistry, and Ranging (MESSENGER) 'fly by' mission conducted during 2011 to 2015 has significantly improved our understanding of the state of isostasy on Mercury through its acquisition of high-resolution gravity and topography data. These data (Fig. 8.17) show, for example, that

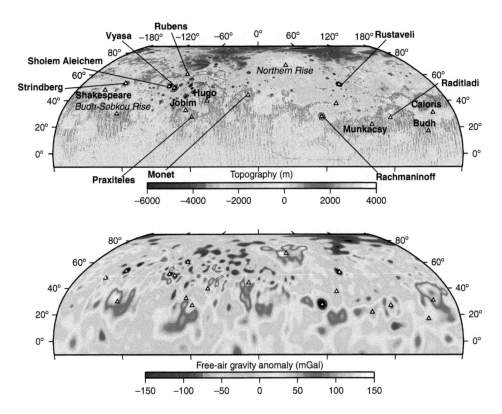

Figure 8.17 Topography and free-air gravity anomaly map of the northern hemisphere of Mercury. Original topography and gravity data were acquired during the MESSENGER 'fly-by' mission and made available by the Planetary Data System (http://pds- geosciences.wustl.edu) and Genova et al. (2019). The topography is based on a 5–10-km grid and so resolves wavelengths down to ~10–20 km, while the gravity is complete to degree and order 90 and so resolves wavelengths down to ~125 km. White filled triangles show selected craters, basins and rises with diameters > 150 km. Labelled features on the topography map show craters associated with gravity anomaly 'lows'. Names locate either craters (bold font) or rises (italic font) associated with gravity anomaly 'highs'.

while most craters with diameters > 150 km are associated gravity anomaly 'lows' (e.g., Vyasa, Rachmaninoff, Strindberg), a few of which are associated with gravity 'highs' (e.g., Caloris, Munkacsy, Shakespeare). The 'lows' reflect the topographic depression of the impact and so are probably craters that have not yet been infilled by volcanoclastic sediments or basaltic lava. The occurrence of 'highs', however, reflects an infill load and its compensation so could potentially yield an estimate of T_e, as was demonstrated in Fig. 8.13 for lunar craters, most of which are infilled. The gravity anomaly 'high' associated with Caloris is significantly wider, and hence of longer wavelength (Fig. 8.17), than that of other craters and may indicate that this large, infilled impact crater is underlain by a broad mass excess (James et al. 2015). Other long-wavelength (wavelength > 1,020 km) 'highs' are associated with the Northern and Bodh-Sobkou rises and are notable because of their high admittance (> 60 mGal km^{-1}). James et al. (2015) have argued that these features are supported either by deep compensation; for example, by some form of flow if viscosities in the Mercurian mantle were low enough, or by uplift due to plate flexure of a relatively rigid lithosphere, or by some combination of these factors – a view generally supported by Kay and Dombard (2019) for the long-wavelength topography that surrounds Caloris. Unfortunately, no spectral analysis has yet been carried out to separate out from the gravity field the relative roles of these two effects.

Nevertheless, there has been some work on crustal thickness. Padovan et al. (2015) and Sori (2018) have used admittance functions derived from geoid and topography data to invert for the crustal thickness, assuming Airy isostasy and densities of the crust derived from mineralogical abundance and grain-density considerations. They found thicknesses of 35±18 km and 26±11 km respectively, similar to those on the Moon and other terrestrial planets. In a related, more recent study, Watters et al. (2021) used the Bouguer gravity anomaly to invert for the crustal thickness, assuming an average thickness of crust of 35 km and densities of the crust and mantle of 2,900 and 3,150 kg m^{-3} respectively. They found thicknesses in the range of 15 to 65 km, with the thickest crust associated with several lobate scarps similar to the one mapped earlier by Strom et al. (1975), while the thinnest crust was characterised by an absence of such features. None of these crustal thickness studies, however, have taken into account the possibility of flexure, and so the state of isostasy and the support of Mercury's relatively subdued topography remains unclear.

8.4 Mars

Mars is a small planet with a light, CO_2-rich atmosphere and a mass that is just 11 per cent of that of the Earth. Like the Moon and Mercury, Mars is heavily cratered. Unlike these two planets, however, Mars shows evidence on its surface of erosion, sedimentation, and mass transport, suggesting that at some time in its past, the planet had both a denser atmosphere and water.

The possibility of water, and hence life, on Mars has generated considerable societal interest in the exploration of the planet. As a result, there have been more 'fly-by' and 'orbiter' missions to Mars than to any other terrestrial planet. The Viking mission, which encountered the planet during 1976, was one of the most complex ever mounted by NASA, and it involved two orbiters and two landers.

Mars has a wide range of topographic features and has considerable relief. Its surface is dominated by the topographically high region of the Tharsis province (Fig. 8.18). Superimposed on Tharsis are several volcanoes and what appear to be extensional structures in the form of rift valleys.

The volcanoes of Tharsis province are unusual in several respects. Firstly, they are among the largest constructional features in the solar system (e.g., Blasius and Cutts, 1981). Olympus Mons (Fig. 8.19), for example, reaches heights of up to 22 km above surrounding regions, and its base is up to 600 km across. Secondly, they consist mainly of shield volcanoes that are composed of thousands of individual lava flows, many of which appear to have been emplaced through tubes and channels. Like their terrestrial counterparts the summits of the Tharsis volcanoes are characterised by calderas, some of which show evidence for repeated vertical collapse and tectonic modification of their steep rims. In addition, the slopes of the volcanoes show

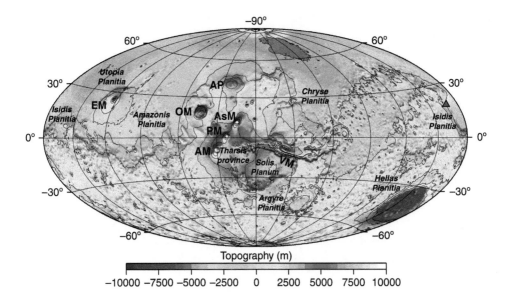

Figure 8.18 Topography map of Mars based on Mars Orbiter Laser Altimeter (MOLA) mission data (Smith et al. 2001). EM = Elysium Mons, OM = Olympus Mons, AP = Alba Patera, AsM = Ascraeus Mons, PM = Pavonis Mons, AM = Arsia Mons, VM = Valles Marineris. The purple-filled triangle on the edge of Isidis Planitia locates the landing site of *Perseverance Rover*, which is currently exploring a possible fan-delta at the edge of the crater.

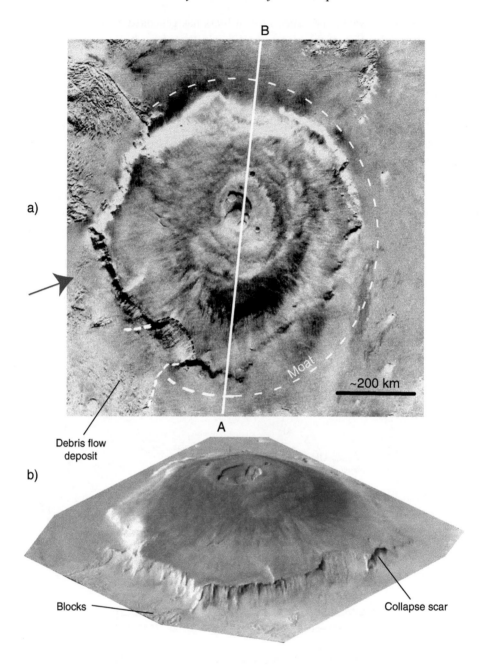

Figure 8.19 Camera image of Olympus Mons on the west flank of the Tharsis Rise. (a) Plan view showing summit caldera, large-scale flank collapses and downslope debris deposits. (b) Perspective view from the south-west. Reproduced with permission from fig. 7.11 of Greely (1994).

evidence of large-scale sector collapse, especially on their flanks where the flexural moat is overfilled.

The large size of Olympus Mons, compared to terrestrial volcanoes,[6] has raised many questions concerning the state of isostasy on Mars. Thurber and Toksoz (1978), for example, argued that Olympus Mons is regionally rather than locally compensated, based on the absence of evidence of normal faulting in the region flanking the volcano. By comparing the topography of the flank of the volcano with the predictions of simple models, they argued that T_e of the Martian lithosphere is high and probably exceeds 150 km.

Olympus Mons is associated with some of the largest positive gravity anomalies on Mars (e.g., Sjogren, 1979). The anomaly at a spacecraft altitude of 275 km, for example, reaches a maximum amplitude of +344 mGal over the crest of the volcano. Sjogren (1979) used the gravity anomaly and Gauss's theorem to estimate the mass anomaly associated with the volcano. He showed that Olympus Mons is associated with a mass anomaly of some 8.7×10^{18} kg, which is similar to the mass inferred from the topography, assuming a uniform density of 3,000 kg m^{-3}. The close similarity between the gravity- and topography-derived estimates suggested to Sjogren that Olympus Mons is essentially uncompensated.

But what did Sjogren mean by this? In a re-examination of the topography of Olympus Mons, Comer et al. (1985) agreed with Thurber and Toksoz's observation of the lack of evidence for normal faulting on the volcano flanks. They pointed out, though, there was evidence for a broad 1–2-km topographic depression or moat flanking the volcano. (Hints of a moat can be seen in Fig. 8.19a.) By comparing the depression with the predictions of simple models, they concluded that the best-fit value of T_e is 200 km.

Despite a high T_e, the large load of Olympus Mons causes a significant flexural depression and, hence, bending stress. Comer et al. (1985), for example, showed that stresses peak at 50–200 MPa at distances of 400–800 km from the load centre. They queried whether stresses of this magnitude could be supported by the strength of the Martian lithosphere. Viscoelastic relaxation, global compressive stresses, and dynamic support were all considered as possible alternative explanation for the flexure. Comer et al. (1985) concluded, however, that none of these factors were likely to be important. They therefore believed the Martian lithosphere to be strong enough that it could support large volcanic loads by flexure: the lack of concentric graben was explained as the result of a superimposed regional compressive stress, due to lithospheric cooling.

Comer et al. (1985) extended their analysis to other volcanoes in the Tharsis province. Unlike Olympus Mons, some of these volcanoes are flanked by faults, including graben, which bear a close resemblance to the linear bend faults observed seaward of deep-sea trenches on Earth (e.g., Figs. 4.61 and 4.62). A particularly good example is observed on the flanks of Pavonis Mons (Fig. 8.20). Comer et al. (1985) attributed the graben to extensional stresses that were generated in the Martian lithosphere by volcano loading. By measuring the radial distance to the graben from the centre of Pavonis Mons, Arsia Mons, Alba Patera and Ascraeus Mons and comparing them with predictions of simple elastic plate models, they estimated T_e values

---

[6] Tenerife, Canary Islands, appears to have the highest volcanoes on Earth, rising ~12 km above the surface of the oceanic crust.

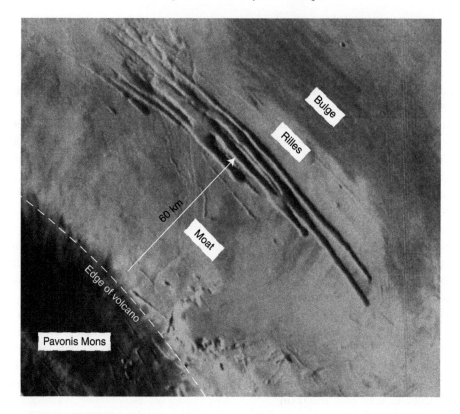

Figure 8.20 Camera image of the north-east side of Pavonis Mons (Fig. 8.18) showing the rilles which are interpreted as flexural bend faults. Reproduced from fig. 5 of Comer et al. (1985), copyright by the American Geophysical Union.

that were in the range of 18–54 km. These values (e.g., Table 8.2)[7] are substantially lower than those deduced at Olympus Mons.

Other constraints on the T_e of the Martian lithosphere have come from studies of impact craters. Some of the largest craters are Hellas, Isidis and Argyre (Fig. 8.18). The craters are circular in outline, are surrounded by concentric arcuate belts of mountainous terrain and closely resemble large impact craters on the Moon and Mercury. Of special interest is Isidis because it is flanked by concentric graben. Comer et al. (1985) argued, by analogy with lunar rilles, that the graben formed in response to stresses that were generated in response to volcanic material infilling and loading the impact basin. By comparing the distance to the graben with the predictions of the stress maxima based on different values of T_e, they estimate that Isidis T_e is 200–300 km.

An outstanding question is what is the origin of the high, plateau-like features such as Tharsis and Elsyium (Fig. 8.18) on which several of the largest volcanoes are superimposed? Carr (1974) attributed radial fractures on Tharsis (e.g., Valles Marineris) to uplift of

---

[7] Table 8.2 is available from the Cambridge University Press server.

the crust by convective upwelling in the underlying mantle. Solomon and Head (1979) suggested, however, that the rises are mainly constructional features that are regionally supported by the Martian lithosphere. They argued that early in the history of Mars, the lithosphere responded locally to volcanic loading. Later, as the planet cooled and became more rigid, the lithosphere responded more regionally. The long time of cooling (4 Gyr?), the relatively young age of loading and therefore the large age of the lithosphere at the time of volcano loading may account for the high values of T_e that have been reported, which, in turn, helps explain the extreme heights of the volcanoes, for example, in the Tharsis region.

Recently, there have been a number of new NASA missions that have significantly enhanced the resolution of the Mars gravity field. These include the 1996–2007 Mars Global Surveyor (MGS) and, since 2001, the Mars Odyssey, and 2006–present Mars Reconnaissance Orbiter (MRO) mission data. Figure 8.21 shows, for example, the Goddard Mars Model-3 (GMM-3) which has been derived from these data and is complete to degree and order $n = 120$ and so resolves features in the Martian gravity field down to wavelengths of ~180 km (Genova et al. 2016).

Fig. 8.18 shows that the Mars free-air gravity anomaly field is dominated by large amplitude 'highs' associated with the volcanoes of the Tharsis province and the Elysium rise and 'lows' associated with Valles Marineris. The large impact basins are associated with 'lows' with central 'highs' (e.g., Isidis, Argyre) or without (e.g. Helles). Other more regional 'highs' characterise the northern and southern highlands and regional 'lows' characterise the basins of Utopia, Amazonis and Chryse. There is no obvious expression in the free-air gravity field of the boundary that separates the ancient highlands of the

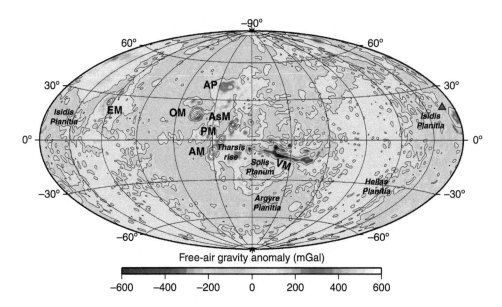

Figure 8.21 Free-air gravity anomaly map based on the Goddard Mars Model-3 (GMM-3) (Genova et al. 2016). See Fig. 8.18 for labels.

Southern Hemisphere and the young-appearing northern lowlands of the Northern Hemisphere of Mars, which is marked by a scarp like feature, extensional and compressional tectonics, and differences[8] in mean elevation up to 6 km and which Watters (2003) has attributed to plate flexure.

Studies of GMM-3 data have yielded several new estimates of T_e, along with estimates of crustal thickness, heat flux and stress state. There is general agreement that T_e at Olympus Mons is high, > 100 km, and in the range 93–300 km (Zuber et al. 2000, Belleguic et al. 2005, Beuthe et al. 2012, Ding et al. 2019). Other volcanoes, however, yield more inconsistent results with estimates at Arsia, Pavonis and Ascraeus Mons, for example, ranging widely from 10– >20, 26–175 and 22–105 km respectively.

The studies referred to previously were based on the admittance and coherence techniques which use the spectra between observed gravity and topography to estimate T_e considering surface and sub-surface loading. Although spectra can be localised to individual features, there are difficulties, as outlined by Beuthe et al. (2005), in regions such as the Tharsis rise where window sizes need, on the one hand, to be small enough to focus on an individual volcano yet, on the other, not so small that they cannot resolve a high T_e or exceed the resolution of the gravity and topography data. Indeed, Beuthe et al. (2012) resolved only minimum or maximum estimates of T_e for the major volcanoes on the Tharsis rise and no estimates at all for Pavonis Mons and Alba Patera.

Another approach to inverse (i.e., spectral) modelling is forward modelling. The application of such an approach to the estimation of T_e at Olympus Mons is illustrated in Fig. 8.22b. This volcano is located on the westernmost flank of the Tharsis rise (Fig. 8.22a). And since the rise resembles mid-plate topographic swells on Earth such as that at Hawaii, for example, and Mège and Masson (1996), among others, have speculated it is supported by a deep mantle plume, we need to first remove it from the observed topography and gravity data. This can most easily be achieved by filtering, for example, with a Gaussian filter of sufficient width (in this case $\sigma = $ ~900 km) to isolate the ~5,000-km wide rise. The difference between the observed and filtered topography yields the surface (topographic) load required to calculate the flexure and the gravity anomaly. Fig. 8.22b compares the observed gravity anomaly based on GMM-3 data at Olympus Mons to the calculated anomaly based on the sum of the gravity effect of the positive anomaly due to the load and the negative anomaly caused by the flexure of low-density crust into high-density mantle, an assumed density of the load of 3,300 kg m^{-3} (which is within the range of estimates based on measurements of Martian meteorites which are believed to have originated from the Tharsis volcanoes), densities of infill, crust and mantle of 2,900, 2,900 and 3,300 respectively and an elastic plate with T_e of 50, 100 and 200 km. Comparisons of observed and calculated gravity anomalies suggest T_e ~ 200 km. This T_e accounts well for the amplitude and wavelength of the observed anomaly. Placing bounds on T_e is, however, difficult. While $T_e \ll 200$ km can be ruled out because such values predict too-small-amplitude positive anomalies over the crest of the volcano and too-large-amplitude negative anomalies over its flanks, it is possible that $T_e > 200$ km but the

---

[8] Referred to as the 'Mars dichotomy'.

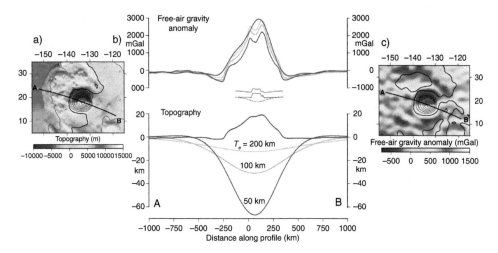

Figure 8.22 Topography and free-air gravity anomaly of the Olympus Mons region. (a) Observed topography based on MOLA data. Contour interval = 2.5 km. (b) Comparison of observed and calculated free-air gravity anomaly along Profile AB. The calculations are based on a three-dimensional elastic plate model, densities of load, infill, crust and mantle of 3,300, 2,900, 2,900 and 3,300 kg m^{-3} respectively and T_e = 50, 100 and 200 km. The inset shows – at the same scale – the bathymetry (solid blue line), free-air gravity anomaly (solid red line) and flexure (dotted blue line) of the top oceanic crust along a profile that intersects the Hawaiian Ridge between Oahu and Molokai. (c) Observed free-air gravity anomaly based on GMM-3 data. Contour interval = 500 mGal.

difference in the calculated anomaly between T_e = 200 km and, say, T_e = 300 km is simply too small to resolve in the observed anomaly. Nevertheless, the results in Fig. 8.22 seem consistent with previous estimates which predict (e.g., Zuber et al. 2000, Belleguic et al. 2005, Beuthe et al. 2012, Ding et al. 2019) uniformly high values for Olympus Mons with $93 < T_e < 300$ km.

The widest range of T_e estimates from previous studies has arisen for Arsia, Pavonis and Ascraeus Mons, which have been emplaced on the north-western edge of Tharsis province (Fig. 8.23). This is surprising, given that the three volcanoes form a linear NE–SW trending chain and, in the absence of plate tectonics on Mars, would be expected to have been emplaced on similar-age lithosphere at the time of loading.

Figure 8.24 compares observed and calculated gravity anomalies along Profile CD, which connects the summits of Arsia, Pavonis and Ascraeus Mons. At first glance it appears the calculated anomaly based on T_e = 200 km provides the best overall fit to the observed anomaly along the profile. However, the Root Mean Square (RMS) difference between observed and calculated gravity anomalies reveals a distinct minimum at T_e = 100 km (Inset, Fig. 8.24b). This arises mainly because the calculated anomaly based on T_e = 100 km better fits the gradients of the observed anomaly than T_e = 200 km, especially over the flanks of Arsia Mons. Another point is that there is a distinct 'bulge' in its flexure curve (by ~1–2 km) due to the superposition of the bulges caused by the 'yoking' of the

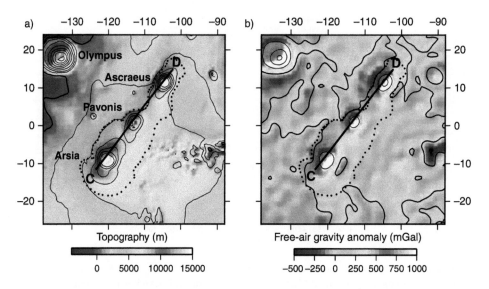

Figure 8.23 Topography and free-air gravity anomaly in the region of Arsia, Pavonis and Ascraeus Mons. (a) Topography. Contour interval = 2.5 km. (b) Free-air gravity anomaly. Contour interval = 500 mGal. Small black filled circles show the region used to isolate the surface topographic 'driving' loads of the three volcanoes. Profile CD is shown in Fig. 8.24.

Pavonis and Ascraeus loads that coincides for the $T_e = 100$ km case with the position of the rilles shown in Fig. 8.20. The rilles have been interpreted as graben which formed in response to extensional bending stresses. The $T_e = 200$ km case reveals no such 'bulge'. Therefore, we may conclude from these considerations that high values of T_e are required for Arsia, Pavonis and Ascraeus Mons; that there is little evidence of any change in T_e along the chain of volcanoes; and that the volcanoes were most probably emplaced on the Tharsis rise during the same period of time.

We pointed earlier that the contribution of the Tharsis rise to the topography and gravity anomaly had been removed by filtering prior to carrying out the flexure analysis. Figure 8.24b shows (dashed lines) the filtered topography and gravity that was removed along Profile CD in the region of the rilles to be 5.1 km and 160 mGal respectively. This corresponds to an admittance of 31.4 mGal km^{-1} which is applicable to features of widths of $\sim 2\,\sigma$ or wavelengths of $\sim 3{,}600$ km. This agrees with the long-wavelength admittance of ~ 30 mGal km^{-1} derived by Williams et al. (2008) for the 'Thaumasia Highlands', which includes Pavonis Mons but is significantly less than that derived by Williams et al. (2008) for 'West Tharsis', which includes Pavonis as well as Arsia, Ascraeus and Olympus Mons, of ~ 60 mGal km^{-1}, and by McKenzie et al. (2010) for convection beneath continental lithosphere of ~ 50 mGal km^{-1}. This is suggestive of different supportive mechanisms for the Tharsis rise with perhaps a more static isostatically balanced flexural supported origin for its eastern region and a more dynamic origin for its western region.

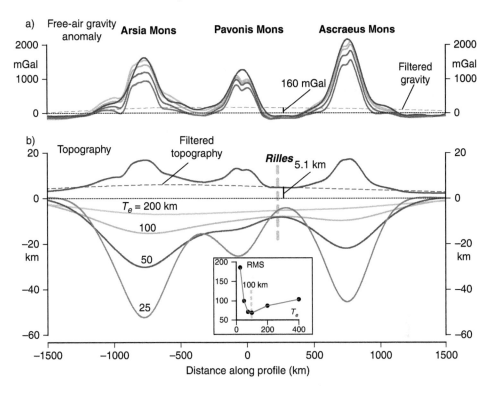

Figure 8.24 Comparison of observed and calculated gravity anomalies along Profile CD (Fig. 8.23), which intersects the crests of Arsia, Pavonis and Ascraeus Mons. (a) Gravity anomaly. The observed gravity is based on the GMM-3 data with Tharsis rise (wavelength ~1,800 km) removed from it. The calculated gravity (solid colour lines) is based on the surface topographic loads, the same elastic plate model, and densities as in Fig. 8.23, and $T_e = 25$, 50, 100 and 200 km. Grey dashed line shows the filtered gravity. (b) Topography and flexure of the top of the crust. Blue dashed line shows the filtered topography. Thick grey vertical line locates the rilles observed in Fig. 8.20 and shows they correlate with a flexural arch between Pavonis and Ascraeus Mons. The inset shows the Root Mean Square (RMS) difference between observed and calculated gravity as a function of T_e.

8.5 Venus

Venus is the terrestrial planet that most resembles Earth. It has a similar diameter and a nearly equivalent mass and density. At the surface, however, Venus is very different. The dominantly CO_2-rich atmosphere of the planet is nearly one hundred times denser than the Earth, and its surface temperature is several hundreds of degrees higher.

We know from 'fly-by' missions that the surface of Venus is remarkably smooth, compared with other terrestrial planets (Petengill et al., 1980). Highlands (i.e., regions with mountains > 10 km) cover only 5 per cent of the surface area and are concentrated in two main regions: Ishtar Terra and Aphrodite Terra (Fig. 8.25a). The remaining part of Venus consists mainly of lowlands with interspersed dome-like regions.

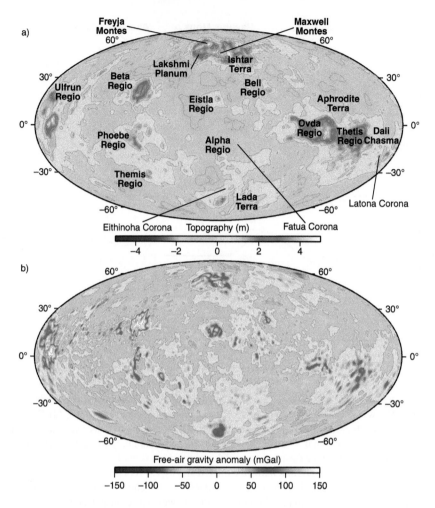

Figure 8.25 Topography and free-air gravity anomaly data derived from Magellan mission spherical harmonic models. (a) Topography. Contour interval: 2 km. (b) Free-air gravity anomaly. Contour interval: 100 mGal. Note the generally close correlation between topography and gravity anomaly. Data provided by the Planetary Data System (http://pds-geosciences.wustl.edu).

The existence of topographic highs and high surface temperatures on Venus has been a paradox since the earliest studies. Weertman (1979) argued, for example, that topographic highs could exist even if the surface temperatures are high, provided the rocks that comprise the outer layers of Venus have a 'dry' rheology. He based this idea on the results of experimental rock mechanics that demonstrate that the onset of ductility in commonly occurring minerals such as quartz occurs at much greater depths for dry than wet samples. If this is the case for Venusian minerals, then Weertman's view was that the outer layers of the planet could be rigid despite the high temperatures and, hence, some form of plate tectonics might exist on Venus.

Data from the Pioneer Venus Orbiter mission showed that gravity anomalies on Venus are not as large as they are on the Moon or Mars (Sjogren et al., 1980). Moreover, anomalies appear to correlate better with topography than on Earth. Sjogren and colleagues found that the gravity effect of the uncompensated topography was significantly larger than the observed gravity anomalies. He therefore concluded that there was 'at least partial isostasy with the possibility of complete compensation'.

Phillips et al. (1981) calculated the gravitational admittance for three features in the equatorial region of Venus (Western Aphrodite and two small features to the north-west) and compared them with the predictions of isostatic models. They showed that the admittance was consistent with either a local model of isostasy with a depth of compensation of about 100 km or a regional, flexural model with T_e in the range of 104–224 km. However, Phillips et al. (1981) dismissed both models: the local model because it was questionable that density differences could be maintained to such great depths on Venus due to viscous flow, and the regional (flexural) model because it was considered unlikely the planet could have a high T_e, given its high surface temperature. They preferred a model in which the topography of Venus was dynamically rather than statically supported.

A dynamic model has also been invoked by McGill et al. (1981) to explain the broad topographic rise at Beta Regio, one of the dome-like regions of Venus. The rise is characterised by rift structures that bear a striking resemblance to terrestrial rift valleys and their flanking rim uplifts (Fig. 8.26). McGill et al. (1981) ruled out a constructional origin for Beta Regio, although they did appear to accept that Theia and Rhea Mons, up to 4-km-high features on its southern and northern flanks respectively, might be shield volcanoes.

Cazenave and Dominh (1981) examined the relationship between topography and gravity along a single profile of the 'rolling plains province' between Eistla and Bell Regio. The province is characterised by many circular features[9] that resemble impacts or volcanic craters. By comparing observed and calculated anomalies based on a simple elastic model, they concluded that T_e of the Venus lithosphere was small and about 7.5 km.

Reasenberg and Bills (1983) criticised the results of Cazenave and Dominh. Their main point was that Cazenave and Dominh had used an unreasonably small crustal thickness of 6 km and that a larger value for the crustal thickness would have yielded a significantly larger value of T_e.

This was confirmed by Bowin (1983), who examined the relationship between topography and gravity along a profile of Aphrodite Terra. In particular, he found that a crustal thickness of 40 km would require a T_e in the range of 180–200 km! He argued that since such large values of T_e were unlikely on Venus, the best-fitting model was one in which Aphrodite Terra was locally compensated.

Bowin et al. (1985) extended their earlier analyses to other topographic features on Venus. They found, for example, a GTR (see Section 7.5.2) of 12–18 mGal km^{-1} over Beta Regio and Atla Regio compared to 5–9 mGal km^{-1} over Aphrodite Terra. They pointed out that these differences could be due to variations in either the depth of compensation (with

---

[9] Later called coronae.

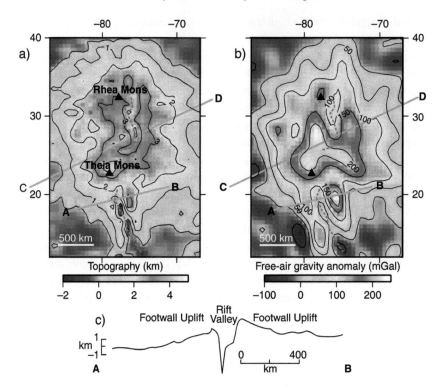

Figure 8.26 Topography and free-air gravity anomaly data in the region of Beta Regio based on Magellan mission data. (a) Topography. Contour interval = 1 km. Dashed line shows interpreted extent of the rift valley. (b) Free-air gravity anomaly. Contour interval = 50 mGal. (c) Topography profile AB reproduced from figs. 2b and 4b of Solomon et al. (1992) which show a 40–110-km-wide rift valley flanked by rim uplifts, copyright of the American Geophysical Union. Profile CD is shown in fig. 16(c) of McKenzie (1994).

Beta Regio and Atla Regio having a greater depth of compensation than Aphrodite Terra) or dynamic support (with Beta Regio and Atla Regio having more dynamic support than Aphrodite). Their preferred explanation, however, was some form of viscous relaxation in the Venusian lithosphere. According to this model, Beta Regio and Atla Regio have a higher GTR than Aphrodite Terra because they are younger and, hence, less relaxed.

The possibility that surface features on Venus have undergone significant viscous relaxation is a reasonable one, given the planet's high surface temperatures. Grimm and Solomon (1988), for example, compared the shapes of craters on Venus with the predictions of viscous models in which it was assumed that their initial topography resembled that of a fresh lunar crater. Their preferred model was one in which the crust and upper mantle behave as a linearly viscous fluid, with a viscosity that depends on the initial stress-difference as well as the temperature and the rheology of the Venusian crust and mantle. They found a 'trade-off' in the model between the crustal thickness and the average thermal gradient. This is because viscosity decreases with an increase in temperature, so relaxation is more rapid at a higher temperature. Also, the crust will be less viscous at the same temperature than the mantle, so the relaxation will be more rapid for a thicker crust. Therefore, significant relaxation (> 50 per cent) can occur if either the temperature gradient is high and the crustal thickness is low, or the temperature gradient is low and the crustal thickness is high. By comparing the observed and predicted crater topography, they were able to place limiting bounds on these parameters. They found, for example, that significant crater relief can be achieved on Venus for the largest craters (i.e., diameter > 100 km) if either the gradient is 20°C km^{-1} and the crustal thickness is < 10 km or the gradient is 10°C km^{-1} and the crustal thickness is < 20 km. Since heat-loss arguments suggest that the mean thermal gradient on Venus is probably not less than 10°C km^{-1}, then their results suggest that the crustal thickness needed to preserve crater relief must be of the order of 10–20 km.

As Grimm and Solomon (1988) point out, these estimates for crustal thickness on Venus are significantly less than the depths of compensation determined earlier by Phillips et al. (1981) and Bowin (1983). The reason for this, they argue, is that dynamic effects may significantly contribute to both the long-wavelength gravity and the topography field of Venus.

One of the most interesting regions of Venus, imaged during the Venera and Pioneer missions, is Ishtar Terra (Fig. 8.25a). This region is characterised by a 3.5–4.5-km-high plateau that is surrounded by a rim of high mountains that rise upwards of 1–3 km above the plateau (Head et al., 1990). One of these mountains, Maxwell Montes, is associated with a banded terrain that resembles that of the Valley and Ridge in the southern Appalachians and, for this reason, has been interpreted as the result of folding (Fig. 8.27). Maxwell

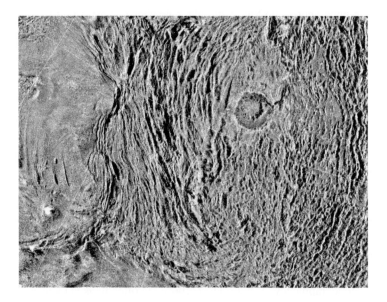

Figure 8.27 Synthetic Aperture Radar image obtained during the Venera mission showing the banded terrain on the west flank of Maxwell Montes in Ishtar Terra (Fig. 8.25a). Image courtesy of the Planetary Data System (http://pds-geosciences.wustl.edu).

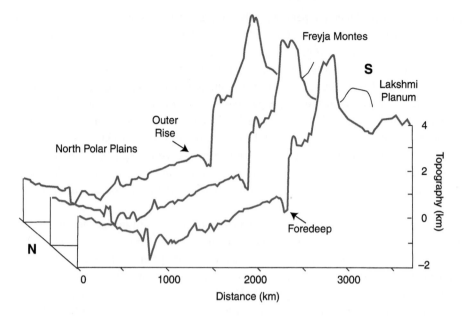

Figure 8.28 Topography profiles of the northern flank of Ishtar Terra in the region of Freyja Montes. Reproduced from fig. 2 of Solomon and Head (1990), copyright by the American Geophysical Union.

Montes and Freyja Montes are flanked on their southern and northern flanks respectively by a 100–150-km-wide depression Fig. 8.20, which reaches depths of up to 1 km below the surrounding plains. The depression resembles in its asymmetric shape the foredeeps that flank mountain belts on Earth (Fig. 8.28). Head et al. (1990) therefore attributed the foredeep to flexure due to underthrusting of the North Polar Plains lithosphere beneath Ishtar Terra.

Solomon and Head (1990) showed that the Freyja Montes foredeep is associated with an outer topographic rise which bears a striking resemblance to the outer rise seaward of deep-sea trenches in the western Pacific Ocean (e.g., Fig. 4.55). They compared topographic profiles of the foredeep and the outer rise with calculated profiles based on a broken-plate model and found a best-fitting T_e that was in the range of 11–18 km.[10]

In 1990, Venus was encountered during the Magellan mission. This mission acquired topographic data, which had a resolution 20 times better than that of the preceding Venera and Pioneer missions. By summer, 1992, Magellan had mapped 98 per cent of the surface of Venus.

One of most spectacular topographic features imaged by Magellan were the coronae. These circular features comprise concentric ridges and a central interior 'plains' region that forms either a topographic low or high. The coronae are often flanked by deep depressions

---

[10] Assumes Young's modulus = 100 GPa and Poisson's ratio = 0.25.

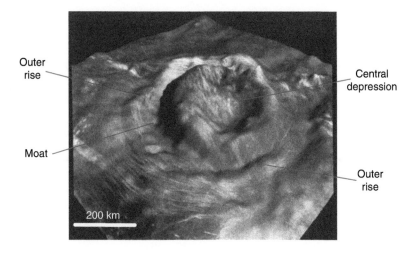

Figure 8.29 Side aperture image draped on the topography of Fatua Corona showing a central depression, flanking moat and outer rise. Reproduced from fig. 1b of Squyres et al. (1992), copyright by the American Geophysical Union.

or moats. Figure 8.29 shows an example from west of Aphrodite Terra and north of Lada Terra (Fig. 8.25a). This corona, known as Fatua Corona, has a diameter of about 300 km and stands approximately 600 m above the level of the surrounding plains. The Corona has a central depression of approximately 300 m and is flanked by a well-developed moat and outer topographic rise.

Sandwell and Schubert (1992) carried out the first systematic analysis of the moats that flank coronae, attributing them to elastic plate flexure. They used the Freyja Montes example studied earlier by Solomon and Head (1990) as a test of their methodology (Fig. 8.30) and then applied it to Eithinoha, Heng-O, Artemis and Latona coronae (Fig. 8.31). Altogether, Sandwell and Schubert (1992) examined > 50 profiles of moats and their flanking outer rises, finding values of T_e that, when corrected for curvature, ranged from a low of 15 km for Eithinoha to a high of 40 km for Heng-O. The low value at Eithinoha is noteworthy because this corona has a well-developed central depression (Fig. 8.32), a feature expected for wide volcanic loads emplaced on weak lithosphere, especially where there is little or no material that infills the surface flexure (see, e.g., Fig. 4.16).

While values of T_e derived in Figs. 8.30 and 8.31 are significantly less than estimates from Mars and the Moon, they are similar to values from oceanic regions on Earth.

Johnson and Sandwell (1994) pointed out, however, that evidence for flexure on Venus is limited. They concluded, for example, that there were only seven areas of the planet that exhibited flexural features, in addition to the five reported previously by Sandwell and Schubert (1992). These seven areas yield $12 < T_e < 34$ km. As Johnson and Sandwell (1994) point out, the existence of flexure implies large bending stresses, and supporting such stresses in the presence

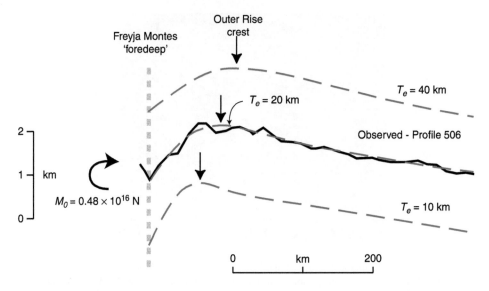

Figure 8.30 Comparison of the observed topography of the 'moat' and 'outer rise' north of Freyja Montes in Ishtar Terra (Fig. 8.25a) with calculated profiles based on a semi-infinite (broken) elastic plate model and T_e values of 10, 20 and 40 km. Reproduced from fig. 3 of Sandwell and Schubert (1992), copyright by the American Geophysical Union.

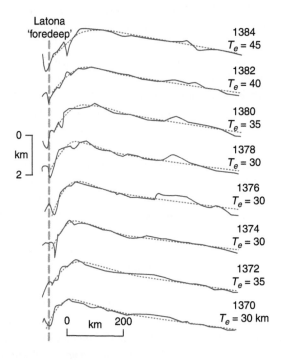

Figure 8.31 Comparison of observed (solid blue line) and calculated (red dashed line) topography profiles of the moat and outer rise associated with the Latona Corona in Aphrodite Terra. Reproduced from fig. 13 of Sandwell and Schubert (1992), copyright by the American Geophysical Union.

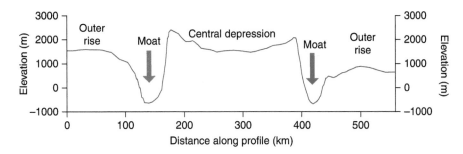

Figure 8.32 Topographic profile of Eithinoha Corona in Lada Terra. Reproduced from fig. 1 of Sandwell and Schubert (1992), copyright by the American Geophysical Union.

of high surface temperatures may be difficult. They therefore considered other models such as that of a viscous rather than elastic lithosphere and schemes in which flexure is cumulative, reaching its maximum role during the later stages of coronae evolution.

Smrekar (1994) and colleagues were among the first to use spectral techniques to determine the relationship between gravity and topography data over Beta Regio, Atla Regio, Bell Regio and Eistla Regio. They compared the observed admittance and coherence with predictions based on models of local isostasy (e.g., Airy and Pratt), surface and subsurface flexural loading (e.g., Forsyth, 1985) and a mantle plume (e.g., McNutt and Shure, 1986) that impinges on the base of the lithosphere. They found depths of compensation that ranged from 125 km for Bell Regio to 250 km for Beta Regio. Admittance and coherence spectra for Atla Regio and Bell Regio could be fit reasonably well by a surface and sub-surface loading model with T_e in the range of 30–50 km, although the plume model provided the best overall fit to the data. The spectral data from Beta Regio and Eistla Regio were too poor, however, to be used to estimate T_e.

One difficulty with the Smrekar (1994) study is the large scatter in her admittance and coherence estimates. In addition, she used a spherical harmonic representation of the gravity field that was only complete to degree and order, n, 60. This limits resolution to wavelengths of 630 km and higher. Although local gravity inversions enabled a resolution of 400–600 km to be obtained locally, these wavelengths are probably still too short to fully define the state of isostasy. On Earth, for example, the flexural effects of the lithosphere only begin to dominate over other effects, such as convection, at wavelengths less than about 600 km. It is questionable, therefore, whether the T_e estimates of Smrekar (1994), which are somewhat higher than those deduced from forward modelling, can be considered reliable.

McKenzie and Nimmo (1997) also used spectral techniques. However, they used a gravity field that was complete to degree and order 120. The resulting admittance was smoother than that of Smrekar and colleagues. Moreover, admittance values were obtained down to wavelengths of 200 km, which more clearly defined the uncompensated and compensated regions and the region of the spectra that might be associated with convection.

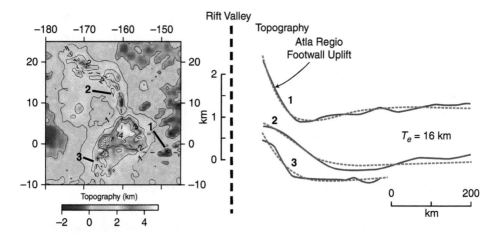

Figure 8.33 Comparison of observed and calculated topography in the region of Atla Regio. The observed topography is based on Magellan mission data. Dashed line shows the rift valley based on McKenzie and Nimmo (1997). The calculated topography is based on a flexure model with T_e = 16 km. Reproduced with permission from fig. 13 of McKenzie and Nimmo (1997).

The main result of the McKenzie and Nimmo study was to show that T_e in the regions of Atla Regio, Ulfrun Regio, Phoebe Regio, Eistla Regio and the Dali Chasma (near Latona Corona) were in the range of 11–38 km. These estimates are lower than those obtained by Smrekar, but similar to the forward modelling results of Sandwell and Schubert (1992) and Johnson and Sandwell (1994), as confirmed by McKenzie and Nimmo who showed, for example, that the footwall uplift that flanks some of the rift systems in Atla Regio (Fig. 8.33) could be modelled by an elastic plate with a T_e = 16 km.

A consensus is therefore emerging that the terrae and, perhaps, some coronae, are regions where compressional tectonics dominate, while the regio are regions where extensional tectonics dominate. These regions are characterised by $10 < T_e < 45$ km (Table 8.3[11]), and a similar range of T_e values are found in oceanic regions and in some compressional and extensional continental settings on Earth.

An outstanding question is why are the T_e values from Venus and Earth so similar if Venus has such a high surface temperature? This question can be best addressed by Yield Strength Envelope (YSE) considerations.

Suppe and Connors (1992), for example, constructed a set of YSE curves that were appropriate for the temperature and pressure conditions on Venus. They showed that if Venus comprises a 6-km-thick diabase-rich crust that overlies a dunite mantle, then the depth to the brittle–ductile transition would be 10–15 km, assuming a strain rate of 10^{-14} s^{-1} and a temperature gradient of 15°C km^{-1}. But this is still at variance, however, with the T_e

---

[11] Table 8.3 is available from the Cambridge University Press server.

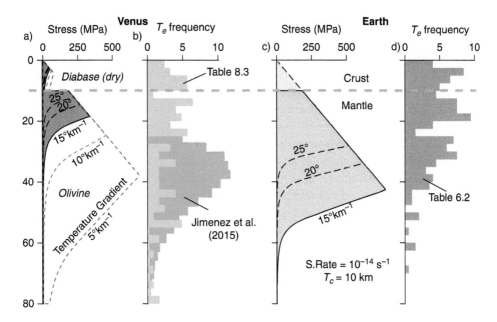

Figure 8.34 Comparison of the Yield Strength Envelope (YSE) curves for Venus and Earth. (a) Venus YSE assuming a surface temperature of 450°C, undried (C = Caristan, 1982) and dry diabase crust (M = Mackwell et al. 1998) and olivine mantle (Goetze, 1978; Goetze and Evans, 1979) and different thermal gradients. A thermal gradient of 15°C km^{-1} corresponds, for example, to ~100 Ma oceanic lithosphere. (b) Histogram of T_e estimates for Venus based on data in Table 8.3 and Jiménez-Diaz et al. (2015). (c) Earth YSE assuming ~100 Ma oceanic lithosphere, a surface temperature of 0°C, an olivine mantle and different temperature gradients. (d) Histogram of all T_e estimates for Earth (oceanic lithosphere) based on data in Table 6.2.

observations. The problem is illustrated in Fig. 8.34, which compares the YSE model for Venus and Earth, the only difference in parameters between them being higher surface temperature for Venus than for Earth. Because of its higher surface temperature, T_e on Venus should be generally smaller than that on Earth. This does not appear to be the case. While Venus is associated with low T_e values, recent studies based on spatially localised admittance studies suggest some high values (e.g., Smrekar and Stofen, 2003; Hoogenboom et al., 2004; Jimensez et al., 2015). Interestingly, Fig. 8.34 suggests that the T_e structure of Venus is more consistent with an Earth YSE than a Venus YSE. The Earth YSE in the figure is based on a ~100 Ma oceanic lithosphere which the cooling plate model predicts is associated with a temperature gradient of ~15°C km^{-1} and a depth of the Brittle-Ductile Transition (BDT) of ~45 km. Higher gradients correlate with younger lithosphere and shallower depths to the BDT, and lower gradients with older lithosphere and deeper depths to the BDT (see, for example, Fig. 6.30). The Venus BDT, however, is ~20 km for a temperature gradient of ~15°C km^{-1} which is about half that for the Earth, so one possible explanation for its high T_e values is a low temperature gradient in the Venusian sub-crustal mantle (say ~5°C km^{-1}) which makes it appear stronger than at the same depth in Earth's sub-crustal mantle.

We need to be cautious, however, before drawing too many conclusions regarding the YSE for Venus. It is possible, for example, that some of the surface features of Venus that have been used to model T_e are not the result of flexure alone. The features that flank coronae, for example, are associated with long-wavelength gravity and topography anomalies, and so it is possible that topography attributed to flexure is caused, at least in part, by processes that are acting below the lithosphere, such as convection. As is known from studies on Earth, it is not always easy to separate flexural effects associated with surface and sub-surface loading from the long-wavelength effects of deep-mantle convection.

The question is: Can the short-wavelength topographic features, which have been interpreted in terms of flexure, also be explained by convection? Moresi and Parsons (1995) for example, argued that they can. According to their models, convection, which is based on a rheological law that includes the possibility of lateral viscosity variations, will produce uncompensated topography at short wavelengths.

McKenzie (1994), however, expressed difficulties with attributing short-wavelength topography to convection. His main argument was that convection is unlikely to explain the admittance values on Venus, which are as high as 100 mGal km^{-1}. On Earth, in mid-plate bathymetric swell regions, the admittance is only about 25 mGal km^{-1}, and it is generally thought that swells are compensated at depths of 60–100 km (see Section 7.5.2). On Venus, in regio regions, the long-wavelength admittance reaches 50–60 mGal km^{-1} (Fig. 8.35) which would suggest depths > 180 km. An admittance of 100 mGal km^{-1} would require such large depths that it is unlikely that convection so deep in the Venusian mantle could contribute to short-wavelength surface topography.

If the short-wavelength topography is caused by flexure rather than convection, then the question that we raised earlier remains: why is T_e on Venus so high, given its high surface temperature?

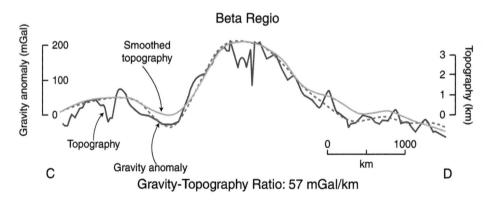

Figure 8.35 Free-air gravity anomaly and topography profile CD of Beta Regio (Fig. 8.26). The gravity anomaly has been plotted at the same scale as the topography, assuming a gravity/topography ratio (admittance) of 57 mGal km^{-1}. Reproduced with permission from fig. 16c of McKenzie (1994).

One possible explanation is the earlier suggestion of Weertman: Venus is a much drier planet than Earth. We know that in commonly occurring materials on Earth, such as quartz, the onset of ductility occurs at greater depths for dry than wet samples. Mackwell et al. (1998), for example, examined the rheology of the more mafic rocks believed to make up the Venusian crust. They have shown that the strength of dry diabase, for example, is much greater than values measured previously under undried or wet conditions. The result would be a much stronger crust. For example, a 10-km-thick undried diabase crust implies a low-strength zone in the lower crust, while a dry diabase crust would not (Fig. 8.34). A dry diabase crust therefore implies no strength contrast between the crust and mantle. Under these conditions, the entire Venusian crust would be rigid – despite its high surface temperatures – and mechanical coupling between the crust and mantle is likely to be strong.

More recent studies have focussed on the origin of coronae and what these features imply about the thermal and mechanical properties of the Venusian mantle. Hoogenboom and Houseman (2006), for example, appealed to basal lithosphere instabilities and upwellings and downwellings caused by variations in heat flow or stress in the asthenosphere beneath the lithosphere. They found the nature of the instability to depend on crustal thickness and long-term strength with high-viscosity crust being flexed downwards over a downwelling and low-viscosity crust being flexed upwards over an upwelling. They predicted a gravity-to-topography ratio (GTR) of 30–80 mGal km^{-1}, which is similar to observations of 35–76 mGal km^{-1} (Schubert and Moore, 1994). Jiménez et al. (2015) and Gulcher et al. (2020), on the other hand, have considered the role of secular cooling and plume-lithosphere as another source of instability contributing to the origin of the coronae.

These considerations suggest that controversy remains about the rheology and the state of isostasy on Venus. On the one hand, there is an excellent correlation between gravity and topography (better than on Earth), which McKenzie suggests reflects vigorous mantle convection. On the other, there is evidence from flexure studies that suggests Venus has a rigid, high-viscosity lithosphere that can support at least some of the planet's present-day topographic relief. The problem is to explain how Venus can have a rigid lithosphere, given its high surface temperatures. One possibility is that Venus is a drier planet than Earth, due perhaps to its closer proximity to the Sun. Therefore, the view is emerging (Ker, 1994) that, despite the planet having been re-surfaced several times in the past by magmatic events, Venus is presently a tectonically quiet planet with a different style of tectonics to Earth and where plate tectonics, as we know it, is absent.

8.6 Earth – Postscript

Earth has been the centrepiece of our study of isostasy, and so it is useful here to review those surface observables (e.g., topography[12] and gravity) made on the Earth and that also can be made with relative ease on the terrestrial planets.

---

[12] Since this is a global discussion we use the term 'topography' here and in the rest of this section for both topography on land and bathymetry at sea.

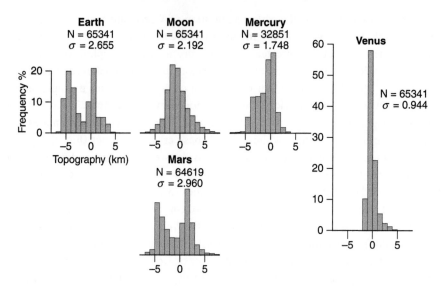

Figure 8.36 Histograms of topography for Earth, Moon, Mercury, Mars and Venus. Numbers show number and one standard deviation of the observations. Note that data on Mercury is limited to its northern hemisphere.

The topography is a key parameter in assessing the state of planetary isostasy. The Earth's topography, for example, is strongly bimodal: the two peaks correspond to the mean height of the continents and the mean depth of the oceans (Fig. 8.36). The bimodality is the consequence of Earth's evolution and the tendency of Earth's topography to approach isostatic equilibrium. The ocean basins are underlain by thin oceanic crust that forms at the mid-ocean ridge and subsides as it cools and increases its density with age. Oceanic crust is eventually subducted at deep-sea trenches, although sufficient remains at the present for there to be a peak in the histogram at ~-4.5 km, which is the depth of ~35 Ma oceanic crust. The continents, in contrast, are the products of a complex amalgam of orogenic belts that have been progressively added to its old cratonic cores, for example, by the closing of the oceans and continent–continent collisions and so are associated with significantly thicker crust than normal oceanic crust. Although continents comprise only ~28 per cent of Earth's surface erosion works to maintain them at or close to sea level, and so there is second peak in Earth's topography at ~+0.5 km.

Mars also is bimodal. The bimodality is at first glance surprising, since Mars is not believed to be currently active, although it may have experienced plate tectonics early in its history. The bimodality reflects the hemispherical 'dichotomy' of the northern plains, which are elevated and may have thin crust, and the southern highlands, which are depressed and may have thick crust. The origin of the dichotomy is not clear, but Zhong and Zuber (2001) speculate it is caused by a large-scale convective upwelling which has elevated at least part if not all of the Tharsis Rise region.

In contrast to the Earth and Mars, the topography of the other terrestrial planets is distinctly unimodal. The Moon, Mercury and Venus, for example, have a single peak at – 0.5, +0.5, and –0.5 km above or below their individual reference surfaces, respectively. The peak is widest for the Moon where there is a range of surface features of different heights and narrowest for Venus where there are a larger number of features of similar height.

Another key parameter is the gravity anomaly. The state of isostasy for the Moon, Mercury, Mars and Venus can be assessed, for example, by comparing the power spectrum of the observed and calculated gravity field, assuming that their topography is uncompensated. Figure 8.37 shows, for example, that the gravity effect of uncompensated topography generally follows the observed gravity field at short wavelengths. This is expected since short-wavelength features will appear uncompensated in the gravity field, irrespective of the state of isostasy. The main exception is Mercury, but we have only computed the spectra for the planet's northern hemisphere and so no doubt have underestimated the power in its observed field at short wavelengths. At long wavelengths, by way of contrast, the uncompensated gravity spectra generally exceed the observed gravity spectra (Fig. 8.28), indicating that some degree of isostatic compensation exists on all the terrestrial planets (Ferrari and Bills, 1979). The wavelength at which the spectra diverge varies, with Earth and Mars showing the clearest divergence at wavelengths > 300 km and >~1,700 km respectively. The different wavelengths reflect the role of flexural isostasy, with Earth characterised by a wide range of loads (e.g., sediment, orogenic and volcanic) on a variable-elastic-thickness lithosphere, and Mars characterised by a more limited range of loads (mainly volcanic) and a uniformly high elastic thickness. There is evidence on both Earth and Mars for some form of mantle dynamics which account for the divergence of their spectra at the longest wavelengths. Venus shows divergence for all wavelengths, with the largest differences at long rather than short wavelengths, and so most closely resembles Earth, while Mercury shows a constant divergence across the spectrum. Bills (1979) explained the latter phenomena, which he had identified in older data on Mars, as the consequence of differences in the

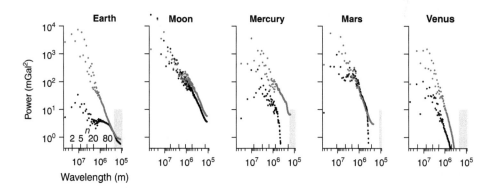

Figure 8.37 Free-air gravity anomaly power spectra for Earth, Moon, Mercury, Mars and Venus. Black dots show the observed gravity anomaly. Red dots show the calculated gravity anomaly assuming topography is uncompensated, the Bouguer 'slab formula' and a density of topography of 2,650 kg m^{-3}.

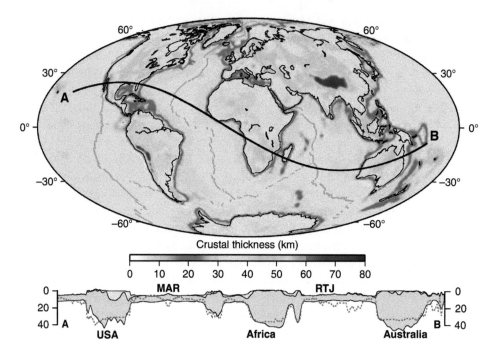

Figure 8.38 Crustal thickness of the Earth based on active source seismic studies as well as earthquake source receiver function studies (CRUST1.0 – Laske et al. 2013). Profile AB shows the regional crustal structure of USA, Africa and Australia and compares the observed seismically constrained depth to Moho (i.e., the base of the purple layer) to the calculated depth to Moho (red dashed line) based on the topography and a standard Airy model of isostasy. Yellow shading shows the sediment distribution. MAR = Mid-Atlantic Ridge. RTJ = Rodrigues Triple Junction.

depth of compensation for each spectral wavelength, with deeper compensation for the longer wavelengths and shallower compensation for the shorter wavelengths. The Moon shows the least divergence of all the terrestrial planets, but this probably reflects its dominance of sub-surface over surface loads which are manifest in the gravity field but not necessarily the topography.

The state of isostasy on the Earth is, without question, the best understood of all the terrestrial planets. On a large scale, the crustal structure of the oceans and continents can be explained remarkably well by a 'standard' Airy model with a zero-elevation crustal thickness of T_c = 31.2 km and a zero-degree crust and mantle density of 2,800 and 3,330 kg m^{-3}, respectively (Fig 8.38).[13] This model also appears to account quite well for the crustal structure in compressional regions where the crust is thicker than normal, and in extensional regions where it is thinner. The main discrepancies are at the mid-Atlantic ridge,

---

[13] Similar parameters were used to calculate the density structure of the 'standard' crust and lithosphere model illustrated in Fig. 2.26.

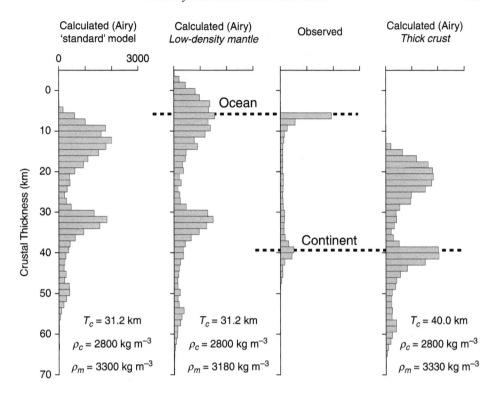

Figure 8.39 Histogram of crustal thickness for the Earth based on CRUST2.0. The calculated crustal is based on the topography and a standard Airy model of isostasy.

where the Airy model predicts too large a crustal thickness, and in the USA, Africa and Australia, where the Airy model generally predicts too small a crustal thickness.

The Earth's crustal thickness, like its topography, is strongly bimodal (Fig. 8.39). However, the crustal thickness 'peaks' deviate from the predictions of a standard Airy model. The ocean basin peak, for example, requires a lower mantle density (i.e., 3,180 kg m^{-3}) to explain it, while the continental peak requires a thicker zero elevation crust (i.e., 40 km). The lower mantle density explains the poor fit seen along Profile AB (Fig. 8.38) between the seismic Moho and the Airy model prediction at the mid-Atlantic ridge, where oceanic crust is thinner than predicted by Airy, and at USA, Africa and Australia, where the continental crust is thicker than predicted by Airy.

While the Airy model therefore explains well large-scale changes in Earth's crustal thickness, it fails on smaller scales. It was shown in Chapter 4, for example, that the crustal structure around volcanic loads cannot be explained by an Airy model of isostasy. These features require a strong lithosphere where the compensation at a topographic or bathymetric feature is spread out over a broad area rather than locally.

The large number of flexure studies on Earth have enabled a significant data base to be compiled on the elastic thickness of the lithosphere, T_e, and how it varies temporally and

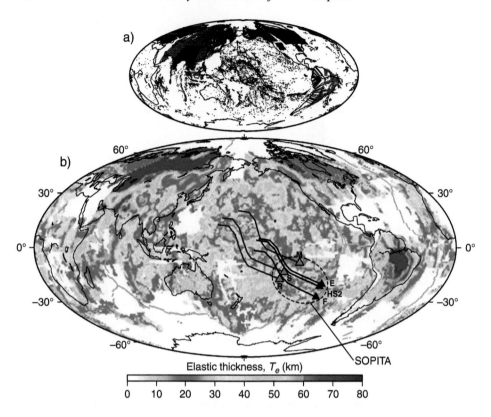

Figure 8.40 Global T_e map based on both forward and spectral and wavelet modelling. (a) Location of T_e estimates. (b) Grid of the T_e data constructed using the GMT (Wessel and Smith, 1991) algorithm nearneighbor with a grid interval of $1 \times 1°$, a $10°$ search radius and a T_e cap set at 80 km. The solid black and grey lines show the backtrack (to 120 Ma) of the Easter (E), HS2 (Searle et al. 1995) and Foundation (F) hotspots, and the Marquesas (M), Society (S) and Rurutu (R) hotspots respectively. The dashed line shows the approximate outline of SOPITA according to Pringle and Duncan (1995).

spatially. Figure 8.40 shows an attempt to use this data base to construct a global T_e map. In the oceans, T_e is based on the forward modelling results compiled in Table 6.1 and the reciprocal admittance (spectral) results of Watts et al. (2006). In the continents, a mixed data set was used that comprised the forward modelling results compiled in Table 6.2, selected admittance, and coherence (spectral) results (including those illustrated in Figs. 5.27 and 5.31), and results from wavelet methods (e.g., Audet and Burgmann (2011) where no other estimates are available, such as for China, Alaska, south-east Asia and parts of South America.

The importance of the global T_e map is that it is an indicator of how strong the Earth's outermost layers appear to past, current, as well as future, long-term loads superimposed on it. The presence of past loads, for example, is well seen in the central Pacific Ocean, where we see that regions of low T_e are not just confined to the East Pacific Rise but extend away from it in a generally north-west direction as far as the West Pacific Seamount Province (Koppers et al. 1995). Moreover, the low T_e does not follow oceanic isochrons, as might be expected, but intersects them at a high angle. The reason for this is that, once oceanic lithosphere has been loaded by a hotspot volcano, it retains a 'memory' of its thermal and mechanical setting,[14] such that if the plate migrates, and the hotspot remains fixed with respect to the deep mantle, then both the volcano and its associated T_e migrate with the plate. This is seen in Fig. 8.40b, which shows the backtracked trails of six present-day hotspots: three on the East Pacific Rise (solid black lines) and three off-ridge (grey solid lines). The six hotspots occupy the region of the so-called SOuth Pacific Isotopic and Thermal Anomaly (SOPITA) which is a broad region of unusually shallow seafloor (sometimes dubbed a 'superswell') and anomalous Sr, Nd and Pb isotope ratios, and so the isotope anomaly would be expected to migrate along with the volcano and the low T_e. Indeed, rock samples from the Magellan, Marshall and Wake seamounts in the West Pacific Seamount Province reveal that the SOPITA signature and the six backtracked hotspot trails generally encompass the region of low T_e.

Current or recent loads and unloads would not, of course, necessarily 'see' the region of low T_e in the central Pacific Ocean. Rather, they would reflect the increase in the strength of oceanic lithosphere as it continues to cool with age following an unloading or loading event. Therefore, loads and unloads such as those associated with, for example, solid Earth and Ocean tides (Mantovani et al. 2005), large-scale coal mining (Klein et al. 1997), and geologically recent phenomena such as tidal flexure at ice margins (Vaughan, 1995) and Plio-Pleistocene sediment accumulations at continental margins (e.g., Ebro – Nelson and Maldonado 1990, Barents Sea – Vogt et al. 1998) will 'see' the present-day age, and hence thermal and mechanical structure, of the lithosphere.

Perhaps the most interesting applications of a global T_e map is the information that it provides on the relative contributions of flexure to the global gravity and topography fields. For example, a histogram of a grid of oceanic T_e values (Fig. 8.41), which resembles a Yield

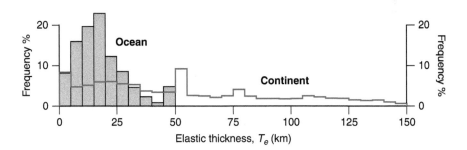

Figure 8.41 Histogram of the T_e data used in the construction of Fig. 8.41b. Note that T_e values > 80 km have been clipped in Fig. 8.40b.

---

[14] In this sense the lithosphere is 'forever', as O'Reilly et al. (2001) have suggested is the case for Archaen lithosphere based on its sub-continental mantle buoyancy and depletion.

Strength Envelope 'sailboat', has a distinct peak at 15–20 km. Continental T_e grids, on the other hand, show no such peak, but smaller, less well-defined peaks at approximately 50–55, 75–80 and 105–110 km. The isostatic response function (e.g., Fig. 5.21) shows that a T_e of 15–20 km will contribute little to the global gravity field for loads with a wavelength, $\lambda_w >$ ~800 km, which corresponds to degree and order $n < 50$. For a T_e of 105–110 km, the maximum continental peak, there is little contribution for $\lambda_w >$ ~3,000 km, which corresponds to $n < 13$.

These estimates of the limiting wavelength of flexural isostasy are generally supported by consideration of the Earth's topography and gravity power spectra. Figure 8.42, for example, compares the power spectra of the gravity field (black filled circles) to calculations based on the gravity effect of uncompensated topography (red filled circles) and its Bouguer-type compensation, which can be considered representative of a topography supported by an infinitely rigid lithosphere (i.e., $T_e = \infty$), and the gravity effect of topography and its Airy-type compensation (green filled circles), which can be considered representative of topography supported by an infinitely weak lithosphere (i.e., $T_e = 0$ km), to the observed free-air gravity anomaly spectra. The comparisons reveal

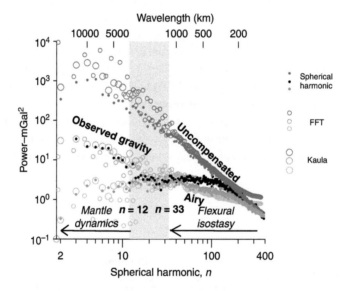

Figure 8.42 Power spectrum of Earth's free-air gravity anomaly field (black and grey symbols). The red and pink symbols show the power of the uncompensated topography and bathymetry, and the green and light green symbols show the gravity field of the topography and bathymetry and its compensation based on an Airy model of isostasy. Small filled symbols show a solution based on spherical harmonics of Watts and Moore (2017). Medium open symbols show a solution based on the GMT algorithm grdfft which is based on the Fast Fourier Transform (FFT). Large open symbols show the solution based on spherical harmonics of Kaula (1967). The grey shaded area shows the region between $12 < n < 33$ (i.e., wavelengths between ~1,210–3,340 km), where the relative role of mantle dynamics and flexural isostasy is uncertain.

a general agreement between the power of observed and calculated spectra at short wavelengths. At intermediate wavelengths, however, the calculated spectra diverge significantly from the observed spectra. The divergence can be directly attributed to isostasy, as was recognised earlier by Kaula (1967). However, the calculated Airy spectra generally underpredict the observed spectra at intermediate wavelengths, as has also been demonstrated by Kaula (1967), Rapp[15] (1989) and Tsoulis (2001). Recently, Watts and Moore (2017) showed that the observed spectra could be explained by the spectra of a flexure model of isostasy with $T_e = 34.0 \pm 4.0$ km, which is a regional rather than local scheme of compensation. At long wavelengths, the observed spectra show a marked increase in power and divergence from the spectra of all isostatic models which tend towards zero at the longest wavelengths. It is this part of the power spectrum that cannot therefore be accounted for by crustal isostasy and so was attributed by McKenzie (1967, 1977), among others, to mantle dynamics.

The significance of Fig. 8.42 is that it quantifies for us the contribution of flexural isostasy and mantle dynamics to the observed topography and gravity field; flexural crustal isostasy dominates the waveband $33 < n < 400$ which corresponds to wavelengths of ~100–1,210 km, while mantle dynamics dominates the waveband $2 < n < 12$ which corresponds to wavelengths of ~3,340–20,040 km. There is still a waveband ($12 < n < 33$) into which flexural isostasy or mantle dynamics may extend, but there is no compelling evidence as to which is dominant in the spectra data.

The spatial extent of where flexural isostasy and mantle dynamics dominate can be illustrated by plotting the free-air gravity anomaly in their respective wavelength bands. Figure 8.43a, for example, shows that flexural isostasy ($33 < n < 400$) dominates along seamount chains (e.g., Hawaiian-Emperor, Louisville, Western Pacific Seamount Province), where the negative-positive-negative patterns in the gravity field are observed, and at island arcs, deep-sea trenches and outer rises (e.g., circum-Pacific), where positive-negative-positive patterns are observed. Positive patterns indicate cases where Moho is shallower than predicted, say, by Airy due to flexure, and negative patterns where Moho is deeper. In the continents flexural isostasy dominates along the borders of Tibet and the Iran 'blocks', where negative-positive patterns are observed; the negative in this case reflecting depression of the crust beneath the foreland (e.g., Ganges), and the positive reflecting flexural support of the adjacent mountain belts (e.g., Himalaya). Figures 8.43b and 8.43c show that mantle dynamics ($2 < n < 12$) dominates in the west-central Pacific Ocean and North Atlantic Ocean regions, where positive gravity anomalies generally correlate with positive topography anomalies (i.e., shallower than expected bathymetry). Elsewhere, at the Middle America and Peru-Chile subduction zones, there are positive gravity anomalies which generally correlate with negative topography anomalies (i.e., deeper than expected bathymetry).

Included in mantle dynamics are processes that induce vertical stresses on the base of the plates, such as those associated with mantle convection. As was recognised earlier by

---

[15] Rapp attributed the departures in the observed and calculated power spectra at intermediate wavelengths as 'defects in the Airy model'.

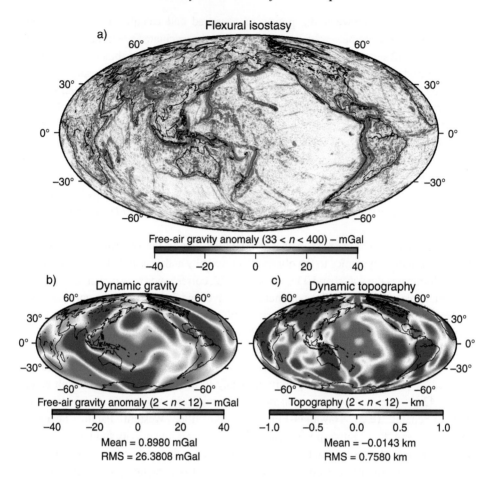

Figure 8.43 Free-air gravity anomaly field for the spherical harmonics that are dominated by flexural isostasy and mantle dynamics. (a) Flexural isostasy ($33 < n < 400$). The map reveals those features where flexure dominates such as along the Hawaiian–Emperor seamount chain, the Louisville Ridge and the Western Pacific Seamount Province. (b) and (c) Dynamic gravity and dynamic topography ($2 < n < 12$) respectively. The maps reveal those features where mantle dynamics dominate, such as beneath the west-central Pacific Ocean, where dynamic gravity and topography show a degree of correlation. Figures reproduced from figs. 4a and 8 in Watts and Moore (2017), copyright by the American Geophysical Union.

Morgan (1965) and McKenzie (1977), one of the strongest arguments for convection is a correlation between gravity and topography. One part of the convective circulation is the accretion and cooling of the oceanic lithosphere at mid-oceanic ridges and its subduction at deep-sea trench-island arc systems. Indeed, the North Atlantic Ocean is associated with a broad dynamic gravity and topography high which appears symmetric about the mid-Atlantic ridge crest and has been interpreted in terms of a convective upwelling associated with the Iceland plume. The East Pacific Rise, however, is characterised by an absence of a long-wavelength gravity and bathymetric high. There is a broad region of dynamic gravity and topography highs in the west-central Pacific Ocean, where, interestingly, it correlates spatially with the 'Western Pacific Seamount Province' – a region of numerous isolated seamounts on the ocean floor (Koppers et al., 1998), and the 'Pacific anomaly' – a low *S*-wave velocity region at the core–mantle boundary (He et al., 2006).

Superimposed on any large-scale convective circulation are likely to be smaller scales of convection such as that associated with individual hotspots. Prominent gravity anomaly highs, for example, that correlate with topographic highs in the region of Hawaii, Marquesas and Cape Verde 'swells' (e.g., Watts, 1976) have been interpreted, for example, as the result of a convective upwelling in the mantle. Prominent lows that correlate with topographic lows in the region of the Australia–Antarctic discordance (e.g., Gurnis and Müller, 2003), on the other hand, have been interpreted as the products of downwelling.

This discussion suggests that it should be relatively easy to separate the effects of flexure from surface observables on Earth and isolate the deep processes such as convection that are occurring beneath the lithosphere. The problem is that while it may be relatively easy to do in the oceans, where a suitable reference surface exists, and where away from volcanic constructs there is a relatively simple relationship between seafloor depth and plate age, it is much more difficult to carry out in the continents. One of the challenges for the future, therefore, is to determine more precisely the T_e structure of the continents.

8.7 Summary

We have shown in this chapter that, in comparison with Earth, our knowledge of the state of isostasy on the Moon, Mercury, Mars and Venus is fragmentary. In many ways, the situation is like it was at the turn of the twentieth century on Earth, when the early isostasists had to rely only on gravity anomaly and topography data. The paucity of data on the physical properties of these planets, and, more importantly, the lack of seismic constraints on their crustal and mantle structure preclude a quantitative assessment of the state of isostasy on the terrestrial planets.

Satellite encounters with the planets during the last three decades of the twentieth century have, however, yielded spectacular new discoveries. There is evidence, for example, from gravity and topography data that some degree of isostatic compensation exists on the Moon, Mercury, Mars and Venus. The significance of this observation is that it suggests that there is a weak zone at some depth in each planet that permits isostatic adjustment to take place in response to shifting loads on the surface. In addition, there is evidence on all planets for a strong outermost layer, or lithosphere, that helps to support these loads on long geological timescales.

The most significant loads on the Moon are the basaltic lavas that infilled its giant impact craters. There is evidence from compressional ridges and extensional grabens that the lunar lithosphere responded to these sub-surface loads by some form of flexure. Comparisons of the distance to the ridges and grabens with predictions of the stress maxima based on simple elastic plate models suggest that the lunar lithosphere has a T_e in the range of 15–100 km. There is evidence from the depth and diameter of large craters that, despite the long time that has elapsed since cratering and load emplacement, viscous relaxation is probably not a major contributor to lunar topography.

Mars has some of the largest volcanoes and impact basins of any planet in the solar system. Remarkably, both types of surface feature appear to be supported by the strength of the lithosphere. The T_e values that are required to explain the gravity anomaly, moat topography and the distance to flanking graben exceed 200 km. Mars therefore has some of the highest T_e values that have been recorded on the terrestrial planets and, hence, one of the strongest lithospheres.

Venus shows a wide range of topographic features that can be attributed to flexure, including rift flank uplifts, outer topographic rises and trench-like depressions and volcanoes. The T_e values that are required to explain these features, as well as other regions of the planet, are in the range of 10–30 km. These estimates are significantly smaller than those determined from flexure studies on Mars and the Moon. They are also smaller than those of the continents, but similar to those of the oceans. The main question is why Venus has such a similar T_e structure to Earth yet has so much higher surface temperatures. Yield Strength Envelope considerations suggest that Venus should have much lower T_e values than the Earth, perhaps as much as a factor of two lower. Yet, values of up to 35 km and higher have been documented. One possibility is that Venus, due to its proximity to the Sun, is a dry planet, such that the onset of ductility of its constitutive minerals occurs at much greater depths.

The Earth is the yardstick against which the state of isostasy on the terrestrial planets will be assessed in the future. The primary data sets will continue to be gravity anomaly and topography data together with seismic data which have the potential to image the surfaces of flexure directly. We are close to defining the relative contributions of plate flexure and mantle dynamics to Earth's topography and gravity fields as well as to its crustal structure and vertical motion history. The acquisition of higher-resolution data will increase the number of estimates of T_e of the planets, which, in turn, will help us to understand better the complexities of their geodynamical evolution.

References

Abelson, P. H., 1998. *Ross Gunn. May 12, 1897–October 15, 1966, Biographical Memoirs*, Washington, DC, National Academy Press, pp. 110–25.

Abercrombie, R. E., and Ekstrom, G., 2007. Earthquake slip on oceanic transform faults, Nature, *v.* 410, pp. 74–7.

Abers, G. A., and Lyon-Caen, H., 1990. Regional gravity anomalies, depth of the foreland basin and isostatic compensation of the New Guinea Highlands, Tectonics, *v.* 9, pp. 1479–93.

Adam, C., and Bonneville, A., 2008. No thinning of the lithosphere beneath northern part of the Cook-Austral volcanic chains, J. Geophys. Res., *v.* 113, B10104. https://doi.org/10.1029/2007JB005313.

Airy, G. B., 1855. On the computation of the effect of the attraction of mountain-masses, as disturbing the apparent astronomical latitude of stations of geodetic surveys, Phil. Trans. R. Soc., *v.* 145, pp. 101–4.

Ali, M. Y., and Watts, A. B., 2009. Subsidence history, gravity anomalies and flexure of the United Arab Emirates (UAE) foreland basin, GeoArabia, *v.* 14, pp. 17–44, https://doi.org/10.2113/geoarabia140217.

Ali, M. Y., Watts, A. B., and Hill, I., 2003. A seismic reflection profile study of lithospheric flexure in the vicinity of the Cape Verde islands, J. Geophys. Res., *v.* 108, https://doi.org/10.1029/2002JB002155.

Ali, M. Y., Watts, A. B., Searle, M. P. et al., 2020. Geophysical imaging of ophiolite structure in the United Arab Emirates, Nat. Comm., *v.* 11, https://doi.org/10.1038/s41467-020-16521-0.

Allen, M. B., Jones, S., Ismail-Zadah, A., Simmons, M., and Anderson, L., 2002. Onset of subduction as the cause of rapid Pliocene-Quaternary subsidence in the south Caspian basin, Geology, *v.* 30, pp. 775–8.

Anderson, D. L., and Minster, J. B., 1980. Seismic velocity, attenuation, and rheology of the upper mantle, in Allègre, C., ed., *Coulomb Volume*, Paris, Centre National de la Recherche Scientifique, pp. 13–22.

Anderson, R. N., and Noltimier, H. C., 1973. A model for the horst and graben structure of mid-ocean ridge crests based upon spreading velocity and basalt delivery to the oceanic crust, Geophys. J. R. Astr. Soc., *v.* 34, pp. 137–47.

Anonymous, 1878. The interior of the Earth. Abstract of an address given at the Cumberland Association for the Advancement of Science by Sir George B. Airy. Revised by the author, Nature, *v.* 18, pp. 41–4, https://doi.org/10.1038/018041b0.

Anonymous, 1919. Joseph Barrell (1869–1919), Am. J. Sci., *v.* 48, pp. 251–80.

Antoine, J. W., Martin, R. G., Pyle, T. G., and Bryant, W. R., 1974. Continental margins of the Gulf of Mexico, in Burk, C. A., and Drake, C. L., eds., *The Geology of Continental Margins*, Springer, Berlin, pp. 683–94.

Apple, R. A., and Macdonald, G. A., 1966. The Rise of Sea Level in Contemporary Times at Honaunau, Hawaii, Pacific Science, *v.* 20, pp. 125–36.

Araki, H., Tazawa, S., Noda, H. et al., 2009. Lunar global shape and polar topography derived from Kaguya-LALT laser altimetry, Science, *v.* 323, pp. 897–900, https://doi .org/10.1126/science.1164146.

Armin, R. A., and Mayer, L., 1983. Subsidence analysis of the Cordillera miogeocline: Implications for timing of late Proterozoic rifting and amount of extension, Geology, *v.* 11, pp. 702–5.

Armstrong, G. D., 1997. Potential field signatures and flexural rigidity of the lithosphere in Ireland, Ph.D. thesis, National University of Ireland, 184 pp.

Armstrong, G. D., and Watts, A. B., 2001. Spatial variations in Te in the southern Appalachians, eastern United States, J. Geophys. Res.,*v.* 106, pp. 22,009–26.

Artemjev, M. E., and Artyushkov, E. V., 1971. Structure and isostasy of the Baikal rift and the mechanism of rifting, J. Geophys. Res., *v.* 76, pp. 1197–211.

Artemjev, M. E., and Kaban, M. K., 1994. Density inhomogeneities, isostasy and flexural rigidity of the lithosphere in the Transcaspian region, Tectonophysics, *v.* 240, pp. 281–97.

Asada, T., and Shimamura, H., 1975. A structure of the oceanic lithosphere as revealed by ocean-bottom seismography. *Abstracts of Papers Presented at the Interdisciplinary Symposia*, Grenoble, 16th General Assembly, International Union of Geodesy and Geophysics, p. 91.

Audet, P., 2011. Directional wavelet analysis on the sphere: Application to gravity and topography of the terrestrial planets, J. Geophys. Res., *v.* 116, https://doi.org/10.1029/ 2010JE003710.

Audet, P., and Burgmann, R., 2011. Dominant role of tectonic inheritance in supercontinent cycles, Nat. Geosci., *v.* 4, pp. 184–7, https://doi.org/10.1038/ NGEO1080.

Audet, P., and Mareschal, J.-C., 2007. Wavelet analysis of the coherence between Bouguer gravity and topography: Application to the elastic thickness anisotropy in the Canadian Shield, Geophys. J. Int., *v.* 168, pp. 287–98, https://doi.org/10.1111/ j.1365-246X.2006.03231.x.

Avouac, J. P., and Burov, E., 1996. Erosion as a driving mechanism of intracontinental mountain growth?, J. Geophys. Res., *v.* 101, pp. 17747–69.

Babbage, C., 1847. Observations on the Temple of Serapis, at Pozzuoli, near Naples, with remarks on certain causes which may produce geological cycles of great extent, Quart. J. Geol. Soc. Lond., *v.* 3, pp. 186–217.

Bai, Y., Dong, D., Kirby, J. F., Williams, S. E., & Wang, Z., 2018. The effect of dynamic topography and gravity on lithospheric effective elastic thickness estimation: a case study, *Geophys. J. Int.*, 214, 623–34, https://doi: 10.1093/gji/ggy162.

Baines, A. G., Cheadle, M. J., Dick, H. J. B. et al., 2003. Mechanism for generating the anomalous of oceanic core complexes: Atlantis Bank, southwest Indian Ridge, Geology, *v.* 31, pp. 1105–8, https://doi.org/10.1130/G19829.1.

Baird, D. J., Knapp, J. H., Steer, D. N., Brown, L. D., and Nelson, K. D., 1995. Upper-mantle reflectivity beneath the Williston basin, phase-change Moho, and the origin of intracratonic basins, Geology, *v.* 23, pp. 431–4.

Baker, B. H., and Wohlenberg, J., 1971. Structure and evolution of the Kenya rift valley, Nature, *v.* 229, pp. 538–42.

Baldwin, R. B., 1971. The question of isostasy on the Moon, Phys. Earth Planet. Interiors, *v.* 4, pp. 167–79.

Banks, R. J., and Swain, C. J., 1978. The isostatic compensation of East Africa, Proc. R. Soc. London, *v.* 364, pp. 331–52.

Banks, R. J., Parker, R. L., and Huestis, S. P., 1977. Isostatic compensation on a continental scale: Local versus regional mechanisms, Geophys. J. R. Astr. Soc., *v.* 51, pp. 431–52.

Bargar, K. E., and Jackson, E. D., 1974, Calculated volumes of individual shield volcanoes along the Hawaiian-Emperor chain, Jour. Research U.S. Geol. Survey, *v.* 2, pp. 545–50.

Barrell, J., 1914a, The strength of the Earth's crust. I. Geologic tests of the limits of strength, J. Geol., *v.* 22, pp. 28–48.

Barrell, J., 1914b, The strength of the Earth's crust. II. Regional distribution of isostatic compensation, J. Geol., *v.* 22, pp. 145–65.

Barrell, J., 1914c, The strength of the Earth's crust. III. Influence of variable rate of isostatic compensation, J. Geol., *v.* 22, pp. 209–36.

Barrell, J., 1914d, The strength of the Earth's crust. IV. Heterogeneity and rigidity of the crust as measured by departures from isostasy, J. Geol., *v.* 23, pp. 289–314.

Barrell, J., 1914e, The strength of the Earth's crust. V. The depth of masses producing gravity anomalies and deflection residuals: Section A, Development of criteria for spheroidal masses. Section B, Applications of criteria to determine the limits of depth, form and mass, J. Geol., *v.* 22, pp. 441–68, 537–55.

Barrell, J., 1914f, The strength of the Earth's crust. VI. Relations of isostatic movements to a sphere of weakness – the asthenosphere, J. Geol., *v.* 22, pp. 655–83.

Barrell, J., 1914g, The strength of the Earth's crust. VII. Variations of strength with depth as shown by the nature of departures from isostasy, J. Geol, *v.* 22, pp. 729–41.

Barrell, J., 1914h, The strength of the Earth's crust. VIII. Physical conditions controlling the nature of lithosphere and asthenosphere: Section A, Relations between rigidity, strength and igneous activity. Section B, Relations with other fields of geophysics, J. Geol., *v.* 23, pp. 424–43 and pp. 499–515.

Barrell, J., 1919a, The nature and bearings of isostasy, Am. J. Sci., *v.* XLVIII, pp. 281–90.

Barrell, J., 1919b, The status of the theory of isostasy, Am. J. Sci., *v.* XLVIII, pp. 291–338.

Barton, P. J., and Wood, R. J., 1984, Tectonic evolution of the North Sea basin: Crustal stretching and subsidence, Geophys. J. R. Astr. Soc., *v.* 79, pp. 987–1022.

Basile, C., Mascle, J., Popoff, M., Bouillin, J. P., and Mascle, G., 1993, The Ivory Coast-Ghana transform margin: A marginal ridge structure deduced from seismic data, Tectonophysics, *v.* 222, pp. 1–19.

Bassi, G., Keen, C. E., and Potter, P., 1993, Contrasting styles of rifting: Models and examples from the eastern Canadian margin, Tectonics, *v.* 12, pp. 639–55.

Bassin, C., Laske, G., and Masters, G., 2000, The Current Limits of Resolution for Surface Wave Tomography in North America, EOS Trans AGU, *v.* 81, http://igppweb.ucsd.edu/~gabi/rem.html.

Bauer, K., Neben, S. Schreckenber, B. et al. 2000, Deep structure of the Namibia continental margin as derived from integrated geophysical studies, J. Geophys. Res., *v.* 105, pp. 25829–53.

Beaumont, C., 1978, The evolution of sedimentary basins on a viscoelastic lithosphere: Theory and examples, Geophys. J. R. Astr. Soc., *v.* 55, pp. 471–97.

Beaumont, C., 1979, On rheological zonation of the lithosphere during flexure, Tectonophysics, *v.* 59, pp. 347–65.

Beaumont, C., 1981, Foreland basins, Geophys. J. R. Astr. Soc., *v.* 65, pp. 291–329.

Bécel, A., Shillington, D. J., Nedimović, M. R., Webb, S. C., and Kuehn, H., 2015, Origin of dipping structures in fast-spreading oceanic lower crust offshore Alaska imaged by multichannel seismic data, Earth Planet. Sci. Lett., *v.* 424, pp. 26–37, https://doi.org/10.1016/j.epsl.2015.05.016.

Bechtel, T. D., Forsyth, D. W., and Swain, C. J., 1987, Mechanisms of isostatic compensation in the vicinity of the East African Rift, Kenya, Geophys. J. R. Astr. Soc., *v.* 90, pp. 445–65.

Bechtel, T. D., Forsyth, D. W., Sharpton, V. L., and Grieve, R. A. F., 1990, Variations in effective elastic thickness of the North American lithosphere, Nature, *v.* 343, pp. 636–8.

Becker, R. H., and Sultan, M., 2009, Land subsidence in the Nile Delta: inferences from radar interferometry, The Holocene, *v.* 19, pp. 949–54, https://doi.org/10.1177/0959683609336558.

Behn, M. D., Lin, J., and Zuber, M. T., 2002, A continuum mechanics model for normal faulting using a strain-rate softening rheology: implications for thermal and rheological controls on continental and oceanic rifting, Earth Planet. Sci. Lett., *v.* 202, pp. 725–40.

Bell, R., Karner, G. D., and Steckler, M. S., 1988, Early Mesozoic rift basins of Eastern North America and their gravity anomalies: The role of detachments during extension, Tectonics, *v.* 7, pp. 447–62.

Bellas, A., and Zhong, S., 2022, Effects of a Weak Lower Crust on the Flexure of Continental Lithosphere, J. Geophys. Res., *v.* 126, https://doi.org/10.1029/2021JB022678

Belleguic, V., Lognonne, P., and Wieczorek, M., 2005, Constraints on the Martian lithosphere from gravity and topography data, J. Geophys. Res., *v.* 110, https://doi.org/10.1029/2005JE002437.

Bendick, R., and Flesch, L., 2007, Reconciling lithospheric deformation and lower crustal flow beneath central Tibet, Geology, *v.* 35, pp. 895–8.

Benjamin, M. T., Johnson, N. M., and Naeser, C. W., 1987, Rapid uplift in the Bolivian Andes, Geology, *v.* 15, pp. 680–3.

Beuthe, M., 2008, Thin elastic shells with variable thickness for lithospheric flexure of one-plate planets, Geophys. J. Int., *v.* 172, 817–41.

Beuthe, M., Le Maistre, S., Rosenblatt, P., Pätzold, M., and V. Dehant, V., 2012, Density and lithospheric thickness of the Tharsis Province from MEX MaRS and MRO gravity data, J. Geophys. Res., *v.* 117, https://doi.org/10.1029/2011JE003976.

Bhattacharji, S., and Singh, R. N., 1984, Thermo-mechanical structure of the southern part of the Indian shield and its relevance to Precambrian basin evolution, Tectonophysics, *v.* 105, pp. 103–20.

Bianco, T. A., Ito, G., Becker, J. M., and Garcia, M. O., 2005, Secondary Hawaiian volcanism formed by flexural arch decompression, Geochem. Geophys. Geosystems, *v.* 6, https://doi.org/10.1029/2005GC000945.

Billen M. I., and Gurnis M., 2005. Constraints on subducting plate strength within the Kermadec trench. J. Geophys. Res., *v.* 110, https://doi.org/10.1029/2004JB003308.

Bills, B. G., 1979, Planetary isostasy: Topographic and gravitational variance spectra, Lunar Sci., *v.* 10, pp. 132–4.

Bird, P., 1979, Continental delamination and the Colorado Plateau, J. Geophys. Res., *v.* 84, pp. 7561–71.

Bishop, P., 2007, Long-term landscape evolution: linking tectonics and surface processes, Earth Surf. Process. Landforms, *v.* 32, pp. 329–65.

Bittencourt, A. C. da S. P., Dominguez, J. M. L., and Ussami, N., 1999, Flexure as a control on the large-scale geomorphic characteristics of the eastern Brazil coastal zone, J. Coastal Research, *v.* 15, pp. 505–19.

Blasius, K. R., and Cutts, J. A., 1981, Topography of Martian central volcanoes, Icarus, *v.* 45, pp. 87–112.

Bloom, A., 1967, Pleistocene shorelines: A new test of isostasy, Geol. Soc. Am. Bull., *v.* 78, pp. 1477–94.

Blum, M. D., and Roberts, H. H., 2009, Drowning of the Mississippi delta due to insufficient sediment supply and global sea-level rise, Nature Geoscience, *v.* 2, pp. 488–91, https://doi.org/10.1038/NGEO553.

Bodine, J. H., 1980, *Numerical Computation of Plate Flexure in Marine Geophysics, Technical Report No. 1,* New York, Lamont-Doherty Geological Observatory of Columbia University.

Bodine, J. H., and Watts, A. B., 1979, On lithospheric flexure seaward of the Bonin and Mariana Trenches, Earth Planet. Sci. Lett., *v.* 43, pp. 132–48.

Bodine, J. H., Steckler, M. S., and Watts, A. B., 1981, Observations of flexure and the rheology of the oceanic lithosphere, J. Geophys. Res., *v.* 86, pp. 3695–707.

Bogdanov, I., Huaman, D., Thovert, J.-F., Genthon, P., and Adler, P. M., 2007. A model for fracturation in the Loyalty Islands, *C. R. Geoscience,* 339, 840–848.

Bohannon, R. G., Naeser, C. W., and Schmidt, D. L., 1989, The timing of uplift, volcanism and rifting peripheral to the Red Sea: A case of passive rifting?, J. Geophys. Res., *v.* 94, pp. 10315–30.

Bohnenstiehl, D. R., and Kleinrock, M. C., 2000, Evidence for spreading-rate dependence in the displacement-length ratios of abyssal hill faults at mid-ocean ridges, Geology, *v.* 28, pp. 395–8.

Boland, J. N., and Tullis, T. E., 1986, Deformation behaviour of wet and dry clinopyroxenite in the brittle to ductile transition region, Geophys. Mon. Am. Geophys. Union, *v.* 36, pp. 35–50.

Bonatti, E., Ligi, M., Gasperini, L. et al. 1994, Transform migration and vertical tectonics at the Romanche fracture zone, equatorial Atlantic, J. Geophys. Res., *v.* 99, pp. 21779–802.

Bond, G. C., and Kominz, M. A., 1984, Construction of tectonic subsidence curves for the early Paleozoic Miogeocline, southern Canadian Rocky Mountains: Implications for subsidence mechanisms, age of breakup and crustal thinning, Geol. Soc. Am. Bull., *v.* 95, pp. 155–73.

Bonneville, A., Barriot, J.-P., and Bayer, R., 1988. Evidence from geoid data of a hot spot origin from the southern Mascarene Plateau and Mascarene Islands (Indian Ocean), *J. Geophys. Res.,* 93, 4199–4212.

Boscovich, R. G., 1755, *De litteraria expeditione per Pontificam ditionem,* Rome, Typographio Palladis, 475 pp.

Bosworth, W., 1985, Geometry of propagating continental rifts, Nature, *v.* 316, pp. 625–7.

Bott, M. H. P., 1960, The use of rapid digital computing methods for direct gravity interpretation of sedimentary basins, Geophys. J. R. Astr. Soc., *v.* 3, pp. 63–7.

Bott, M. H. P., 1971, *The Interior of the Earth,* London, Edward Arnold, 316 pp.

Bott, M. H. P., 1976, Formation of sedimentary basins of graben type by extension of the continental crust, Tectonophysics, *v.* 36, pp. 77–86.

Bott, M. H. P., 1997, Modeling the formation of a half graben using realistic upper crustal rheology, J. Geophys. Res., *v.* 102, pp. 24,605–17.

Bott M. H. P., 1996, Flexure associated with planar faulting, Geophys. J. Int. *v.* 126, pp. F21–F24.

Bott, M. H. P., and Dean, D. S., 1973, Stress diffusion from plate boundaries, Nature, *v.* 243, pp. 339–41.

Bouguer, P., 1749, *La figure de la terre*, Paris, Quay des Augustins, 364 pp.

Bowie, W., 1922, The Earth's crust and isostasy, Geograph. Rev., *v.* XII, no. 4, pp. 613–627.

Bowie, W., 1926, Isostasy in western Siberia, Am. J. Sci., *v.* 211, pp. 113–18.

Bowie, W., 1927, *Isostasy – The Science of the Equilibrium of the Earth's Crust*, New York, E. P. Dutton, 275 pp.

Bowie, W., 1929, Possible origin of oceans and continents, Gerlands Beitrage zur Geophysik, *v.* XXI, pp. 178–82.

Bowie, W. (Compiler), 1932, *Comments on Isostasy Made by Authors of Geological and Geophysical Books and Papers*, National Research Council, Washington, DC, 49 pp.

Bowin, C., 1983, Gravity, topography, and crustal evolution of Venus, Icarus, v. 56, pp. 345–71.

Bowin, C., and Milligan, J., 1985, Negative gravity anomaly over spreading rift valleys: Mid-Atlantic ridge at 26° N, Tectonophysics, *v.* 113, pp. 233–56.

Bowin, C., Abers, G., and Shure, L., 1985, Gravity field of Venus at constant altitude and comparisons with Earth, J. Geophys. Res., *v.* 90, pp. C757–C770.

Brace, W. F., and Kohlstedt, D. L., 1980, Limits on lithospheric stress imposed by laboratory experiments, J. Geophys. Res., *v.* 85, pp. 6248–52.

Braga, L. F. S., 1991, Isostatic evolution and crustal structure of the Amazon continental margin determined by admittance analyses and inversion of gravity data, Ph.D thesis, Oregon State University, 197 pp.

Braitenberg, C., Wang, Y., Fang, J., and Hsu, H. T., 2003. Spatial variations of flexure parameters over the Tibet-Quinghai plateau, Earth Planet. Sci. Lett., *v.* 205, pp. 211–24.

Braun, J., and Beaumont, C., 1987, Styles of continental rifting: results from dynamic models of lithospheric extension, in Beaumont, C., and Tankard, A. J., eds., *Sedimentary Basins and Basin-Forming Mechanisms*, Calgary, Alberta, Canadian Society of Petroleum Geologists, pp. 241–58.

Braun, J., and Willett, S. D., 2013, A very efficient O(n), implicit and parallel method to solve the stream power equation governing fluvial incision and landscape evolution, Geomorphology, *v.* 180–181, pp. 170–9, https://doi.org/10.1016/j.geomorph.2012.10.008.

Brigham, E. O., 1974, *The Fast Fourier Transform*, Englewood Cliffs, NJ, Prentice-Hall, 251 pp.

Broecker, W. S., 1966, Glacial rebound and the deformation of the shorelines of proglacial lakes, J. Geophys. Res., *v.* 71, pp. 4777–83.

Broecker, W. S., Kennett, J. P., Flower, B. P. et al. 1989, Routing of meltwater from the Laurentide Ice Sheet during the Younger Dryas cold episode, Nature, *v.* 341, pp. 318–21.

Brooks, M., 1970, Positive Bouguer anomalies in some orogenic belts, Geol. Mag., *v.* 111, pp. 399–400.

Brotchie, J. F., and Silvester, R., 1969, On crustal flexure, J. Geophys. Res, v. 74, pp. 5240–5.

Brown, C. D., and Phillips, R. J., 1999, Flexural rift flank uplift at the Rio Grande rift, New Mexico, Tectonics, *v.* 18, pp. 1275–91.

Brown, I. C., 1872, The Venerable Archdeacon Pratt, Archdeacon of Calcutta: A Sketch, *Mission Life*, v. 3, Part 1 (New Series), pp. 163–9.

Brune, S., Heine, C., Clift, P. D., and Pérez-Gussinyé, M., 2017, Rifted margin architecture and crustal rheology: Reviewing Iberia-Newfoundland, Central South Atlantic, and South China Sea, Marine and Petroleum Geology, *v.* 79, pp. 257–281, https://doi .org/10.1016/j.marpetgeo.2016.10.018.

Brunet, M.-F., 1986, The influence of the evolution of the Pyrenees on adjacent basins, Tectonophysics, *v.* 129, pp. 343–54.

Brunet, M.-F., and Le Pichon, X., 1982, Subsidence of the Paris Basin, J. Geophys. Res., *v.* 87, pp. 8547–60.

Bry, M., and White, N., 2007, Reappraising elastic thickness variation at oceanic trenches, J. Geophys. Res., *v.* 112, https://doi.org/10.1029/2005JB004190.

Buck, W. R., 1991, Modes of continental lithospheric extension, J. Geophys. Res., *v.* 96, pp. 20,161–78.

Buck, W. R., 1988, Flexural rotation of normal faults, Tectonics, v. 7, pp. 959–73.

Buck, W. R., 2001, Accretional curvature of lithosphere at magmatic spreading centers and flexural support of axial highs, J. Geophys. Res., *v.* 106, pp. 3953–60.

Buck, W. R., 2017, The role of magmatic loads and rift jumps in generating seaward dipping reflectors on volcanic rifted margins, Earth Planet. Sci. Lett., *v.* 466, pp. 62–9, https://doi.org/10.1016/j.epsl.2017.02.041.

Bullard, E. C., 1936, Gravity measurements in East Africa, Phil. Trans. R. Soc., *v.* 235, pp. 445–531.

Burbank, D. W., 1992, Causes of recent Himalayan uplift deduced from deposited patterns in the Ganges basin, Nature, *v.* 357, pp. 680–3.

Burgess, P. M., Gurnis, M., and Moresi, L., 1997, Formation of sequences in the cratonic interior of North America by interaction between mantle, eustatic, and stratigraphic processes, Geol. Soc. Am. Bull., *v.* 108, pp. 1515–35.

Burov, E. B., and Diament, M., 1992, Flexure of the continental lithosphere with multi-layered rheology, Geophys. J., v. 109, pp. 449–68.

Burov, E. B., and Diament, M., 1995, The effective elastic thickness (Te) of continental lithosphere: What does it really mean?, J. Geophys. Res., *v.* 100, pp. 3905–27.

Burov, E. B., and Molnar, P., 1998, Gravity anomalies over the Ferghana Valley (central Asia) and intracontinental deformation, J. Geophys. Res., *v.* 103, pp. 18,137–52.

Burov, E. B., and Watts, A. B., 2006. The long-term strength of continental lithosphere: jelly sandwich or creme brulee?, *GSA Today*, 16, 4–10, https://doi: 10.1130/1052-5173(2006)016<4:TLT SOC>2.0.CO;2

Burov, E. B., Kogan, M. G., Lyon-Caen, H., and Molnar, P., 1990, Gravity anomalies, the deep structure, and dynamic processes beneath the Tien Shan, Earth Planet. Sci. Lett, v. 96, pp. 367–83.

Burrard, S., 1920, A brief review of the evidence upon which the theory of isostasy has been based, Geog. J., *v.* 56, pp. 47–52.

Byerlee, J. D., 1978, Friction of rocks, Pageoph, v. 116, pp. 615–26.

Caldwell, J. G., and Turcotte, D. L., 1979, Dependence of the elastic thickness of the oceanic lithosphere on age, J. Geophys. Res., *v.* 84, pp. 7572–6.

Caldwell, J. G., Haxby, W. F., Karig, D. E., and Turcotte, D. L., 1976, On the applicability of a universal elastic trench profile, Earth Planet. Sci. Lett., *v.* 31, pp. 239–46.

Calmant, S., 1987, The elastic thickness of the lithosphere in the Pacific Ocean, Earth Planet. Sci. Lett., *v.* 85, pp. 277–88.

Calmant, S., and Cazenave, A., 1986, The effective elastic lithosphere under the Cook-Austral and Society islands, Earth Planet. Sci. Lett., *v.* 77, pp. 187–202.

Calmant, S., Francheteau, J., and Cazenave, A., 1990, Elastic layer thickening with age of the oceanic lithosphere, Geophys. J., *v.* 100, pp. 59–67.

Campanile, D., Nambiar, C. G., Bishop, P., Widdowson, M., and Brown, R., 2008. Sedimentation record in the Konkan-Kerala basin: Implications for the evolution of the Western Ghats and the Western Indian passive margin, Basin Research, *v.* 20, pp. 3–22, https://doi.org/10.1111/j.1365-2117.2007.00341.x.

Canals, M., Puig, P., Durrieu de Madron, X. et al. 2006, Flushing submarine canyons, Nature, *v.* 444, pp. 354–7, https://doi.org/10.1038/nature05271.

Canales, J. P., Dañobeitia, J. J., Detrick, R. S. et al. 1997, Variations in axial morphology along the Galápagos spreading center and the influence of the Galápagos hotspot, J. Geophys. Res., *v.* 102, pp. 27341–54.

Canales, J. P., Danobeitia, J. J., and Watts, A. B., 2000, Wide-angle seismic constraints on the internal structure of Tenerife, Canary Islands, J. Volcan. Geotherm. Res., *v.* 103, pp. 65–81.

Caporali, A., 1995, Gravity anomalies and the flexure of the lithosphere in the Karakoram, Pakistan,J. Geophys. Res., *v.* 100, pp. 15075–85.

Caress, D. W., McNutt, M. K., Detrick, R. S., and Mutter, J. C., 1995, Seismic imaging of hotspot-related crustal underplating beneath the Marquesas Islands, Nature, *v.* 373, pp. 600–3.

Caristan, Y., 1982, The transition from high-temperature creep to fracture in Maryland diabase, J. Geophys. Res., *v.* 87, pp. 6781–90.

Carlson, R. L., and Raskin, G. S., 1984, The density of the ocean crust: Nature, *v.* 311, pp. 555–8.

Carr, M. H., 1974, Tectonism and volcanism of the Tharsis region of Mars, J. Geophys. Res., *v.* 79, pp. 3943–9.

Carslaw, H. S., and Jaeger, J. C., 1959, *Heat Conduction in Solids*, Oxford, Oxford University Press.

Carter, N. L., and Tsenn, M. C., 1987, Flow properties of continental lithosphere, Tectonophysics, *v.* 136, pp. 27–63.

Cathles, L. M., 1975, *The Viscosity of the Earth's Mantle*, Princeton, Princeton University Press, 386 pp.

Cattin, R., Martelet, G., Henry, P. et al. 2001. Gravity anomalies, crustal structure and thermo-mechanical support of the Himalaya of Central Tibet, Geophys. J. Int., *v.* 147, pp. 381–92.

Cazenave, A., and Dominh, K., 1981, Elastic thickness of the Venus lithosphere, Geophys. Res. Lett., *v.* 8, pp. 1039–42.

Cazenave, A., and Dominh, K., 1984, Geoid heights over the Louisville Ridge (South Pacific), J. Geophys. Res., *v.* 89, pp. 11171–9.

Cazenave, A., Lago, B., Dominh, K., and Lambeck, K., 1980, On the response of the ocean lithosphere to seamount loads from Geos 3 satellite radar altimeter observations, Geophys. J. R. Astr. Soc., *v.* 63, pp. 233–52.

Cederborn, C. E., Sinclair, H. D., Schlunegger, F., and Rahn, M. K., 2004, Climate-induced rebound and exhumation of the European Alps, Geology, *v.* 32, pp. 709–12, https://doi.org/10.1130/G20491.1.

Chamberlin, R. T., 1931, Isostasy from the geological point of view, J. Geology, *v.* 39, pp. 1–23.

Champagnac, J. D., Molnar, P., and Anderson, R. S., 2007, Quaternary erosion-induced isostatic rebound in the western Alps, Geology, *v.* 35, pp. 195–8, https://doi.org/10.1130/G23053A.1.

Champagnac, J. D., van der Beek, P., Diraison, G., and Dauphin, S., 2008, Flexural isostatic response of the Alps to increased Quaternary erosion recorded by foreland basin remnants, SE France, Terra Nova, *v.* 20, pp. 213–20, https://doi.org/10.1111/j.1365-3121.2008.00809.x.

Chapman, L. G., 1954, An outlet of Lake Algonquin at Fossmill, Ontario, Proc. Geol. Assoc. Can., *v.* 6, pp. 61–8.

Chapman, M. E., and Bodine, J. H., 1984, Considerations of the indirect effect in marine gravity modelling, J. Geophys. Res., *v.* 84, pp. 3889–92.

Chappell, J., 1974, Late Quaternary glacio- and hydro-isostasy, on a layered Earth, Quatern. Res., *v.* 4, pp. 429–40.

Chapple, W. M., 1978, Mechanics of thin-skinned fold-and-thrust belts, Geol. Soc. Am. Bull., *v.* 89, pp. 1189–98.

Charvis, P., Laesanpura, A., Gallart, J. et al. 1999, Spatial distribution of hotspot material added to the lithosphere under La Réunion, from wide-angle seismic data, J. Geophys. Res., *v.* 104, pp. 2875–93.

Chen, W. P., and Molnar, P., 1983, Focal depths of intracontinental and intraplate earthquakes and their implications for the thermal and mechanical properties of the lithosphere, J. Geophys. Res., *v.* 88, pp. 4183–214.

Chian, D., Louden, K. E., and Reid, I., 1995, Crustal structure of the Labrador Sea conjugate margin and implications for the formation of nonvolcanic continental margins, J. Geophys. Res., *v.* 100, pp. 24239–53.

Chopra, P. N., and Paterson, M. S., 1981, The experimental deformation of dunite, Tectonophysics, *v.* 78, pp. 453–73.

Cisne, J. L., 1984, Depth-dependant sedimentation and the flexural edge effect in epeiric seas: Measuring water depth relative to the lithosphere's flexural wavelength, J. Geology, *v.* 93, pp. 567–76.

Cita, M. B., and Zocchi, M., 1978, Distribution patterns of benthic foraminifera on the floor of the Mediterranean Sea, Oceanologica Acta, *v.* 1, pp. 445–62.

Clague, D. A., and Jarrard, R. D., 1973, Tertiary Pacific Plate motion deduced from the Hawaiian – Emperor Chain, Geol. Soc. Am. Bull., *v.* 84, pp. 1135–54.

Clague, D. A., and Dalrymple, G. B., 1987, The Hawaiian-Emperor volcanic chain. Part I. Geologic evolution, in Decker, R. W., Wright, T. L., and Stauffer, P. H., eds., *Volcanism in Hawaii*, Chapter 1, U.S. Geological Survey Professional Paper 1350, pp. 5–54.

Clark, J. A., 1980, A numerical model of worldwide sea level changes on a viscoelastic Earth, in Morner, N.-A., ed., *Earth Rheology, Isostasy and Eustasy*, New York, John Wiley & Sons, pp. 525–34.

Clark, J. A., Farrell, W. E., and Peltier, W. R., 1978, Global changes in postglacial sea level, Quatern. Res., *v.* 9, pp. 265–87.

Clift, P. D., Turner, J., and Ocean Drilling Program Leg 152 Scientific Party, 1995, Dynamic support by the Icelandic plume and vertical tectonics of the northeast Atlantic continental margins, J. Geophys. Res., *v.* 100, pp. 24,473–86.

Clift, P. D., and Lorenzo, J. M., 1999, Flexural unloading and uplift along the cote d'Ivoire-Ghana transform margin, equatorial Atlantic, J. Geophys. Res., *v.* 104, pp. 25257–74.

Cloetingh, S., and Burov, E. B., 1996, Thermomechanical structure of European continental lithosphere: Constraints from rheological profiles and EET estimates, Geophys. J. Int., *v.* 124, pp. 695–723.

Cloetingh, S. A. P. L., Wortel, M. J. R., and Vlaar, N. J., 1982, Evolution of passive continental margins and initiation of subduction zones, Nature, *v.* 297, pp. 139–42.

Cloetingh, S., Van Wees, J. D., van der Beek, P. A., and Spadini, G., 1995, Role of pre-rift rheology in kinematics of extensional basin formation: Constraints from thermomechanical models of Mediterranean and intracratonic basins, Mar. Pet. Geol., *v.* 12, pp. 793–807.

Cloetingh, S., and Burov, E., 2011, Lithospheric folding and sedimentary basin evolution: a review and analysis of formation mechanisms, Basin Research, *v.* 23, pp. 257–90, https://doi.org/10.1111/j.1365-2117.2010.00490.x.

Clouard, V., Campos, J., Lemoine, A., Perez, A., and Kausel, E., 2007. Outer rise stress changes related to the subduction of the Juan Fernandez Ridge, central Chile, *J. Geophys. Res.*, *v.* 112, doi: https://doi:10.1029/2005JB003999.

Coakley, B. J., and Gurnis, M., 1995, Far-field tilting of Laurentia during the Ordovician and constraints on the evolution of a slab under an ancient continent, J. Geophys. Res., *v.* 100, pp. 6313–27.

Coakley, B. J., and Watts, A. B., 1991, Tectonic controls on the development of unconformities: The North Slope, Alaska, Tectonics, *v.* 10, pp. 101–30.

Cochran, J. R., 1973, Gravity and magnetic investigations in the Guiana Basin, Western Equatorial Atlantic, Geol. Soc. Am. Bull., *v.* 84, pp. 3249–68.

Cochran, J. R., 1979, An analysis of isostasy in the world's oceans 2. Midocean ridge crests, J. Geophys. Res., *v.* 84, pp. 4713–29.

Cochran, J. R., 1980, Some remarks on isostasy and the long-term behavior of the continental lithosphere, Earth Planet. Sci. Lett., *v.* 46, pp. 266–74.

Cochran, J. R., 1981, Simple models of diffuse extension and the pre-seafloor spreading development of the continental margin of the northwestern Gulf of Alden, *Proceedings of the 26th International Cong Symposium on Continental Margins, Oceanologica Acta*, pp. 154–65.

Cochran, J. R., and Talwani, M., 1977, Free-air gravity anomalies in the world's oceans and their relationship to residual elevation, Geophys. J. Roy. Astr. Soc., *v.* 50, pp. 495–552.

Cogan, J., Rigo, L., Grasso, M., and Lerche, I., 1989, Flexural tectonics of southeastern Sicily, J. Geodyn., *v.* 11, pp. 189–241.

Cohen, S. C., and Darby, D. J., 2003, Tectonic plate coupling and elastic thickness derived from the inversion of a steady state viscoelastic model using geodetic data: Application to southern North Island New Zealand, J. Geophys. Res., *v.* 108, https://doi.org/10.1029/2002JB001687.

Cohen, T. J., and Meyer, R. P., 1966, The midcontinent gravity high: Gross crustal structure, in Steinhardt, J. S., and Smith, T. J., eds., *The Earth beneath the Continents*, Monograph 10, Washington, DC, American Geophysical Union, pp. 141–65.

Coleman, J. M., 1988, Dynamic changes and processes in the Mississippi River delta, Geol. Soc. Amer. Bull., *v.* 100, pp. 999–1015.

Collette, B. J., 1974, Thermal contraction joints in a spreading seafloor as origin of fracture zones, Nature, v. 251, pp. 299–300.

Collier J. S., and Watts A. B. 2001. Lithospheric response to volcanic loading by the Canary Islands: Constraints from seismic reflection data in their flexural moats, Geophys. J. Int. *v.* 147, pp. 660–76

Comer, R. P., 1983, Thick plate flexure, Geophys. J. R. Astr. Soc., *v.* 72, pp. 101–13.

Comer, R. P., 1986, Comments on: 'Thick-plate flexure re-examined', Geophys. J. R. Astr. Soc., *v.* 85, pp. 467–8.

Comer, R. P., Solomon, S. C., and Head, J. W., 1985, Mars: Thickness of the lithosphere from the tectonic response to volcanic loads, Rev. Geophys. and Space Phys., *v.* 23, pp. 61–92.

Conel, J. E., and Holstrom, G. B., 1968, Lunar mascons: A near-surface interpretation, Science, *v.* 162, pp. 1403–5.

Conte, S. D., and De Boor, C., 1965, *Elementary Numerical Analysis, An Algorithmic Approach*, New York, McGraw-Hill.

Contreras-Reyes, E., Grevemeyer, I., Watts, A. B. et al. 2010a, Crustal intrusion beneath the Louisville hotspot track, Earth and Planet. Sci. Lett., *v.* 289, pp. 323–33, https:// doi.org/10.1016/j.epsl.2009.11.020.

Contreras-Reyes, E., and Osses, A., 2010b, Lithospheric flexure modelling seaward of the Chile trench: Implications for oceanic plate weakening in the trench outer rise region, Geophys. J. Int., *v.* 182, pp. 97–112, https://doi.org/10.1111/j.1365-246X.2010.04629.x.

Corfield, R. I., Watts, A. B., and Searle, M. P., 2005, Subsidence of the North Indian Continental Margin, Zanskar Himalaya, NW India, J. Geol. Soc. London, *v.* 162, pp. 135–46.

Courtney, R. C., and Beaumont, C., 1983, Thermally-activated creep and flexure of the oceanic lithosphere, Nature, *v.* 305, pp. 201–4.

Cox, K. G., 1980, A model for flood basalt volcanism, J. Petrol., *v.* 21, pp. 629–50.

Cox, K. G., 1993, Continental magmatic underplating, Phil. Trans. R. Soc. Lond., *v.* 342, pp. 155–66.

Crisp, J. A. 1984, Rates of magma emplacement and volcanic output, Journal of Volcanology and Geothermal Research, *v.* 20, pp. 177–211.

Crittenden, M. D., 1963, Effective viscosity of the Earth derived from isostatic loading of Pleistocene Lake Bonneville, J. Geophys. Res., *v.* 68, pp. 5517–30.

Crittenden, M. D., 1967, Viscosity and finite strength of the mantle as determined from water and ice loads, Geophys. J. R. Astr. Soc., *v.* 14, pp. 261–79.

Crosby, A. G., 2007, An assessment of the accuracy of admittance and coherence estimates using synthetic data, Geophys. J. Int., *v.* 171, pp. 25–54, https://doi.org/10.1111/j.1365 -246X.2007.03520.x.

Crosby, A. G., McKenzie, D., and Sclater, J. G., 2006, The relationship between depth, age and gravity in the oceans, Geophys. J. Int., *v.* 166, pp. 553–73, https://doi.org/10.1111/ j.1365-246X.2006.03015.x.

Crucifix, M., Loutre, M.-F., Lambeck, K., and Berger, A., 2001, Effect of isostatic rebound on modelled ice volume variations during the last 200 kyr, Earth Planet. Sci. Letts., *v.* 184, pp. 623–33.

Curray, J. R., and Moore, D. G., 1971, Growth of the Bengal deep-sea fan and denudation in the Himalayas, Geol. Soc. Am. Bull., *v.* 82, pp. 563–72.

D'Agostino, A., and McKenzie, D., 1999, Convective support of long-wavelength topography in the Apennines (Italy), Terra Nova, *v.* 11, pp. 234–8.

Dalloubeix, C., Fleitout, L., and Diament, M., 1988, A new analysis of gravity and topography data over the Mid-Atlantic Ridge: Non-compensation of the axial valley, Earth Planet. Sci. Lett., *v.* 88, pp. 308–20.

Dalwood, R. E. T., 1996, A seismic study of lithospheric flexure in the vicinity of the Canary Islands, D.Phil. thesis, Oxford University.

Daly, E., Brown, C., Stark, C. P., and Ebinger, C. J., 2004, Wavelet and multitaper coherence methods for assessing the elastic thickness of the Irish Atlantic margin, Geophys. J. Int., *v.* 159, pp. 445–59, https://doi.org/10.1111/j.1365-246X.2004.02427.x.

Daly, M. C., Lawrence, S. R., Diemu-Tshiband, K., and Matouana, B., 1992, Tectonic evolution of the Cuvette Centrale, Zaire, J. Geol. Soc. Lond., *v.* 149, pp. 539–46.

Daly, R. A., 1934, *The Changing World of the Ice Age*, New Haven, CT, Yale University Press, 271 pp.

Daly, R. A., 1939, Regional departures from ideal isostasy, Bull. Geol. Soc. Am., *v.* 50, pp. 387–420.

Damuth, J. E., 1975, Amazon Cone: Morphology, sediments, age and growth pattern, Geol. Soc. Am. Bull., *v.* 86, pp. 863–78.

Dana, J. D., 1896, *Manual of Geology*, New York, American Book Co, 1087 pp.

Daňobeitia, J. J., Canales, J. P., and Dehghani, G. A., 1994, An estimation of the elastic thickness of the lithosphere in the Canary archipelago using admittance function, Geophys. Res. Lett., *v.* 21, pp. 2649–52.

Darwin, C. R., 1842, *The Structure and Distribution of Coral Reefs. Being the First Part of the Geology of the Voyage of the Beagle, under the Command of Capt. Fitzroy, R. N. during the Years 1832 to 1836*, London, Smith Elder and Co., https://bit.ly/3eFnA5g.

Darwin, C. R., 1844, *Geological Observations on the Volcanic Islands Visited during the Voyage of H.M.S. Beagle, Together with Some Brief Notices of The Geology of Australia and the Cape of Good Hope. Being the Second Part of the Geology of the Voyage of the Beagle, under the Command of Capt. Fitzroy, R.N. during the Years 1832 to 1836*, London, Smith Elder and Co., https://bit.ly/3euiMzq.

Darwin, G. H., 1908, On the stresses caused in the interior of the Earth by the weight of continents and mountains, in *Scientific Papers, v. II, Tidal Friction and Cosmogony*, Cambridge, Cambridge University Press, pp. 459–514. https://bit.ly/3L7bEWc.

Davies, G. F., 1981, Regional compensation of subducted lithosphere: Effects on geoid, gravity and topography from a preliminary model, Earth Planet. Sci. Lett., *v.* 54, pp. 431–41.

De Bremaecker, J. C., 1977, Is the oceanic lithosphere elastic or viscous?, J. Geophys. Res., *v.* 82, pp. 2001–4.

De Rito, R. F., Cozzarelli, F. A., and Hodge, D. S., 1986, A forward approach to the problem of viscoelasticity and the thickness of the mechanical lithosphere, J. Geophys. Res., *v.* 91, pp. 8295–313.

DeCelles, P. G., and Giles, K. A., 1996, Foreland basin systems, Basin Research, *v.* 8, pp. 105–23.

DeCelles, P. G., Robinson, D. M., and Zandt, G., 2002, Implications of shortening in the Himalayan fold-thrust belt for uplift of the Tibetan Plateau, *Tectonics*, *v.* 21, https://doi.org/10.1029/2001TC001322.

Delaney, J. P., 1940, Leonardo da Vinci on isostasy, Science, *v.* 91, pp. 546–7.

de Lyndon, F., 1932, Discussion of 'The isostasy of the Uinta Mountains' by Andrew C. Lawson, J. Geology, *v.* 40, pp. 664–9.

Densmore, A. L., Yong, L., Ellis, M. A., and Rongjun, Z., 2005, Active tectonics and erosional unloading at the eastern margin of the Tibetan plateau, Journal of Mountain Science, *v.* 2, pp. 146–54.

Desegaulx, P., Kooi, H., and Cloetingh, S., 1991, Consequences of foreland basin development on thinned continental lithosphere: Application to the Aquitaine basin (SW France), Earth Planet. Sci. Lett., *v.* 106, pp. 116–32.

Detrick, R. S., 1981, An analysis of geoid anomalies across the Mendocino fracture zone: Implications for thermal models of the lithosphere, J. Geophys. Res., *v.* 86, pp. 11751–62.

Detrick, R. S., and Watts, A. B., 1979, An analysis of isostasy in the world's oceans, 3. Aseismic ridges, J. Geophys. Res., *v.* 84, pp. 3637–53.

Dewey, J. F., and Bird, J. M., 1970, Mountain belts and the new global tectonics, J. Geophys. Res., *v.* 75, no. 14, pp. 2625–47.

Dewey, J. F., and Pitman, W. C., 1982, Late Palaeozoic basins of the southern U. S. continental interior, Phil. Trans. R. Soc., *v.* 305, pp. 145–8.

de Voogd, B., Palome, S. P., Hirn, A. et al., 1999. Vertical movements and material transport during hotspot activity: Seismic reflection profiling offshore La Reunion, *J. Geophys. Res.*, 104, 2855–74.

Diament, M., Sibuet, J. C., and Hadaoui, A., 1986, Isostasy of the Northern Bay of Biscay continental margin, Geophys. J. R. Astr. Soc., *v.* 86, pp. 893–907.

Dickinson, W. R., 1998, Geomorphology and geodynamics of the Cook-Austral Island-Seamount Chain in the South Pacific Ocean: Implications for hotspots and plumes, Int. Geol. Rev., *v.* 40, pp. 1039–75, https://doi.org/10.1080/00206819809465254.

Dickinson, W. R., 2001, Paleoshoreline record of relative Holocene sea levels on Pacific islands, Earth. Sci. Rev., *v.* 55, pp. 191–234.

Dickinson, W. R., 2004, Picture essay of Pacific Island coasts, J. of Coastal Research, *v.* 204, pp. 1012–34.

Dickinson, W. R., and Green, R. C., 1998, Geoarchaeological context of Holocene subsidence at the ferry berth Lapita site, Mulifanua, Upolu, Somoa, Geoarchaeol., *v.* 13, pp. 239–63.

Dietz, R. S., 1961, Continent and ocean basin evolution by spreading of the sea floor, Nature, *v.* 190, pp. 854–7.

Ding, M., Lin, J., Gu, C., Huang, Q., and Zuber, M. T., 2019, Variations in Martian lithospheric strength based on gravity/topography analysis, J. Geophys. Res. Planets, *v.* 124, pp. 3095–118, https://doi.org/10.1029/2019JE005937.

Dixon, J. E., Dixon, T. H., Bell, D. R., and Malservisi, R., 2004, Lateral variation in upper mantle viscosity: role of Water, Earth Planet. Sci. Lett., *v.* 222, pp. 451–67.

Dorman, L. M., and Lewis, B. T. R., 1970, Experimental Isostasy 1: Theory of determination of the Earth's response to a concentrated load, J. Geophys. Res., *v.* 75, pp. 3357–65.

Dorman, L. M., and Lewis, B. T. R., 1972, Experimental Isostasy 3: Inversion of the Isostatic Green Function and Lateral Density Changes, J. Geophys. Res., *v.* 77, pp. 3068–77.

Doucouré, C. M., de Wit, M. J., and Mushayandebru, M. F., 1996, Effective elastic thickness of the continental lithosphere in South Africa, J. Geophys. Res., *v.* 101, pp. 11291–303.

Downey, N. J., and Gurnis, M., 2009, Instantaneous dynamics of the cratonic Congo basin, J. Geophys. Res., *v.* 114, https://doi.org/10.1029/2008JB006066.

Driscoll, N. W., and Karner, G. D., 1994, Flexural deformation due to Amazon Fan loading: A feedback mechanism affecting sediment delivery to margins, Geology, *v.* 22, pp. 1015–18.

Dubois, J., Launay, J., and Recy, J., 1974, Uplift movements in New Caledonia – Loyalty Islands area and their plate tectonics interpretation, Tectonophysics, *v.* 24, pp. 133–50.

Dubois, J., Launay, J., and Recy, J., 1975, Some new evidence on lithospheric bulges close to island arcs, Tectonophysics, *v.* 26, pp. 189–96.

Duroy, Y., Farah, A., and Lillie, R. J., 1989, Subsurface densities and lithospheric flexure of the Himalayan foreland in Pakistan, in Malinconico, L. L., and Lillie, R. J., eds., *Tectonics of the Western Himalaya*, Special Paper 232, Geological Society of America, pp. 217–36.

Dutton, C. E., 1882, *Physics of the Earth's Crust* by the Rev. Osmond Fisher, Am. J. Science, *v.* 23, no. 136, pp. 283–90.

Dutton, C. E., 1889, On some of the greater problems of physical geology, Bull. Phil. Soc. Washington, *v.* 2, pp. 51–64.

Dykstra, M., 2005, Dynamics of submarine sediment mass-transport, from the shelf to the deep-sea, Ph.D. thesis, University of California, Santa Barbara.

Dyment, J., 1990, Some complimentary approaches to improve the deep seismic reflection studies in sedimentary basin environment, The Celtic Sea basin, in Pinet, B., and Bois, C., eds., *The Potential of Deep Seismic Profiling for Hydrocarbon Exploration*, Paris, Editions Technip, pp. 403–23.

Ebinger, C. J., and Hayward, N. J., 1996, Soft plates and hot spots: Views from Afar, J. Geophys. Res., *v.* 101, pp. 21,859–76.

Ebinger, C. J., Bechtel, T. D., Forsyth, D. W., and Bowin, C. O., 1989, Effective elastic plate thickness beneath the East African and Afar plateaus and dynamic compensation of the uplifts, J. Geophys. Res., *v.* 94, pp. 2883–901.

Ebinger, C. J., Crow, M. J., Rosendahl, B. R., Livingstone, D. A., and LeFournier, J., 1984, Structural evolution of Lake Malawi, Africa, Nature, *v.* 308, pp. 627–9.

Ebinger, C. J., Karner, G. D., and Weissel, J. K., 1991, Mechanical strength of extended continental lithosphere: Constraints from the western rift system, Tectonics, *v.* 10, pp. 1239–56.

Eddy, D. R., Van Avendonk, H. J. A., Christeson, G. L. et al. 2014, Deep crustal structure of the northeastern Gulf of Mexico: Implications for rift evolution and seafloor spreading, J. Geophys. Res., *v.* 119, pp. 6802–22, https://doi.org/10.1002/2014JB011311.

Eddy, D. R., Van Avendonk, H. J. A., Christeson, G. L., and Norton, I. O., 2018, Structure and origin of the rifted margin of the northern Gulf of Mexico Geosphere, *v.* 14, pp. 1804–17, https://doi.org/10.1130/GES01662.1.

Eldholm, O., Skogseid, J., Planke, S., and Gladczenko, T. P., 1995, Volcanic margin concepts, in Banda, E., Torné, M., and Talwani, M., eds., *Rifted Ocean-Continent Boundaries*, Dordrecht, Kluwer Academic Publishers, pp. 1–16.

Elliott, D., 1976, The motion of thrust sheets, J. Geophys. Res., *v.* 81, pp. 949–62.

Elliott, T., 1986, Deltas, in Reading, H. G., ed., *Sedimentary Environments and Facies*, Oxford, Blackwell Scientific Publications, pp. 113–54.

Emery, K. O., and Garrison, L. E., 1967, Sea level 7000 to 20,000 years ago, Science, *v.* 157, pp. 684–7.

England, P. C., 1981, Metamorphic pressure estimates and sediment volumes for the Alpine orogeny: An independent control on geobarometers, Earth Planet. Sci. Lett., *v.* 56, pp. 387–97.

England, P. C., and McKenzie, D. P., 1982, A thin viscous sheet model for continental deformation, Geophys. J. R. Astr. Soc., *v.* 70, pp. 295–321.

Erickson, S. G., 1993, Sedimentary loading, lithospheric flexure, and subduction initiation at passive margins, Geology, *v.* 21, pp. 125–8.

Evangelidis, C. P., Minshull, T. A., and Henstock, T. J., 2004, Three-dimensional crustal structure of Ascension Island from active source seismic tomography, Geophys. J Int., *v.* 159, pp. 311–25, https://doi.org/10.1111/j.1365-246X.2004.02396.x.

Everest, G., 1847, *An Account of the Measurement of Two Sections of the Meridional Arc of India*, London, The Court of Directors of the Hon. East-India Company, Great Queen Street.

Farr, H. K., 1980, Multibeam bathymetric sonar: Sea beam and hydro chart, Marine Geodesy, *v.* 4, pp. 77–93, https://doi.org/10.1080/15210608009379375.

Fan, G., Wallace, T. C., Beck, S. L., and Chase, C. G., 1996, Gravity anomaly and flexural model: Constraints on the structure beneath the Peruvian Andes, Tectonophysics, v. 255, pp. 99–109.

Feighner, M. A., and Richards, M. A., 1994, Lithospheric structure and compensation mechanism of the Galápogos Arc, J. Geophys. Res., *v.* 99, pp. 6711–29.

Fernandez J., Tizzani P., Manzo M. et al., 2009. Gravity-driven deformation of Tenerife measured by InSAR time series analysis, Geophysical Res. Lett., *v.* 36, https://doi.org/10.1029/2008GL036920.

Ferrari, A. J., 1977, Lunar gravity: A harmonic analysis, J. Geophys. Res., *v.* 82, pp. 3065–84.

Ferrari, A. J., and Bills, B. G., 1979, Planetary geodesy, Rev. Geophys. Space Phys., *v.* 17, pp. 1663–77.

Fielding, E., Isaacs, B., Barazangi, M., and Duncan, C., 1994, How flat is Tibet?, Geology, *v.* 22, pp. 163–7.

Filmer, P. E., and McNutt, M. K., 1989, Geoid anomalies over the Canary Islands group, Mar. Geophys. Res., *v.* 11, pp. 77–87.

Fischer, K. M., McNutt, M. K., and Shure, L., 1986, Thermal and mechanical constraints on the lithosphere beneath the Marquesas swell: Nature, *v.* 322, pp. 733–6.

Fisher, O., 1881, *Physics of the Earth's Crust*, London, Macmillan and Co.

Flemings, P. B., and Jordan, T. E., 1990, Stratigraphic modelling of foreland basins: Interpreting thrust deformation and lithosphere rheology, Geology, *v.* 18, pp. 430–4.

Flinch, J. F., Bally, A. W., and Wu, S., 1996, Emplacement of a passive-margin evaporitic allochthon in the Betic Cordillera of Spain, Geology, *v.* 24, pp. 67–70.

Fluck, P., Hyndman R. D., and Lowe C., 2003. Effective elastic thickness Te of the lithosphere in western Canada, J. Geophys. Res., *v.* 108, https://doi.org/10.1029/2002JB002201.

Forsyth, D. W., 1975, The early structural evolution and anisotropy of the oceanic upper mantle, Geophys. J. R. Astr. Soc., *v.* 43, pp. 103–62.

Forsyth, D. W., 1985, Subsurface loading and estimates of the flexural rigidity of continental lithosphere, J. Geophys. Res., *v.* 90, pp. 12,623–32.

Foucher, J. P., and Pichon, X. L., 1972, Comments on 'Thermal effects of the formation of Atlantic continental margins by continental break up' by N. H. Sleep, *Geophys J. Roy. Astr. Soc., v.* 29, 43–46.

Fowler, C. M. R., and Nisbet, E. G., 1985, The subsidence of the Williston basin, Can. J. Earth Sci., *v.* 22, pp. 408–15.

Fowler, S. R., White, R. S., Spence, G. D., and Westbrook, G. K., 1989, The Hatton Bank continental margin – II. Deep structure from two-ship expanding spread seismic profiles, Geophys. J. Int., *v.* 96, pp. 295–309.

Franke, D., 1968. Curvature of island arcs, Nature, *v.* 220, pp. 363–4.

Freedman A. P., and Parsons B., 1986. Seasat-derived gravity over the Musician seamounts, J. Geophys. Res., *v.* 91, pp. 8325–40

Frey, F. A., Clague, D., Mahoney, J. J., and Sinton, J. M., 2000, Volcanism at the edge of the Hawaiian Plume: Petrogenesis of submarine alkalic lavas from the North Arch volcanic field, J. of Petrology, *v.* 41, pp. 667–91.

Fulton, R. J., and Walcott, R. I., 1975, Lithospheric flexure as shown by deformation of glacial lake shorelines in southern British Columbia, Geol. Soc. Am. Memoir, *v.* 142, pp. 163–73.

Gadd, S. A., and Scrutton, R. A., 1997, An integrated thermomechanical model for transform continental margin evolution, Geo-Mar. Lett., *v.* 17, pp. 21–30.

Galgana, G. A., McGovern, P. J., and Grosfils, E. B., 2009. Coupled models of lithospheric flexure and magma chamber pressurization at large volcanoes on Venus, Proceedings of the COMSOL Conference 2009 Boston, www.comsol.com/paper/download/44556/Galgana.pdf.

Gallart, J., Driad, L., Charvis, P. et al. 1999, Pertubation to the lithosphere along the hotspot track of La Réunion from an off-shore on-shore seismic transect, J. Geophys. Res., *v.* 104, pp. 2895–908.

Gangopadhyay, A., and Talwani, P., 2003, Symptomatic features of intraplate earthquakes, Seismological Research Letters, *v.* 74, pp. 864–82.

Garcia, M. O., Haskins, E. H., Stolper, E. M., and Baker, M., 2007, *Stratigraphy of the Hawai'i Scientific Drilling Project core (HSDP2): Anatomy of a Hawaiian shield volcano*, Geochem. Geophys., *v.* 8, https://doi.org/10.1029/2006GC001379.

Garcia-Castellanos, D., 2002. Interplay between lithospheric flexure and river transport in foreland basins, *Basin Research*, 14, 89–104.

Garcia-Castellanos, D., Fernandez, M., and Torné, M., 1997, Numerical modeling of foreland basin formation: A program relating thrusting, flexure, sediment geometry and lithosphere rheology, Computers & Geosciences, *v.* 23, pp. 993–1003.

Garcia-Castellanos, D., Vergés, J., Gaspar-Escribano, J., and Cloetingh, S., 2003, Interplay between tectonics, climate, and fluvial transport during the Cenozoic evolution of the Ebro Basin (NE Iberia), J. Geophys. Res., *v.* 108, https://doi.org/10.1029/2002JB002073.

Garner, D. L., and Turcotte, D. L., 1984, The thermal and mechanical evolution of the Anadarko basin, Tectonophysics, *v.* 107, pp. 1–24.

Gaspar-Escribano, J. M., van Wees, J. D., ter Voorde, M. et al. 2001. Three-dimensional flexural modelling of the Ebro Basin (NE Iberia), Geophys. J. Int., *v.* 145, pp. 349–67.

Genova, A., Goossens, S., Lemoine, F. G. et al. 2016, Seasonal and static gravity field of Mars from MGS, Mars Odyssey and MRO radio science, Icarus, *v.* 272, pp. 228–45, http://dx.doi.org/10.1016/j.icarus.2016.02.050.

Genova, A., Goossens, S., Mazarico, E. et al., 2019, Geodetic evidence that Mercury has a solid inner core, Geophys. Res. Letters, *v.* 46, pp. 3625–33, https://doi.org/10.1029/2018GL081135.

Gerrard, I., and Smith, G. C., 1982, Post-Paleozoic succession and structure of the south-western African continental margin, in Watkins, J. S., and Drake, C. L., eds., *Studies in Continental Margin Geology,* Tulsa, OK, American Association of Petroleum Geologists, pp. 49–74.

Gibb, R. A., and Thomas, M. D., 1976, Gravity signature of fossil plate boundaries in the Canadian Shield, Nature, *v.* 262, pp. 199–200.

Gibson, I. L., 1966, Crustal flexures and flood basalts, Tectonophysics, *v.* 3, pp. 447–56.

Gilbert, G. K., 1889, The strength of the Earth's crust, Bull. Geol. Soc. Am., *v.* 1, pp. 23–7.

Gilbert, G. K., 1890, *Lake Bonneville,* United States Geological Survey Memoir *v.* 1, Washington, DC, Government Printing Office, 438 pp.

Gilbert, G. K., 1893, The Moon's face, Bull. Phil. Soc. Washington, *v.* 12, pp. 241–92.

Gilbert, G. K., 1895a, New light on isostasy, J. Geol., *v.* 3, pp. 331–4.

Gilbert, G. K., 1895b, Notes on the gravity determinations reported by G. R. Putnam, Bull. Phil. Soc. Washington, *v.* 13, pp. 61–75.

Gilbert, G. K., 1913, Interpretation of anomalies of gravity, U.S. Geol. Surv. Prof. Paper, *v.* 85, pp. 29–37.

Gilchrist, A. R., and Summerfield, M. A., 1990, Differential denudation and flexural isostasy in formation of rifted-margin upwarps, Nature, *v.* 346, pp. 739–42.

Gilchrist, A. R., Summerfield, M. A., and Cockburn, H. A. P., 1994, Landscape dissection, isostatic uplift, and the morphologic development of orogens, Geology, *v.* 22, pp. 963–6.

Gilvarry, J. J., 1969, Nature of the lunar mascons, Nature, *v.* 221, pp. 732–6.

Girdler, R. W., 1963, Geophysical studies of rift valleys, Phys. Chem. Earth, *v.* 5, pp. 121–56.

Goetze, C., 1978, The mechanisms of creep in olivine, Phil. Trans. R. Soc. Lond., *v.* 288, pp. 99–119.

Goetze, C., and Evans, B., 1979, Stress and temperature in the bending lithosphere as constrained by experimental rock mechanics, Geophys. J. R. Astr. Soc., *v.* 59, pp. 463–78.

Gögüs, O. H., and Pysklywec, R. N., 2008, Near-surface diagnostics of dripping or delaminating lithosphere, J. Geophys. Res., *v.* 113, https://doi.org/10.1029/2007JB005123.

Golombek, M. P., and McGill, G. E., 1983, Grabens, basin tectonics, and maximum total expansion of the Moon, J. Geophys. Res., *v.* 88, pp. 3563–78.

Gong, S., Wieczorek, M. A., Nimmo, F. et al. 2016, Thicknesses of mare basalts on the Moon from gravity and topography, J. Geophys. Res. Planets, *v.* 121, pp. 854–70, https://doi.org/10.1002/2016JE005008.

Goodbred, S. L., and Kuehl, S. A., 2000, The significance of large sediment supply, active tectonism, and eustasy on margin sequence development: Late Quaternary stratigraphy and evolution of the Ganges–Brahmaputra delta, Sedimentary Geology, *v.* 133, pp. 227–48.

Goodwillie, A. M., and Watts, A. B., 1993, An altimetric and bathymetric study of elastic thickness in the central Pacific Ocean, Earth Planet. Sci. Lett., *v.* 118, pp. 311–26.

Goossens, S., Sabaka, T. J., Wieczorek, M. A. et al., 2020, High-resolution gravity field models from GRAIL data and implications for models of the density structure of the Moon's crust, J. Geophys. Res. Planets, *v.* 125, https://doi.org/10.1029/2019JE006086.

Gordon, R. G., 1998, The plate tectonic approximation: Plate nonrigidity, diffuse plate boundaries, and global plate reconstructions, Ann. Rev. Earth Planet. Sci., *v.* 26, pp. 615–42.

Got, J.-L., Monteiller, V., Monteux, J., Hassani, R., and Okubo, P., 2008, Deformation and rupture of oceanic crust may control growth of Hawaiian volcanoes, Nature, 451, 453–6, https://doi.org/10.1038/nature06481.

Govers, R., Meijer, P., and Krijgsman, W., 2009, Regional isostatic response to Messinian Salinity Crisis events, Tectonophysics, *v.* 463, pp. 109–29, https://doi.org/10.1016/j.tecto.2008.09.026.

Grand, S. P., and Helmberger, D. V., 1984, Upper mantle shear structure beneath the Northwest Atlantic Ocean, J. Geophys. Res., *v.* 89, pp. 11465–75.

Greely, R., 1994, *Planetary Landscapes*, New York, Chapman & Hall, 286 pp.

Greene, M. T., 1982, *Geology in the Nineteenth Century*, Ithaca, NY, Cornell University Press, 324 pp.

Greenlee, S. M., and Moore, T. C., 1988, Recognition and interpretation of depositional sequences and calculation of sea-level changes from stratigraphic data – offshore New Jersey and Alabama Tertiary, in Wilgus, C. K., Hastings, B. S., Kendall, C. G. S. C., Posamentier, H. W., Ross, C. A., and Wagoner, J. C. V., eds., *Sea-Level Changes: An Integrated Approach*, Broken Arrow, OK, Society of Economic Paleontologists and Mineralogists, pp. 329–53.

Greenroyd, C. J., Peirce, C., Rodger, M., Watts, A. B., and Hobbs, R. W., 2008, Demerara Plateau – the structure and evolution of a transform passive margin, Geophys. J. Int., *v.* 172, pp. 549–64, https://doi.org/10.1111/j.1365-246X.2007.03662.x.

Grevemeyer I., Weigel W., Schussler S., and Avedik F., 2001. Crustal and upper mantle structure and lithospheric flexure along the Society Island hotspot chain, Geophys. J. Int., *v.* 147, pp. 123–40.

Grevemeyer, I., Kaul, N., Diaz-Naveas, J. L. et al., 2005, Heat flow and bending-related faulting at subduction trenches: Case studies offshore of Nicaragua and Central Chile, Earth Planet. Sci. Lett., *v.* 236, pp. 238–48, https://doi.org/10.1016/j.epsl.2005.04.048.

Grevemeyer, I., Ranero, C. R., and Ivandic, M., 2018, Structure of oceanic crust and serpentinization at subduction trenches, Geosphere, *v.* 14, pp. 395–418, https://doi.org/10.1130/GES01537.1.

Grevemeyer, I., Rüpke, L. H., Morgan, J. P., Iyer, K., and Devey, C. W., 2021, Extensional tectonics and two-stage crustal accretion at oceanic transform faults, Nature, *v.* 591, pp. 402–7, https://doi.org/10.1038/s41586-021-03278-9.

Grigg, R. W., and Jones, A. T., 1997, Uplift caused by lithospheric flexure in the Hawaiian Archipelago as revealed by elevated coral deposits, Mar. Geol., *v.* 141, pp. 11–25.

Grimm, R. E., and Solomon, S. C., 1988, Viscous relaxation of impact crater relief on Venus: Constraints on crustal thickness and thermal gradient, J. Geophys. Res., *v.* 93, pp. 11191–929.

Grotzinger, J., and Royden, L., 1990, Elastic strength of the Slave craton at 1.9 Gyr and implications for the thermal evolution of the continents, Nature, *v.* 347, pp. 64–6.

Gubler, E., Kahle, H.-G., Klingelé, E., Mueller, S., and Oliver, R., 1981, Recent crustal movements in Switzerland and their geophysical interpretation, Tectonophysics, *v.* 71, pp. 125–52.

Gülcher, A. J. P., Gerya, T. V., Montési, L. G. J., and Munch, J., 2020, Corona structures driven by plume–lithosphere interactions and evidence for ongoing plume activity on Venus, Nature Geoscience, *v.* 13, pp. 547–54, https://doi.org/10.1038/s41561-020-0606-1.

Gunn, R., 1937, A quantitative study of mountain building on an unsymmetrical Earth, J. Franklin Inst., *v.* 224, pp. 19–53.

Gunn, R., 1943a, A quantitative evaluation of the influence of the lithosphere on the anomalies of gravity, J. Franklin Institute, *v.* 236, pp. 47–65.

Gunn, R., 1943b, A quantitative study of isobaric equilibrium and gravity anomalies in the Hawaiian Islands, J. Franklin Institute, *v.* 236, pp. 373–90.

Gunn, R., 1944, A quantitative study of the lithosphere and gravity anomalies along the Atlantic coast, J. Franklin Inst., *v.* 237, pp. 139–54.

Gunn, R., 1947, Quantitative aspects of juxtaposed ocean deeps, mountain chains and volcanic ranges, Geophysics, *v.* 12, pp. 238–55.

Gunn, R., 1949, Isostasy extended, J. Geol., *v.* 57, pp. 263–79.

Gunn, R., 1962, *Autobiographical Notes*, New York, American Institute of Physics.

Gurnis, M., 1990, Ridge spreading, subduction, and sea level fluctuations, Science, *v.* 250, pp. 970–72.

Gurnis, M., and Müller, R. D., 2003, Origin of the Australian-Antarctic Discordance from an ancient slab and mantle wedge, Geol. Soc. Amer. Special Papers, *v.* 372, pp. 417–29, https://doi.org/10.1130/0-8137-2372-8.417.

Haddad, D., and Watts, A. B. 1999, Subsidence history, gravity anomalies, and flexure of the northeast Australian margin in Papua New Guinea, Tectonics, *v.* 18, pp. 827–42.

Haddad, D., Watts, A. B., and Lindsay, J., 2001, Evolution of the intracratonic Officer Basin, central Australia: implications from subsidence analysis and gravity modelling, Basin Research, *v.* 13, pp. 217–38.

Hager, B. H., 1984, Subducted slabs and the geoid: Constraints on mantle rheology and flow, J. Geophys. Res., *v.* 89, pp. 6003–15.

Hager, B. H., Clayton, R. A., Richards, M. A., Cromer, R. P., and Dziewonski, A. M., 1985, Lower mantle heterogeneity, dynamic topography and the geoid, Nature, v. 313, pp. 541–5.

Hajnal, Z., Fowler, C. M. R., Mereu, R. F. et al., 1984, An initial analysis of the Earth's crust under the Williston Basin: 1979 CO-CRUST experiment, J. Geophys. Res., *v.* 89, pp. 9381–400.

Hammer, P. T. C, Dorman, L. M., Hildebrand, J. A., and Cornuelle, B. D., 1994, Jasper seamount structure: Seafloor seismic refraction tomography, J. Geophys. Res., *v.* 99, pp. 6731–52.

Hampel, A., Karow, T., Maniatis, G., and Hetzel, R., 2010, Slip rate variations on faults during glacial loading and post-glacial unloading: implications for the viscosity structure of the lithosphere, J. Geol. Soc. London, *v.* 167, pp. 385–99, https://doi.org/10.1144/0016-76492008-137.

Hanks, T. C., 1971, The Kuril trench-Hokkaido rise system: Large shallow earthquakes and simple models of deformation, Geophys. J. R. Astr. Soc., v. 23, pp. 173–89.

Hansen, F. D., 1982, Semibrittle creep of selected crustal rocks at 1000 MPa, Ph.D. thesis, Texas A&M University, College Station.

Hansen, F. D., and Carter, N. L., 1983, Semibrittle creep of dry and wet Westerly granite at 1000 MPa, *US Symposium on Rock Mechanics, 24th, Texas A&M University, College Station*, pp. 429–47.

Haq, B. U., Harbenbol, J., and Vail, P. R., 1987, Chronology of fluctuating sea levels since the Triassic, Science, v. 235, pp. 1156–67.

Harris, R. N., and Chapman, D. S., 1994, A comparison of mechanical thickness estimates from trough and seamount loading in the southeastern Gulf of Alaska, J. Geophys. Res., v. 99, pp. 9297–317.

Harrison, J. C., and Brisbin, W. C, 1959, Gravity anomalies off the west coast of North America, 1: Seamount Jasper, Bull. Geol. Soc. Am., v. 70, pp. 929–34.

Harry, D. L., and Mickus, K. L., 1998, Gravity constraints on lithosphere flexure and the structure of the late Paleozoic Ouachita orogen in Arkansas and Oklahoma, south central North America, Tectonics, v. 17, pp. 187–202.

Harry, D. L., Oldow, J. S., and Sawyer, D. S., 1995, The growth of orogenic belts and the role of crustal heterogeneities in decollement tectonics, Bull. Geol. Soc. Am., v. 107, pp. 1411–26.

Hartley, A. J., 2003, Andean uplift and climate change, J. Geol. Soc. London, v. 160, pp. 7–10.

Hartley, R., 1995, Isostasy of Africa: Implications for the thermo-mechanical behaviour of the continental lithosphere, D.Phil. thesis, Oxford University.

Hartley, R. W., and Allen, P. A., 1994, Interior cratonic basins of Africa: Relation to continental break-up and role of mantle convection, Basin Res., v. 6, pp. 95–113.

Hartley, R., Watts, A. B., and Fairhead, J. D., 1996, Isostasy of Africa, Earth Planet. Sci. Lett., v. 137, pp. 1–18.

Haskell, N. A., 1935, The motion of a fluid under the surface load, Physics, v. 6, pp. 265–9.

Haskell, N. A., 1937, The viscosity of the asthenosphere, Am. J. Sci., v. 33, pp. 22–8.

Haxby, W. F., and Parmentier, E. M., 1988, Thermal contraction and the state of stress in the oceanic lithosphere, J. Geophys. Res., v. 93, pp. 6419–29.

Haxby, W. F., and Turcotte, D. L., 1978, On isostatic geoid anomalies, J. Geophys. Res. v. 83, pp. 5473–8.

Haxby, W. F., Turcotte, D. L., and Bird, J. M., 1976, Thermal and mechanical evolution of the Michigan Basin, Tectonophysics, v. 36, pp. 57–75.

Haxby, W. F., and Weissel, J. K., 1986, Evidence for small-scale mantle convection from Seasat altimeter data, J. Geophys. Res., v. 91, 3507–20.

Hayford, J. F., 1909, *The Figure of the Earth and Isostasy from Measurements in the United States*, Washington, DC, Government Printing Office, 178 pp.

Hayford, J. F., and Bowie, W., 1912, *The Effect of Topography and Isostatic Compensation upon the Intensity of Gravity*, Washington DC, Coast and Geodetic Service, Government Printing Office, 132 pp.

He, Y., Wen, L., and Zheng, T., 2006, Geographic boundary and shear wave velocity structure of the "Pacific anomaly" near the core-mantle boundary beneath the western Pacific, Earth Planet. Sci. Lett., v. 244, pp. 302–14, https://doi.org/10.1016/j.epsl.2006.02.007.

Head, J. W., Vorder Bruegge, R. W., and Crumpler, L. S., 1990, Venus orogenic belt environments: Architecture and origin, Geophys. Res. Lett., v. 17, pp. 1337–400.

Heard, H. C., and Carter, N. L., 1968, Experimentally induced 'natural' intragranular flow in quartz and quartzite, Am. J. Sci., v. 266, pp. 1–42.

Heezen, B. C, Tharp, M., and Ewing, M., 1959, *The floors of the oceans 1. The North Atlantic Ocean,* Geol. Soc. Am. Spec. Papers, *v.* 65, p. 122.

Hegarty, K. A., Weissel, J. K., and Mutter, J. C., 1988, Subsidence history of Australia's Southern margin: Constraints on basin models, Am. Assoc. Pet. Geol., *v.* 72, pp. 615–633.

Heine, C., Müller, R. D., Steinberger, B., and Torsvik, T. H., 2008, Subsidence in intracontinental basins due to dynamic topography, Phys. Earth Planet. Inter., *v.* 171, pp. 252–64, https://doi.org/10.1016/j.pepi.2008.05.008.

Heiskanen, W. A., 1931, Isostatic tables for the reduction of gravimetric observations calculated on the basis of Airy's hypothesis, Bull. Géodésique, *v.* 30, pp. 110–29.

Heiskanen, W. A., 1966, Isostasy, in Hedgpeth, J. W., ed., *Encyclopedia of Science,* New York, McGraw-Hill, pp. 776–82.

Heiskanen, W. A., and Vening Meinesz, F. A., 1958, *The Earth and Its Gravity Field,* New York, Toronto and London, McGraw-Hill, 470 pp.

Herschel, J., 1836, Letter to C. Lyell, in Babbage, C., ed., *The Ninth Bridgewater Treatise, A Fragment,* London, John Murray, pp. 202–17.

Hertz, H., 1884, *On the Equilibrium of Floating Elastic Plates: Wiedmann's Annalen,* v. 22, pp. 449–55.

Hess, H., 1962, History of ocean basins, in Engel, A. E. J., James, H. L., and Leonard, B. F., eds., *Petrologic Studies: A Volume in Honor of A. F. Buddington,* Boulder, CO, Geological Society of America, pp. 599–620.

Hetényi, M., 1979, *Beams on Elastic Foundations,* Ann Arbor, MI, University of Michigan Press, 255 pp.

Hetényi, G., Cattin, R., Vergne, J., and Nábêlek, J. L., 2006. The effective elastic thickness of the India plate from receiver function imaging, gravity anomalies and thermomechanical modelling, Geophys. J. Int., *v.* 167, pp. 1106–18, https://doi.org/10.11/j.365-246X.2006.03198.x.

Hieronymus, C. F., and Bercovici, D., 2000, Non-hotspot formation of volcanic chains: control of tectonic and flexural stresses on magma transport, Earth Planet. Sci. Lett., *v.* 181, pp. 539–54.

Higgins, S. A., Overeem, I., Steckler, M. S. et al., 2014, InSAR measurements of compaction and subsidence in the Ganges-Brahmaputra Delta, Bangladesh, *v.* 119, pp. 1768–81, https://doi.org/10.1002/2014JF003117.

Hildenbrand, A., Gillot, P.-Y., and Le Roy, I., 2004, Volcano-tectonic and geochemical evolution of an oceanic intra-plate volcano: Tahiti-Nui (French Polynesia), Earth Planet. Sci. Lett., *v.* 217, 349–65, https://doi.org/10.1016/S0012-821X(03)00599-5.

Hill, K. C., 1991, Structure of the Papuan Fold Belt, Papua New Guinea, Am. Assoc. Pet. Geol., *v.* 75, pp. 857–72.

Hillaire-Marcel, C., and Fairbridge, R. W., 1978, Isostasy and eustasy in Hudson Bay, Geology, *v.* 7, pp. 117–22.

Hillier, J. K., and Watts, A. B., 2007, Global distribution of seamounts from ship-track bathymetry data, Geophys. Res. Letts., *v.* 34, https://doi.org/10.1029/2007GL029874.

Hinojosa, J. H., and Mickus, K. L., 2002, Thermoelastic modeling of lithospheric uplift: A finite difference numerical solution, Computers & Geosciences, *v.* 28, pp. 155–67.

Hinz, K., 1981, A hypothesis on terrestrial catastrophes: Wedges of very thick oceanward dipping layers beneath passive margins – their origin and paleoenvironment significance, *Geol. Jahrb.,* v. E22, pp. 345–63.

Hinze, W. J., Allen, D. J., Fox, A. J. et al. 1992. Geophysical investigations and crustal structure of the North American Midcontinent Rift system, *Tectonophysics,* v. 213, 17–32.

Hirano, M., 1976, Mathematical model and the concept of equilibrium in connection with slope shear ratio, Z. Geomorph. N. F., Suppl. Bd, *v.* 25, pp. 50–71.

Hirano, N. Takahashi, E., Yamamoto, J. et al., 2006, Volcanism in response to plate flexure, Science, *v.* 313, https://doi.org/10.1126/science.1128235.

Hiroa, T. R., 1957, *Arts and Crafts of Hawaii*, Honolulu, HI, B. P. Bishop Museum Press, pp. 369–72.

Hirth, G., and Kohlstedt, D. L., 2003. The rheology of the upper mantle and the mantle wedge: a view from the experimentalists, in Eiler, J., ed., *Inside the Subduction Factory*, AGU Geophysical Monograph Series, *v.* 138, Washington, DC, American Geophysical Union, pp. 83–105.

Hoernle, K., Hauff, F., Werner, R. et al., 2011, Origin of Indian Ocean Seamount Province by shallow recycling of continental lithosphere, Nature Geoscience, 4, 883–87, https://doi.org/10.1038/NGEO1331.

Hoffmann, G., Silver, E., Day, S., Driscoll, N., and Appelgate, B., 2010, Drowned carbonate platforms in the Bismarck Sea, Papua New Guinea, Marine Geophysical Researches, https://doi.org/10.1007/s11001-010-9079-8.

Holeman, J. H., 1968, The sediment yield of the major rivers of the world, Water Resources Res., *v.* 4, pp. 737–47.

Holmes, A., 1931. Radioactivity and earth movements. Trans. Geol. Soc. Glasgow, *v.* 18, pp. 559–606.

Holt, W. E., and Stern, T. A., 1991, Sediment loading on the western platform of the New Zealand continent: Implications for the strength of a continental margin, Earth Planet. Sci. Lett., *v.* 107, pp. 523–38.

Holt, W. E., and Stern, T. A., 1994, Subduction, platform subsidence, and foreland thrust loading: The late Tertiary development of Taranaki Basin, New Zealand, Tectonics, *v.* 13, pp. 1068–92.

Hood, L. L., Mitchell, D. L., Lin, R. P., Acuna, M. H., and Binder, A. B., 1999, Initial measurements of the lunar induced magetic dipole moment using lunar prospector magnetometer data, Geophys. Res. Letters, *v.* 26, pp. 2327–30.

Hoogenboom, T., and Houseman, G. A., 2006, Rayleigh-Taylor instability as a mechanism for corona formation on Venus, Icarus, *v.* 180, pp. 292–307, https://doi.org/10.1016/j.icarus.2005.11.001.

Hoogenboom, T., Smrekar, S. E., Anderson, F. S., and Houseman, G., 2004. Admittance survey of type 1 coronae on Venus, *J. Geophys. Res.*, *109*, https://doi.org/10.1029/2003JE002171.

Hospers, J., 1965, Gravity field and structure of the Niger Delta, West Africa, Geol. Soc. Am. Bull., *v.* 76, pp. 407–22.

Houtz, R. E., Ludwig, W. J., Milliman, J. D., and Grow, J. A., 1978, Structure of the northern Brazilian continental margin, Geol. Soc. Am. Bull., *v.* 88, pp. 711–19.

Huang, P. Y., and Solomon, S. C., 1988, Centroid depths of mid-ocean ridge earthquakes; dependance on spreading rate, J. Geophys. Res., *v.* 93, pp. 13445–77.

Huang, J., and Zhong, S., 2005, Sublithospheric small-scale convection and its implications for the residual topography at old ocean basins and the plate model, J. Geophys. Res., *v.* 110, https://doi.org/10.1029/2004JB003153.

Huang, Q., Xiao, Z., and Xiao, L., 2014, Subsurface structures of large volcanic complexes on the nearside of the Moon: A view from GRAIL gravity, Icarus, *v.* 243, pp. 48–57, https://doi.org/10.1016/j.icarus.2014.09.009.

Hulme, G., 1972, Mascons and isostasy, Nature, *v.* 238, pp. 448–50.

References

Hunter, J., and Watts, A. B., 2016, Gravity anomalies, flexure and mantle rheology seaward of circum-Pacific trenches, Geophys. J. Int., *v.* 207, pp. 288–316, https://doi .org/10.1093/gji/ggw275.

Hutchinson, D. R., Golmshtok, A. J., Zonenshain, L. P. et al., 1992, Depositional and tectonic framework of the rift basins of Lake Baikal from multichannel seismic data, Geology, *v.* 20, pp. 589–92.

Hyndman, R. D., Christensen, N. I., and Drury, M. J., 1979, Seismic velocities, densities, electrical resistivities, porosities and thermal conductivities of core samples from boreholes into the islands of Bermuda and the Azores, in Talwani, M., Harrison, C. G., and Hayes, D. E., eds., *Deep Drilling Results in the Atlantic Ocean: Oceanic Crust*, Maurice Ewing Series 2, Washington, DC, American Geophysical Union Monograph, pp. 94–112.

Illies, J. H., 1970, Graben tectonics as related to crust-mantle interaction, in Illies, J. H., and Mueller, S., eds., *Graben Problems*, Stuttgart, Schweizerbart'sche Verlagsbuchhandlung, pp. 4–26.

Ito, G., McNutt, M. K., and Gibson, R. L., 1995, Crustal structure of the Tuamotu Plateau, 15°S, and implications for its origin, J. Geophys. Res., *v.* 100, pp. 8097–114.

Ito, G., and Taira, A., 2000, Compensation of the Ontong Java Plateau by surface and subsurface loading, J. Geophys. Res., *v.* 105, pp. 11171–83.

Iwasaki, T., and Matsu'ura, M., 1982, Quasi-static crustal deformations due to a surface load: Rheological structure of the Earth's crust and upper mantle, J. Phys. Earth, *v.* 30, pp. 469–508.

Jackson, J., 2002, Strength of the continental lithosphere: Time to abandon the jelly sandwich? *GSA Today*, *v.* September, pp. 4–10, doi: https://doi.org/10.1130/1052-5173(2002)012<0004:SOTCLT>2.0.CO;2.

Jackson, M., and Bilham, R., 1994, Constraints on Himalayan deformation inferred from vertical velocity fields in Nepal and Tibet, J. Geophys. Res., *v.* 99, pp. 13,897–912, https://doi.org/10.1029/94JB00714.

Jacobi, R. D., 1981, Peripheral bulge – a causal mechanism for the Lower/Middle Ordovician unconformity along the western margin of the Northern Appalachians, Earth Planet. Sci. Lett., *v.* 56, pp. 245–51.

Jaeger, J. C., 1969, *Elasticity, Fracture and Flow with Engineering and Geological Applications*, Frome and London, Menthuen & Co. Ltd and Science Paperbacks, 268 pp.

James, D. E., 1971, Andean crustal and upper mantle structure, J. Geophys. Res., *v.* 76, pp. 3246–71.

James, E. A., and Evans, P. R., 1971, The stratigraphy of the offshore Gippsland basin, Aust. Pet. Exp. Assoc. J., *v.* 11, pp. 71–4.

James, P. B., Zuber, M. T., Phillips, R. J., and Solomon, S. C., 2015, Support of long-wavelength topography on Mercury inferred from MESSENGER measurements of gravity and topography, J. Geophys. Res. Planets, *v.* 120, pp. 287–310, https://doi .org/10.1002/2014JE004713.

James, T. S., 1992, The Hudson Bay free-air gravity anomaly and glacial rebound, Geophys. Res. Lett., *v.* 19, pp. 861–4.

Janle, P., 1981, A crustal model of the mare Serenitatis, J. Geophys., *v.* 49, pp. 57–65.

Jarrard, R. D., and Turner, D. L., 1979, Comments on 'Lithospheric flexure and uplifted atolls' by M. K. McNutt and H. W. Menard, J. Geophys. Res., *v.* 84, pp. 5691–4.

Jeffreys, H., 1926, On the nature of isostasy, Beitrage zur Geophysik, v. XV, no. Heft 2, pp. 153–74.

Jeffreys, H., 1928, Isostasy by William Bowie, Geol. Mag., *v.* 65, 279–82, https://doi.org /210.1017/S0016756800107745.

Jeffreys, H., 1954, Types of isostatic adjustment, Am. J. Sci., *v.* 243-A (Daly volume), pp. 352–9.

Jiang, X., Jin, Y., and McNutt, M. K., 2004. Lithospheric deformation beneath the Altyn Tagh and West Kunlun faults from recent gravity surveys, J. Geophys. Res., *v.* 109, https://doi.org/10.1029/2003JB002444.

Jiménez-Díaz, A., Ruiz, J., Kirby, J. F. et al., 2015, Lithospheric structure of Venus from gravity and topography, Icarus, *v.* 260, pp. 215–31, https://doi.org/10.1016/j.icarus.2015.07.020.

Jiménez-Munt, I., Fernàndez, M., Saura, E., Vérges, J., and Garcia-Castellanos, D., 2012, 3-D lithospheric structure and regional/residual Bouguer anomalies in the Arabia–Eurasia collision (Iran), Geophys. J. Int., *v.* 190, pp. 1311–24, https://doi.org/10.1111/j.1365-246X.2012.05580.x.

Jin, Y., McNutt, M. K., and Zhu, Y., 1994, Evidence from gravity and topography data for folding in Tibet: Nature, *v.* 371, pp. 669–74.

Johnson, C. L., and Sandwell, D. T., 1994, Lithospheric flexure on Venus, Geophys. J. Int., *v.* 119, pp. 627–47.

Johnson, M. E., Baarli, B. G., Cachão, M. et al., 2012, Rhodoliths, uniformitarianism, and Darwin: Pleistocene and Recent carbonate deposits in the Cape Verde and Canary archipelagos, Paleogeography, Paleoclimatology, Paleoecology, *v.* 329–330, pp. 83–100, https://doi.org/10.1016/j.palaeo.2012.02.019.

Johnson, M. R. W., 1981, The erosion factor in the emplacement of the Keystone thrust sheet (South East Nevada) across a land surface, Geol. Mag., *v.* 118, pp. 501–7.

Jordan, T. A., and Watts, A. B., 2005, Gravity anomalies, flexure and the elastic thickness structure of the India-Eurasia collisional system, Earth Planet. Sci. Lett., *v.* 236, 732–50, https://doi.org/10.1016/j.epsl.2005.05.036.

Jordan, T. E., 1981, Thrust loads and foreland basin evolution, Cretaceous, western United States, Am. Assoc. Pet. Geol. Bull, *v.* 65, pp. 2506–20.

Judge, A. V., and McNutt, M. K., 1991, The relationship between plate curvature and elastic plate thickness: A study of the Peru-Chile trench, J. Geophys. Res., *v.* 96, pp. 16625–40.

Jurkowski, G., Ni, J., and Brown, L., 1984, Modern uparching of the Gulf coastal plain, J. Geophys. Res., *v.* 89, 6247–55.

Kalnins, L. M., and Watts, A. B., 2009, Spatial variations in effective elastic thickness in the western Pacific Ocean and their implications for Mesozoic volcanism, Earth and Planet. Sci. Lett., *v.* 286, pp. 89–100, https://doi.org/10.1016/j.epsl.2009.06.018.

Kamp, P. J. J., and Tippett, J. M., 1993, Dynamics of Pacific Plate crust in the South Island (New Zealand) zone of oblique continent-continent convergence, J. Geophys. Res., *v.* 98, pp. 16,105–18.

Kanamori, H., 1971, Great earthquakes at island arcs and the lithosphere, Tectonophysics, *v.* 12, pp. 187–98.

Karato, S., and Wu, P., 1993, Rheology of the upper mantle, Science, *v.* 260, pp. 771–8.

Karato, S., Paterson, M. S., and FitzGerald, J. D., 1986, Rheology of synthetic olivine aggregates: Influence of grain size and water, J. Geophys. Res., *v.* 91, pp. 8151–76.

Karner, G. D., 1982, Spectral representation of isostatic models, BMR J. Australian Geol. Geophys., *v.* 7, pp. 55–62.

Karner, G. D., and Watts, A. B., 1982, On isostasy at Atlantic-type continental margins, J. Geophys. Res., *v.* 87, pp. 2923–48.

Karner, G. D., and Watts, A. B., 1983, Gravity anomalies and flexure of the lithosphere at mountain ranges, J. Geophys. Res., *v.* 88, pp. 10449–77.

Karner, G. D., and Weissel, J. K., 1990, Factors controlling the location of compressional deformation of oceanic lithosphere in the Central Indian Ocean, J. Geophys. Res., *v.* 95, pp. 19,795–810.

Karner, G. D., Egan, S. S., and Weissel, J. K., 1992, Modeling the tectonic development of the Tucano and Sergipe-Alagonas rift basins, Brazil, Tectonophysics, *v.* 215, pp. 133–60.

Karner, G. D., Steckler, M. S., and Thorne, J., 1983, Long-term mechanical properties of the continental lithosphere, Nature, *v.* 304, pp. 250–3.

Katayama, I., and Karato, S.-I., 2008. Low-temperature, high stress deformation of olivine under water-saturated conditions, Phys. Earth Planet. Inter., 168, 125–33, https://doi.org/10.1016/j.pepi.2008.05.019.

Kaula, W. M., 1967, Geophysical implications of satellite determinations of the Earth's gravitational field, Space Science Reviews, *v.* 7, pp. 769–94.

Kawakatsu, H., Kumar, P., Takei, Y. et al., 2009, Seismic evidence for sharp lithosphere-asthenosphere boundaries of oceanic plates, Science, *v.* 324, pp. 499–502, https://doi.org/10.1126/science.1169499.

Kay, J. P., and Dombard, A. J., 2019, Long-wavelength topography on Mercury is not from folding of the lithosphere, Icarus, *v.* 319, pp. 724–8, https://doi.org/10.1016/j.icarus.2018.09.040.

Kay, M., 1947, Geosynclinal nomenclature and the Craton, Bull. Amer. Asoc. Pet. Geologists, *v.* 31, pp. 1289–93.

Kearney, J., 2011, George Rockwell Putnam, Lighthouse DIGEST, July/August, East Machias, ME, Foghorn Publishing, https://shop.foghornpublishing.com/product-category/magazine-subscription.

Keen, C. E., and Dehler, S. A., 1997, Extensional styles and gravity anomalies at rifted continental margins: Some North Atlantic examples, Tectonics, *v.* 16, pp. 744–54.

Keen, C. E., Keen, M. J., Barrett, D. L., and Heffler, D. G., 1975, Some aspects of the ocean-continent transition at the continental margin of eastern North America, in *Offshore Geology of Eastern Canada*, Ottawa, Geological Survey of Canada, Department of Energy, Mines and Resources, pp. 189–97.

Kelemen, P. B., and Holbrook, W. S., 1995, Origin of thick, high-velocity igneous crust along the East Coast Margin, J. Geophys. Res., *v.* 100, pp. 10077–94.

Ker, R. A., 1994, A new portrait of Venus: thick-skinned and decrepit, Science, *v.* 263, pp. 759–60.

Kim C. H., Park C. H., Jeong E. Y. et al., 2009. Flexural isostasy and loading sequence of the Dokdo seamounts on the Ulleung Basin in the East Sea (Sea of Japan), J. of Asian Earth Sciences, *v.* 35, pp. 459–68, https://doi.org/10.1016/j.jseaes.2009.02.009.

Kim, S.-S., and Wessel, P., 2010, Flexure modelling at seamounts with dense cores, Geophysical J. Int., *v.* 182, Issue 2, https://doi.org/10.1111/j.1365-1246X.2010.04653.x.

King-Hubbert, M., and Melton, F. A., 1930, Isostasy: A critical review, J. Geol., *v.* 38, pp. 673–95.

Kinsman, D. J. J., 1975, Rift valley basins and sedimentary history of trailing continental margins, in Fischer, A. G., and Judson, S., eds., *Petroleum and Global Tectonics*, Princeton, Princeton University Press, pp. 83–126.

Kirby, S. H., and Kronenberg, A. K., 1987, Rheology of the lithosphere: Selected topics, Rev. Geophys., *v.* 25, pp. 1219–44.

Kirby, J. F., and Swain, C. J., 2009, A reassessment of spectral Te estimation in continental interiors: The case of North America, J. Geophys. Res., *v.* 114, https://doi.org/10.1029/2009JB006356.

Kirby, J. F., and Swain, C. J., 2011, Improving the spatial resolution of effective elastic thickness estimation with the fan wavelet transform, Computers & Geosciences, *v.* 37, pp. 1345–54, https://doi.org/10.1016/j.cageo.2010.10.008.

Kirby, J. F., and Swain, C. J., 2014, The long-wavelength admittance and effective elastic thickness of the Canadian Shield, J. Geophys. Res., *v.* 119, pp. 5187–214, https://doi.org/10.1002/2013JB010578.

Kjeldstad, A., Skogseid, J., Langtangen, H. P., Bjørlykke, K., and Høeg, K., 2003, Differential loading by prograding sedimentary wedges on continental margins: An arch-forming mechanism, J. Geophys. Res., *v.* 108, https://doi.org/10.1029/2001JB001145.

Klein, A., Jacoby, W., and Smilde, P., 1997, Mining-induced crustal deformation in northwest Germany: modelling the rheological structure of the lithosphere, Earth Planet. Sci. Lett., *v.* 147, pp. 107–23.

Klemann, V., Martinec, Z., and Ivins, E. R., 2008, Glacial isostasy and plate motion, Journal of Geodynamics, *v.* 46, pp. 95–103, https://doi.org/10.1016/j.jog.2008.04.005.

Kley, J., and Monaldi, C. R., 1998, Tectonic shortening and crustal thickness in the Central Andes, Geology, *v.* 26, pp. 723–6.

Klitgord, K. D., Hutchinson, D. R., and Schouten, H., 1988, US Atlantic continental margin: Structural and tectonic framework., in Sheridan, R. E., and Grow, J. A., eds., *The Atlantic Continental Margin*, Boulder, CO, Geological Society of America, pp. 19–56.

Kobayashi, K., Nakanishi, M., Tamaki, K., and Ogawa, Y., 1998, Outer slope faulting associated with the western Kuril and Japan trenches, Geophys. J. Int., *v.* 134, pp. 356–72.

Kocks, W. F., 1975. *Thermodynamics and Kinetics of Slip*, Progress in Materials Science Thermodynamics and Kinetics of Slip, *v.* 19, Oxford, UK, Pergamon Press.

Kogan, M. G., and McNutt, M. K., 1987, Isostasy in the USSR I: Admittance data, in Fuchs, K., and Froidevaux, C., eds., *Composition, Structure, and Dynamics of the Lithosphere-Asthenosphere System*, Geodynamics Series, *v.* 16, Washington, DC, American Geophysical Union, pp. 301–7.

Kogan, M. G., Fairhead, J. D., Balmino, G., and Makedonskii, E. L., 1994, Tectonic fabric and lithospheric strength of northern Eurasia based on gravity data, Geophys. Res. Lett., *v.* 21, pp. 2653–6.

Kooi, H., Cloetingh, S., and Burrus, J., 1992, Lithospheric necking and regional isostasy at extensional basins 1. Subsidence and gravity modeling with an application to the Gulf of Lions Margin (SE France), J. Geophys. Res., *v.* 97, pp. 17553–71.

Koppers, A. A. P., Staudigel, H., Christie, D. M., Dieu, J. J., and Pringle, M. S., 1995, Sr-Nd-Pb isotope geochemistry of leg 144 West Pacific guyots: implications for the geochemical evolution of the "SOPITA" mantle anomaly, in Haggerty, L. A., Premoli Silva, I., Rack, F., and McNutt, M. K., eds., *Proceedings of the Ocean Drilling Project Scientific Results*, *v.* 144, pp. 535–45.

Koppers, A. A. P., Staudigel, H., Wijbrans, J. R., and Pringle, M. S., 1998, The Magellan seamount trail: implications for Cretaceous hotspot volcanism and absolute Pacific plate motion, Earth Planet. Sci. Lett., *v.* 163, pp. 53–68.

Korstgard, J. A., and Lerche, I., 1992, Flexural plate representation of Danish Central Graben evolution, J. Geodyn., *v.* 16, pp. 181–209.

KRISP Working Party, 1991, Large-scale variation in lithospheric structure along and across the Kenya rift, Nature, *v.* 354, pp. 223–7.

Kristoffersen, Y., and Talwani, M., 1977, Extinct triple junction south of Greenland relative to North America, Geol. Soc. Am. Bull., *v.* 88, pp. 1037–49.

Kruse, S. E., and McNutt, M. K., 1988, Compensation of Paleozoic orogens: A comparison of the Urals to the Appalachians, Tectonophysics, *v.* 154, pp. 1–17.

Kruse, S. E., and Royden, L. H., 1994, Bending and unbending of an elastic lithosphere: The Cenozoic history of the Apennine and Dinaride foredeep basins, Tectonics, *v.* 13, pp. 278–302.

Kruse, S. E., Liu, Z. J., Naar, D. F., and Duncan, R. A., 1997, Effective elastic thickness of the lithosphere along the Easter Seamount Chain, J. Geophys. Res., *v.* 102, pp. 27305–17.

Kuckes, A. F., 1977, Strength and rigidity of the elastic lunar lithosphere and implications for present-day mantle convection in the Moon, J. Phys. Earth Planet. Int., *v.* 14, pp. 1–12.

Kuhn, M., 2003, Geoid determination with density hypotheses from isostatic models and geological information, J. Geodesy, *v.* 77, pp. 50–65, https://doi.org/10.1007/s00190-002-0297-y.

Kumar, N., Leyden, R., Carvalho, J., and Francisconi, O., 1979, *Sediment Isopach Map: Brazilian Continental Margin*, Tulsa, OK, American Association of Petroleum Geologists.

Kunze, A. W. G., 1976, Evidence for isostasy in the lunar mascon maria, The Moon, *v.* 15, pp. 415–19.

Kuo, B. Y., and Forsyth, D. W., 1988, Gravity anomalies of the ridge-transform system in the south Atlantic between 31 and 34. 5°S: Upwelling centers and variations in crustal thickness, Mar. Geophys. Res., *v.* 10, pp. 205–32.

Kuo, B. Y., and Parmentier, E. M., 1986, Flexure and thickening of the lithosphere at the East Pacific Rise, Geophys. Res. Lett., *v.* 13, pp. 681–4.

Kusznir, N. J., and Egan, S. S., 1990, Simple-shear and pure-shear models of extensional sedimentary basin formation: Application to the Jean d'Arc basin, Grand Banks of Newfoundland, in Tankard, A. J., and Blackwill, H. R., eds., *Extensional Tectonics of the North Atlantic margins*, Memoir 46, Tulsa, OK, American Association of Petroleum Geologists, pp. 305–22.

Kusznir, N. J., Marsden, G., and Egan, S. S., 1991, A flexural-cantilever simple-shear/pure-shear model of continental lithosphere extension: Applications to the Jeanne d'Arc basin, Grand Banks and Viking Graben, North Sea, in Roberts, A. M., Yielding, G., and Freeman, B., eds., *The Geometry of Normal Faults*, v. 56, Spec. Publ. Geological Society of London, pp. 41–60.

Lago, B., and Cazenave, A., 1981, State of stress in the oceanic lithosphere in response to loading, Geophys. J. R. Astr. Soc, *v.* 64, pp. 785–99.

Lago, B., and Rabinowicz, M., 1984, Admittance for a convection in a layered spherical shell, Geophys. J. R. Astr. Soc., *v.* 77, pp. 461–82.

Lallemand, S., Culotta, R., and von Huene, R., 1989, Subduction of the Daiichi Kashima seamount in the Japan trench, Tectonophysics, *v.* 160, pp. 231–47.

Lamb, S., 2002, Is it all in the crust? Nature, *v.* 420, pp. 130–1.

Lamb, S., and Hoke, L., 1997, Origin of the high plateau in the Central Andes, Bolivia, South America, Tectonics, *v.* 16, pp. 623–49.

Lamb, S., Moore, J. D. P., Pérez-Gussinyé, M., and Stern, T., 2020, Global whole lithosphere isostasy: Implications for surface elevations, structure, strength, and densities of the continental lithosphere, geochemistry, geophysics, Geosystems, *v.* 21, https://doi.org/10.1029/2020GC009150.

Lambeck, K., 1972, Gravity anomalies over Ocean Ridges, Geophys. J. R. Astr. Soc., *v.* 30, pp. 37–53.

Lambeck, K., 1981a, Flexure of the ocean lithosphere from island uplift, bathymetry and geoid height observations: The Society Islands, Geophys. J. R. Astr. Soc., *v.* 67, pp. 91–114.

Lambeck, K., 1981b, Lithospheric response to volcanic loading in the Southern Cook Islands, Earth Planet. Sci. Lett., *v.* 55, pp. 482–96.

Lambeck, K., 1983, Structure and evolution of the intra-cratonic basins of central Australia, Geophys. J. R. Astr. Soc., *v.* 74, pp. 843–86.

Lambeck, K., 1984, Structure and evolution of the Amadeus, Officer and Ngalia Basins of Central Australia, Aust. J. Earth Sci., *v.* 31, pp. 25–48.

Lambeck, K., and Nakiboglu, S. M., 1981, Seamount loading and stress in the ocean lithosphere 2. Viscoelastic and elastic-viscoelastic models, J. Geophys. Res., *v.* 86, pp. 6961–84.

Lambert W. D., 1930, The form of the geoid on the hypothesis of complete isostatic compensation, Bulletin Geodesique (1922–1941), *v.* 26, pp. 98–106.

Lane, N., Watts, A. B., and Farrant, A., 2007, An analysis of the Cotswolds topography: Insights into the landscape response to denudational isostasy, J. Geological Soc. London, *v.* 165, pp. 85–103, https://doi.org/110.1144/0016-76492006-76492179.

LASE Study Group, 1986, The Structure of the US East Coast Passive Margin from Large Aperture Seismic Experiments (LASE), Mar. Pet. Geol., *v.* 3, pp. 234–42.

Lash, G. G., and Engelder, T., 2007, Jointing within the outer arc of a forebulge at the onset of the Alleghanian Orogeny, J. Struct. Geol., *v.* 29, pp. 774–86, https://doi.org/10.1016/j.jsg.2006.12.002.

Laske, G., Masters, G., Ma, Z., and Pasyanos, M., 2013, Update on CRUST1.0 – A 1-degree global model of Earth's crust, Geophys. Res. Abstr., *v.* 15, pp. EGU2013–2658, http://igppweb.ucsd.edu/~gabi/rem.html.

Latychev, K., Mitrovica, J. X., Tamisiea, M. E., and Tromp, J., 2005, Influence of lithospheric thickness variations on 3-D crustal velocities due to glacial isostatic adjustment, Geophys. Res. Letts., *v.* 32, https://doi.org/10.1029/2004GL021454.

Lavier, L. L., and Steckler, M. S., 1997, The effect of sedimentary cover on the flexural strength of continental lithosphere, Nature, *v.* 389, pp. 476–9.

Lavier, L. L., Buck, W. R., and Poliakov, A. N. B., 1999, Self-consistent rolling-hinge model for the evolution of large-offset low-angle normal faults, Geology, *v.* 27, pp. 1127–30.

Lawrence, D. T., Doyle, M., and Aigner, T., 1990, Stratigraphic simulation of sedimentary basins: Concepts and calibration, Am. Assoc. Pet. Geol., *v.* 74, pp. 273–95.

Lawson, A. C., 1938, The isostasy of large deltas, Geol. Soc. Am. Bull., *v.* 149, pp. 401–16.

Lawson, A. C., 1942, Mississippi Delta – A study in isostasy, Bull. Geol. Soc. of Am., *v.* 53, pp. 1231–54.

Le Pichon, X., 1968, Sea-floor spreading and continental drift, J. Geophys. Res., *v.* 73, pp. 3661–97.

Le Pichon, X., Francheteau, J., and Bonnin, J., 1973, *Plate Tectonics*, Amsterdam, Elsevier Scientific.

Lee T.-G., Moon J.-W., and Jung M.-S., 2009. Three-dimensional flexure modelling of seamounts near the Ogasawara Fracture Zone in the western Pacific, Geophys. J. Int., *v.* 177, pp. 247–58, https://doi.org/10.1111/j.1365-246X.2008.04054.x.

Leeder, M. R., 1991, Denudation, vertical crustal movements and sedimentary basin infill, Geologische Rundschau, *v.* 80, pp. 441–58.

Leeds, A. R., Knopoff, L., and Kausel, E. G., 1974, Variations of upper-mantle structure under the Pacific Ocean, Science, *v.* 186, pp. 141–3.

Lefeldt, M., Grevemeyer, I., Gosler, J., and Bialas, J., 2009. Intraplate seismicity and related mantle hydration at the Nicaraguan trench outer rise, *Geophys. J. Int., v.* 178, pp. 742–752. https://doi.org/10.1111/j.1365.1246X.2009.04167.x.

Leier, A. L., DeCelles, P. G., and Pelletier, J. D., 2005, Mountains, monsoons and megafans, Geology, *v.* 33, pp. 289–92, https://doi.org/10.1130/G21228.1.

Lerner-Lam, A. L., and Jordan, T. H., 1987, How thick are the continents?, J. Geophys. Res., *v.* 92, pp. 14,007–26.

Levitt, D. A., and Sandwell, D. T., 1995, Lithospheric bending at subduction zones based on depth soundings and satellite gravity, J. Geophys. Res., *v.* 100, pp. 379–400.

Lewis, B. T. R., and Dorman, L. M., 1970, Experimental isostasy 2: An isostatic model for the U. S. A. derived from gravity and topographic data, J. Geophys. Res., *v.* 75, pp. 3367–86.

Li, F., Dyt, C., and Griffiths, C., 2004, 3D modelling of flexural isostatic deformation, Computers & Geosciences, *v.* 30, pp. 1105–15, https://doi.org/10.1016/j.cageo.2004.08.005.

Li, X., and Götze, H.-J., 2001, Tutorial: Ellipsoid, geoid, gravity, geodesy and geophysics, Geophysics, *v.* 66, pp. 1660–8.

Lillie, R. J., 1985, Tectonically buried continent/ocean boundary, Ouachita Mountains, Arkansas, Geology, *v.* 13, pp. 18–21.

Lin, A.-T., and Watts, A. B., 2002. Origin of the West Taiwan basin by orogenic loading and flexure of a rifted continental margin, J. Geophys. Res., *v.* 107, https://doi.org/10.1029/2001JB000669.

Lindsay, J. F., and Leven, J. H., 1996, Evolution of a Neoproterozoic to Paleozoic intracratonic setting, Officer Basin, South Australia, Basin Res., *v.* 8, pp. 403–24.

Lipman, P. W., Clague, D. A., Moore, J. G., and Holcomb, R. T., 1989, South Arch volcanic field – Newly identified young lava flows on the sea floor south of the Hawaiian Ridge, Geology, *v.* 17, 611–14.

Lithgow-Bertelloni, C., and Silver, P. G., 1998, Dynamic topography, plate driving forces and the African superswell, Nature, *v.* 395, pp. 269–72.

Little, T. A., Van_Dissen, R., Schermer, E., and Carne, R., 2009, Late Holocene surface ruptures on the southern Wairarapa fault, New Zealand: Link between earthquakes and the uplifting of beach ridges on a rocky coast, Lithosphere, *v.* 1, https://doi.org/10.1130/L7.1.

Liu, L., Spasojevic, S., and Gurnis, M., 2008, Reconstructing Farallon plate subduction beneath North America back to the Late Cretaceous, Science, *v.* 322, pp. 934–7, https://doi.org/10.1126/science.1162921.

Liu, S.-F., and Nummedal, D., 2004, Late Cretaceous subsidence in Wyoming: Quantifying the dynamic component. Geology, *v.* 32, pp. 397–400, https://doi.org/10.1130/G20318.1.

Loncarevic, B. D., and Ewing, G. N., 1963, Geophysical study of the Orpheus gravity anomaly, The Proceedings of the Seventh World Petroleum Congress, pp. 827–35.

Londoño, J., and Lorenzo, J. M., 2004. Geodynamics of continental plate collision during late tertiary foreland basin evolution in the Timor Sea: constraints from foreland sequences, elastic flexure and normal faulting, Tectonophysics, *v.* 392, pp. 37–54, https://doi.org/10.1016/j.tecto.2004.04.007.

Lorenzo, J. M., and Vera, E. E., 1992, Thermal uplift and erosion across the ocean-continent boundary of the southern Exmouth Plateau, Earth Planet. Sci. Lett., *v.* 108, pp. 79–92.

Lorenzo J. M., and Wessel P., 1997. Flexure across a continent–ocean fracture zone: the northern Falkland/Malvinas Plateau, South Atlantic, Geo-Marine Letters, *v.* 17, pp. 110–18.

Lorenzo, J. M., O'Brien, G. W., Stewart, J., and Tandon, K., 1998, Inelastic yielding and forebulge shape across a modern foreland basin: North West Shelf of Australia, Timor Sea, Geophys. Res. Lett., *v.* 25, pp. 1455–8.

Louden, K. E., and Forsyth, D. W., 1976, Thermal conduction across fracture zones and the gravitational edge effect, J. Geophys. Res., *v.* 81, pp. 4869–74.

Louden, K. E., and Forsyth, D. W., 1982, Crustal structure and isostatic compensation near the Kane fracture zone from topography and gravity measurements – I. Spectral analysis approach, Geophys. J. R. astr. Soc., *v.* 68, pp. 725–50.

Lowry, A. R., and Smith, R. B., 1994, Flexural rigidity of the Basin and Range-Colorado Plateau-Rocky Mountain transition from coherence analysis of gravity and topography, J. Geophys. Res., *v.* 99, pp. 20,123–40.

Lowry, A. R., and Smith, R. B., 1995, Strength and rheology of the western U. *S.* Cordillera, J. Geophys. Res., *v.* 100, pp. 17,947–63.

Ludwig, K. R., Szabo, B. J., Moore, J. G., and Simmons, K. R., 1991, Crustal subsidence rate of Hawaii determined from 234U/238U ages of drowned coral reefs, Geology, *v.* 19, pp. 171–4.

Luis, J. F., and Neves, M. C., 2006, The isostatic compensation of the Azores Plateau: A 3D admittance and coherence analysis, J. Volcan. Geothermal Res., *v.* 156, pp. 10–22, https://doi.org/10.1016/j.jvolgeores.2006.03.010.

Luttrell, K., and Sandwell, D., 2012, Constraints on 3-D stress in the crust from support of mid-ocean ridge topography, J. Geophys. Res., *v.* 117, pp. 1–19, https://doi.org/10.1029/2011JB008765.

Lyell, C., 1832–33, *Principles of Geology*, London, John Murray, 1322 pp.

Lyon-Caen, H., and Molnar, P., 1983, Constraints on the structure of the Himalaya from an analysis of gravity anomalies and a flexural model of the lithosphere, J. Geophys. Res., *v.* 88, pp. 8171–91.

Lyon-Caen, H., and Molnar, P., 1984, Gravity anomalies and the structure of the western Tibet and the southern Tarim basin, Geophys. Res. Lett., *v.* 11, pp. 1251–4.

Lyon-Caen, H., Molnar, P., and Suarez, G., 1985, Gravity anomalies and flexure of the Brazilian shield beneath the Bolivian Andes, Earth Planet. Sci. Lett., *v.* 75, pp. 81–92.

Lyons, S. N., Sandwell, D. T., and Smith, W. H. F., 2000. Three-dimensional estimation of elastic thickness under the Louisville Ridge, J. Geophys. Res. *v.* 105, pp. 13,239–52.

Macario, A., Malinverno, A., and Haxby, W. F., 1995, On the robustness of elastic thickness estimates using the coherence method, J. Geophys. Res., *v.* 100, pp. 15,163–72.

MacCurdy, E., 1928, *The Mind of Leonardo da Vinci*, London, Jonathan Cape, 360 pp.

MacCurdy, E., 1956, *The Notebooks of Leonardo da Vinci*, 2 volumes, translated by Edward MacCurdy, London, Jonathan Cape, 566 and 610 pp.

Mackwell, S. J., Zimmerman, M. E., and Kohlstedt, D. L., 1998, High-temperature deformation of dry diabase with applications to tectonics on Venus, J. Geophys. Res., *v.* 103, pp. 975–84.

Maclennan, J., and Lovell, B., 2002, Control of regional sea level by surface uplift and subsidence caused by magmatic underplating of Earth's crust, Geology, *v.* 30, pp. 675–8.

Madon, M., 2007, Overpressure development in rift basins: an example from the Malay Basin, offshore Peninsular Malaysia, Petroleum Geoscience, *v.* 13, pp. 169–80, https://doi.org/10.1144/1354-079307-744.

Madsen, J. A., Forsyth, D. W., and Detrick, R. S., 1984, A new isostatic model for the East Pacific Rise Crest, J. Geophys. Res., *v.* 89, pp. 9997–10,016.

Magde, L. S., Detrick, R. S., and the TERRA Group, 1995, Crustal and upper mantle contribution to the axial gravity anomaly at the southern East Pacific Rise, J. Geophys. Res., *v.* 100, pp. 3747–66.

Magnavita, L. P., Davison, I., and Kusznir, N. J., 1994, Rifting, erosion, and uplift history of the Recôncavo-Tucano-Jatobá Rift, northeast Brazil, Tectonics, *v.* 13, pp. 367–88.

Manea M., Manea V. C., Ferrari, L., Kostoglodov V., and Bandy, W. L., 2005. Elastic thickness of the oceanic lithosphere beneath the Tehuantepec ridge, Earth Planet Sci. Lett., *v.* 238, pp. 64–77, https://doi.org/10.1016/j.epsl.2005.06.060.

Manriquez, P., Contreras-Reyes, E., and Osses, A., 2014, Lithospheric 3-D flexure of the oceanic plate seaward of the trench using variable elastic thickness, Geophys. J. Int., *v.* 196, 681–93, https://doi.org/10.1093/gji/ggt464.

Mantovani, M. S., Shukowsky, M. W., de Freitas, S. R. C., and Neves, B. B. B., 2005, Lithosphere mechanical behaviour inferred from tidal gravity anomalies: a comparison of Africa and South America, Earth Planet. Sci. Lett., *v.* 230, pp. 397–412, https://doi.org/10.1016/j.epsl.2004.12.007.

Mareschal, J. C., and Jaupart, C., 2004. Variations of surface heat flow and lithospheric thermal structure beneath the North American craton, Earth Planet. Sci. Lett., *v.* 223, 65–77.

Masek, J. G., Isacks, B. L., and Fielding, E. J., 1994, Rift flank uplift in Tibet: Evidence for a viscous lower crust, Tectonics, *v.* 13, pp. 659–67.

Mathews, K. J., Maloney, K. T., Zahirovic, S. et al. 2016, Global plate boundary evolution and kinematics since the late Paleozoic, Global and Planetary Change, *v.* 146, pp. 226–50, https://doi.org/10.1016/j.gloplacha.2016.10.002.

McAdoo, D. C., 1981, Geoid anomalies in the vicinity of subduction zones, J. Geophys. Res., *v.* 86, pp. 6073–90.

McAdoo, D. C., 1982, On the compensation of geoid anomalies due to subducting slabs, J. Geophys. Res., *v.* 87, pp. 8684–92.

McAdoo, D. C., and Martin, C. F., 1984, Seasat observations of lithospheric flexure seaward of trenches, J. Geophys. Res., *v.* 89, pp. 3201–10.

McAdoo, D. C., Caldwell, J. G., and Turcotte, D. L., 1978, On the elastic-perfectly plastic bending of the lithosphere under generalized loading with application to the Kuril Trench, Geophys. J. Int., *v.* 54, pp. 11–26.

McAdoo, D. C., Martin, C. F., and Poulouse, S., 1985, Seasat observations of flexure: Evidence for a strong lithosphere, Tectonophysics, *v.* 116, pp. 209–22.

McConnell, R. K., 1965, Isostatic adjustment in a layered earth, J. Geophys. Res., *v.* 70, pp. 5171–88.

McConnell, R. K., 1968, Viscosity of the mantle from relaxation time spectra of isostatic adjustment, J. Geophys. Res., *v.* 73, pp. 7089–105.

McGill, G. E., Steenstrup, S. J., Barton, C., and Ford, P. G., 1981, Continental rifting and the origin of Beta Regio, Venus, Geophys. Res. Lett., *v.* 8, pp. 737–40.

McGinnis, L. D., 1970, Tectonics and the gravity field in the Continental Interior, J. Geophys. Res., *v.* 75, pp. 317–31.

McGinnis, J. P., Driscoll, N. W., Karner, G. D., and Brumbaugh, W. D., 1993, Flexural response of passive margins to deep-sea erosion and slope retreat: Implications for relative sea-level, Geology, *v.* 21, pp. 893–6.

McGovern, P. J., 2007, Flexural stresses beneath Hawaii: Implications for the October 15, 2006, earthquakes and magma ascent, Geophys. Res. Letts., *v.* 34, https://doi.org/10.1029/2007GL031305.

McGovern P. J., Grosfils, E. B., Galgana, G. A. et al. 2014, Lithospheric flexure and volcano basal boundary conditions: keys to the structural evolution of large volcanic edifices on the terrestrial planets. Geol. Soc. London Special Publications, *v.* 401, https://doi.org/10.1144/SP401.7.

McKenzie, D. P., 1967, Some remarks on heat flow and gravity anomalies: J. Geophys. Res. *v.* 72, pp. 6261–73.

McKenzie, D. P., 1977, Surface deformation, gravity anomalies and convection, Geophys. J. Int., *v.* 48, pp. 211–38.

McKenzie, D. P., 1978, Some remarks on the development of sedimentary basins, Earth Planet. Sci. Lett., *v.* 40, pp. 25–32.

McKenzie, D. P., 1994, The relationship between topography and gravity on Earth and Venus, Icarus, *v.* 112, pp. 55–88.

McKenzie, D. P., 2003, Estimating T_e in the presence of internal loads, J. Geophys. Res., *v.* 108, https://doi.org/10.1029/2002JB001766.

McKenzie, D. P., 2010, The influence of dynamically supported topography on estimates of T_e, Earth Planet. Sci. Lett., *v.* 95, pp. 127–38, https://doi.org/10.1016/j.epsl.2010.03 .033.

McKenzie, D. P., and Bowin, C. O., 1976, The relationship between bathymetry and gravity in the Atlantic Ocean, J. Geophys. Res., *v.* 81, pp. 1903–15.

McKenzie, D. P., and Fairhead, J. D., 1997, Estimates of the effective elastic thickness of the continental lithosphere from Bouguer and free-air gravity anomalies, J. Geophys. Res., *v.* 102, pp. 27,523–52.

McKenzie, D. P., and Nimmo, F., 1997, Elastic thickness estimates for Venus from line of sight accelerations, Icarus, *v.* 130, pp. 198–216.

McKenzie, D. P., and Parker, R. L., 1967, The North Pacific: An example of tectonics on a sphere, Nature, *v.* 216, pp. 1276–80.

McKenzie, D. P., Roberts, J., and Weiss, N., 1974, Numerical models of convection in the Earth's mantle, Tectonophysics, *v.* 19, pp. 89–103.

McKenzie, D., Jackson, J., and Priestley, K., 2005, Thermal structure of oceanic and continental lithosphere, Earth Planet. Sci. Lett., *v.* 233, pp. 337–49, https://doi.org/10 .1016/j.epsl.2005.02.005.

McKenzie, D., Daly, M. C., and Priestley, K., 2015, The lithospheric structure of Pangea, Geology, *v.* 43, pp. 783–6, https://doi.org/10.1130/G36819.1.

McMurtry, G. M., Campbell, J. F., Fryer, G. J., and Fietzke, J., 2010, Uplift of Oahu, Hawaii, during the past 500 k.y. as recorded by elevated reef deposits, Geology, *v.* 38, pp. 27–30, https://doi.org/10.1130/G30378.1.

McNutt, M. K., 1979, Compensation of oceanic topography: An application of the response function technique to the Surveyor area, J. Geophys. Res., *v.* 84, pp. 7589–98.

McNutt, M. K., 1984, Lithospheric flexure and thermal anomalies, J. Geophys. Res., *v.* 89, pp. 11,180–94.

McNutt, M. K., 1988, Thermal and mechanical properties of the Cape Verde Rise, J. Geophys. Res., *v.* 93, pp. 2784–94.

McNutt, M. K., and Kogan, M., 1987, Isostasy in the USSR II: Interpretation of admittance data, in Fuchs, K., and Froidevaux, C., eds., *Composition, Structure, Dynamics of the Lithosphere-Asthenosphere System*, Geodynamics Series 16, Washington, DC, American Geophysical Union, pp. 309–27.

McNutt, M. K., and Menard, H. W., 1978, Lithospheric flexure and uplifted atolls, J. Geophys. Res., *v.* 83, pp. 1206–12.

McNutt, M. K., and Menard, H. W., 1982, Constraints on yield strength in the oceanic lithosphere derived from observations of flexure, Geophys. J. R. Astr. Soc., *v.* 71, pp. 363–94.

McNutt, M. K., and Parker, R. L., 1978, Isostasy in Australia and the evolution of the compensation mechanism, Science, *v.* 199, pp. 773–5.

McNutt, M. K., and Shure, L., 1986, Estimating the compensation depth of the Hawaiian swell with linear filters, J. Geophys. Res., *v.* 91, pp. 13915–23.

McNutt, M. K., Diament, M., and Kogan, M. G., 1988, Variations of elastic plate thickness at continental thrust belts: J. Geophys. Res., *v.* 93, pp. 8825–38.

McQueen H. W. S., and Lambeck K., 1989. The accuracy of some lithospheric bending parameters, Geophys. J., *v.* 96, pp. 401–13.

Mège, D., and Masson, P., 1996, A plume tectonics model for the Tharsis province, Mars, Planet. Space Sci., *v.* 44, pp. 1499–546.

Mei, S., Suzuki, A. M., Kohlstedt, D. L., Dixon, N. A., and Durham, W. B., 2010, Experimental constraints on the strength of the lithospheric mantle, J. Geophys. Res., *v.* 115, https://doi.org/10.1029/2009JB006873.

Meissner, R., and Wever, T., 1986. Intracontinental seismicity, strength of crustal units, and the seismic signature of fault zones, *Phil. Trans Roy. Soc. London*, 317, 45–61.

Mello, U. T., and Bender, A. A., 1988, On isostasy at the equatorial margin of Brazil, Revista Brasileira de Geociencias, *v.* 18, pp. 237–46.

Melosh, H. J., 1978, Dynamic support of the outer rise, Geophys. Res. Lett., *v.* 5, pp.321–4.

Menard, H. W., 1964, *Marine Geology of the Pacific*, New York, McGraw-Hill, 271 pp.

Menard, H. W., 1973. Depth anomalies and the bobbing motion of drifting islands, *J. Geophys. Res.*, 78, 5128–37.

Menard, H. W., 1986, *Islands*, Scientific American Library Series, Oxford, W. H. Freeman and Co, 230 pp.

Miall, A. D., 1990, *Principles of Sedimentary Basin Analysis*, New York, Springer-Verlag, 668 pp.

Miller, K. G., Kominz, M. A., Browning, J. V. et al., 2005, The Phanerozoic Record of Global Sea-Level Change, Science, *v.* 310, pp. 1293–8, https://doi.org/10.1126/science.1116412.

Minshull, T. A., and Brozena, J. M., 1997, Gravity anomalies and flexure of the lithosphere at Ascension Island, Geophys. J. Int., *v.* 131, pp. 347–60.

Minshull T. A., Ishizuka O., Mitchell N. C., and Evangelidis C., 2003, Vertical motions and lithosphere rheology at Ascension Island. EOS Trans. Amer. Geophys. Union, *v.* 84, http://eprints.soton.ac.uk/id/eprint/1376

Minshull T. A., Ishizuka O., and Garcia-Castellanos D., 2010. Long-term growth and subsidence of Ascension Island: Constraints on the rheology of young oceanic lithosphere, Geophys. Res. Letters, *v.* 37, https://doi.org/10.1029/2010GL045112.

Mohriak, W. U., Hobbs, R., and Dewey, J. F., 1990, Basin-forming processes and the deep structure of the Campos basin, offshore Brazil, Mar. Pet. Geol., *v.* 7, pp. 94–122.

Molnar, P., and England, P. C., 1990, Late Cenozoic uplift of mountain ranges and global climate change: Chicken or egg?, Nature, *v.* 346, pp. 29–34.

Monsalve G., McGovern P., and Sheehan A. 2009. Mantle fault zones beneath the Himalayan collision: Flexure of the continental lithosphere, Tectonophysics, *v.* 477, pp. 66–76, https://doi.org/10.1016/j.tecto.2008.12.014.

Montadert, L., de Charpel, O., Roberts, D. G., Guennoc, P., and Sibuet, J., 1979, Northeast Atlantic passive continental margins: Rifting and subsidence processes, in Talwani, M., Hey, W., and Ryan, W. B. F., eds., *Deep Drilling Results in the Atlantic Ocean: Continental Margins and Paleoenvironment*, Maurice Ewing Series 3: Washington, DC, American Geophysical Union, pp. 154–86.

Monteverde, D. H., Mountain, G. S., and Miller, K. G., 2008, Early Miocene sequence development across the New Jersey margin, Basin Res., *v.* 20, pp. 249–67, https://doi.org/10.1111/j.1365-2117.2008.00351.x.

Montgomery, D. R., and Stolar, D. B., 2006, Reconsidering Himalayan River anticlines, Geomorphology, *v.* 82, pp. 4–15.

Mooney, W. D., Laske, G., and Masters, T. G., 1998, CRUST 5.1: A global crustal model at 5° × 5°, J. Geophys. Res., *v.* 103, pp. 727–47.

Moore, J. G., 1970, Relationship between Subsidence and Volcanic Load, Hawaii, Bull. Volcanologique, *v.* 34, pp. 562–76.

Moore, J. G., and Campbell, J. F., 1987, Age of tilted reefs, Hawaii, J. Geophys. Res., *v.* 92, pp. 2641–6.

Moore, J. G., Clague, D. A., Holcomb, R. T. et al., 1989, Prodigous submarine landslides on the Hawaiian Ridge, J. Geophys. Res., *v.* 94, pp. 17,465–84.

Moore, J. M., and Fornari, D. J., 1984, Drowned reefs as indicators of the rate of subsidence of the island of Hawaii, J. Geol., *v.* 92, pp. 752–9.

Moresi, L., and Parsons, B., 1995, Interpreting gravity, geoid, and topography for convection with temperature dependent viscosity, Application to surface features on Venus, J. Geophys. Res., *v.* 100, pp. 21155–71.

Moretti, I., and Royden, L., 1988, Deflection, gravity anomalies and tectonics of doubly subducted continental lithosphere: Adriatic and Ionian Seas, Tectonics, *v.* 7, pp. 875–93.

Moretti, I., and Turcotte, D. L., 1985, A model for erosion, sedimentation, and flexure with application to New Caledonia, J. Geodyn., *v.* 3, pp. 155–68.

Morgan, J. K., Moore, G. F., and Clague, D. A., 2003, Slope failure and volcanic spreading along the submarine south flank of Kilauea volcano, Hawaii, J. Geophys. Res., *v.* 108, https://doi.org/10.1029/2003JB002411.

Morgan, R. L., and Watts, A. B., 2018, Seismic and gravity constraints on flexural models for the origin of seaward dipping reflectors, Geophys. J. Int., *v.* 214, pp. 2073–83, https://doi.org/10.1093/gji/ggy243.

Morgan, W. J., 1965, Gravity anomalies and convection currents. 1. A sphere and cylinder sinking beneath the surface of a viscous fluid, J. Geophys. Res., *v.* 70, pp. 6175–87.

Morgan, W. J., 1968, Rises, trenches, great faults and crustal blocks, J. Geophys. Res., *v.* 73, no. 6, pp. 1959–82.

Morner, N.-A., 1969, The Late Quaternary history of the Kattegatt Sea and the Swedish West Coast, deglaciation, isostasy and eustasy, Sveriger Geologiska Underoskning, Arsbok 63 Ser. C, *v.* 640, no. 3, pp. 404–53.

Morris, A., and Maffione, M., 2016, Is the Troodos ophiolite (Cyprus) a complete, transform fault–bounded Neotethyan ridge segment?, Geology, *v.* 44, pp. 199–202, https://doi.org/10.1130/G37529.1.

Moucha, R., Forte, A. M., Mitrovica, J. X. et al., 2008b, Dynamic topography and long-term sea-level variations: there is no such thing as a stable continental platform, Earth Planet. Sci. Lett., *v.* 271, pp. 101–8, https://doi.org/10.1016/j.epsl.2008.03.056.

Moucha, R., Forte, A. M., Rowley, D. B. et al., 2008a, Mantle convection and the recent evolution of the Colorado Plateau and the Rio Grande rift valley, Geology, *v.* 36, pp. 439–42, https:/doi.org/10.1130/G24577A.1.

Muller, P. M., and Sjogren, W. L., 1968, Mascons: Lunar mass concentrations, Science, *v.* 161, pp. 680–4.

Müller, R. D., Roest, W. R., Royer, J.-Y., Gahagan, L. M., and Sclater, J. G., 1997, Digital isochrons of the world's ocean floor, J. Geophys. Res., *v.* 102, pp. 3211–14.

Müller, R. D., Sdrolias, M., Gaina, C., Steinberger, B., and Heine, C., 2008, Long-term sea-level fluctuations driven by ocean basin dynamics, Science, *v.* 319, pp. 1357–62, https://doi.org/10.1126/science.1151540.

Munk, W. H., and Cartwright, D. E., 1966, Tidal spectroscopy and prediction, J. Geophys. Res., *v.* 259, pp. 533–81.

Mussman, W. J., and Read, J. F., 1986, Sedimentology and development of a pasive- to a convergent-margin unconformity: Middle Ordovician Knox-Beekmantown unconformity, Virginia Appalachians, Bull. Geol. Soc. Am., *v.* 97, pp. 282–95.

Mutter, C. Z., and Mutter, J. C., 1993, Variations in thickness of layer 3 dominate oceanic crustal structure, Earth Planet. Sci. Lett., *v.* 117, pp. 295–317.

Mutter, J. C., Talwani, M., and Stoffa, P. L., 1982, Origin of seaward-dipping reflectors in oceanic crust off the Norwegian Margin by "subaerial seafloor spreading", Geology, *v.* 10, pp. 353–7.

Mutter, J. C., Talwani, M., and Stoffa, P. L., 1984, Evidence for a thick oceanic crust adjacent to the Norwegian margin, J. Geophys. Res., *v.* 89, pp. 483–502.

Nabelek, J., Hetényi, G., Vergne, J., Sapkota, S., Kafle, B., and the Hi-CLIMB team, 2009, Underplating in the Himalaya-Tibet Collision Zone revealed by the Hi-CLIMB experiment, Science, *v.* 325, pp. 1371–4, www.sciencemag.org/cgi/content/full/325/5946/1371/DC1.

Nadai, A., 1963, *Theory of Flow and Fracture of Solids*, New York, McGraw-Hill, 705 pp.

Nadirov, R. S., Bagirov, E., and Tagiyev M., 1997. Flexural plate subsidence, sedimentation rates, and structural development of the super-deep South Caspian basin, Mar. Pet. Geology, *v.* 14, pp. 383–400.

Nakiboglu, S. M., and Lambeck, K., 1983, A reevaluation of the isostatic rebound of Lake Bonneville, J. Geophys. Res., *v.* 88, pp. 439–47.

Nakiboglu, S. M., and Lambeck, K., 1985, Comments on thermal isostasy, J. of Geodynamics, *v.* 2, pp. 51–65.

Nelson, C. H., and Maldonado, A. A., 1990, Factors controlling late Cenozoic margin growth from the Ebro delta to the western Mediterranean deep sea, Marine Geol., *v.* 95, pp. 419–40.

Nettleton, L. L., 1939, Determination of density for reduction of gravity observations, Geophysics, *v.* 4, pp. 176–83.

Neumann, E.-R., Olsen, K. H., Baldridge, W. S., and Sundvoll, B., 1992, The Oslo rift: a review, Tectonophysics, *v.* 208, pp. 1–18.

Neumann, G. A., and Zuber, M. T., 1996, Coherence of lunar mare basins, Proc. Lunar Planet. Sci. Conf. 27th, pp. 953–4.

Newman, R., and White, N., 1997, Rheology of the continental lithosphere inferred from sedimentary basins, Nature, *v.* 385, pp. 621–4.

Nimmo, F., and Watters, T. R., 2004, Depth of faulting on Mercury: Implications for heat flux and crustal and effective elastic thickness, Geophys. Res. Lett., *v.* 31, https://doi.org/10.1029/2003GL018847.

Nishimura, C. E., and Forsyth, D. W., 1989, The anisotropic structure of the upper mantle in the Pacific, Geophys. J., *v.* 96, pp. 203–29.

Nishimura, T., and Thatcher, W., 2003, Rheology of the lithosphere inferred from postseismic uplift following the 1959 Hebgen Lake earthquake, J. Geophys. Res., *v.* 108, https://doi.org/10.1029/2002JB002191.

Nunn, J. A., and Aires, J. R., 1988, Gravity anomalies and flexure of the lithosphere at the middle Amazon Basin, Brazil, J. Geophys. Res., *v.* 93, pp. 415–28.

Nunn, J. A., Czerniak, M., and Pilger, R. H., 1987, Constraints on the structure of Brooks Range and Colville Basin, Northern Alaska, from flexure and gravity analysis, Tectonics, *v.* 6, pp. 603–17.

Nunn, P. D., 1994, *Oceanic Islands,* Oxford, Blackwell, 413 pp.

Nyquist, J. E., and Wang, H. F., 1988, Flexural modeling of the Midcontinent rift, J. Geophys. Res., *v.* 93, pp. 8852–68.

O'Connor, J. M., Steinberger, B., Regelous, M. et al. 2013, Constraints on past plate and mantle motion from new ages for the Hawaiian-Emperor Seamount Chain, Geochem. Geophys., *v.* 14, https://doi.org/10.1002/ggge.20267.

O'Keefe, J. A., 1968, Isostasy on the moon, Science, *v.* 162, pp. 1405–6.

O'Reilly, S. Y., Griffin, W. L., Poudjom_Djomani, Y. H., and Morgan, P., 2001, Are lithospheres forever? Tracking changes in Subcontinental Lithospheric Mantle Through Time, GSA Today, *v.* 11, pp. 4–9.

Ojeda, G. Y., and Whitman, D., 2002, Effect of windowing on lithosphere elastic thickness estimates obtained via the coherence method: Results from northern South America, J. Geophys. Res., *v.* 107, https://doi.org/10.1029/2000JB00014.

Olive, J.-A., Behn, M. D., and Malatesta, L. C., 2014, Modes of extensional faulting controlled by surface processes, Geophys. Res. Letters, *v.* 41, pp. 6725–33, https://doi.org/10.1002/2014GL061507.

Olive, J.-A., Malatesta, L. A., Behn, M. D., and Buck, W. R., 2020, Sensitivity of rift tectonics to global variability in the efficiency of river erosion, PNAS *v.* 119, https://doi.org/10.1073/pnas.2115077119.

Opdyke, N. D., Spangler, D. P., Smith, D. L., Jones, D. S., and Lindquist, D. C., 1984, Origin of the epeirogenic uplift of Pliocene-Pleistocene beach ridges in Florida and development of the Florida Karst, Geology, *v.* 12, pp. 226–8.

Orme, A. R., 2007, Clarence Edward Dutton (1841–1912): soldier, polymath and aesthete, Geological Society of London, Special Publications, *v.* 287, pp. 271–86, https:/doi.org/10.1144/SP287.21.

Owens, R., 1996, *The Morphology and Tectonics of the Rekyjanes Ridge,* D.Phil. thesis, Oxford University.

Padovan, S., Wieczorek, M. A., Margot, J.-L., Tosi, N., and Solomon, S. C., 2015, Thickness of the crust of Mercury from geoid-to-topography ratios, Geophys. Res. Letters, *v.* 42, pp. 1029–38, https://doi.org/10.1002/2014GL062487.

Panasyuk, S. V., and Hager, B. H., 2000, Inversion for mantle viscosity profiles constrained by dynamic topography and the geoid, and their estimated errors, Geophys. J. Int., *v.* 143, pp. 821–36.

Pang, M., and Nummedal, D., 1995, Flexural subsidence and basement tectonics of the Cretaceous Western Interior basin, United States, Geology, *v.* 23, pp. 173–6.

Pari, G., and Peltier, W. R., 1996, The free-air gravity constraint on subcontinental mantle dynamics, J. Geophys. Res., *v.* 101, pp. 28105–32.

Pari, G., and Peltier, W. R., 2000, Subcontinental mantle dynamics: A further analysis based on the joint constraints of dynamic surface topography and free-air gravity, J. Geophys. Res., *v.* 105, pp. 5635–62.

Parker, R. L., 1972, The rapid calculation of potential anomalies, Geophys. J. R. Astr. Soc., *v.* 31, pp. 447–55.

Parmentier, E. M., and Forsyth, D. W., 1987, Three-dimensional flow beneath a slow-spreading ridge axis: A dynamic contribution to deepening of the median valley toward fracture zones, J. Geophys. Res., *v.* 90, pp. 678–84.

Parmentier, E. M., and Haxby, W. F., 1986, Thermal stresses in the oceanic lithosphere: Evidence from geoid anomalies at fracture zones, J. Geophys. Res., *v.* 91, pp. 7193–204.

Parsiegla, N., Stankiewicz, J., Gohl, K., Ryberg, T., and Uenzelmann-Neben, G., 2009, Southern Africa continental margin: Dynamic processes of a transform margin, Geochem. Geophys., *v.* 10, https://doi.org/10.1029/2008GC002196.

Parsons, B. E., and McKenzie, D. P., 1978, Mantle convection and the thermal structure of plates, J. Geophys. Res., *v.* 83, pp. 4485–96.

Parsons, B. E., and Molnar, P., 1976, The origin of outer topographic rises associated with trenches, Geophys. J. R. Astr. Soc., *v.* 45, pp. 707–12.

Parsons, B. E., and Sclater, J. G., 1977, An analysis of the variation of ocean floor bathymetry and heat flow with age, J. Geophys. Res., *v.* 82, pp. 803–27.

Patton, T. L., and O'Connor, S. J., 1988, Cretaceous flexural history of northern Oman Mountain foredeep, Am. Assoc. Pet. Geol., *v.* 72, pp. 797–809.

Paxman, G. J. G., 2015, Quantifying tectonic and erosion-driven uplift in the Gamburtsev Subglacial Mountains of East Antarctica, MESc. thesis, Oxford University.

Paxman, G. J. G., Watts, A. B., Ferraccioli, F. et al. 2016, Erosion-driven uplift in the Gamburtsev Subglacial Mountains of East Antarctica, *Earth Planet. Sci. Lett.*, *v.* 452, pp. 1–14, https://doi.org/10.1016/j.epsl.2016.07.040.

Pazzaglia, F. J., and Gardner, T. W., 1994, Late Cenozoic flexural deformation of the middle U.S. Atlantic passive margin, J. Geophys. Res., *v.* 99, pp. 12143–57.

Pearson, W. C., and Lister, C. R. B., 1979, The gravity signatures of isostatic, thermally-expanded ridge crests, Phys. Earth Planet. Inter., *v.* 19, pp. 73–84.

Pedoja, K., Husson, L., Regard, V. et al. 2011, Relative sea-level fall since the last interglacial stage: Are coasts uplifting worldwide?, Earth Sci. Rev., *v.* 108, pp. 1–15, https://doi.org/10.1016/j.earscirev.2011.05.002.

Peirce, C., Whitmarsh, R. B., Scrutton, R. A. et al. 1996, Cote d'Ivoire-Ghana margin: seismic imaging of passive rifted crust adjacent to a transform continental margin, Geophys. J. Int., *v.* 125, pp. 781–95.

Pelletier, J. D., 2004, Estimate of three-dimensional flexural-isostatic response to unloading: Rock uplift due to late Cenozoic glacial erosion in the western United States, Geology, *v.* 32, pp. 161–4.

Peltier, W. R., 1974, The impulse response of a Maxwell Earth, Rev. Geophys. Space Phys., *v.* 12, pp. 649–69.

Peltier, W. R., 1980, Models of glacial isostasy and relative sea level, in Bally, A. W. Bender, P. L., McGetchin, T. R. and Walcott, R. I. eds., *Dynamics of the Plate Interiors*, Geodynamics Series, *v.* 1, Washington, DC, American Geophysical Union, pp. 111–28.

Peltier, W. R., and Andrews, J. T., 1976, Glacial-isostatic adjustment – I. The forward problem, Geophys. J. R. Astr. Soc., *v.* 46, pp. 605–46.

Peper, T., and Cloetingh, S., 1992, Lithosphere dynamics and tectono-stratigraphic evolution of the Mesozoic Betic rifted margin (southeastern Spain), Tectonophysics, *v.* 203, pp. 345–61.

Pérez-Gussinyé, M., and Reston, T. J., 2001, Rheological evolution during extension at nonvolcanic rifted margins: Onset of serpentinization and development of detachments leading to continental breakup, J. Geophys. Res., *v.* 106, pp. 3961–75.

Pérez-Gussinyé, M., Lowry, A. R., Watts, A. B., and Velicogna, I., 2004, On the recovery of effective elastic thickness using spectral methods: Examples from synthetic data and from the Fennoscandia Shield, J. Geophys. Res., *v.* 109, https://doi.org/10.1029/2003JB002788.

Pérez-Gussinyé M., and Watts, A. B., 2005. The long-term strength of Europe and its implications for plate-forming processes, Nature, *v.* 436, https://doi.org/10.1038/nature03854.

Pérez-Gussinyé, M., Lowry, A. R., and Watts, A. B., 2007, Effective elastic thickness of South America and its implications for intracontinental deformation, Geochem. Geophys., *v.* 8, https://doi.org/10.1029/2006GC001511.

Pérez-Gussinyé, M., Metois, M., Fernandez, M. et al., 2009, Effective elastic thickness of Africa and its relationship to other proxies for lithospheric structure and surface tectonics, Earth Planet. Sci. Lett., *v.* 287, pp. 152–67, https://doi.org/10.1016/j.epsl.2009.08.004.

Pérez-Gussinyé, M., Andrés-Martínez, M., Araújo, M. et al. 2020, Lithospheric strength and rift migration controls on synrift stratigraphy and breakup unconformities at rifted margins: Examples from numerical models, the Atlantic and South China Sea Margins, Tectonics, *v.* 39, https://doi.org/10.1029/2020TC006255.

Perrier, R., and Quiblier, J., 1974, Thickness changes in sedimentary layers during compaction history: methods for quantitative evaluation, Amer. Assoc. Pet. Geologists Bulletin, *v.* 53, pp. 507–20.

Petengill, G. H., Eliason, E., Ford, P. G. et al. 1980, Venus Pioneer radar results: Altimetry and surface properties: J. Geophys. Res., *v.* 85, pp. 8261–70.

Peterson C., and Roy M., 2005. Gravity and flexure models of the San Luis, Albuquerque, and Tularosa basins in the Rio Grande rift, New Mexico, and southern Colorado, *New Mexico Geological Society, 56th Field Conference Guidebook*, pp. 105–14. https://doi.org/10.56577/FFC-56.105.

Petrunin, A., and Sobolev, S. V., 2006, What controls thickness of sediments and lithospheric deformation at a pull-apart basin?, Geology, *v.* 34, pp. 389–92, https://doi.org/310.1130/G22158.22151.

Phillips, R. J., 1994, Estimating lithospheric properties at Atla Regio, Venus, Icarus, *v.* 112, pp. 147–70.

Phillips, R. J., Kaula, W. M., McGill, G. E., and Malin, M. C., 1981, Tectonics and evolution of Venus, Science, *v.* 212, pp. 879–87.

Phipps Morgan, J., Parmentier, E. M., and Lin, J., 1987, Mechanisms for the origin of mid-ocean ridge axial topography: Implications for the thermal and mechanical structure of accreting plate boundaries, J. Geophys. Res., *v.* 92, pp. 12,823–36.

Pilkington, M., 1990, Lithospheric flexure and gravity anomalies at Proterozoic plate boundaries in the Canadian Shield, Tectonophysics, *v.* 176, pp. 277–90.

Pim, J., Peirce, C., Watts, A. B., Grevemeyer, I., and Krabbenhoeft, A., 2008, Crustal structure and origin of the Cape Verde Rise, Earth Planet. Sci. Lett., 272, 422–8, https://doi.org/10.1016/j.epsl.2008.05.012.

Pinet, C., Jaupart, C., Mareschal, J.-C. et al. 1991, Heat flow and lithospheric structure of the eastern Canadian shield, J. Geophys. Res., *v.* 96, pp. 19,941–63.

Pirazzoli, P. I., 1983, Mise en evidence d'une flexure active de la lithosphere, dans l'archipel de la Societe (Polynesie francais), d'apres la position des rivages de la fin de l'Holocene, C. R. Acad. Sci. Paris (Ser. II), *v.* 296, pp. 695–8.

Pitman, W. C., 1978, The relationship between eustasy and stratigraphic sequences of passive margins, Geol. Soc. Am. Bull., *v.* 89, pp. 1389–1403.

Pitman, W. C., and Heirtzler, J. R., 1966, Magnetic anomalies over the Pacific-Antarctic Ridge, Science, *v.* 154, pp. 1164–6.

Pockalny, R. A., Gente, A. P., and Buck, R., 1996, Oceanic transverse ridges: A flexural response to fracture-zone-normal extension, Geology, *v.* 24, pp. 71–4.

Poudjom Djomani, Y. H., Diament, M., and Albouy, Y., 1992, Mechanical behaviour of the lithosphere beneath the Adamawa uplift (Cameroon, West Africa) based on gravity data, J. African Earth Sci., *v.* 15, pp. 81–90.

Poudjom Djomani, Y. H., Fairhead, J. D., and Griffin, W. L., 1999, The flexural rigidity of Fennoscandia: Reflection of the tectonothermal age of the lithospheric mantle, Earth Planet. Sci. Lett., *v.* 174, pp. 139–54.

Poudjom Djomani, Y. H., Nnange, J. M., Diament, M., Ebinger, C. J., and Fairhead, J. D., 1995, Effective elastic thickness and crustal thickness variations in west central Africa inferred from gravity data, J. Geophys. Res., *v.* 100, pp. 22047–70.

Pratt, J. H., 1855, On the attraction of the Himalaya mountains, and of the elevated regions beyond them, upon the plumb line in India, Phil. Trans. R. Soc., *v.* 145, pp. 53–100.

Pratt, J. H., 1859, On the deflection of the plumb-line in India, caused by the attraction of the Himalaya mountains and of the elevated regions beyond; and its modification by the compensating effect of a deficiency of matter below the mountain mass, Phil. Trans. R. Soc., *v.* 149, pp. 745–96.

Pratt, J. H., 1864, Speculations on the constitution of the Earth's crust, Proc. R. Soc. Lond., *v.* XIII, pp. 253–76.

Pratt, J. H., 1871, Speculations on the constitution of the Earth's crust, Phil. Trans. R. Soc., *v.* 161, pp. 335–58.

Prest, V. K., 1970, Quaternary geology of Canada, in Douglas, R. J. W., ed., *Geology and Economic Minerals of Canada*, Ottawa, Department of Energy, Mines and Resources, pp. 677–766.

Price, N. J., and Audley-Charles, M. G., 1983, Plate rupture by hydraulic fracture resulting in overthrusting, Nature, *v.* 306, pp. 572–5.

Price, R. A., 1973, Large-scale gravitational flow of supracrustal rocks, southern Canadian Rockies, in DeJong, K. A., and Scholten, R., eds., *Gravity and Tectonics*, New York, Wiley-Interscience, pp. 491–501.

Priestley, K., and Debayle, E., 2003, Seismic evidence for a moderately thick lithosphere beneath the Siberian Platform, Geophys. Res. Letters, *v.* 30, https://doi.org/10.1029/2002GL015931.

Priestley, K., and McKenzie, D., 2013, The relationship between shear wave velcoity, temperature, attenuation and viscosity in the shallow part of the mantle, Earth Planet. Sci. Lett., *v.* 381, 78–91, http://dx.doi.org/10.1016/j.epsl.2013.08.022.

Priestley, K., McKenzie, D., and Ho, T., 2018, A lithosphere–asthenosphere boundary – a global model derived from multimode surface-wave tomography and petrology, in Yuan, H., and Romanowicz, B., eds., *Lithospheric Discontinuities*, American Geophysical Union and John Wiley and Sons Inc., pp. 111–24.

Pringle, M. S., and Duncan, R. A., 1995, Radiometric ages of basement lavas recovered at Loen, Wodejebato, MIT, and Takuyo-daisan guyots, northwestern Pacific Ocean, in Haggerty, J. A., Silva, I. P., Rack, F., and McNutt, M. K., eds., *Proceedings of the Ocean Drilling Program, Scientific Results*, v. 144, pp. 547–57.

Pullen, S., and Lambeck, K., 1981, Mascons and loading of the lunar lithosphere, Proc. Lunar Planet. Sci. Conf. 12th, pp. 853–65.

Purdy, G. M., and Detrick, R. S., 1986, Crustal structure of the Mid-Atlantic Ridge at 23°N from seismic refraction, J. Geophys. Res., *v.* 91, pp. 3739–62.

Purdy, G. M., Kong, L. S. L., Christeson, G. L., and Solomon, S. C., 1992, Relationship between spreading rate and the seismic structure of mid-ocean ridges, Nature, v. 355, pp. 815–17.

Putnam, G. R., 1895, Results of a transcontinental series of gravity measurements, Bull. Phil. Soc. Washington, *v.* 13, pp. 31–60.

Putnam, G. R., 1912, Condition of the Earth's crust, Science, *v.* 36, pp. 869–71.

Putnam, G. R., 1922, Condition of the Earth's crust and the earlier American gravity observations, Bull. Geol. Soc. Am., *v.* 33, pp. 287–302.

Putnam, G. R., 1929, Isostasy, Nature, *v.* 123, pp. 298–9.

Putnam, G. R., 1930, Isostatic compensation in relation to geological problems, J. Geol., *v.* 38, pp. 590–9.

Quidelleur, X., Hildenbrand, A., and Samper, A., 2008, Causal link between Quaternary paleoclimatic changes and volcanic islands evolution, Geophys. Res. Lett., *v.* 35, https://doi.org/10.1029/2007GL031849.

Quinlan, G. M., 1987, Models of subsidence mechanisms in intracratonic basins, and their applicability to North American examples, in Beaumont, C., and Tankard, A. J., eds., *Sedimentary Basins and Basin-Forming Mechanisms*, Memoir 12, Calgary, Canadian Society of Petroleum Geology, pp. 463–81.

Quinlan, G. M., and Beaumont, C, 1984, Appalachian thrusting, lithospheric flexure and the Paleozoic stratigraphy of the eastern Interior of North America, Can. J. Earth Sci., *v.* 21, pp. 973–96.

Qureshi, I. R., 1976, Two-dimensionality on a spherical Earth – A problem in gravity reductions, Pageoph, *v* 114, pp. 81–95.

Qureshy, M. N., 1969, Thickening of a basalt layer as a possible cause for the uplift of the Himalayas – A suggestion based on gravity data, Tectonophysics, *v.* 7, pp. 137–57.

Rajesh, R. S., and Mishra, D. C., 2004. Lithospheric thickness and mechanical strength of the Indian shield, Earth Planet. Sci. Lett., *v.* 225, pp. 319–28.

Rajesh, R. S., Stephen J., and Mishra, D. C., 2003. Isostatic response and anisotropy of the eastern Himalyan-Tibetan Plateau: A reappraisal using multitaper spectral analysis, Geophys. Res. Lett., *v.* 30, https://doi.org/10.29/2002GL016104.

Ranero, C. R., Phipps_Morgan, J., McIntosh, K., and Reichert, C., 2003, Bending related faulting and mantle serpentinisation at the Middle America trench, Nature, *v.* 425, pp. 367–73.

Ranero, C. R., Villaseñor, A., Morgan, J. P., and Weinrebe, W., 2005, Relationship between bend-faulting at trenches and intermediate-depth seismicity, Geochem. Geophys., *v.* 6, https://doi.org/10.1029/2005GC000997.

Rapp, R. H., 1989, The decay of the spectrum of the gravitational potential and the topography for the Earth, Geophys. J. Int., *v.* 99, pp. 449–55.

Rapp, R. H., and Pavlis, N. K., 1990, The development and analysis of geopotential coefficient models to spherical harmonic degree 360, J. Geophys. Res., *v.* 95, pp. 21885–911.

Raterron, P., Wu, Y., Weidner, D. J., and Chen, J., 2004. Low-temperature olivine rheology at high pressure, Phys. Earth Planet. Inter., *v.* 145, 149–59.

Ravaut, P., Al Yahya'ey, A., Bayer, R., and Lesquer, A., 1993, Response isostatique de la plate-forme arabique au chargement ophiolitique en Oman, Compte Rendue de l'Academie Science Paris, v. 317, pp. 463–70.

Rayleigh, C. B., Kirby, S. H., Carter, N. L., and Avé Lallement, H. G., 1971, Slip and the clinoenstatite transformation as competing rate processes in enstatite, J. Geophys. Res., *v.* 76, pp. 4011–22.

Reading, H. G., and Collinson, J. D., 1996, Clastic coasts (Chapter 6), in Reading, H. G., ed., *Sedimentary Environments: Processes, Facies and Stratigraphy*, pp. 154–228, Blackwell Science.

Reasenberg, R. D., and Bills, B. G., 1983, Critique of "Elastic thickness of the Venus lithosphere estimated from topography and gravity" by A. Cazenave and K. Dominh, Geophys. Res. Lett., *v.* 10, pp. 93–6.

Rees, B. A., Detrick, R. S., and Coakley, B. J., 1993, Seismic stratigraphy of the Hawaiian flexural moat, Geol. Soc. Am. Bull, *v.* 105, pp. 189–205.

Regan, J., and Anderson, D. L., 1985, Anisotropic models of the upper mantle, Phys. Earth Planet. Int., *v.* 35, pp. 227–63.

Reid, H. F., 1922, Isostasy and earth movements, Bull. Geol. Soc. Am., *v.* 33, pp. 317–18.

Reid, H. F., 1932, in Bowie, W., ed., *Comments on Isostasy*, Washington, DC, National Research Council, 49 pp.

Reston, T. J., Krawczyk, C., and Klaeschen, D., 1996, The S reflector west of Galicia (Spain); evidence from pre-stack depth migration for detachment faulting during continental breakup, J. Geophys. Res., *v.* 101, pp. 8075–91.

Reynolds, D. J., Steckler, M. S., and Coakley, B. J., 1991, The role of the sediment load in sequence stratigraphy: The influence of flexural isostasy and compaction, J. Geophys. Res., *v.* 96, pp. 6931–49.

Richter, F. M., and Parsons, B., 1975, On the interaction of two scales of convection in the mantle, J. Geophys. Res., *v.* 80, pp. 2529–41.

Ritzwoller, M. H., Shapiro, N. M., and Zhong, S., 2004, Cooling history of the Pacific lithosphere, Earth Planet. Sci. Lett., *v.* 226, pp. 69–84.

Robertson, A. H. F., 1998, Tectonic significance of the Eratosthenes seamount: A continental fragment in the process of collision with a subduction zone in the eastern Mediterranean (Ocean Drilling Project Leg 160), Tectonophysics, *v.* 298, pp. 63–82.

Robinson, A. H., Peirce, C., and Funnell, M. J., 2018, Construction and subduction of the Louisville Ridge, SW Pacific – insights from wide-angle seismic data modelling, Geophys. J. Int., *v.* 215, pp. 2222–45, https://doi.org/10.1093/gji/ggy397.

Robinson, N. H., 1980, *The Royal Society Catalogue of Portraits*, London, The Royal Society, 343 pp.

Roddaz, M., Baby, P., Brusset, S., Hermoza, W., and Darrozes, J. M., 2005, Forebulge dynamics and environmental control in Western Amazonia: The case study of the Arch of Iquitos (Peru), Tectonophysics, *v.* 399, 87–108, https://doi.org/10.1016/j.tecto.2004.12.017.

Ross, J. V., and Nielsen, K. C, 1978, High-temperature flow of wet polycrystalline enstatite, Tectonophysics, *v.* 44, pp. 233–61.

Royden, L., 1988a, Flexural behaviour of the continental lithosphere in Italy: Constraints imposed by gravity and deflection data, J. Geophys. Res., *v.* 93, pp. 7747–66.

Royden, L. H., 1988b, Late Cenozoic tectonics of the Pannonian basin system, in Royden, L. H., and Horvath, F., eds., *The Pannonian Basin: A Study in Basin Evolution*, Memoir 45: Tulsa, OK, American Association of Petroleum Geologists, pp. 27–48.

Royden, L., 1993, The tectonic expression of slab pull at convergent plate boundaries, Tectonics, *v.* 12, pp. 303–25.

Royden, L., 1996, Coupling and decoupling of crust and mantle in convergent orogens: Implications for strain partitioning in the crust, J. Geophys. Res., *v.* 101, pp. 17679–705.

Royden, L., and Karner, G. D., 1984, Flexure of the lithosphere beneath the Apennine and Carpathian foredeep basins: Evidence for an insufficient topographic load, Amer. Assoc. Pet. Geol., *v.* 68, pp. 704–12.

Royden, L., and Keen, C. E., 1980, Rifting process and thermal evolution of the continental margin of eastern Canada determined from subsidence curves, Earth Planet. Sci. Lett., *v.* 51, pp. 343–61.

Royden, L., Sclater, J. G., and Von Herzen, R. P., 1980, Continental Margin Subsidence and Heat flow: Important parameters in formation of petroleum hydrocarbons, Am. Assoc. Petrol. Geol. Bull., *v.* 64, pp. 173–87.

Ruppel, C., and McNutt, M. K., 1990, Regional compensation of the Greater Caucasus mountains based on an analysis of Bouguer gravity data, Earth Planet. Sci. Lett., *v.* 98, pp. 360–79.

Rutter, E. H., and Brodie, K. H., 1991, Lithosphere rheology – A note of caution, J. Struct. Geol., *v.* 13, pp. 363–7.

Sacek, V., and Ussami, N., 2009, Reappraisal of the effective elastic thickness for the sub-Andes using 3-D finite element flexural modelling, gravity and geological constraints, Geophys J. Int., *v.* 179, pp. 778–6, https://doi.org/10.1111/j.1365-246X.2009.04334.x.

Sandwell, D. T., 1986, Thermal stress and the spacing of transform faults, J. Geophys. Res., *v.* 91, pp. 6405–17.

Sandwell, D. T., 1982, Thermal isostasy: Response of a moving lithosphere to a distributed heat source, J. Geophys. Res., *v.* 87, pp. 1001–14.

Sandwell, D. T., 2022, *Advanced Geodynamics: The Fourier Transform Method,* Cambridge, Cambridge University Press, 272 pp.

Sandwell, D. T., and Schubert, G., 1982, Lithospheric flexure at fracture zones, J. Geophys. Res., *v.* 87, pp. 4657–67.

Sandwell, D. T., and Schubert, G., 1992, Flexural ridges, trenches, and outer rises around coronae on Venus, J. Geophys. Res., *v.* 97, pp. 16069–83.

Sandwell, D. T., and Smith, W. H. F., 1997, Marine gravity anomaly from Geosat and ERS-1 satellite altimetry, J. Geophys. Res., *v.* 102, pp. 10039–54

Sandwell, D. T., Müller, R. D., Smith, W. H. F., Garcia, E., and Francis, R., 2014, New global marine gravity model from CryoSat-2 and Jason-1 reveals buried tectonic structure, Science, *v.* 346, pp. 65–7, https://doi.org/10.1126/science.1258213.

Sandwell, D. T., Harper, H., Tozer, B., and Smith, W. H. F., 2019, Gravity field recovery from geodetic altimeter missions, Advances in Space Research, https://doi.org/10.1016/j.asr.2019.09.011.

Sanford, B. V., 1987, Paleozoic geology of the Hudson Platform, in *Sedimentary Basins and Basin-Forming Mechanisms: Intracratonic Basins,* Amer. Assoc. Pet. Geol. Memoir, *v.* 12, pp. 483–505.

Sauramo, M., 1955, Die Geschichte der Ostsee, Ann. Acad. Sci. Fennicae, *v.* 51, pp. 1–522.

Schedl, A., and Wiltschko, D. V., 1984, Sedimentological effects of a moving terrain, J. of Geology, *v.* 92, pp. 273–87.

Schofield, J. C., 1967, 1-Post glacial sea-level maxima a function of salinity?. 2-Pleistocene sealevel evidence from Cook Islands, J. of Geosciences, *v.* 10, pp. 115–20.

Schofield, J. C., and Nelson, C. S., 1978, Dolomitisation and Quaternary climate of Niue Island, Pacific Ocean: Pacific Geol., *v.* 13, pp. 37–48.

Schubert, G., and Moore, W. B., 1994, Gravity over Coronae and Chasmata on Venus, Icarus, *v.* 112, pp. 130–46.

Schultz, P. H., and Gault, D. E., 1975, Seismic effects from major basin formation on the Moon and Mercury, Moon, *v.* 12, pp. 159–77.

Schumm, S. A., 1963, The disparity between present rates of denudation and orogeny, U.S. Geol. Surv. Prof. Papers, *v.* 454–H, pp. H1–H13.

Sclater, J. G., and Francheteau, J. 1970, The implications of terrestrial heat flow observations on current tectonic and geochemical models of the crust and upper mantle of the Earth, Geophys. J. R. Astr. Soc., *v.* 20, pp. 509–42.

Sclater, J. G., Jaupart, C., and Galson, D., 1980, The heat flow through oceanic and continental crust and the heat loss of the Earth, Rev. Geophys. Space Phys., *v.* 18, pp. 269–311.

Sclater, J. G., Lawver, L. A., and Parsons, B. E., 1975, Comparison of long wavelength residual elevation and free air gravity anomalies in the north Atlantic and possible implications for the thickness of the lithosphere plate, J. Geophys. Res, *v.* 80, pp. 1031–52.

Scruton, P. C., 1960, Delta building and the deltaic sequence, in Shepard, F. C., Phleger, F. B., and van Andel, T. H., eds., *Recent Sediments, Northwest Gulf of Mexico,* Tulsa, OK, American Association Petroleum Geologists, pp. 82–102.

Scrutton, R. A., 1982, Crustal structure and development of sheared passive continental margins, in Scrutton, R. A., ed., *Dynamics of Passive Margins,* Geodynamics Series 6, Washington, DC, American Geophysical Union, pp. 133–40.

Searle, M. P., Law, R. D., and Jessup, M. J., 2006, Crustal structure, restoration, and evolution of the Greater Himalaya in Nepal-South Tibet: implications for channel flow and ductile extrusion of the middle crust, J. Geol. Soc. London, *v.* 268, pp. 355–78.

Searle, R. C., 1970, Evidence from gravity anomalies for thinning of the lithosphere beneath the Rift Valley in Kenya, Geophys. J. R. Astr. Soc., *v.* 21, pp. 13–31.

Searle, R. C., Francheteau, J., and Cornaglia, B., 1995, New observations on mid-plate volcanism and the tectonic history of the Pacific plate, Tahiti to Easter microplate, Earth Planet. Sci. Lett., *v.* 131, pp. 395–421.

Searle, R. C., Keeton, J. A., Owens, R. B. et al., 1998, The Reykjanes Ridge: Structure and tectonics of a hot-spot-influenced slow-spreading ridge, from multibeam bathymetry, gravity and magnetic investigations, Earth Planet. Sci. Lett., *v.* 160, pp. 463–78.

Seno, T., and Yamanaka, Y., 1996, Double seismic zones, compressional deep trench-outer rise events, and superplumes, in Bebout, G. E., Scholl, D. W., Kirby, S. H., and Platt, J. P., eds., *Subduction Top to Bottom,* Monograph 36, Washington, DC, American Geophysical Union, pp. 347–55.

Shaber, G. G., Boyce, J. M., and Trask, N. J., 1977, Moon-Mercury: Large impact structures, isostasy and average crustal viscosity, Phys. Earth Planet. Int., *v.* 15, pp. 189–201.

Shah, A. K., and Buck, W. R., 2003, Plate bending stresses at axial highs, and implications for faulting behavior, Earth Planet. Sci. Lett., *v.* 211, pp. 343–56, https://doi.org/10.1016/S0012-821X(03)00187-0.

Shapiro, N. M., and Ritzwoller, M. H., 2002, Monte-Carlo inversion for a global shear-velocity model of the crust and upper mantle, Geophys. J. Int., *v.* 151, pp. 88–105.

Sharp, W. D., and Clague, D. A., 2006, 50-Ma Initiation of Hawaiian-Emperor Bend Records Major Change in Pacific Plate Motion, Science, *v.* 313, pp. 1281–4.

Sharp, W. D., and Renne, P. R., 2005, The 40Ar/39Ar dating of core recovered by the Hawaii Scientific Drilling Project (phase 2), Hilo, Hawaii, Geochem. Geophys., *v.* 6, https://doi.org/10.1029/2004GC000846.

Sheehan, A. F., and McNutt, M. K., 1989, Constraints on thermal and mechanical structure of the oceanic lithosphere at the Bermuda Rise from geoid height and depth anomalies, Earth Planet. Sci. Lett., *v.* 93, pp. 377–91.

Sheffels, B., and McNutt, M., 1986, Role of subsurface loads and regional compensation in the isostatic balance of the Transverse Ranges, California: Evidence for intracontinental subduction, J. Geophys. Res., *v.* 91, pp. 6419–31.

Shelton, G., and Tullis, J., 1981, Experimental flow laws for crustal rocks, EOS Trans. American Geophysical Union, *v.* 62, p. 396.

Shepard, F. P., 1923, Isostasy as a result of Earth shrinkage, J. Geol., *v.* 31, pp. 208–16.

Shi, X., Burov, E., Leroy, S., Qiu, X., and Xia, B., 2005, Intrusion and its implication for subsidence: A case from the Baiyun Sag, on the northern margin of the South China Sea, Tectonophysics, *v.* 407, pp. 117–34, https://doi.org/10.1016/j.tecto.2005.07.004.

Shillington, D. J., Bécel, A., Nedimović, M. R. et al., 2015, Link between plate fabric, hydration and subduction zone seismicity in Alaska, Nature Geoscience, *v.* 8, pp. 961–4, https://doi.org/910.1038/ngeo2586.

Shudofsky, G. N., Cloetingh, S., Stein, S., and Wortel, R., 1987, Unusually deep earthquakes in east Africa: Constraints on the thermo-mechanical structure of a continental rift system, Geophys. Res. Lett., *v.* 14, pp. 741–4.

Sibuet, J.-C., Le Pichon, X., and Goslin, J., 1974, Thickness of lithosphere deduced from gravity edge effects across the Mendocino Fault, Nature, *v.* 252, pp. 676–9.

Sibuet, J.-C., and Veyrat-Peinet, B., 1980, Gravimetric model of the Atlantic Equatorial Fracture Zones, J. Geophys. Res., *v.* 85, pp. 943–54.

Simons, F. J., Zuber, M. T., and Korenaga, J., 2000, Isostatic response of the Australian lithosphere: Estimation of effective elastic thickness and anisotropy using multitaper spectral analysis, J. Geophys. Res., *v.* 105, pp. 19163–84.

Simpson, G., 2004, Role of river incision in enhancing deformation, Geology, *v.* 32, pp. 341–4, https://doi.org/10.1130/G20190.1.

Simpson, R. W., Jachens, R. C., Blakeley, R. J., and Saltus, R. W., 1986, A new isostatic residual gravity map of the conterminous United States with a discussion on the significance of isostatic residual anomalies, J. Geophys. Res., *v.* 91, pp. 8348–72.

Sinclair, H. D., Coakley, B. J., Allen, P. A., and Watts, A. B., 1991, Simulation of foreland basin stratigraphy using a diffusion model of mountain belt uplift and erosion: An example from the central Alps, Switzerland, Tectonics, *v.* 10, pp. 599–620.

Sjogren, W. L., 1979, Mars gravity: High resolution results from Viking orbiter 2, Science, *v.* 203, pp. 1006–9.

Sjogren, W. L., Phillips, R. J., Birkeland, P. W., and Wimberly, R. N., 1980, Gravity anomalies on Venus, J. Geophys. Res., *v.* 85, pp. 8295–302.

Sleep, N. H., 1971, Thermal effects of the formation of Atlantic continental margins by continental breakup, Geophys. J. R. Astr. Soc., *v.* 24, pp. 325–50.

Sleep, N. H., 1973, Crustal thinning on Atlantic continental margins: evidence from older margins, in *Implications of Continental Drift to the Earth Sciences*, pp. 685–92, New York, Academic Press.

Sleep, N. H., and Rosendahl, B. R., 1979, Topography and tectonics of mid-ocean ridge axes, J. Geophys. Res., *v.* 84, pp. 6831–9.

Sleep, N. H., and Snell, N. S., 1976, Thermal contraction and flexure of mid-continent and Atlantic marginal basins, Geophys. J. R. Astr. Soc., *v.* 45, pp. 125–54.

Sloss, L., 1963, Sequences in the cratonic interior of North America, Geol. Soc. Am. Bull., *v.* 74, pp. 93–113.

Sloss, L. L., and Scherer, W., 1975, Geometry of sedimentary basins: Applications to the Devonian of North America and Europe, Geol. Soc. Am. Memoir 142, pp. 71–88.

Smallwood, J. R., 2010. Bouguer redeemed: The successful 1737–1740 gravity experiments on Pichincha and Chimborazo, Earth Sciences History, *v.* 29, pp. 1–29.

Smith, D. E., Zuber, M. T., Frey, H. V. et al. 1998, Topography of the Northern hemisphere of Mars from Mars Orbiter Laser Altimeter, Science, *v.* 279, pp. 1686–92.

Smith, D. E., Zuber, M. T., Solomon, S. C. et al. 1999, The global topography of Mars from MOLA, Science, *v.* 284, pp. 1495–503.

Smith, D. E., Zuber, M. T., Frey, H. V. et al., 2001, Mars Orbiter Laser Altimeter – Experiment summary after the first year of global mapping of Mars., J. Geophys. Res., *v.* 106, pp. 23689–722, https://doi.org/10.1029/2000JE001364.

Smith, D. E., Zuber, M. T., Neumann, G. A. et al., 2010, Initial observations from the Lunar Orbiter Laser Altimeter (LOLA), Geophys. Res. Letters, *v.* 37, 6 pp., https://doi .org/10.1029/2010GL043751.

Smith, J. R., and Wessel, P., 2000, Isostatic consequences of giant landslides on the Hawaiian Ridge, Pure and Appl. Geophys., *v.* 157, pp. 1097–114.

Smith, W. H. F., Staudigel, H., Watts, A. B., and Pringle, M. S., 1989, The Magellan Seamounts: Early Cretaceous record of the South Pacific isotopic and thermal anomaly, J. Geophys. Res., *v.* 94, pp. 10,501–23.

Smith, W. H. F., and Sandwell, D. T., 1997, Global Sea Floor Topography from Satellite Altimetry and Ship Depth Soundings, Science, *v.* 277, pp. 1956–62.

Smrekar, S. E., 1994, Evidence for active hotspots on Venus from analysis of Magellan gravity data, Icarus, *v.* 112, pp. 2–26.

Smrekar, S. E., and Stofan, E. R., 2003. Effects of lithospheric properties on the formation of Type 2 coronae on Venus, *J. Geophys. Res.*, 108, https://doi.org/1029/ 2002JE001930.

Snyder, D. B., and Barazangi, M., 1986, Deep crustal structure and flexure of the Arabian plate beneath the Zagros collisional mountain belt as inferred from gravity observations, Tectonics, *v.* 5, pp.361–73.

Soha, J. M., Lynn, D. J., Lorre, J. J. et al. 1975, IPL processing of the Mariner 10 images of Mercury, J. Geophys. Res., *v.* 80, pp. 2394–414.

Sohn, R. A., and Sims, K. W. W., 2005, Bending as a mechanism for triggering off-axis volcanism on the East Pacific Rise, Geology, *v.* 33, pp. 93–6, https://doi.org/10.1130/ G21116.1.

Solomon, S. C., 1976, Geophysical constraints on radial and lateral temperature variations in the upper mantle: Am. Mineralogist, *v.* 61, pp. 788–803.

Solomon, S. C., and Head, J. W., 1979, Vertical movement in mare basins: Relation to mare emplacement, basin tectonics, and lunar thermal history, J. Geophys. Res., *v.* 84, pp. 1667–82.

Solomon, S. C., and Head, J. W., 1980, Lunar mascon basins: Lava filling, tectonics and evolution of the lithosphere, Rev. Geophys. and Space Phys., *v.* 18, pp. 107–41.

Solomon, S. C., and Head, J. W., 1982, Evolution of the Tharsis province of Mars: The importance of heterogeneous lithospheric thickness and volcanic construction, J. Geophys. Res., *v.* 87, pp. 9755–74.

Solomon, S. C., and Head, J. W., 1990, Lithospheric flexure beneath the Freyja Montes foredeep, Venus: Constraints on lithospheric thermal gradient and heat flow, Geophys. Res. Lett., *v.* 17, pp. 1393–403.

Solomon, S. C., Comer, R. P., and Head, J. W., 1982, The evolution of impact basins: Viscous relaxation of topographic relief, J. Geophys. Res., *v.* 87, pp. 3975–92.

Solomon, S. C., Smrekar, S. E., Bundschadler, D. L. et al. 1992, Venus tectonics: An overview of Magellan observations, J. Geophys. Res., *v.* 97, pp. 13199–255.

Sori, M. M., 2018, A thin, dense crust for Mercury, *Earth Planet. Sci. Lett.*, *v.* 489, pp. 92–9, https://doi.org/10.1016/j.epsl.2018.02.033.

Spadini, G., Cloetingh, S., and Berlotti, G., 1995, Thermo-mechanical modeling of the Tyrrhenian Sea, Tectonics, *v.* 14, pp. 629–44.

Squyres, S. W., Janes, D. M., Baker, G. et al. 1992, The morphology and evolution of coronae on Venus, J. Geophys. Res., *v.* 97, pp. 13611–34.

Stanley, J.-D., 2001, Dating modern deltas: Progress, problems, and prognostics, Annu. Rev. Earth Planet. Sci., *v.* 29, pp. 257–94.

Stark, C. P., Stewart, J., and Ebinger, C. J., 2003, Wavelet transform mapping of the effective elastic thickness and plate loading: Validation using synthetic data and application to the study of the South African tectonics, J. Geophys. Res., *v.* 108, https://doi.org/10.1029/2001JB000609.

Staudigel, H., and Schmincke, H.-U., 1984, The Pliocene seamount series of La Palma/ Canary Islands, J. Geophys. Res., *v.* 89, pp. 11195–215.

Steckler, M. S., 1985, Uplift and extension in the Gulf of Suez: Indications of induced mantle convection, Nature, *v.* 317, pp. 135–9.

Steckler, M. S., and Watts, A. B., 1980, The Gulf of Lion: Subsidence of a young continental margin, Nature, *v.* 287, pp. 425–9.

Steckler, M. S., Mountain, G. S., Miller, K. G., and Christie-Blick, N., 1999, Reconstruction of Tertiary progradation and clinoform development on the New Jersey passive margin by 2-D backstripping, Mar. Geol., *v.* 154, pp. 399–420.

Stephen, J., Singh, S. B., and Yedekar, D. B., 2003. Elastic thickness and isostatic coherence anisotropy in the South Indian Peninsular Shield and its implications, Geophys. Res. Lett., *v.* 30, https://doi.org/10.1029/2003GL017686.

Stern, R. J., Reagan, M., Ishizuka, O., Ohara, Y., and Whattam, S., 2012, To understand subduction initiation, study forearc crust: To understand forearc crust, study ophiolites, Lithosphere, *v.* 4, pp. 469–83, https://doi.org/10.1130/L183.1.

Stern, T. A., and McBride, J. H., 1998, Seismic exploration of continental strike-slip zones, Tectonophysics, *v.* 286, pp. 63–78.

Stern, T. A., and ten Brink, U. S., 1989, Flexural uplift of the Transantarctic Mountains, J. Geophys. Res., *v.* 94, pp. 10,315–30.

Stern, T. A., Baxter, A. K., and Baxter, P. J., 2005, Isostatic rebound due to glacial erosion within the Transantarctic Mountains, Geology, *v.* 33, pp. 221–4.

Stewart, J., 1998, Gravity anomalies, flexure and the thermal and mechanical properties of the continental lithosphere, D.Phil thesis, Oxford University.

Stewart, J., and Watts, A. B., 1997, Gravity anomalies and spatial variations of flexural rigidity at mountain ranges, J. Geophys. Res., *v.* 102, pp. 5327–52.

Stewart, I. S., Sauber, J., and Rose, J., 2000, Glacio-seismotectonics: Ice sheets, crustal deformation and seismicity, Quat. Science Reviews, *v.* 19, pp. 1367–89.

Stockmal, G. S., Beaumont, C., and Boutilier, R., 1986, Geodynamic models of covergent margin tectonics: Transition from rifted margin to overthrust belt and consequences for foreland-basin development, Am. Assoc. Pet. Geol., *v.* 70, pp. 181–90.

Stocks, T., and Wüst, G., 1935, *Die Tiefenverhaltnisse des offenen Atlantischen Ozeans,* Berlin und Leipzig, Verlag von Walter de Gruyter & Co, 31 pp.

Stoffa, P., and Buhl, P., 1980, Two-ship multichannel seismic experiments for deep crustal studies, J. Geophys. Res., *v.* 84, pp. 7645–60.

Stolper, E. M., DePaolo, D. J., and Thomas, D. M., 2009, Deep drilling into a mantle plume volcano: The Hawaii Scientific Drilling Project, Scientific Drilling, *v.* 7, pp. 4–14.

Strom, R. G., 1971, Lunar mare ridges, rings and volcanic ring complexes, Mod. Geol., *v.* 2, pp. 133–57.

Strom, R. G., Terrile, R. J., and Guest, J. E., 1975, Tectonism and volcanism on Mercury, J. Geophys. Res., *v.* 80, pp. 2478–507.

Sugano, T., and Heki, K., 2004. Isostasy of the Moon from high-resolution gravity and topography data: Implication for its thermal history, *Geophys. Res. Lett.*, *v.* 31, https://doi.org/10.1029/2004GL022059.

Suguio, K., Martin, L., and Flexor, J., 1980, Sea level fluctuations during the past 6000 years along the coast of the state of Sao Paulo, Brazil, in Morner, N.-A., ed., *Earth Rheology, Isostasy and Eustasy*, New York, John Wiley & Sons, pp. 471–86.

Suppe, J., and Connors, C., 1992, Critical taper wedge mechanics of fold-and-thrust belts on Venus: Initial results from Magellan, J. Geophys. Res., *v.* 97, pp. 13545–57.

Suyenaga, W., 1977, Earth deformation in response to surface loading: Application to the formation of the Hawaiian Ridge, PhD. thesis, University of Hawaii.

Swain, C. J., and Kirby, J. F., 2003, The effect of 'noise' on estimates of the elastic thickness of the continental lithosphere by the coherence method, Geophys. Res. Letts., *v.* 30, https://doi.org/10.1029/2003GL017070.

Swain, C. J., and Kirby, J. F., 2006, An effective elastic thickness map of Australia from wavelet transforms of gravity and topography using Forsyth's method, Geophys. Res. Letts., *v.* 33, https://doi.org/10.1029/2005GL025090.

Sweeney, J. F., 1977, Subsidence of the Sverdrup basin, Canadian Arctic Islands, Geol. Soc. Am. Bull., *v.* 88, pp. 41–8.

Sykes, L. R., 1970, Seismicity of the Indian Ocean and a possible nascent island arc between Ceylon and Australia, J. Geophys. Res., *v.* 75, pp. 5041–55.

Sykes, L. R., 1978, Intraplate seismicity, reactivation of pre-existing zones of weakness, alkaline magmatism and other tectonism postdating continental fragmentation, Rev. Geophys., *v.* 16, pp. 621–88.

Talwani, M., 1972, Lunar Gravity Traverse Experiment, The Moon, *v.* 4, p. 307, https://doi.org/10.1007/BF00561998.

Talwani, M., Worzel, J. L., and Landisman, M., 1959, Rapid gravity computations for two-dimensional bodies with applications to the Mendocino submarine fracture zones, J. Geophys. Res., *v.* 54, pp. 49–59.

Tang, J., Lerche, I., and Cogan, J., 1992, An inverse method for calculating basement geometry, J. Geodynamics, *v* 15, pp. 85–106.

Tanimoto, T., 1997, Bending of spherical lithosphere – axisymmetric case, Geophys. J. Int., *v.* 129, pp. 305–10.

Tanimoto, T., 1998, State of stress within a bending spherical shell and its implications for subducting lithosphere, Geophys. J. Int., *v.* 134, pp. 199–206.

Tanner, J. G., and Uffen, R. J., 1960, Gravity anomalies in the Gaspé Peninsula, Publ. Dom. Obs., *v.* 21, pp. 221–60.

Tapponier, P., and Francheteau, J., 1978, Necking of the lithosphere and the mechanics of slowly accreting plate boundaries, J. Geophys. Res., *v.* 83, pp. 3955–70.

Tarduno, J. A., Duncan, R. A., Scholl, D. W. et al., 2003, The Emperor Seamounts: Southward motion of the Hawaiian hotspot plume in Earth's mantle, Science, *v.* 301, pp. 1064–9, https://doi.org/10.1126/science.1086442.

Tassara, A., Swain, C., Hackney, R., and Kirby, J., 2006, Elastic thickness structure of South America estimated using wavelets and satellite-derived gravity data, Earth Planet. Sci. Lett., *v.* 253, pp. 17–36, https://doi.org/10.1016/j.epsl.2006.10.008.

Tate, M., White, N., and Conroy, J.-J., 1993, Lithospheric extension and magmatism in the Porcupine Basin West of Ireland, J. Geophys. Res., *v.* 98, pp. 13905–23.

Taylor, S. R., 1989, Geophysical framework of the Appalachians and adjacent Grenville Province, in Pakiser, L., and Mooney, W., eds., *Geophysical Framework of the Continental United States*, Geological Society of America Memoir, *v.* 1, pp. 317–47.

ten Brink, U., 1991, Volcano spacing and rigidity, Geology, *v.* 19, pp. 397–400.

ten Brink, U. S., and **Brocher, T. M.**, 1987, Multichannel seismic evidence for a subcrustal intrusive complex under Oahu and a model for Hawaiian volcanism, J. Geophys. Res., *v.* 92, pp. 13687–707.

ten Brink, U. S., and Stern, T., 1992, Rift flank uplifts and hinterland basins: Comparison of the Transantarctic Mountains with the Great Escarpment of Southern Africa, J. Geophys. Res., *v.* 97, pp. 569–85.

ten Brink, U. S., and Watts, A. B., 1985, Seismic stratigraphy of the flexural moat flanking the Hawaiian Islands, Nature, *v.* 317, pp. 421–4.

ten Brink, U. S., Be-Avraham, Z., Bell, R. E. et al. 1993, Structure of the Dead Sea pull-apart basin from gravity analysis, J. Geophys. Res., *v.* 98, pp. 21887–94.

ten Brink, U. S., Hackney, R. I., Bannister, S., Stern, T. A., and **Makovsky, Y.**, 1997. Uplift of the Transantarctic Mountains and the bedrock beneath the East Antarctic ice sheet, J. Geophys. Res., *v.* 102, pp. 27,603–21.

Tessema, A., and Antoine, L. A. G., 2003. Variation in effective elastic plate thickness of the East Africa lithosphere, J. Geophys. Res. *v.* 108, https://doi.org/10.1029/2002JB002200.

Thom, B. G., and Chappell, J., 1978, Holocene sea level change: An interpretation, Phil. Trans. R. Soc. Lond., *v.* 291, pp. 187–94.

Thurber, C. H., and Toksoz, M. N., 1978, Martian lithospheric thickness from elastic flexure theory, Geophys. Res. Lett., *v.* 5, pp.977–80.

Tiley, R., McKenzie, D, and White, N., 2003. The elastic thickness of the British Isles, J. Geol. Soc. Lond., *v.* 160, pp. 499–502.

Timoshenko, S., 1958a, *Strength of Materials. Part I. Elementary Theory and Problems*, New York, D. Van Nostrand Co, 442 pp.

Timoshenko, S., 1958b, *Strength of Materials. Part II. Advanced Theory and Problems*, New York, D. Van Nostrand Co, 572 pp.

Timoshenko, S. P., and Woinowsky-Krieger, S., 1959, *Theory of Plates and Shells*, 2nd ed., New York, McGraw-Hill, 575 pp.

Tiwari, V. M., and Mishra, D. C., 1999. Estimation of effective elastic thickness from gravity and topography data under the Deccan volcanic province, India, Earth Planet. Sci. Lett., *v.* 171, pp. 289–99.

Todd, B. J., and Keen, C. E., 1989, Temperature effects and their geological consequences at transform margins, Can. J. Earth Sci., *v.* 26, pp. 2591–603.

Tornqvist, T. E., Wallace, D. J., Storms, J. E. A. et al. 2008, Mississippi Delta subsidence primarily caused by compaction of Holocene strata, Nature Geoscience, *v.* 1, pp. 173–6, https://doi.org/10.1038/ngeo129.

Tozer, B., Watts, A. B., and Daly, M. C., 2017, Crustal structure, gravity anomalies and subsidence history of the Parnaíba cratonic basin, Northeast Brazil, J. Geophys. Res., *v.* 122, pp. 5591–621, https://doi.org/10.1002/2017JB014348.

Trask, N. J., and Guest, J. E., 1975, Preliminary terrain map of Mercury, J. Geophys. Res., *v.* 80, pp. 2461–77.

Tsoulis, D., 2001, A comparison between the Airy/Heiskanen and the Pratt/Hayford isostatic models for the computation of potential harmonic coefficients, J. Geodesy, *v.* 74, pp. 637–43.

Turcotte, D. L., 1974, Are transform faults thermal contraction cracks?, J. Geophys. Res., *v.* 79, pp. 2573–7.

Turcotte, D. L., and McAdoo, D. C., 1979, Thermal subsidence and petroleum generation in the southwestern block of the Los Angeles basin, California, J. Geophys. Res., *v.* 84, pp. 3460–4.

Turcotte, D. L., and Schubert, G., 1982, *Geodynamics: Applications of Continuum Physics to Geological Problems,* New York, John Wiley, 450 pp.

Umino S., Lipman P. W., and Obata S., 2000, Subaqueous lava flow lobes, observed on ROV KAIKO dives off Hawaii, Geology *v.* 28, pp. 503–6.

Upcott, N. M., Mukasa, R. K., Ebinger, C. J., and Karner, G. D., 1996, Along-axis segmentation and isostasy in the Western Rift, East Africa, J. Geophys. Res., *v.* 101, pp. 3247–68.

Urbancic, N., Ghent, R., Johnson, C. L. et al., 2017, Subsurface density structure of Taurus-Littrow Valley using Apollo 17 gravity data, J. Geophys. Res. Planets, *v.* 122, pp. 1181–94, https://doi.org/10.1002/2017JE005296.

Ussami, N., Cogo de Sa, N., and Molina, E. C., 1993, Gravity map of Brazil. 2. Regional and residual isostatic anomalies and their correlation with major tectonic provinces, J. Geophys. Res., *v.* 98, pp. 2199–208.

Ussami, N., Karner, G. D., and Bott, M. H. P., 1986, Crustal detachment during South Atlantic rifting and formation of Tucano-Gabon basin system, Nature, *v.* 322, pp. 629–32.

Vai, G. B., 2006, Isostasy in Luigi Ferdinando Marsili's manuscripts, in *The Origins of Geology in Italy,* edited by Vai, G. B. and Caldwell, W. G. E., Geological Society of America Special Papers, *v.* 411, pp. 95–127, https://doi.org/10.1130/2006.2411(07).

Vail, P. R., Mitchum, R. M., and Thompson, S., 1977, Relative sea-level from coastal onlap, in Payton, C. E., ed., *Seismic Stratigraphy – Applications to Hydrocarbon Exploration,* Memoir 26, Tulsa, OK, American Association of Petroleum Geologists, pp. 63–82.

Van Avendonk, H. J. A., Christeson, G. L., Norton, I. O., and Eddy, D. R., 2015, Continental rifting and sediment infill in the northwestern Gulf of Mexico, Geology, *v.* 43, pp. 631–4, https://doi.org/10.1130/G36798.1.

van den Berg, J., van de Wal, R. S. W., and Oerlemans, J., 2006, Recovering lateral variations in lithospheric strength from bedrock motion data using a coupled ice sheet-lithosphere model, J. Geophys. Res., *v.* 111, https://doi.org/10.1029/2005JB003790.

van den Berg, J., van de Wal, R. S. W., Milne, G. A. and Oerlemans, J., 2008, Effect of isostasy on dynamical ice sheet modeling: A case study for Eurasia, J. Geophys. Res., *v.* 113, https://doi.org/10.1029/2007JB004994.

van der beek, P. A., and Cloetingh, S., 1992, Lithospheric flexure and the tectonic evolution of the Betic Cordilleras (SE Spain), Tectonophysics, *v.* 203, pp. 325–44.

van Wees, J. D., and Cloetingh, S., 1994, A finite-difference technique to incorporate spatial variations in rigidity and planar faults into 3–D models for lithospheric flexure, Geophys. J. Int., *v.* 117, pp. 179–95.

van Wyk de Vries, B., and Matela, R., 1998, Styles of volcano-induced deformation: numerical models of substratum flexure, spreading and extrusion, J. Volcanology and Geothermal Research, *v.* 81, pp.1–18.

Vaughan, D. G., 1995, Tidal flexure at ice shelf margins, J. Geophys. Res., *v.* 100, pp. 6213–24.

Vening Meinesz, F. A., 1929, *Theory and Practise of Pendulum Observations at Sea,* Technische boekhandel en drukkerij: Delft, J. Waltman Jr, 95 pp.

Vening Meinesz, F. A., 1931, Une nouvelle methode pour la réduction isostatique régionale de l'intensité de la pesanteur, Bull. Géodésique, *v.* 29, pp. 33–51.

Vening Meinesz, F. A., 1932, in Bowie, W., ed., *Comments on Isostasy*, Washington, DC, National Research Council, 49 pp.

Vening Meinesz, F. A., 1941a, *Gravity Expeditions at Sea 1934–1939, the Expeditions, the Computations and the Results*, v. III, Delft, The Netherlands Geodetic Commission, 97 pp.

Vening Meinesz, F. A., 1941b, *Gravity over the Hawaiian Archipelago and over the Madeira Area; Conclusions about the Earth's Crust*, Proceedings, Koninklijke Nederlandse Akok, v. 44, Wetensiag, 41 pp.

Vening Meinesz, F. A., 1948, *Gravity Expeditions at Sea 1923–1938, Complete Results with Isostatic Reduction Interpretation of the Results*, v. IV, Delft, The Netherlands Geodetic Commission, 24 pp.

Vening Meinesz, F. A., 1950, Les graben africains, résultat de compression ou de tension dans la croûte terrestre?, Inst. R. Colonial Belge, Bull, v. 21, pp. 539–52.

Ventsel, E., and Krauthammer, T., 2001, *Thin Plates and Shells: Theory, Analysis, and Applications*, New York, Marcel Dekker, 666 pp.

Vera, E. E., Mutter, J. C, Buhl, P. et al. 1990, The structure of 0- to 0.2-m.y.-old oceanic crust at 9°N on the East Pacific Rise from expanded spread profiles, J. Geophys. Res., v. 95, pp. 15,529–56.

Verhoef, J., and Jackson, H. R., 1991, Admittance signatures of rifted and transform margins: Examples from eastern Canada, Geophys. J. Int., v. 105, pp. 229–39.

Vogt, N., Pinedo-Vasquez, M., Brondızio, E. S. et al. 2016, Local ecological knowledge and incremental adaptation to changing flood patterns in the Amazon delta, Sustainability Science, v. 11, pp. 611–623, https://doi.org/10.1007/s11625-015-0352-2.

Vogt, P. R., Jung, W. Y., and Brozena, J., 1998, Arctic margin gravity highs: Deeper meaning for sediment depocenters?, Mar. Geophysical Res., v. 20, pp. 459–77.

Wager, L. R., and Deer, W. A., 1938, A dyke swarm and crustal flexure in East Greenland, Geol. Mag., v. 75, pp. 39–46.

Walcott, R. I., 1970a, Flexural rigidity, thickness, and viscosity of the lithosphere, J. Geophys. Res., v. 75, pp. 3941–53.

Walcott, R. I., 1970b, Flexure of the lithosphere at Hawaii, Tectonophysics, v. 9, pp. 435–46.

Walcott, R. I., 1970c, Isostatic response to loading of the crust in Canada, Can. J. Earth Sci., v. 7, pp. 716–27.

Walcott, R. I., 1970d, An isostatic origin for basement uplifts, Can. J. Earth Sci., v. 7, pp. 931–7.

Walcott, R. I., 1972a, Gravity, flexure, and the growth of sedimentary basins at a continental edge, Geol. Soc. Am. Bull., v. 83, pp. 1845–8.

Walcott, R. I., 1972b, Late Quaternary vertical movements in Eastern North America: Quantitative evidence of glacio-isostatic rebound, Rev. Geophys. Space Phys., v. 10, pp. 849–84.

Walcott, R. I., 1973, Structure of the Earth from glacio-isostatic rebound, Annu. Rev. Earth Planet Sci., v. 1, pp. 15–37.

Walcott, R. I., 1976, Lithospheric flexure, analysis of gravity anomalies, and the propagation of seamount chains, in Sutton, G. H., Manghnani, M. H., and Moberly, R., eds., *The Geophysics of the Pacific Ocean Basin and Its Margin*, Geophysical Monograph 19, Washington, DC, American Geophysical Union, pp. 431–8.

Wang, H., Wright, T. J., Yu, Y. et al. 2012, InSAR reveals coastal subsidence in the Pearl River Delta, China, Geophys. J. Int., v. 191, pp. 1119–28, https://doi.org/10.1111/j.1365-246X.2012.05687.x.

Wang, X., and Cochran, J. R., 1993, Gravity anomalies, isostasy and mantle flow at the East Pacific Rise crest, J. Geophys. Res., *v.* 98, pp. 19505–31.

Wang, Y., and Mareschal, J.-C., 1999, Elastic thickness of the lithosphere in the Central Canadian Shield, Geophys. Res. Lett., *v.* 26, pp. 3033–6.

Waschbusch, P. J., and Royden, L. H., 1992, Spatial and temporal evolution of foredeep basins: Lateral strength variations and inelastic yielding in continental lithosphere, Basin Res., *v.* 4, pp. 179–96.

Watters, T. R., 2003, Lithospheric flexure and the origin of the dichotomy boundary on Mars, Geology, *v.* 31, pp. 271–4.

Watters, T. R., James, P. B., and Selvans, M. M., 2021, Mercury's crustal thickness and contractional strain, Geophys. Res. Letters, *v.* 48, https://doi.org/10.1029/2021GL093528.

Watters, T. R., Schultz, R. A., Robinson, M. S., and Cook, A. C., 2002, The mechanical and thermal structure of Mercury's early lithosphere, Geophys. Res. Letters, *v.* 29, doi: https://doi.org/10.1029/2001GL014308.

Watts, A. B., 1972, Geophysical investigations east of the Magdalen Islands, southern Gulf of St. Lawrence, Can. J. Earth Sci., *v.* 9, pp. 1504–28.

Watts, A. B., 1976, Gravity and bathymetry in the Central Pacific Ocean, J. Geophys. Res., *v.* 81, pp. 1533–53.

Watts, A. B., 1978, An analysis of isostasy in the world's oceans: 1. Hawaiian-Emperor Seamount Chain, J. Geophys. Res., *v.* 83, pp. 5989–6004.

Watts, A. B., 1981, The U.S. Atlantic continental margin: Subsidence history, crustal structure and thermal evolution, in Bally, A. W., ed., *The Geology of Continental Margins*, Education Course Notes #19, Tulsa, OK, American Association of Petroleum Geologists, pp. 1–75.

Watts, A. B., 1982a, *Gravity Anomalies over Oceanic Rifts, Continental and Oceanic Rifts*, Geodynamics Series 8, Washington, DC, American Geophysical Union, pp. 99–106.

Watts, A. B., 1982b, Tectonic subsidence, flexure and global changes of sea-level, Nature, *v.* 297, pp. 469–74.

Watts, A. B., 1988, Gravity anomalies, crustal structure and flexure of the lithosphere at the Baltimore Canyon Trough, Earth Planet. Sci. Lett., *v.* 89, pp. 221–38.

Watts, A. B., 1989, Lithospheric flexure due to prograding sediment loads: Implications for the origin of offlap/onlap patterns in sedimentary basins, Basin Res., *v.* 2, pp. 133–44.

Watts, A. B., 1992, The effective elastic thickness of the lithosphere and the evolution of foreland basins, Basin Res., *v.* 4, pp. 169–78.

Watts, A. B., 1993, The formation of sedimentary basins, in Brown, G. C., Hawkesworth, C. J., and Wilson, R. C. L., eds., *Understanding the Earth*, Cambridge, Cambridge University Press, pp. 301–24.

Watts, A. B., 1994, Crustal structure, gravity anomalies and flexure of the lithosphere in the Canary Islands, Geophys. J. Int., *v.* 119, pp. 648–66.

Watts, A. B., 2007. An overview, in Watts, A. B., ed., *Treatise of Geophysics. Volume 6: Crust and Lithosphere Dynamics*, Amsterdam, Elsevier, pp. 1–48.

Watts, A. B., 2012, Models for the evolution of passive margins, in Roberts, D. G., and Bally, A. W., eds., *Regional Geology and Tectonics: Phanerozoic Rift Systems and Sedimentary Basins*, Amsterdam, Elsevier, pp. 33–57.

Watts, A. B., and Burov, E. B., 2003, Lithospheric strength and its relationship to the elastic and seismogenic layer thickness, Earth Planet. Sci. Lett., *v.* 213, pp. 113–31, https://doi.org/10.1016/S0012-821X(03)00289-9.

Watts, A. B., and Cochran, J. R., 1974, Gravity anomalies and flexure of the lithosphere along the Hawaiian-Emperor seamount chain, Geophys. J. R. Astr. Soc., *v.* 38, pp. 119–41.

Watts, A. B., and Fairhead, J. D., 1997, Gravity anomalies and magmatism at the British Isles continental margin, J. Geol. Soc. London, *v.* 154, pp. 523–9.

Watts, A. B., and Marr, C., 1995, Gravity anomalies and the thermal and mechanical structure of rifted continental margins, in Banda, E., Talwani, M., and Torné, M., eds., *Rifted Ocean-Continent Boundaries*, Dordrecht, Kluwer Academic Publishers, pp. 65–94.

Watts, A. B., and Moore, J. D. P., 2017, Flexural isostasy: Constraints from gravity and topography power spectra, J. Geophys. Res., *v.* 122, https://doi.org/10.1002/2017JB014571.

Watts, A. B., and Ribe, N. M., 1984, On geoid heights and flexure of the lithosphere at sea-mounts, J. Geophys. Res., *v.* 89, pp. 11152–70.

Watts, A. B., and Ryan, W. B. F., 1976, Flexure of the lithosphere and continental margin basins, Tectonophysics, *v.* 36, pp. 25–44.

Watts, A. B., and Steckler, M. S., 1979, Subsidence and eustasy at the continental margin of eastern North America, in Talwani, M., Hay, W., and Ryan, W. B. F., eds., *Deep Drilling Results in the Atlantic Ocean: Continental Margins and Paleoenvironment*, Maurice Ewing Series 3, Washington, DC, American Geophysical Union, pp. 218–34.

Watts, A. B., and Stewart, J., 1998, Gravity anomalies and segmentation of the continental margin offshore West Africa, Earth Planet. Sci. Lett., *v.* 156, pp. 239–52.

Watts, A. B., and Talwani, M., 1974, Gravity anomalies seaward of deep-sea trenches and their tectonic implications, Geophys. J. R. Astr. Soc., *v.* 36, pp. 57–92.

Watts, A. B., and Talwani, M., 1975a, Gravity effect of downgoing lithospheric slabs beneath island arcs, Geol. Soc. Am. Bull, *v.* 86, pp. 1–4.

Watts, A. B., and Talwani, M., 1975b, *Gravity Field of the Northwest Pacific Ocean Basin and Its Margin: Hawaii and Vicinity*, Boulder, CO, Geological Society of America.

Watts, A. B., and ten Brink, U. S., 1989, Crustal structure, flexure and subsidence history of the Hawaiian Islands, J. Geophys. Res., *v.* 94, pp. 10473–500.

Watts, A. B., and Thorne, J. A., 1984, Tectonics, global changes in sea-level and their relationship to stratigraphic sequences at the U.S. Atlantic continental margin, Mar. Pet. Geol., *v.* 1, pp. 319–39.

Watts, A. B., and Torné, M., 1992a, Crustal structure and the mechanical properties of extended continental lithosphere in the Valencia trough (Western Mediterranean), J. Geol. Soc. Lond., *v.* 149, pp. 813–27.

Watts, A. B., and Torné, M., 1992b, Subsidence history, crustal structure and thermal evolution of the Valencia trough: A young extensional basin in the western Mediterranean, J. Geophys. Res., *v.* 97, pp. 20021–41.

Watts, A. B., and Zhong, S., 2000, Observations of flexure and the rheology of oceanic lithosphere, Geophys. J. Int., *v.* 142, pp. 855–75.

Watts, A. B., Bodine, J. H., and Steckler, M. S., 1980, Observations of flexure and the state of stress in the oceanic lithosphere, J. Geophys. Res., *v.* 85, pp. 6369–76.

Watts, A. B., Cochran, J. R., and Selzer, G., 1975, Gravity anomalies and flexure of the lithosphere: A three-dimensional study of the Great Meteor Seamount, N.E. Atlantic, J. Geophys. Res., *v.* 80, pp. 1391–8.

Watts, A. B., Grevemeyer, I., Shillington, D. J. et al. 2021, Seismic structure, gravity anomalies and flexure along the Emperor seamount chain, J. Geophys. Res., *v.* 126, https://doi.org/10.1029/2020JB021109.

Watts, A. B., Karner, G. D., and Steckler, M. S., 1982, Lithospheric flexure and the evolution of sedimentary basins, in Kent, P., Bott, M. H. P., McKenzie, D. P., and

574   *References*

Williams, C. A., eds., *The Evolution of Sedimentary Basins*, Philosophical Transactions of the Royal Society of London, *v.* 305A, pp. 249–81.

Watts, A. B., Lamb, S. H., Fairhead, J. D., and Dewey, J. F., 1995, Lithospheric flexure and bending of the Central Andes, Earth Planet. Sci. Lett., *v.* 134, pp. 9–21.

Watts, A. B., Peirce, C., Collier, J. et al. 1997, A seismic study of lithospheric flexure in the vicinity of Tenerife, Canary Islands, Earth Planet. Sci. Lett., *v.* 146, pp. 431–47.

Watts, A. B., Rodger, M., Peirce, C., Greenroyd, C. J., and Hobbs, R. W., 2009, Seismic structure, gravity anomalies, and flexure of the Amazon continental margin, NE Brazil, J. Geophys. Res., *v.* 114, https://doi.org/10.1029/2008JB006259.

Watts, A. B., Sandwell, D. T., Smith, W. H. F., and Wessel, P., 2006, Global gravity, bathymetry, and the distribution of submarine volcanism through space and time, J. Geophys. Res., *v.* 111, https://doi.org/10.1029/2005JB004083.

Watts, A. B., ten Brink, U., Buhl, P., and Brocher, T., 1985, A multichannel seismic study of lithospheric flexure across the Hawaiian-Emperor seamount chain, Nature, *v.* 315, pp. 105–11.

Watts, A. B., Tozer, B., Daly, M. C., and Smith, J., 2018, A comparative study of the Parnaíba, Michigan and Congo cratonic basins, in Daly, M. C., Fuck, R. A., Julia, J., Macdonald, D. I., and Watts, A. B., eds., *Cratonic Basin Formation: A Case Study of the Parnaíba Basin of Brazil*, Geological Society, London, Special Publications, *v.* 472, pp. 45–66, https://doi.org/10.1144/SP472.6.

Weber, R. C., Lin, P.-Y., Garnero, E. J., Williams, Q., and Lognonné, P., 2011, Seismic Detection of the Lunar Core, Science, *v.* 331, pp. 309–12, https://doi.org/10.1126/science.1199375.

Webster, J. M., Clague, D. A., and Braga, J. C., 2007, Support for the giant wave hypothesis: Evidence from submerged terraces off Lanai, Hawaii, Int. J. Earth Sci. (Geol. Rundsch), *v.* 96, pp. 517–24.

Webster, J. M., Clague, D. A., Braga, J. C. et al., 2006, Drowned coralline algal dominated deposits off Lanai, Hawaii; carbonate accretion and vertical tectonics over the last 30 ka, Marine Geology, *v.* 225, pp. 223–46.

Weertman, J., 1979, Height of mountains on Venus and the creep properties of rock, Phys. Earth Planet. Int., *v.* 19, pp. 197–207.

Wegener, A., 1966, *The Origin of Continents and Oceans* (Translated from the Fourth Revised German Edition by John Biram). New York, Dover Publications, 246 pp.

Weigel, W., and Grevemeyer, I., 1999, The Great Meteor seamount seismic structure of a submerged intraplate volcano, Geodynamics, *v.* 28, pp. 27–40.

Weissel, J. K., and Karner, G. D., 1989, Flexural uplift of rift flanks due to mechanical unloading of the lithosphere during extension, J. Geophys. Res., *v.* 94, pp. 13,919–50.

Weissel, J. K., and Watts, A. B., 1979, Tectonic evolution of the Coral Sea Basin, J. Geophys. Res., *v.* 84, pp. 4572–82.

Weissel, J. K., Anderson, R. N., and Geller, C. A., 1980, Deformation of the Indo-Australian plate, Nature, *v.* 287, pp. 284–91.

Wellman, P., 1978, Gravity evidence for abrupt changes in mean crustal density at the junction of Australian crustal blocks, BMR J. Aust. Geol. Geophys., *v.* 3, pp. 153–62.

Wernicke, B., 1985, Uniform sense normal simple shear of the continental lithosphere, Can. J. Earth Sci., *v.* 22, pp. 108–25.

Wernicke, B., and Axen, G. J., 1988, On the role of isostasy in the evolution of normal fault systems, Geology, *v.* 16, pp. 848–51.

Wessel, P., 1993, A re-examination of the flexural deformation beneath the Hawaiian Islands, J. Geophys. Res., *v.* 98, pp. 12,177–90.

Wessel, P., 2016, Regional-residual separation of bathymetry and revised estimates of Hawaii plume flux, Geophys. J. Int., *v.* 204, pp. 932–47, https://doi.org/10.1093/gji/ggv472.

Wessel, P., and Haxby, W. F., 1990, Thermal stresses, differential subsidence, and flexure at oceanic fracture zones, J. Geophys. Res., *v.* 95, pp. 375–91.

Wessel, P., and Keating, B. H., 1994, Temporal variations of flexural deformation in Hawaii, J. Geophys. Res., *v.* 99, pp. 2747–56.

Wessel, P., and Lyons, S. 1997, Distribution of large Pacific seamounts from Geosat/ ERS-1: Implications for the history of intraplate volcanism, J. Geophys. Res., *v.* 102, pp. 22459–75.

Wessel, P., and Smith, W. H. F., 1991, Free software helps map and display data, EOS Trans. Amer. Union, *v.* 72, pp. 441–6.

White, N., and McKenzie, D. P., 1988, Formation of the "Steer's Head" geometry of sedimentary basins by differential stretching of the crust and mantle, Geology, *v.* 16, pp. 250–3.

White, R. S., 1992, Crustal structure and magmatism of North Atlantic continental margins, J. Geol. Soc. Lond., *v.* 149, pp. 841–54.

White, R. S., Smith, L. K., Roberts, A. W. et al., 2008, Lower-crustal intrusion on the North Atlantic continental margin, Nature, *v.* 452, pp. 460–4, https://doi.org/10.1038 /nature06687.

Whitehouse, P., Latychev, K., Milne, G. A., Mitrovica, J. X., and Kendall, R., 2006, Impact of 3-D Earth structure on Fennoscandian glacial isostatic adjustment: Implications for space-geodetic estimates of present-day crustal deformations, Geophys. Res. Lett., *v.* 33, https://doi.org/10.1029/2006GL026568.

Whitman, D., 1994, Moho geometry beneath the eastern margin of the Andes, northwest Argentina, and its implications for the elastic thickness of the Andean foreland, J. Geophys. Res., *v.* 99, pp. 15277–89.

Wieczorek, M. A., Neumann, G. A., Nimmo, F. et al., 2013, The crust of the moon as seen by GRAIL, Science, *v.* 339, pp. 671–5, https://doi.org/10.1126/science.1231530.

Wieczorek, M. A., and Zuber, M. T., 2001. A Serenitatis origin for the Imbrian grooves and South Pole-Aitken thorium anomaly, *J. Geophys. Res.*, 106, 27,853–864.

Wiens, D. A., and Stein, S., 1983, Age dependence of oceanic intraplate seismicity and implications for lithospheric evolution, J. Geophys. Res., *v.* 88, pp. 6455–68.

Wiens, D. A., and Stein, S., 1984, Intraplate seismicity and stresses in young oceanic lithosphere, J. Geophys. Res., *v.* 89, pp. 11,442–64.

Willett, S. D., Chapman, D. S., and Neugebauer, H. J., 1985, A thermo-mechanical model of continental lithosphere, Nature, *v.* 314, pp. 520–3.

Williams, J.-P., Nimmo, F., Moore, W. B., and Paige, D. A., 2008, The formation of Tharsis on Mars: What the line-of-sight gravity is telling us, J. Geophys. Res., *v.* 113, https://doi.org/10.1029/2007JE003050.

Williams, J.-P., Ruiz, J., Rosenburg, M. A., Aharonson, O., and Phillips, R. J., 2011, Insolation driven variations of Mercury's lithospheric strength, J. Geophys. Res., *v.* 116, doi: https://doi.org/10.1029/2010JE003655.

Williams, K. K., and Zuber, M. T., 1998, Measurement and analysis of lunar basin depths from Clementine altimetry, Icarus, *v.* 131, pp. 107–22.

Willis, B., 1919, Joseph Barrell and his work, J. Geol., *v.* 27, pp. 664–72.

Wilson, C., 2004, Long wavelength gravity and topography anomalies in the Pacific, MESc. thesis, University of Oxford.

Wilson, D. S., 1992, Focused mantle upwelling beneath mid-ocean ridges: Evidence from sea-mount formation and isostatic compensation of topography, Earth Planet. Sci. Lett., *v.* 113, pp. 41–55.

Wilson, J. T., 1965, A new class of faults and their bearing on continental drift, Nature, *v.* 207, pp. 343–7.

Withjack, M. O., Schlische, R. W., and Olsen, P. E., 1998, Diachronous rifting, drifting, and inversion on the passive margin of central eastern North America: An analog for other passive margins, Amer. Assoc. Pet. Geol. Bull., *v.* 82, pp. 817–35.

Wolf, D., 1985, Thick-plate flexure re-examined, Geophys. J. R. Astr. Soc., *v.* 80, pp. 265–73.

Wolf, D., 1986, Reply to comments by Robert P. Comer, Geophys. J. R. Astr. Soc., *v.* 85, pp. 469–70.

Wolf, D., 1993, The changing role of the lithosphere in models of glacial isostasy: A historical review, Global and Planetary Change, *v.* 8, pp. 95–106.

Wolfson-Schwehr, M., Boettcher, M. S., and Behn, M. D., 2017, Thermal segmentation of mid-ocean ridge-transform faults, Geochem. Geophys., *v.* 18, pp. 3405–18, https://doi.org/10.1002/2017GC006967.

Woodroffe, C. D., 1988, Vertical movement of isolated oceanic islands at plate margins: Evidence from emergent reefs in Tonga (Pacific Ocean) Cayman Islands (Caribbean Sea) and Christmas Island (Indian Ocean), Zeitschrift für Geomorphologie, *v.* Suppl. Bd. 69, pp. 17–37.

Woodroffe, C. D., McLean, R., Polach, H., and Wallensky, E., 1990, Sea level and coral atolls: Late Holocene emergence in the Indian Ocean, Geology, *v.* 18, pp. 62–6.

Wooler, D. A., Smith, A. G., and White, N., 1992, Measuring lithospheric stretching on Tethyan passive margins, J. Geol. Soc. Lond., *v.* 149, pp. 517–32.

Woollard, G. P., 1943, Transcontinental gravitational and magnetic profile of North America and its relation to geologic structure, Bull. Geol. Soc. Am., *v.* 54, pp. 747–90.

Woollard, G. P., 1951, A gravity reconnaisance of the island of Oahu, Trans. Am. Geophys. Union, *v.* 32, pp. 358–67.

Woollard, G. P., 1966, *Principal Facts for Gravity Observations in the Hawaiian Archipelago, Johnston Island, American Samoa and Society Islands*. Data Report No. 3, HIG-66–20, 10 tables. University of Hawaii at Manoa, Hawaii Institute of Geophysics, 5 pp.

Worzel, J. L., and Shurbet, G. L., 1965, Gravity interpretation from standard oceanic and crustal sections, Geol. Soc. Am. Sp. Paper, *v.* 62, pp. 87–100.

Wright, T. J., Elliot, J. R., Wang, H., and Ryder, I., 2013, Earthquake cycle deformation and the Moho: Implications for the rheology of continental lithosphere., Tectonophysics, *v.* 609, pp. 504–23, https://doi.org/10.1016/j.tecto.2013.07.029.

Wyer, P. P. A., 2003, Gravity anomalies and segmentation of the eastern USA passive continental margin, Ph.D thesis, University of Oxford.

Wyer, P., and Watts, A. B., 2006, Gravity anomalies and segmentation at the East Coast, USA continental margin, Geophys. J. Int., *v.* 166, pp. 1015–38.

Xie, X., and Heller, P. L., 2009, Plate tectonics and basins subsidence history, Geol. Soc. Am. Bull., *v.* 121, pp. 55–64, https://doi.org/10.1130/B26398.1

Xu, C., Dunn, R. A., Watts, A. B. et al., 2022. A seismic tomography, gravity, and flexure study of the crust and upper mantle structure of the Emperor Seamounts at Jimmu guyot, *J. Geophys. Res.*, *v.* 127, e2021JB023241. https://doi.org/10.1029/2021JB023241.

Yamasaki, T., and Houseman, G. A., 2012, The crustal viscosity gradient measured from post-seismic deformation: A case study of the 1997 Manyi (Tibet) earthquake, Earth Planet. Sci. Lett., *v.* 351–352, pp. 105–14, https://doi.org/10.1016/j.epsl.2012.07.030.

Yang, Z., and Chen, W.-P., 2010, Earthquakes along the East African Rift System: A multiscale, system-wide perspective, J. Geophys. Res., *v.* 115, https://doi.org/10.1029/2009JB006779.

Yong, L., Allen, P. A., Densmore, A. L., and Qiang, X., 2003. Evolution of the Longmen Shan foreland basin (western Sichuan, China) during the Late Triassic Indosinian orogeny. Basin Research, *v.* 15, pp. 117–38.

Zhang, F., Lin, J., and Zhan, W., 2014, Variations in oceanic plate bending along the Mariana trench, Earth Planet. Sci. Lett., *v.* 401, pp. 206–14, https://doi.org/10.1016/j.epsl.2014.05.032.

Zhang, F., Lin, J., Zhou, Z., Yang, H., and Zhan, W., 2018, Intra- and intertrench variations in flexural bending of the Manila, Mariana and global trenches: implications on plate weakening in controlling trench dynamics, Geophys. J. Int., *v.* 212, pp. 1429–49, https://doi.org/10.1093/gji/ggx488.

Zhang, N., Zhong, S., and Flowers, R. M., 2012, Predicting and testing continental vertical motion histories since the Paleozoic, Earth Planet. Sci. Lett., *v.* 317–318, pp. 426–35, https://doi.org/10.1016/j.epsl.2011.10.041.

Zheng Y., and Arkani-Hamed J., 2002. Rigidity of the Atlantic oceanic lithosphere beneath New England seamounts, Tectonophysics, *v.* 359, pp. 359–69.

Zhong, S., 1992, Viscous flow model of a subduction zone with a faulted lithosphere: Long and short wavelength topography, gravity and geoid, Geophys. Res. Lett., *v.* 19, pp. 1891–4.

Zhong, S., 1997, Dynamics of crustal compensation and its influences on crustal isostasy, J. Geophys. Res., *v.* 102, pp. 15,287–99.

Zhong, S., and Gurnis, M., 1994, Controls on trench topography from dynamic models of subducted slabs, J. Geophys. Res., *v.* 99, pp. 15,683–95.

Zhong, S., and Zuber, M. T., 2001, Degree-1 mantle convection and the crustal dichotomy on Mars, Earth Planet. Sci. Lett., *v.* 189, pp. 75–84.

Zhong, S., and Watts, A. B., 2002, Constraints on the dynamics of mantle plumes from uplift of the Hawaiian Islands, Earth Planet. Sci. Lett., *v.* 203, pp. 105–16.

Zhong, S., and Watts, A. B., 2013, Lithospheric deformation induced by loading of the Hawaiian Islands and its implications for mantle rheology, J. Geophys. Res., *v.* 118, pp. 6025–48, https://doi.org/10.1002/2013JB010408.

Zhong, Z., Yan, J., Rodriguez, A. P., and Dohm, J. M., 2018, Ancient selenophysical structure of the Grimaldi basin: Constraints from GRAIL gravity and LOLA topography, Icarus, *v.* 309, pp. 411–21, https://doi.org/10.1016/j.icarus.2017.11.030.

Zhou, S., 1991, A model of thick plate deformation and its application to the isostatic movements due to surface, subsurface and internal loadings, Geophys. J. Int., *v.* 105, pp. 381–95.

Zoetemeijer, R., Desegaulx, P., Cloetingh, S., Roure, F., and Moretti, I., 1990, Lithospheric dynamics and tectono-stratigraphic evolution of the Ebro basin, J. Geophys. Res., *v.* 95, pp. 2701–11.

Zuber, M. T., Bechtel, T. D., and Forsyth, D. W., 1989, Effective elastic thickness of the lithosphere and mechanisms of isostatic compensation in Australia, J. Geophys. Res., *v.* 94, pp. 13919–30.

Zuber, M. T., Solomon, S. C., Phillips, R. J. et al., 2000, Internal structure and early thermal evolution of Mars from Mars Global Surveyor Topography and Gravity, Science, *v.* 287, pp. 1788–93.

Zuber, M. T., Smith, D. E., Watkins, M. M. et al., 2013, Gravity field of the moon from the Gravity Recovery and Interior Laboratory (GRAIL) Mission, Science, *v.* 339, pp. 668–71, https://doi.org/10.1126/science.1231507.

Index